An Unexpected End to the Journey

An introduction to international accidents on and around railways

Peter van

To Peter:
As said, don't read it in one go!
Hope you like it.
From Peter!

About myself

Following drama training at Utrecht in The Netherlands, I was trained as a road vehicle driver whilst fulfilling obligatory military service (1976/5th recruitment, ID No 530320337, 109 AfdVa), which gave me a taste for driving serious vehicles. After 18 interesting months of nursing vehicles often older than myself (crash-gearboxes and the like) I studied in Leiden to become a transport history orientated museum educational officer, followed after graduation by a stint as a business-to-business advertiser. For about 12 years during this period I worked evenings in the tourist business in and around Amsterdam. In May 1989, I moved to England and, having decided I did not want to drag my heels on the way to work, initially became a London Transport bus driver (badge No N 138790), and then a British Rail train driver in 1992. I was taken off trains early in 2002 following diagnosis of coronary problems as well as serious loss of hearing. As a result, I worked in positions with more direct customer contact, in traincrew management and at a regional traffic control centre before retiring in March 2013.

This book is dedicated to Stanley Hall
MBE, FIRO, FCILT, HonFIRSE,

who provided an ear when needed and a light to illuminate the way.

First published 2016

ISBN 978 0 9575898 3 4

Published by the author through **Shrewdale Publishing**,
25 Monkmoor Road, Shrewsbury SY2 5AG.

Distributed by **Bookworld Wholesale Ltd,** Unit 10, Hodfar Road, Sandy Lane Industrial Estate, Stourport-on-Severn, Worcestershire DY13 9QB; Tel: 01299 823330.

Printed and bound by CPI Group (UK) Ltd, Croydon, CR0 4YY

Contents

Introduction

Everyone more or less instinctively knows how and why accidents happen. This book has its genesis in my curiosity as to why accidents continue to happen in an environment as professional and committed to safety as public passenger transport is. Whilst you may know about some of these accidents already, I hope I am able to provide you with fresh food for thought.

Any errors and omissions in this book are mine. If you would like to comment, please feel free to email me at petervandermark3@iCloud.com. Despite the grim and often tragic subject matter, I hope that you nevertheless enjoy reading the book. Perhaps it is superfluous to point out that the book is not really meant to be read in one go, it is no novel. Please take your time and, for instance, read about one or two incidents per time only.

Acknowledgements

May I first of all express my sincere gratitude to the transport folk I met at work and on my travels, for sharing their knowledge, experiences and opinions. Unfortunately, there are too many to mention individually – it would take another chapter. I thank Bill Bryson for his books on the use of the English language, which made me believe I can write in English. I also thank the various language teachers throughout my life who persevered with me. As a result of their effort I am conversant with a number of European languages, which greatly facilitated access to cabs, control centres, literature and investigation results. I would like to pay respect to those National Accident Investigation Boards that publish their reports on the internet. Similarly, this extends to those organisations, run by volunteers, who put in a lot of work to catalogue and publish PDF files of ancient accident reports on the internet. Finally, my thanks go to those people who have uploaded videos of maritime, road, air and rail accidents to websites. Nothing makes you more aware of the violence of a crash than to see it unfold in front of your eyes.

Individual thanks must go to:

- **Stanley Hall**. Ex-BR safety manager and, in retirement, writer on aspects of rail safety. Stan persuaded me to start writing about rail accidents, and I gladly dedicate this book to him.
- **Wim Coenraad**. Past president of IRSE, senior consultant railway signalling engineer and good friend, who is deeply conversant with all international aspects of signalling and with other facets of railway operations. Despite his lack of spare time, Wim read and commented on what was written, edited the book and helped with getting it into the open.
- **Jim Vine**. Former BR (Southern) traction engineer and project manager, latterly traction and rolling stock consultant for Halcrow plc. Jim gave me the best possible grounding in

practical railway operational and technical matters as well as in the basics and the finer points of rail safety. Sadly, Jim passed away in June 2002.

- **Tony Telford**. Successfully managed the onerous task of clarifying the influence of politics, for better or worse, on virtually every aspect of public transport operation, particularly with regard to the weird yet fascinating political force fields that govern the inordinately expensive rail transport policies in Britain. Tony passed away most unexpectedly in the summer of 2011.

Last but not least, a big thank you to my wife Linda for her forbearance. And to Mop and Mick, our affectionate tomcat twins, who between them graciously allowed me the occasional use of one hand to type up the manuscript.

1: About this Book

The set-up

The incidents described in this book have been selected on the basis of what I already knew and from what sources that knowledge came, what material was available and, above all, how, in my opinion, the accidents illustrate why situations in the daily grind develop into serious problems. From a British point of view the book predominantly describes foreign incidents, but it includes a few British examples that demonstrate particular types of event.

The main accidents are presented according to date, except for a few that are grouped by subject. As far as the information is concerned, my sources include people I have known who were directly or indirectly involved in accidents in Germany, The Netherlands and Britain. Generally I have used accident investigation reports as well as books and articles from specialist magazines. I have tried to touch on the economics behind accidents as one of the reasons why they continue to occur, despite our power to diminish their impact on our lives. You will find that the point of view expressed is that of a person behind the controls, because they were in charge of the equipment involved. Using my own experience and that of a number of colleagues from around the world to analyse the occurrences colours the narrative. It might be partisan to some extent as I have felt free to state things that the official reports could not, due to lack of proof. I hope to make sense of what, in my view, happened and why, even if a few people may have failed in upholding good practice and compromised safety.

You will find some names of accident locations appear in ***bold italics***. These can be found in the appropriate section of Chapter 4 with a full description of the accident that occurred there. Accident locations used to illustrate issues elsewhere appear in normal *italics*. See also the index of accidents.

Nationalities and languages

In order to enhance understanding, a number of railway terms and abbreviations are listed alphabetically and explained in Chapter 3, whilst for the benefit of foreign language readers roughly equivalent terms are given in French, German and Dutch (as well as American English where deemed necessary). Before reading the book it might be a good idea to have a look at this glossary to get an idea of the terms and names used and what they mean. Furthermore, this book is written (I hope, being of foreign extraction) in British English. The railway terms and abbreviations, therefore, are those used in Britain and might not be readily understandable to foreign readers, or indeed in some cases to British readers themselves. Please note that the official World English indications for equipment are used. For example, a set of points in Britain is called a switch in North America (in fact, the apparatus originally was called a switchpoint)! It is indicated as a turnout in this book. Again, if you cannot work out from the text what something is about, have a look in the glossary.

Note that the foreign terms are not exhaustive equivalents of such terms in their own language; many networks have a score of terms where Britain has just one (e.g. distant signal). The French and German railways employ signal aspects that act as distant signals but have a specific function pertaining to the type of following stop signals of which in daily practice several types are used, again with specific functions. The idea of this exercise is to give a foreign reader the chance to get an idea of what is meant by the British English terms used in this book and so better understand the information given.

Countries are indicated by the abbreviations used on road vehicles: A-Austria, AUS-Australia, B-Belgium, CDN-Canada, CH-Switzerland, CZ-Czech Republic, D-Germany, DK-Denmark, E-Spain, F-France, GB-Great Britain, I-Italy, IRL-Irish Republic, J-Japan, N-Norway, NL-The Netherlands, NZ-New Zealand, P-Portugal, PL-Poland, RO-Romania, RUS-Russia, S-Sweden, SF-Finland, USA-United States of America, ZA-South Africa. NA is also used for North America and includes the USA and Canada.

Imperial and metric conversions

The basis for distance and speed calculation is 80 km/h = 50 mph. This easily enables other approximate speeds to be worked out: 40 km/h = 25 mph, 20 km/h = 12.5 mph, 10 km/h = 6.25 mph and 5 km/h = 3.125 mph. In turn this gives 75 km/h = approximately 46 mph, 100 km/h = 62.5 mph, 120 km/h = 75 mph, etc. I am sure that the odd couple of miles per hour difference will not distort your appreciation of what occurred.

A yard approximates to just shorter than a metre. According to the UK train driver manuals 180 metres = 200 yards as an indication of where the AWS magnet is located before reaching a signal. Thus 90 metres = 100 yards, 45 metres = 50 yards, 9 metres = 10 yards, 4.5 metres = 5 yards, etc.

Regarding conversions between litres and gallons, 4.5 litres = 1 British gallon and 5 litres = 1 US gallon.

Names of towns and cities

Whilst the English names of nations are used in this book, names of towns and cities are written in the language used at their actual location. The rationale behind this apparent anomaly is the anticipated number of international readers of this book, who in most cases will know the names of nations in English but often have their own spelling for local names. In the case of an officially multilingual nation (e.g. Canada, Switzerland, Belgium) the name in the book is as used in its actual location, but where a very different local language (Basque, Sorb, Fries, Welsh, Romansch, etc) is spoken, as no one but the local speaker would recognise it, the name used is the one in the main national language (e.g. German, Dutch, English, French, Spanish). In the case of a city like Brussels in Belgium (itself divided into language areas) this still gives problems, in which case the name used is that which is most likely to be understood by English-speaking readers (e.g. the Flemish name Brussel instead of the Walloon Bruxelles). An additional benefit of this approach is that the names as used in this book are more likely to be found in local literature or on local maps. The proper names of counties or provinces, where these differ from English, are also given below to enable checking of accident information and locations on the internet as well as local high-definition maps (1:10,000 and 1:25,000 scale).

With regard to the pronunciation of local names in their own language, I will leave that to your ingenuity. With several languages involved, it would add another chapter and the issue would probably still not be solved satisfactorily.

Listing of some local names used in the book

Aasta	(N): *Åsta*
Aix (la Chapelle)	(D): *Aachen*
Antwerp	(B): *Antwerpen* (Flemish; *Anvers* in Walloon French)

Bavaria	(D): *Bayern*
Berne	(CH): *Bern*
Bruges	(B): *Brugge* (Flemish; *Bruges* in Walloon French)
Brussels	(B): *Brussel* (Flemish; *Bruxelles* in Walloon French)
Cologne	(D): *Köln*
Copenhagen	(DK): *København*
Flanders	(B): *Vlaanderen* (Dutch-speaking Belgium and SW Netherlands)
Flushing	(NL): *Vlissingen*
Geneva	(CH): *Genève* (French; *Genf* in German Swiss)
Genoa	(I): *Genova*
Ghent	(B): *Gent* (Flemish; *Gand* in Walloon French)
Hanover	(D): *Hannover*
Hook of Holland	(NL): *Hoek van Holland*
Lake Constance	(CH, A & D): *Bodensee*
Lucerne	(CH): *Luzern*
Lyons	(F): *Lyon*
Moscow	(RUS): *Moskva*
Munich	(D): *München*
Naples	(I): *Napoli*
Nuremberg	(D): *Nürnberg*
Prague	(CZ): *Praha*
Rome	(I): *Roma*
The Hague	(NL): *Den Haag* (officially *'s-Gravenhage*)
Turin	(I): *Torino*
Tuscany	(I): *Toscana*
Tyrolia	(A): *Tirol*
Walloon	(B): *Wallon* (French-speaking part of Belgium)
Warsaw	(PL): *Warszawa*

2: The Nature of Accidents

The point of departure

1 *Accidents, like sickness and natural disasters, are among those less pleasant aspects of life on planet Earth; mishap is included in the deal at all times*. Accidents and precursor incidents, and their subsequent investigations, are essential components in the ongoing quest to eradicate them. It is, however, impossible to completely eliminate accidents; they will continue to happen.

2 *Transport and work accidents typically involve either a) failure by staff to operate equipment properly, b) failure of equipment, c) unintended destructive actions by the public, or d) unforeseen severe ambient circumstances*. Intentional wrecking, e.g. vandalism and belligerent actions, are not by nature accidents.

3 *A major contributory factor to unsafe practices is insufficiently educated and monitored staff*. Incidents can arise as a result of inadequate proficiency, insufficient appreciation of risk and a lack of technical knowledge.

4 *Our view of the social impact of transport risks is culturally determined*. What, for example, Europeans, Japanese or North Americans consider to be an unacceptable risk would hardly raise an eyebrow in lesser technically developed countries.

5 *We are technically able to prevent most transport and work accidents*. Relevant equipment, however, tends to come at a cost that operators are normally unwilling to pay until another, similar, catastrophe occurs – by then classed as 'avoidable'. Any substantial funds spent on improving safety are usually the result of government action in response to media pressure or, very rarely, a private company boosting its competitive position.

6 *Wholly profit-driven organisations, whether governmental or private, are not usually the most suitable custodians of public transport safety*. Public land transport is not normally sufficiently

profitable. Up-to-date safety, however, important for business as it is, comes at a high cost. When, commercially speaking, purchase and operational costs of safety equipment are deemed too costly, investment can be achieved by making accident liability expensive. Increase of the potential cost of accidents in the courts, and consequently that of insurance, changes the outlook on alternative safety cost estimates and so boosts safety as a result.

7 *Safety measures implemented to counter one type of danger may be a contributory factor to a different type of risk.* Fatal crashes can be caused by imperfectly developed or risk-assessed new methods and equipment, with the wrong balance being struck between positive and negative effects. Additionally there is the misuse of new generations of safety features.

Some practical safety history

The basis of present-day transport and work safety

Well-nigh 250 years of mechanised maritime activities, 200 years of mechanised land transport and 110 years of air transport have given us an insight into what can go wrong if things are not properly managed in order to avoid accidents. Transport from the start of mechanisation in the late 19th century onwards, with the rapidly increasing size, weight and speed of ships, cars and trains, was worked with scant experience of the new equipment and little knowledge of risk prevention. Much depended on intuition and all too often a good dose of luck. Lessons learned from transport disasters, increasingly cross-fed between transport modes, brought about measures to avoid repeat occurrences, which over time have resulted in our present store of practical knowledge of transport safety issues. Such measures later became subject to enforcement by national and international authorities. Nowadays the process of maximising safety is based on this store of knowledge and experience, from which risk assessment and the removing of foreseeable dangers, judged on the basis of so-called ALARP (As Low As Reasonably Practicable) conditions, have been developed. Whilst safety standards must be high and consistent, it should not cost the earth to address all, often infinitesimally small, risk factors. Funds are always short and might well be better used to improve safety elsewhere. Here, incidentally, we encounter different interpretations of ALARP between the various parties that make up the transport industry.

To assess why things go wrong on the railway it is not useful merely to study railway accidents, as no cause is unique to any particular transport mode even if combinations of causal factors may be unlike each other. Transport disasters have common characteristics that provide material to compare and determine patterns in the chain of precursor events, to the benefit of all.

An interesting aspect of transport safety is the history of its improvement. Maritime transport, for instance, might be considerably less safe if it were not for the (usually) professional character of the people who work on ships. Coupled with the increasing sophistication of equipment employed to make ship handling and navigation easier this avoids unsafe situations in the first place. However, on cruise vessels that often carry elderly passengers, some of whom are no longer able to use stairs with ease, emergency evacuation can pose a problem. Watch a mandatory lifeboat drill with its muster exercise on a big cruise ship and imagine this is for real.

Airline safety matters were treated with increasing vigour from the 1920s onwards, as soon as accidents began to involve the regular public rather than just dashing aviators with the occasional oddball guest. Air safety regulation concerning construction, equipment and the interior was already evident well before World War 2. Today's aircraft benefit from the ongoing evolution of just about every aspect of safety regulations, construction, materials used and operation, as well as the rapid development of telecommunications, control and navigation equipment.

Railways started to develop safety protocols seriously from around the 1870s following a number of serious accidents. This was increasingly pushed by the law from around 1880. The big issue initially was the interlocking of signals and turnouts to prevent signallers' mistakes, followed by development of fast and reliable telecommunications equipment like the electric telegraph. The next issue was to ensure that locomotive crews adhered to signal aspects shown, especially the restricted and danger aspects, through the introduction of electric colour-light signals and train protection equipment. On the vehicle construction and equipment side the introduction of capable automatic train braking equipment from 1865 onwards was very important, together with electric traction, lighting and heating and the introduction of all-steel integrally constructed coach bodies. By and large well-developed railways became the safest means of transport until the late 1990s, when European governmental privatisation issues induced ill-conceived profitability experiments aimed at diminishing the cost of operations. A number of accidents followed.

In comparison, private car safety was always different from public transport safety. For a long time no external authority interfered with car construction and equipment. The vehicles were regarded as purely private property in which the idea of safety played no discernible role. When the car later became a versatile and economical means of private transport, regard for safety matters remained absent, even when in the 1920s speeds of 160 km/h (100 mph) became a reality. This changed, however, in the early 1960s when American lawyer Ralph Nader investigated road vehicle accident causes and published his book *Unsafe At Any Speed* in 1965. To its credit, the US Government took note, probably because its armed forces lost more personnel to road accidents than to fighting communism in Asia. The US introduced relevant legislation in 1966, European legislation following soon after. Over time this led to much improved car construction and fitting of safety equipment, but it is of note that improvements in handling and safety features of the private car almost invariably lead to higher road speed and keeping of even smaller safety margins by motorists. Pre-1960s car crashworthiness issues may be perused in the pictures of American photographer Mell Kilpatrick, although it is not a particularly pleasant way to spend time.

Human fallibility

Human fallibility plays a decisive role at some point during any accident. Accidents happen because a) someone does not prepare a trip properly, b) risks are overlooked or underestimated, c) there is no relevant knowledge or insight to forestall or react properly to a deteriorating situation, d) someone is distracted or misjudges the situation or the fitness for work of vehicles and equipment, or e) someone, often not involved in controlling the vehicle, is being stupid or making mischief. Distraction or wrongly gauging the situation is especially dangerous during operation in more extreme ambient circumstances or during equipment failures. Distraction among professional vehicle operators is often caused by failures.

Equipment failure

This is characterised by three issues:

- Equipment components (their resistance against wear from operating circumstances being insufficiently robust) are not designed, tested, manufactured, maintained or checked to the extent required. This causes an initial failure.
- This initial failure causes more comprehensive malfunctioning that takes the situation beyond designed safety margins.

- The operator reacts incorrectly once failure occurs and takes the situation beyond safe margins. This indicates the importance of reliable equipment.

Corporate need to save on (financial) resources
The need to save money is usually the reason for not putting more effort into the design of and research into new equipment and its maintenance. This leads to insufficient reliability, flawed operating environments (e.g. cabs and cockpits) and flawed vehicle operating characteristics. Whilst inexcusable when major accidents occur as a result, this may actually stem from ignorance due to being the first to introduce a new generation of equipment (see, for example, the *Comet* crashes in Chapter 5).

Nevertheless, whatever the cause of an accident, bar some unforeseen natural disaster it always revolves around decisions made by people involved in the run-up to or during the actual occurrence.

What is safety?
In the public environment of medical, sports, transport and leisure venues, safety often appears to be whatever 'safety experts' say it is. This usually concerns reducing the operator's legal responsibilities when providing public facilities or services, or not wanting the bother that accompanies certain types of group behaviour. Hence the often irritating restrictions that arise 'in the interest of safety'.

A good transport and work safety performance refers to the absence of potential and actual harmful incidents during a previous period of travel or work; all the trips or the jobs done were completed as intended. Safety, then, is a statistical notion that is concerned with:

- How often does something negative actually happen?
- How often is this likely to happen again, looking at precursor incidents?
- Over what period?
- For what reasons?
- Under what circumstances?
- With what negative results, and how bad are they?
- How many people were involved?
- How many of those were affected?
- In what way?
- At what cost?

The outcome of these calculations decides whether an activity can be declared sufficiently safe or not, and in case not might trigger measures to reduce risk. Be under no illusions – the economic aspect of providing a certain level of safety is always a consideration! In the economy of work and transport the cost of provision of maximum safety is a high yet normally unproductive cost that you either can or cannot afford. If, as in many less developed countries, pressure to provide the best possible safety is not strong because it would make the cost of providing public transport commercially no longer interesting to the usually private operators, less safety is considered a risk worth taking. Otherwise there would be no public transport, which in turn impacts negatively on the economy of the nation.

Transport safety on your journey

Walking or using your car, motorbike or bicycle on public roads on your way to the station, harbour or airport is statistically the most unsafe part of your journey. Once on the bus, train, ship or aircraft you are considerably safer. Analysing differences between the various modes of private and public transport when getting from A to B shows that public transport employs bigger vehicles that are operated by relatively few staff, compared to the masses of individual motorists and other road users who would otherwise have undertaken the journey. However, the better safety performance of public transport is obtained primarily through effective enforcement of good traffic behaviour in the operatives. Public transport safety varies to a surprising extent in different nations as influenced by national habits, the financial welfare and the powers of transport safety authorities to implement and enforce a relevant safety culture, but as a rule public transport safety is markedly better than that among private motorists from the same environment.

Public transport staff

In the vast majority of cases public transport staff are professional specialists; they earn their income from expertly operating the vehicles on which you travel and therefore are usually expensive employees with a shaky hold on their jobs for reasons of performance quality, health issues and economic circumstances (hence the existence of unions with real clout). It is an important point. It means that the operation of public passenger and freight transport is in the hands of people with a vested interest in delivering a safe performance as their income depends on it. In developed nations independent authorities regularly check on the work performance of operatives with sophisticated equipment that leaves no possibility of hiding mistakes. Alcohol and drug abuse, checked by specialised medical agencies at various points in their careers, for example after an incident, leads to instant dismissal. Misbehaviour at the controls, such as aberrant driving, the use of mobile phones or being less than fully awake during work is recorded in detail. If it is found that public transport vehicle operators do not perform as prescribed, they are taken off operations and put through obligatory improvement courses, to be kept under observation to check on their progress. As a result, public transport is operated by people with a robust motivation to deliver a safe trip.

Additionally they work in an operating culture with a similar incentive, as accidents destroy public goodwill, are extremely costly and might well eliminate the source of income through company failure. Consequently, public transport is safer than private transport and, despite claims to the contrary, when working in a professional passenger transport environment it is clear that the pursuit of operational safety is the explanation for most that happens. Safety comes second only to earning a profitable return on investment, as generally without some profit there is no transport undertaking in the first place. But in my 33-year period of work experience within tourism in The Netherlands and with public passenger transport operators in Britain, the urge for profit never knowingly and with management consent stood in the way of pursuing safe modes of work.

Safe operation of motor vehicles by private individuals

In 1989 in Bexley, Kent, it was my experience that only one out of ten applicants for bus driver positions with what was still then London Transport Buses passed the initial intake tests. Then only six out of thirty people who passed the initial tests were allowed to start their actual bus driver training. Yet around 85% of those people who were rejected held a private driver's licence. This gives an indication of the problems with private motorists, who in comparison with professionals are not seriously tested for aptitude and mental suitability. A crucial difference between private and professional drivers is that the need for transport in private motorists is a derived demand from engagement in other activities, e.g. travel for work to earn a living. The quality of controlling a vehicle for private motorists is not based on published standards but on what the motorist thinks or feels it should be, and it is not possible to enforce safe behaviour through termination of employment. Most private motorists are indifferently trained, after which they are not periodically checked on their current skill and knowledge. There are often unreported problems with the already minimal physical and mental health requirements and, last but not least, motorists are too often in charge of indifferently or badly maintained vehicles. Many private road users tend to look on motoring as an ego-boosting birthright, experienced as a competitive sport rather than a potentially dangerous activity. This approach is reflected in driving under the influence of alcohol or drugs and in giving in to distractions caused by telephones, satellite navigation, music equipment, video gadgets and often hardly necessary warning lights and bleepers. Correction of negative driver behaviour by authorities is at best sketchy, and adherence to the rules is weakly enforced. Checking of private motorists' performance issues may take place until after accidents that involved fatalities or serious damage, but offenders are then not necessarily hit in the law courts with appropriate corrective measures that reflect the seriousness of their offences.

So why do rail accidents happen?

The main reasons for the occurrence of rail accidents caused by rail staff are fatigue, physical ill health, preoccupation/distraction, or a mix of the three. No different, in fact, from what happens during the vast majority of maritime, road and air accidents. As far as any 24/7 occupation like the railways is concerned, the ever-present fatigue is generated by changing shift patterns at all times of the day, which in due course upsets the circadian body clock. In turn this causes sleep, immunity, gastric and coronary problems that result in a perpetual state of feeling unwell, tiredness, inhibition of mental application and concentration on the job. I personally did not realise how tired I was until I was taken off trains in 2002 and resumed the normal course of daily life.

Fatigue, a main component of distraction from concentration on the job, may cause periods of micro-sleep behind the controls that last only seconds but can influence the course of quite a few incidents. Train drivers can become confused, unreliable in their assessments and reactions and, especially in darkness or fog, can lose track of their whereabouts. A problem might then be the location of unexpected danger signals that the driver has never stopped at before, or incomplete knowledge of less familiar signals, especially along lines that are not often used, such as diversionary or seasonal routes. If a line is worked regularly a difficult-to-spot signal or one with unusual indications is not normally a problem, but it may be so on less familiar lines. Since foreign train drivers work on other networks now, timekeeping stress may then become dangerous.

As far as distraction is concerned, rail accidents have been caused by drivers attempting to keep an ailing train going in order to reach a station or siding to clear the main line. The positive influence of reliable equipment on rail safety is proven, but its importance is not always reflected in the rolling stock reliability figures. Vandalism and other forms of upset also cause distraction, especially in darkness, when, for example, the windscreen is hit by a brick, people trespass on the track, or road users blunder across a level crossing at the last second. A really disturbing experience is to have a suicide use your train, or to pass a suicide location before any measures to stop your train and remove the mutilated remains have been taken. It has been likened to the shock military personnel suffer when first seeing death on the battlefield. In private there may be family tensions, with a partner and children unable to cope with the irregular shifts and the inevitable social alienation (as indeed is the case with other 24/7 workers). You are always tired on rest days and hardly ever available or willing to visit friends or relatives, unless you are granted annual leave days. Furthermore, driving trains has become tedious to many. Privatisation has led to the splitting up of formerly diverse work into a few core routes with just one type of traction, so keeping up attention becomes a problem. In fact, on many air, road and shipping routes similar problems play a role, and it could be argued that tedium was a contributory factor in the *Herald of Free Enterprise* car-ferry disaster at Zeebrugge in Belgium on 6 March 1987 in which 193 passengers and crew died (see Chapter 5).

Safety aspects of recruitment, training and crew management

An issue with an impact on safety is traincrew training and management. It is tied in with company reluctance to take on new staff due to the high cost of recruitment and training in the light of uncertainty about future work. In addition, suitable staff, as pointed out earlier, are hard to come by. Under the present passenger rail transport franchising and management contract systems in Europe work can go to another operator, and staff taken on and trained at great expense are then productive for someone else. Politicians have also discovered public transport as a means to parade their mettle and justify political ideology to influence decisions. Consequent ill-considered results have negatively influenced safety on more than one occasion.

Retention of traincrew is also a problem. Besides natural staff turnover, people are forever looking for more interesting, less demanding or better paid work. There are age-related health problems and temporary staffing problems caused by high exposure to seasonal illnesses. Then there is unexpectedly increasing business. As it is impossible to recruit and train new staff expediently, older people, with their more complete traction knowledge, route knowledge and experience, tend to get detailed to such new work and vacate their normal places on the rosters. Their replacement with other people, however, is rarely a like-for-like issue but triggers cascades with urgent additional training for the replacement staff. This reduces the ability of a traincrew depot to cover the allocated services with the available complement, resulting in stress on other traincrew to work the uncovered services in overtime and on booked rest days. People then cannot take up holidays or, in the worst cases, have reserved leisure time cancelled. This in turn causes family upset on top of the extra work stress and in such circumstances the mental resilience of railway staff is sorely taxed.

An additional traincrew management issue during recurring periods of staff shortage is that it is difficult to release individuals for training, monitoring and refreshing turns. It is hard to track performance and to ensure that necessary support is available if needed. In the situations described the occurrence of accidents caused by rule breaches and distraction is quite understandable, especially when it involves newly trained people who, through lack of experience and support, run into traps set by invariably less than perfect day-to-day operating conditions.

Rail operators long relied on gradual on-the-job training of new traincrew. The recruits progressed through the years from starter jobs – traditionally locomotive cleaners and signal box booking boys were for many Britons the starting point of their railway career – to the more involved positions. Staff trainers were usually former operational staff taken off due to health problems etc. Although these people had no training in education and no longer had a direct link with the 'real' job, from their experiences the recruits picked up what was needed, especially during their extended time with various practical tutors. It fostered understanding of relevant rules and procedures when they experienced incidents and saw their tutors take action and then explain, rulebook in hand, why things were done the way they were. Dealing with incidents that way became second nature before they took over the job. However, for various reasons, there is no longer time to train people slowly and grind basic railway attitudes, skills, knowledge and experience into them. Strictly prearranged, objective-referenced training sessions within a drastically reduced period have become important for cost reasons. Whilst nowadays it has legally to be demonstrated that traincrew are instructed according to pre-set standards, many issues they learn in the classroom hardly ever occur in real life and are probably forgotten by the time they are needed. People need time to experience incidents and so gain insight. Operators failed to pick up on that fact and neglected to provide sufficient refresher opportunities, so mishaps consequently occurred or were unduly exacerbated, as driving trains safely is as much a matter of experience as it is of theoretical knowledge. Furthermore, whilst understanding of safety issues did not always receive sufficient attention, the importance of timekeeping often did. This has been found to be a contributory cause in accidents.

The ever-present chain of events

Errors on the railway occur a number of times each day, as they do everywhere else. Yet accidents rarely happen, because the railway has developed a safety-layered operation resulting from previous experience that prevents a single mistake from causing a mishap or allowing events to spin out of control. Research into public transport accidents shows that it takes a chain of precursor incidents and mistakes to bring about a mishap, as normally no accident involving professional operators occurs because of a single root cause. Accidents that initially seem straightforward (e.g. a train driver passing a red signal) turn out to have causal strands that, taken individually, were of no direct consequence but together and in the circumstances of the actual accident played a decisive role. Distraction, fatigue, weather, time of day, signalling characteristics, rolling stock and other people can all play a part.

A typical example is the ***Ladbroke Grove*** accident close to London Paddington that took place on an October morning in 1999. It was this crash that first sparked my interest in writing about such matters. The signalling and layout from Paddington to Ladbroke Grove Junction were known to pose unusual risk. Some signals had a bad SPAD (Signal Passed At Danger) record, and due to the east-west orientation of the Great Western main line low early morning or late afternoon sunlight could refract in the coloured-glass Fresnel signal lenses, causing them to glow up as if all the signal aspects were lit. This is officially known as phantom aspects and as ghosting in jargon. Consider the following issues surrounding this accident:

- A high local passenger driving staff turnover, caused after privatisation by intercity and freight train operators poaching qualified staff with more pleasant depot environments, higher wages and more interesting work, occasionally forced local passenger operators to urgently recruit train drivers at London suburban depots.
- Consequently, new driving staff were taken on without sufficient background checks, which in the case of the driver involved in this accident led to his acceptance even though he had a conviction that would normally have disqualified him. Although

it had no influence whatever on the incident, it does show that it did not involve a first-rate recruit by the standards that the railways themselves had set.

- This recruit was quickly trained by an organisation which, post accident, showed a proven lack of educational competence. Route knowledge also received less than satisfactory attention. As a result, the train driver was undereducated when he passed out. His understanding of basic rail safety appears questionable.
- After passing out, and only a few weeks into his job, he approached the locally well-known multi-SPAD signal SN109 under restricted aspects through the fairly complex layout of six bi-directional tracks merging into four normal Up and Down tracks through a set of high-speed connections at Ladbroke Grove Junction when *potentially* the abovementioned sun-induced phantom aspects occurred at signal SN109, mounted at gantry eight with five other junction signals that additionally may have confused him.
- But in any case, and for whatever reason, *a ninth Category A SPAD within a few years occurred at junction signal SN109 and this was the root cause of the accident*. A similar SPAD event had occurred about a year before, but in that case the offending driver had stopped his train.
- Worse, however, was that not only had the SPAD gone unnoticed by the driver, but instead of stopping he accelerated. *The logical inference is that he considered it appropriate to do so. Based on interpretation of signal aspects? Did SN109 present phantom aspects?* It was proven to have shown a red aspect without route indication. Even in case of phantom aspects the driver should have noticed, stopped and queried, adhering to the relevant rules.
- This incident in turn wrong-footed signallers who had previously experienced a multitude of other SPADs in this area, but the drivers involved had always stopped and reported. *Therefore, the signallers were hampered by their expectations of what was going to happen*. According to accusations aired in the media, this caused them to fail to react sufficiently urgently to avert or mitigate the impending collision.
- Signallers had not regularly been taken through emergency procedures covering various scenarios, either theoretically or practically on simulators. They handled electronic equipment that was not set up efficiently to deal with emergencies, especially of this particular type. Furthermore, the rules and regulations

governing these situations were written in three different publications, one of which was a sheaf of photocopied and stapled A4 pages.

- It did not help that two different types of radio telephone equipment, National Radio Network on the HST express trains and Cab Secure Radio on the locals, were in operation, especially as NRN was not conducive to stopping traffic quickly in any emergency. CSR was better, but was not available on the express. It involved calling up a screen through rarely used menus on a computer.
- The signallers worked under a rail network management that reacted to reports on SPAD problems in the area (from the safety director of the lead train operator) with a conspicuous lack of urgency. This apparently was based on the irrational observation that train drivers should simply stop at red signals. Therefore no research into the reason for the surprising amount of SPADs in this small area was initiated.
- Flank protection at turnouts was not provided in the layout, as it had been designed with full Automatic Train Protection (ATP) in mind to provide the necessary safety. Application of ATP by then was abandoned in Britain on cost grounds, which resulted in the situation that on-board ATP equipment was not provided on all trains (such as the offending DMU) working in and out of Paddington, even though the track was fitted and the system was operational.
- Only the obsolete Automatic Warning System (AWS) was available on all trains. This system was not designed to intervene in the case of a driver reacting erroneously on a cautionary signal aspect but cancelling the warning. Therefore it did not stop the offending train when that was what was needed to prevent this accident.
- On a world scale uniquely high speeds of up to 100 mph (160 km/h) were allowed through this modern layout just a few miles from a major European rail terminus. As a result, especially on the main line side of the layout at Ladbroke Grove Junction, fast-running trains frequently crossed each other's pathways. The high frequency of crossing trains increased the likelihood of a disaster, and the speed exacerbated the result of the crash. Following this crash, TPWS (a British ATP substitute) was fairly soon fitted to both track and trains.
- A major fuel fire occurred after the collision because the trains

involved were diesel powered rather than electric. This substantially worsened the results of the crash. Electrification of track and trains on the Great Western main line will address this issue from 2017 onwards.

- Not initially obvious but demonstrable is the negative influence of the privatisation imposed on previously vertically integrated national public rail operators. Accidents that happened throughout Europe due, for example, to staffing issues of private operators would not have occurred under a single national operator.

The above issues conspired within a short moment that morning to cause this high-speed head-on collision and fire. Whilst a new driver was involved here, even experienced staff with years of problem-free work behind them might still be caught out, and that is when the need for a train protection system such as ATP or PTC is most convincingly demonstrated.

By the late 1990s, British train drivers were among the last frontline rail staff in western Europe able to cause massive high-speed train crashes. That situation largely ended with the long-awaited implementation of the Train Protection & Warning System (TPWS) in 2002 and 2003. TPWS is a system, however, that covers only half the signals on the rail network, as the majority of automatic plain line signals are not felt to carry enough risk to warrant the cost of installing this equipment. The foreseen restricted lifetime of TPWS (ETCS being on the horizon) is another motivator to avoid spending. Following fitment of TPWS in Britain the incident and accident rates have dropped only marginally, as predicted. The railways are very safe already, so the beneficial impact of such systems is of necessity limited (which is the argument usually made when disputing the need for such systems). At present this issue is being played out in the US where PTC is due to be installed. Nevertheless, in Britain, due to only part fitting of signals with TPWS, Murphy's Law still operates on the railways. The rear-end collision between two intermodal freight trains following a SPAD at *Shallowford* near Stafford on 16 October 2003 points to what is still possible; it was not prevented by TPWS, although its fitting had been finished along the West Coast main line and the system was operational on both trains involved. Luckily, no one was killed, but had this accident involved a passenger train, deaths might well have occurred and the start of the TPWS train protection period would have been tarnished straight away. On the other hand, some of the older ATP systems elsewhere in Europe did not prevent accidents from occurring either.

The technical side of a transport accident

In an accident with motorised vehicles on land, in the air or at sea a massive amount of kinetic energy, determined by mass (weight) and velocity (speed), has to be spent before the vehicles involved come to rest. The destructive effects on vehicles and their contents are furthermore determined by terrain, cargo and vehicle construction features. Trains may weigh from fifty to many thousands of tonnes, travel at speeds of 300 km/h (185 mph) or more and will usually take an appreciable time to come to rest. The process of losing kinetic energy after an accident is indicated by the damage caused to the vehicles and their environment; the character and the extent of the damage in turn gives an indication of the speeds and weights involved. A rough but gradual deceleration of rail vehicles is usually found in derailment and vehicle positions, missing bogies (trucks) and damage to bodies and underbody equipment. Passenger coaches may roll or come to a stop at considerable distances from the track, but nowadays with bodies and windows intact, offering integer survival space for the occupants. In earlier days, however, vehicles tended to telescope, break up and throw their occupants out through shattered windows and under the still moving vehicles. (Freight vehicles may do all these things but additionally rear up against each other or buckle.) A sudden violent stop may result in massive structural damage to vehicle bodies; they may well disintegrate. Such damage is caused not only by the actual collision but also by heavy pieces of equipment (e.g. the 4- to 6-tonne bogies) that detach themselves from vehicles and fly around.

A heavy head-on or rear-end accident usually causes a pile of wreckage from different vehicles somewhere at or near the impact point. Less damaged rolling stock that has broken away and is still on its wheels may climb or even move past this impact point. If the front vehicles of the colliding trains have 'sacrificed' themselves through, for example, weaker construction and have thus caused much of the collision energy to be spent whilst breaking up, the following coaches may be found with comparatively less damage. If there is 'no way out', individual rolling stock may end up in very strange positions such as on top of each other, extremely deformed or telescoped.

There is a line of thought that automatic couplers hold a derailing or colliding train better in line than the old type of buffers and screw couplers common in Europe. I query that. A modern articulated train such as the French TGV is more likely to stay in line on derailment or after collision than a train made up of individual vehicles, whatever the type of coupling used. This is safer for the occupants, as such a train tends to stay within or near its kinetic envelope, so the danger of a secondary collision with a train approaching on an adjacent track is somewhat diminished. Having said that, the electro-diesel Alvia Talgo articulated train involved in the high-speed *Santiago de Compostela* crash in Spain on 24 July 2013 was excessively damaged and broke apart, killing 79 and injuring 140 of the 222 people on board. Furthermore, what is certain is that a train coupled with central buffing and coupling gear has a far stronger tendency to jack-knife (concertina) than a train with the classic side-buffers and therefore may foul other tracks, leading to collisions. This phenomenon was clear in the ***Eschede*** accident, involving a train with central couplers between the vehicles of an ICE 1 set. Rolling stock involved in accidents in the Americas also tends to end up in such positions.

The root causes

Accidents can be grouped into a number of categories. To me there are three main groups:

- Those caused by railway staff, usually gathered under the heading **human error**.
- Those that occur to railway staff and the people or goods on the train they occupy, usually ascribed to **equipment failure**.
- Those caused by malicious intentions or erratic behaviour by the **general public**. This includes level crossing incidents.

All other categories, notably those concerning the design and construction of trains (e.g. the use of wood as construction material and the use of gas stored under coaches for the Pintsch coal-gas lighting) only come into play after the incident has been caused by one of the three main categories listed above.

Human error

This covers such instances as erratic (over)loading, erratic dispatch and handling of trains, erroneous control of train traffic such as through illegal use of interlock override equipment, leaving doors and hatches open, badly stowed and dislodged cargo causing derailments and vehicle failures, items reaching outside the loading gauge, Category A passing of signals at danger (SPADs), unauthorised moves such as entering or leaving a track engineering possession without permission, setting back following a SPAD or a platform overshoot without permission, running at excessive speed through speed restrictions, not doing brake tests as prescribed, failing to apply parking brakes after berthing a train causing runaways, and not adhering to other rules and standards such as harbouring expectations about signals coming off on time at junctions or stations ahead.

Much of this is down to train driver error and much has been countered by fitting automatic train protection equipment to signalling and trains. More strictly enforced procedures in yards and stations have also done their bit to ensure safer rail traffic, although things occasionally still go wrong but rarely catastrophically. Under normal operating circumstances signallers have in fact been prevented from make serious mistakes for a long time; from about 1870 onwards their equipment has been designed to ignore erroneous inputs at increasing levels of sophistication. Nevertheless, accidents have occurred relatively recently, partly due to obsolescent signalling equipment. In particular, the radio dispatch procedures after World War 2 were at the root of head-on crashes throughout the world. When the full protection of signalling interlocks has to be suspended to work around signalling faults or after disconnections during engineering work, potentially fatal mistakes can still be made. If train speeds are low and the traincrew involved are alert and conversant with the route, the consequences of such mistakes, albeit time-consuming, are not dangerous to the travelling public. However, serious accidents as a result of signalling issues have happened throughout the centuries. Such traffic control mistakes tend to occur when there is more traffic than the layout or traffic controller can comfortably handle. Not everyone can work under stress when attention to procedural detail is required (see ***Quintinshill***). This can safely be observed on model railways during demonstrations, when dealing with non-interlocked signals and turnouts sees trains end up in places where they should not be.

A contributory factor to accidents which is too little understood is that of language used during telephone and radio conversations. Staff are trained to use only certain expressions to avoid misunderstanding and to keep telecommunications conversations short, along prescribed lines. Yet 'real' telephone conversations develop when there is stress, instead of the businesslike exchange of information that enables the parties to assess the situation together and agree a method of working around problems whilst taking into account the provisions of the rulebook. In too many cases the railway has still to learn from air traffic control in that respect, but even in the air I have encountered some surprising deviations from what is required or even permitted, especially in multilingual areas. With regard to air traffic control telecommunications issues, who can forget what caused the terrible disaster with the two Boeing 747s on the runway at *Tenerife Los Rodeos Airport*? (see Chapter 5).

Equipment failure

Equipment failure on the railway covers various levels of seriousness, most of which have no bearing at all on the safety of the train. However, when a number of issues – not in themselves individually significant – come together suddenly a safety threshold may be passed, resulting in a dangerous situation. On trains extreme situations can arise, such as couplers failing, or wheels, axles or bogies coming apart under a speeding train, together with any number of other external influences that can determine the gravity of the situation.

In 1998, at ***Eschede*** in Germany, equipment failure in the form of the disintegration of a single wheel tyre set up a crash that resulted in 101 deaths, together with 105 badly injured casualties and 88 less seriously injured. During a similar wheel-failure incident with an East Coast Mk IV trainset on 16 June 1998 at *Sandy,* Bedfordshire, only a few weeks later, a vehicle derailed at 160 km/h (100 mph) but the train was safely stopped. Apart from nine slightly hurt passengers no one else on the train was aware of the seriousness of the situation until reading about it in the media.

In this book you will read about railway equipment failures, notably on freight wagons. Some curious things happened with tank wagons in the US, more precisely with tanks constructed of the wrong kind of steel with regard to the cargo they were expected to transport, and with disintegration of so-called stub-sill types of tank wagons, which at one time were popular in continental Europe as well. Incompetent loading of heavy, compact freight such as rock boulders for sea defences, steel slabs and steel coils have caused wagon frames to collapse en route. Overloading of bulk aggregates or bad locking of equipment has caused unloading doors to open and spillage of cargo over the top of wagons, and using tank wagons fitted with insufficiently effective internal slosh-baffles, in combination with bad track, has caused the derailment and overturning of wagons. On intermodal trains carrying ISO containers or swap bodies, absence of twist-locks has caused empty containers to be blown off a moving train during gales. Invariably these have ended up alongside the track or on neighbouring tracks, but luckily in most cases the steel bodies have caused track circuits to go down and have put signals back to danger. Bad weather has also been the cause of a number of accidents.

Signalling equipment failures nowadays undoubtedly lead to delay but almost never to accidents. A failed signal shows a red aspect and if there is no aspect at all, the signal has to be treated as showing a red aspect (stop) until the driver and controlling signaller have worked their way around it. When track failures such as buckled or broken rails occur, especially during extremes of temperature, signals are set to red and train traffic stops. Nevertheless, such failures occasionally cause undesirable situations with respect to train safety, but work on the elimination of this problem is ongoing. Turnouts can be problematic and lethal in cases of bad maintenance.

Collapses of railway bridges, embankments and tunnels are extremely rare but have unfortunately happened on more than one occasion. The tunnel collapse at *Soissons* on 16 June 1972 took 108 lives and was the third worst rail accident in France after ***St Michel de Maurienne*** and ***Lagny***. A few of the bridge collapses (such as the *Dee Bridge* collapse in 1847, *Ashtabula Creek* in 1876, the *Tay Bridge* disaster in 1879 and *Birs Bridge* in 1891) caused a great many deaths indeed.

Virtually all bridge designers and engineers in the world have had to gain their knowledge of design, construction parameters and methods, dynamic strains and stresses, properties of construction material, manufacturing and construction quality control as well as methods of maintenance the hard way. However, they have learned from each other's mistakes; international exchange of information regarding accidents in the civil, electrical and metal engineering professions has always been liberal, so lessons have been learned quickly. Consequently, since the start of the 20th century serious railway bridge failures under a train have been rare but have nevertheless occurred for reasons such as scoured piers and abutments during excessive and violent floods. Viaducts brought down by derailing trains ramming abutments or piers have without exception led to high-damage/high-casualty accidents. However, there is very little that can be done to avoid these events apart from properly maintaining trains and track, fitting check (guard) rails on sharp curves and at other potentially dangerous places, avoiding turnouts in the track in the immediate vicinity of a viaduct, and protecting the abutments and piers of viaducts against impact.

Railway bridges carry yet another risk: a derailment on a bridge may quickly turn into a full-blown disaster if the train consequently falls off the bridge or impacts with and weakens its structure to the extent that it collapses. Track on a bridge deck is usually constructed with check rails so that even a derailed train should be guided across without moving towards parts of the bridge or viaduct that are crucial to its strength or not constructed to carry it or – equally dangerous – leaving its kinetic envelope towards another track.

Last but not least, open railway drawbridges across shipping channels were involved in accidents more than once in the days before automatic train protection, as the root cause of the accident was invariably a SPAD at the protecting signals. Those in the US, with its 41 large railway drawbridges across deepwater shipping channels, have proved the most lethal. Similar incidents in Germany and The Netherlands, for example, have not resulted in quite such disastrous outcomes; pictures of the events tend to show steam locomotives or wagons in awkward positions above the water, although in some cases people were fatally injured. Vessels ramming bridges across navigable waters have proved lethal, even quite recently, although in Europe, Asia and the US, ATP and PTC play a role in preventing trains being involved in such accidents.

Even what to a lay person may look like fairly simple things to build, embankments constructed from aggregate such as crushed rock or soil, have claimed their toll in deaths and injuries. Washouts of embankments and bridges due to violent flooding after heavy monsoon rains in tropical climates continue to occur.

Equipment failure due to maintenance mistakes is classed as human error and does play a role in incidents when equipment reliability problems cause, for example, driver distraction resulting in a SPAD. Reliable track and signalling is safe infrastructure, and a reliable train is a safe train.

Aspects of public behaviour

Accidents caused by members of the public are often unintentional, as is the case in more than half of level crossing accidents (which nowadays are the main cause of railway mishaps). Sometimes, however, incidents – not accidents – are staged for personal gain through robberies, extortion of railway operators and theft of vital equipment (e.g. copper signalling cables) or for ideological purposes. Extreme examples, but with a long and sad history, are the planting of explosive or incendiary devices on trains and in stations and airports, as well as sabotage of track and signalling equipment. War is obviously another situation in which many crashes occur or trains are badly damaged by belligerent actions.

Far more run-of-the-mill are the type of incidents intended to provide some sort of personal satisfaction. These usually involve feckless juvenile and adolescent males, influenced by daring peer behaviour that gets out of control (especially when there are females around) and in the majority of cases is aggravated by being under the influence of alcohol or drugs. Internationally these situations have resulted in acts of inexcusably wanton and vicious vandalism. In one of my earliest contacts with a senior British Rail operations manager I experienced his late evening call-out to a site where a train driver had been killed by a steel construction bar that had been suspended across the tracks from a tunnel mouth and which speared his cab. Other examples include the throwing of projectiles at speeding trains, the suspending of metal items at windscreen level from overhead wires causing electric flashovers inside the cab, and depositing objects ranging from concrete blocks to wrecked cars on railway lines with the intention of causing derailment. More commonplace are things like ignoring stop lights and warning signs at level crossings, pulling emergency brake handles on trains without good reason, wrecking and soiling train interiors ('sports' fans) and exteriors (graffiti 'artists'), males lecherously attacking females, wrecking lineside equipment such as signals, and playing chicken on the track. These really are daily events throughout Europe and beyond, resulting in railway folk shrugging their shoulders and getting on with their jobs as virtually no one appears seriously interested in hitting the perpetrators with appropriate correctional measures. It is usually a matter of clearing the mess, repairing the damage and putting it down to a case of 3D (drunks, druggies and the deranged) in incident reports. Trains are also used for suicidal purposes surprisingly often, which puts a heavy toll on train drivers and lineside witnesses. The incident is often very hard to stomach due to the method used by the suicide and to the resulting mess in places accessible to the public, such as between platforms at stations or at level crossings. All the above goes to illustrate the at times rather strained interface between public transport operations and society with its various problems. Obviously, the infrastructure authorities and rail transport operators have very little leverage on influencing this situation.

Another serious issue is the way in which, on an international scale, many heavy road transport operators appear unaware of the dangers of level crossings to their precious transports, as exemplified by the few crashes that are mentioned in this book or the many to be found in international accident reports or on websites such as YouTube. The degree of ignorance they represent is truly shocking. Furthermore, bear in mind that many train derailments occur due to level crossing incidents with road vehicles. If the derailing train moves over into the kinetic envelope of the opposite track in the case of double or multi-track railway lines, what initially is no more than a bad fright to the people on the train quickly becomes a full-blown disaster if a train on the other track hits the derailed rolling stock. There are many such examples throughout the world, the very worst probably being a triple train crash caused by a truck driver that involved a freight train and two crush-loaded commuter trains on 9 November 1963 near *Yokohama* in Japan, killing 161 people on the two passenger trains. Very often major destruction is avoided only by the quick thinking of the train driver or on-board traincrew in stopping traffic on the other tracks, but there isn't always time and it is of course impossible when the traincrew themselves have been killed or badly injured. Incidentally, the risk of derailment after level crossing collisions is much mitigated by fitting obstacle deflectors under the head-ends of trains.

Initial conclusions

When a train crash happens, some of the observations published in the media may initially give an idea as to what has actually occurred.

Head-on, rear-end and sideswipe collisions

In most cases, with the exception of rare and serious signalling or train equipment failures, a crash results from one of the trains being involved in a SPAD at a protecting signal. If a crash happens when a train is crossing over between parallel running main lines and another train along one of those main lines runs into it, then the train on the main line is in 99% of the cases at fault, as it overran a protecting signal. Signalling systems will not allow the position of turnouts to be changed if the route is set, locked and cleared for another train, so the fact that the turnouts were set for the crossing train indicates that the responsibility for the crash lies with the colliding main-line train running into it. These are the kind of accidents where decent automatic train protection will stop both trains. Fortunately, widespread introduction of such systems in the more highly developed nations of Europe and Asia has dramatically reduced the number of such accidents.

Brake failures

Train brakes, especially the traditional type of air or vacuum brakes, need air pressure or vacuum power to be kept released, as in normal working order brake application is their default mode on failure. Throughout more than a century of widespread use, this automatic train brake equipment has proved reliable if kept properly maintained. It follows from this that whenever brake failure is mentioned as the cause of a train crash there is reason to be sceptical. In many cases brake equipment on vehicles involved in a crash has turned out to have been isolated, brake equipment was poorly maintained or brake blocks were allowed to be worn beyond the acceptable (hence, for example, the racks with replacement brake blocks between the platform tracks at mountain locations) and in other cases staff sloppiness led to the necessary brake tests not having been performed to check brake functioning, before moving off with freight or passengers.

There is another exception on the brake failure issue and that is slippery track. Slipperiness is caused throughout the year by humidity (possibly reinforced by frosty conditions) or by chemical pollution from industry. However, the best known reason for slipperiness is the seasonal leaf-fall period in the autumn; major problems are encountered internationally with the braking and accelerating of trains during this period. Among the treatments for this condition are the easing of train service schedules, drivers taking even greater care than normal, regular application of adhesion increasing gels as well as rail head cleaning with infrared heat or high-pressure water jets. Radio-telephone systems used to warn drivers of slippery spots was another great improvement.

Derailments

Many severe crashes result from a) a fault in the running gear of a train, b) a track alignment fault, c) turnouts shifting under a moving train due to technical issues or a mistake caused by a signaller or shunter, d) excessive speed through sections with speed restrictions of any type, curves, turnouts or faulty track, e) track failing under a train due to construction or maintenance issues, or f) hitting a road vehicle on a level crossing or other heavy objects somewhere along the track.

Fires

Fires, even disregarding deliberately ignited ones, are an ever-present risk on board all means of transport. During train fires, stopping quickly, except in tunnels or on high and long viaducts, is imperative, as there is the risk of feeding the fire with oxygen due to train movement. In the days of steam locomotives and wooden coaches there was the additional risk of fire started by scattered burning coal from the firebox, from the heating stoves in North American carriages and from similar equipment in European sleeping and restaurant vehicles, which was then often fuelled by escaping coal gas from damaged tanks for the Pintsch carriage gas lighting, suspended under the vehicle frames. It did, however, give impetus to the development of electric lighting with batteries and axle-driven dynamos to recharge them when the train was moving, and of steam heating. Yet on many European networks gas-lit passenger vehicles would remain in service well into the 1950s.

In the case of diesel traction, with tanks holding many tonnes of diesel fuel, fire is an ever-present risk. It can start when the fuel tanks are deformed violently as a result of a crash, and fuel, normally quite difficult to light when it is a cold liquid at rest, escapes as a well-aerated mist onto hot engines or is exposed to severe sparking. Such fuel fires then propagate themselves as the heat generates fuel vapours and often initially lesser-damaged parts of the wrecked train will also catch fire. On the road too, incidentally, fuel fires are a well-known risk during violent collisions involving heavy freight vehicles whose tanks are suspended externally beneath the vehicle frame. On electric rolling stock of any type there is a risk of short circuits causing fires, although the quality of maintenance plays a decisive role in this sort of occurrence, as does the speed with which automatic current breakers cut the traction current in the event of short circuits and damage to the overhead system.

What can be learned when looking at crashes?

When looking at pictures of a crash, a number of issues concerning the behaviour of rolling stock can often be identified. Rail rolling stock behaves in a number of recognisable ways under crash conditions: complete destruction, concertinaing or jack-knifing, telescoping, climbing on top of each other or ending up in positions well away from the track on which they travelled depend on the speed and violence with which the accident took place. Nowadays, designated vehicle crumple zones dissipate forces and keep the passenger spaces intact and the colliding trains in or as close to their kinetic envelopes as possible to avoid secondary collisions on neighbouring tracks. Before the 1960s such mitigating measures were technically impossible. In head-on collisions at speed with loco-hauled trains, the heavy locomotives tended to stay lined up against each other, cabs crushed back to the engine room bulkheads, often well entangled and with broken frames, the coaches having folded and jumped on top of each other and on the locomotives.

A typical characteristic of DMU/EMU rail-car trains and modern coaches is that as long they keep the shape of a tube they remain strong and will protect their occupants in violent circumstances. However, as soon as the tube shape is breached, the coaches tend to fold up quickly. As a result, most fatalities these days occur either due to a violent stop from high speed or to the outer wall of the coaches being breached, throwing the occupants out. Concertinaing or jack-knifing occurs when coaches of a crashing train end up in a zigzag across the track, losing their momentum through friction from scraping the ballast, soil and track, stretching their coupler springs and buckling their headstocks. The effect can be seen in many accidents with coaches that had centrally mounted couplers and no side-buffers; the latter appear to hinder jack-knifing to an appreciable degree.

Aluminium rolling stock of heat-welded construction introduced an accident characteristic all of its own: the vehicles rip apart cleanly at the welding seams or the aluminium body comes apart completely from the steel underframe. This mode of destruction characterises itself in very straight-edged pieces of wreckage often containing fully recognisable items such as windows or doors still in it. This problem is caused by the heating of the aluminium construction alloys during welding which weakens the alloy to some extent due to the various melting and burning points of the alloyed metals. Some de-alloying takes place, which changes the alloy-characteristics. This results in a diminished resistance against dynamic structural overloads as they occur during a crash, with resultant weld failure. It is for this reason that aluminium aircraft are riveted or nowadays chemically bonded rather than welded. A recent fabrication method for aluminium constructions, friction stir welding, avoids this problem.

Under crash impact conditions wooden and early types of steel rolling stock with lightly built bodies on separate wooden or steel underframes often showed extensive destruction of the body, or telescoping of one vehicle into another. Under the influence of the sudden impact deceleration the frames of the vehicles came out of horizontal alignment due to weight-transfer effects (such as a car dipping its nose when braking hard), allowing the rear end of the frame of the front vehicle to climb on top of the front end of the following vehicle frame, destroying the body of the other vehicle and maiming all within it as it slid along. In the present manner of so-called integral or stressed-skin construction of a passenger vehicle, the body is designed like a tube with strong end-stoppers called collision posts. This largely (but by no means completely) prevents telescoping, especially when the vehicles also have dedicated crumple zones to assist in dissipating the collision momentum in a controlled way. This means, however, that people in the corridors or using toilets etc may be trapped in the crumple zone on impact. In many accidents on electrified lines the stanchions of the overhead line equipment have caused serious damage to vehicle sides and roofs, endangering people on the train when derailments occurs. There is no solution to this problem at this moment, other than avoiding the initial incident in the first place.

In the case of so-called sideswipes, whereby one train hits another at a sharp angle, windows are shattered and the sides buckle, but the amount of damage usually depends strongly on direction of travel (with or against each other) and speed. Many low-speed sideswipes in the same direction result in nothing worse than two trains leaning at crazed angles against each other. In most cases of violent vehicle impact, deaths and casualties are severe, not only because of damage to vehicles and consequent intrusion into survival space but also due to the sudden and severe impact forces acting on the bodies of the passengers themselves. In a train running at a speed of 200 km/h (125 mph) that is suddenly stopped dead, a body is thrown forward and hits interior walls at an impact speed similar to falling freely from a height of 160 m (525 ft). Death in these circumstances is caused by organ damage and consequent internal bleeding. Princess Diana's death in Paris, travelling at approximately 120 km/h (75 mph) in one of the safest cars available, was caused in this way. She was not wearing a seat belt and was thrown about on impact as a consequence.

3: Glossary of Abbreviations and Terms

Items in *italics* have their own separate entry in the listing below.

AHB Automatic Half Barrier level crossing; ADB Automatic Double Barrier crossing

On an AHB the *level crossing* is protected by train-operated automatically closing and opening half barriers that cover only the entry points for road traffic to the crossing deck (also known as a two-quadrant *level crossing*). A derived and more extensively protected version is the Automatic Double Barrier crossing (or four-quadrant *level crossing*) where in addition to protection with half barriers at the entry side an extra set of delayed-closure half barriers on the exit side prevent zigzag manoeuvres by road users. In order to detect road users that get locked in between the closing barriers often some form of obstacle detection has to be provided, usually working with radar and infrared scanners or pressure-sensitive detectors in the crossing deck that will open the exit barriers to let the vehicle escape off the track.

AOC Automatic Open level Crossing

An automatic open *level crossing* is protected only by train-operated automatic light and audible signals to warn road users against approaching trains. In Britain an additional distinction is made between locally and remotely monitored AOCs. In the former case it is the train driver of the approaching train who checks if the crossing deck is clear of road users; in the latter case a crossing keeper, usually based in a remote *signal box* (*interlocking tower*), monitors the crossing with a CCTV link. In neither case has the AOC proven itself as effective as the AHB in separating road and rail traffic. In fact, many networks are now replacing AOCs with AHBs.

ARS Automatic Route Setting

This is the British generic name for automatic systems that manage train traffic control through recognising train services from the *train describer* equipment and then setting the relevant route through operating *turnouts* and *signals* for them. To that end the system has an electronic memory loaded with continuously updated timetables. It has a link to the *train describer* that reports the identity and detected position of various trains in its control area, and has override facilities for a human traffic controller/signaller to deal with ad hoc train movements that have not been entered into the memory or that are not recognised (e.g. through running in an emergency schedule or one that is badly out of sequence).

The main benefit of ARS systems is that they take care of the repetitive and tedious train regulation jobs and leave the traffic controllers to concentrate on monitoring services and initiate adjustment measures (if necessary) for individual trains. Software identifies development of individual delays well before these cause problems. These systems offer schedule adjustment facilities (such as order of passage through junctions, re-platforming at stations, etc) that after satisfactory rescheduling of the delayed service can be uploaded into the ARS. This is a major enhancement in the ability of a traffic controller to deal in a timely way with disturbances within the services in heavily trafficked areas and around large hub stations. From the ARS, other systems will then be alerted to the change (e.g. platform indicators and public announcements at stations).

ATP Automatic Train Protection; ATO Automatic Train Operation

ATP is the British generic name for any system that is primarily designed to monitor and, if necessary, stop trains in a safe place when they are going too fast or are about to go past a signal at danger. On many other networks (e.g. in Scandinavian nations) the North American acronym ATC is used for Automatic Train Control. System names are in [(B) TBL; (A, D) PZB/LZB; (F) KVB/TVM; (GB) AWS/ATP/TPWS; (NL) ATBEG/ATBNG; (CH) Signum/ZUB].

ETCS (European Train Control System) is the acronym for the various ATP functions of the pan-European ERTMS traffic management and train control system. In the US modern electronic Automatic Train Protection systems are being developed and implemented under the name Positive Train Control (PTC). ATP (or ATC) should not be confused with ATO, Automatic Train Operation, which is the British generic name for systems that automatically drive the train, independent of input from a human driver. ATP functions are of necessity included in such an operation.

Decent ATP should display its function parameters in the cab such as its status (on or off) and health; the nature of its failures; present, permitted and target speed restriction limits ahead; signal aspects ahead; and its reasons for intervention. Furthermore, these issues should be recorded on the *OTDR*. ATP supervises a driver regarding adherence to indications from lineside signals and also to permitted *speed limits*. In the case of an approaching red signal or speed restriction, ATP uses two ways of monitoring driver performance: 1) in older systems it monitors the fall in brake-pipe pressure (called the brake application criterion) to ensure that brakes are applied in order to come to a timely stop or reduce speed appropriately; 2) based on pre-programmed parameters it calculates three braking curves – a normal curve, a warning curve and an intervention curve, usually about 3 km/h (2 mph) apart, but in Britain 5 km/h (3 mph) apart – and then monitors the actual fall of train speed based on those curves. If the retardation does not satisfy the normal braking curve, a warning will be triggered on reaching the warning curve, and if the intervention curve is reached, an emergency brake application is initiated. Most systems then allow a driver to regain control over his brake again if the speed has fallen below the calculated normal curve, but in the case of full air-operated brake systems any speed below 40 km/h (25 mph) will result in a stop anyway, due to inherent lag between brake controller action and kinetic result.

Note that ATP is tied in with the signalling system. Unless it is linked with further monitoring equipment that causes the signals to be reset to danger, ATP will not react to, for example, cattle, vehicles, landslips, rock-falls or trees on the track ahead, vehicles violating warnings at level crossings, derailments or other mishaps. Other systems to deal with these issues are available or are being developed.

At present the various European ATP systems are the main inhibitors to traction roaming freely across national network borders. In order to cross borders, international traction must have all these systems on board, which apart from interfering with traction reliability and increasing the purchasing cost dramatically, also occupies on-board vehicle space. This issue is now being addressed with the introduction of ETCS, which at present is just another expensive system that has to be added to the mix already on board. It is hoped that at some time in the future ETCS will be the only necessary ATP system in Europe, but with the present rate of technological development, ETCS may yet become obsolete within a few years.

AWS Automatic Warning System
An obsolete British system to warn a driver that he is about to pass a *signal* that does not show a clear proceed aspect (green) or that he is approaching a more severe speed restriction. In the case of a warning, a horn sounds on which a button has to be depressed and released within 2.7 seconds for trains authorised to run at maximally 100 mph (160 km/h) and 2 seconds for trains authorised to run at 160-200 km/h (100-125 mph), to avoid the application of emergency braking. A visual indicator then shows a yellow and black segmented disc as a reminder of this cancellation.

If a clear signal is passed, a bell sounds and the visual indicator shows black. Other comparable systems are the old-style German Indusi (nowadays in an improved version known as PZB), the Swiss Signum/ZUB, and the French/Belgian 'Crocodile'.

Bi-directional signalling (reversible signalling)
A system of signalling for double or multi-track railway lines whereby all tracks are signalled for both directions in the way single-track lines are. Consequently, trains can use either track fully signalled in both directions. It allows overtaking manoeuvres or bypassing of obstructions. There are two main variants: 1) full reversible signalling where both tracks are split in *block sections* with a steady array of signals in both directions along both lines; 2) simplified reversible signalling where a 'wrong' line entry signal takes a train on to the reversible line and then has no block signals until the next crossovers are reached, which are protected by a distant and a stop signal. The latter system is used primarily for emergency moves during blockades etc, as it allows only one train at a time in the extended single-line section between the crossover locations. Trains cannot follow each other at block distances as normal, which causes delays in the case of two trains following each other, with no service in the opposite direction. In the event of booked single-line working during track engineering work this is of no great consequence, but otherwise it may compound delays that over time can become severe. This is obviously better than nothing in an emergency situation, but is something to be avoided if not strictly necessary.

Nations such as The Netherlands, Belgium and Switzerland (and the North American continent in the case of fully signalled double or multi-track routes) are extensively signalled in this manner, whilst it is fairly common in Germany and France for example. It is not as common in the UK, and British train drivers are therefore less conversant with certain characteristics of reversible signalling. [(B, F) banalisation; (D) Gleiswechselbetrieb; (NL) Dubbel-Enkelspoor]

Block section
This is often referred to as a signalling block or signal section. The word 'block' is internationally used and indicates a section of *track* that is protected by a *signal* at its beginning and ends at the next signal that protects its next block section. Basic railway safety dictates that under normal operating conditions only one train may be in any one block section at any time (train-separation signalling as opposed to time-interval signalling). Block sections must be of a length that enables a train to comfortably come to a halt at a red *stop signal* from *line speed* at the yellow *distant signal* with a normal service brake application. If, however, shorter block sections are applied then the *signalling* is made capable of slowing a train down to stop over two or even more sections. In Britain this explains the 4-aspect green/double yellow/single yellow/red sequence a driver will see; on many continental systems it is usually a green/yellow plus digit or yellow-green/yellow/red sequence, but in Belgium, for example, it is a green/single yellow/diagonal double yellow/red sequence. In the US on many networks you may see a green/flashing yellow/yellow/red sequence in this situation.

Where, in stations or yards, more than one train has to be allowed in a section, e.g. to attach to an available train in a platform, this is known as permissive block. In all cases this can only be done at very low speeds on the authority of a *signal aspect* that shows that the section is not clear to the next signal. The driver must proceed at a speed that enables him to stop before hitting any obstruction in his path. Permissive block is also used on many networks on congested routes, whereby a train approaching a permissive signal that shows a red aspect is allowed to pass that signal after having stopped and to continue at a speed that enables the train to be stopped for any obstruction ahead. This principle is known as 'Stop & Proceed' in Britain, but is no longer normally used there except in the case of an automatic signal from which the signaller cannot be reached due to complete telecommunications failure. In the age of the mobile phone this is a very unlikely scenario, although there are still circumstances where a driver might decide to use it.

The term moving block signalling refers to the train receiving a safety section ahead of and behind it, which moves with the train within its *distance to go authority*. As each train along the line has such a moving exclusion zone/block section, which is known to a central traffic control computer system in continuous radio-contact with all trains, the computer ensures that no train safety exclusion zone is allowed to come closer to another than the calculated safety distance, which inherently functions as a form of *ATP*. Equally important is that in the event of a sudden emergency stop, all trains behind the one that has stopped receive the emergency stop order at the same time and are made to stop as well, pretty much in the way in which a line of cars on a road come to a stop provided all car drivers are alert. It will be clear that moving block is not a system that can be safely used to its maximum potential under human control; it is designed for centrally controlled trains being operated under *Automatic Train Operation* conditions.

Bogie

The assembly of *wheels*, bearings, suspension, *brake* and motion damping gear in a moving sub-frame under a railway vehicle. This sub-frame can rotate in the horizontal plane under the vehicle body to follow curves and also allows movement in other planes to follow rise and fall in the *track*. In the great majority of bogies the suspension between wheel bearings (axleboxes) and the bogie frame is called the primary suspension, and the suspension between bogie frame and vehicle body is called secondary suspension. There are a few examples with third and even fourth suspension stages (e.g. some of the German pre-war Görlitz types). On any train ready to run one will find *traction vehicles* and *trailer vehicles*. Traction vehicles have bogies with powered axles and can move the train; trailer vehicles have only non-powered axles. On many networks this equipment is known by its North American name of truck. [(B, F) bogie; (B, NL) draaistel; (D) Drehgestell]

Brake testing

Far more important to traincrew than that a train will depart is the certainty that it will stop, and for that reason various brake tests are prescribed before a train first comes off depot, when setting up cabs, after vehicles have been removed from or added to the train, and when brake system isolation features (usually air cocks) have been operated.

A full brake test involves someone walking along the train checking for isolations and application and release of the brake blocks and pads whilst listening for possible air leaks in the brake and main reservoir air pipes and the vehicle's own air systems. A brake continuity test is employed to establish whether all vehicles have an open connection to the brake pipe to enable the brakes to function under the control of the driver's brake valve. There are also tests to check that the various safety systems in the cab, designed to stop the train should problems arise, work as they should. After departure the driver will perform a running brake test to check that the train brakes retard as prescribed, to avoid the unpleasant situation of finding a fault when approaching his first station stop or a first red signal at a junction. The classical air-operated train brake is a very reliable piece of equipment, being fail-safe in the event of brake-pipe air loss or loss of electrical charge in the more modern brake control equipment. If an accident were caused as a result of so-called brake failure, I would be ready to bet that one or some of the brake pipe inter-carriage valve cocks were not open and that a full brake test or brake continuity test had not been performed, resulting in part of the train not braking as required (or at all) and very possibly the train not coming to a stand at the intended location. Such runaways have led to accidents happening as a result. [(B, F) essai de freinage; (B, NL) remproef; (D) Bremsprobe]

Buffers
In Europe this term is used for the two sprung contraptions at the lower end of the *headstock* at either end of the buffer beam. The buffers are fitted to keep the individual vehicles from colliding; the *coupler* (also sprung) holds them together. The buffers of a passenger train should just firmly touch, whereas those of a freight train should be slightly apart.

Burrowing junction (dive-under)
See *Flyover (flying junction)*.

Cab (footplate)
This is the place from where a train is controlled by a *train driver* (NA: *engineer*). For that reason there usually is a cab at the front of the train (in the case of steam locomotives, usually set well back from the front). With coupled-up locomotives, push-pull trainsets and *multiple-units* there may be cabs at either end or mid-train as well. However, the location of a driver's cab should never be used as an indication of the direction in which a train will move; always look for white *headlights,* red *tail-lights* or tailboards. Multiple red and white or white-only light indications normally show that the train is being shunted (switched). [(B, F) cabine; (B, NL) kabine; (D), Führerstand]

Cant
Cant is a feature of track on curves whereby the outer rail is laid higher than the inner rail, causing the train to lean inwards. This compensates

to a certain extent for the centrifugal forces generated as a result of the speed of the train whilst negotiating the curve. As this often happens at widely varying speeds, the cant can rarely be ideal for all trains and is therefore a compromise. The rate of compromise is known as cant deficiency. To accelerate certain services on lines with many curves, tilting trains may be used, which increase the compensating effect of track cant by leaning even further inwards either through a passive inertia effect or, more usually, through action of electronically controlled electric or hydraulic rams in the *bogies*. Such active tilt equipment on trains must be accompanied by safety measures to ensure that failed tilt equipment does not allow the train to maintain the higher speed through the curves. [(B, F) dévers; (B, NL) verkanting; (D) Spurneigung]

CD/RA Close Doors/Right Away indicator
The British equivalent of indicators on many networks with which platform staff indicate to a driver to close the doors and give permission to move away from the platform stop. A nice touch is that these indicators usually are interlocked with the starting signal aspect; they will either repeat the red main aspect or not work if that signal is at red. [(B) marguérite; (D) Abfahrtsignal]

Changing ends
This involves a train driver moving from the front cab to the rear cab in order to change direction, whereby what was previously the rear cab becomes the front cab. A locomotive-hauled train may be worked this way under push-pull conditions. [(B, NL) kop maken; (D) Kopf machen; (B, F) changement de cabine]

Check rails
These are extra rails fitted between the running rails on sharp curves and bridges in order to assist in preventing derailment. However, if derailment does occur, check rails are intended to keep the derailed vehicle in line with the track and the rest of the train.

Coaches, cars, carriages (see also *Bogie*)
Railway passenger vehicles are normally referred to as carriages in British English, although vehicles that make up an EMU or an Underground train are invariably referred to as cars, which points to the strong US influence on multiple-unit and Underground vehicle construction and operation in Britain. A coach is a later name for a passenger vehicle with an open saloon layout, and in daily use has now

reached parity with the expression carriage as most passenger vehicles have a coach layout anyway.

Nowadays a rail passenger vehicle normally consists of an integrally constructed steel or aluminium body mounted on two *bogies* that contain the running gear and brake equipment. The carriage body – the place inside which passengers are carried during a trip – consists of a frame, headstocks at either end, two sides with cantrails (top edges to the roof) and solebars (bottom edges to the underframe), and a roof. The frame, originally the strong and rigid underfloor member upon which the body was built, is now an integral part of the carriage body construction, welded together with headstocks, sides and roof. The fitting together of all these parts gives the carriage body its strength. The headstocks are the end bulkheads of the body, where the *coupler* and (in Europe) *buffers*, brake pipe and main reservoir air connectors, traction, communication and heating electricity jumper cable and tail-lights are located together with the gangways connecting the coaches with each other. [(B, F) voiture, wagon; (B, NL) rijtuig; (D) Wagen; (NA) car]

Coasting
Coasting occurs when a train rolls under its own momentum without using power. This is often employed downhill and on the approach to stations in order to conserve energy if running on time. Coasting boards are positioned at locations where a driver can shut off power.

Collar (reminder device)
A simple piece of equipment in a mechanical or *NX* signal box, attached to operating levers or put over buttons on a panel to remind a signaller that the equipment concerned should not be operated. In mechanical signal boxes these collars are metal rings that can be slid down the handle to block operation of the unlocking lever, making hurried or ill-considered moves with that handle impossible. In boxes with NX equipment they are usually partly hollowed out magnetic discs that can be applied over the buttons that should not be operated. Incidents have occurred when the signaller has omitted to put a collar over the levers or buttons to remind him that he should not allow trains to enter the track concerned. Computerised systems employ something similar.

Concertinaing (jack-knifing)
Concertinaing is one of the effects caused by the shedding of energy that results from a severe collision or other abrupt stop of a train. Coaches derail and shed their kinetic energy through folding up in a zigzag along

the track, *derailing*, losing *bogies*, etc but often staying upright. Although centrally mounted automatic *buffer/coupling* devices tend to produce this behaviour more than the screw-couplers with side-buffers still prevalent in Europe, there have also been similar accidents in Europe when the vehicles were coupled with centrally mounted *couplers* without side-buffers.

Couplers, couplings and buffers

These devices are used to attach the various vehicles that make up a train to each other. There are hundreds of different types of couplers but only a very few are interchangeable. The *buffers* are separate sprung devices that keep the vehicles from smashing into each other; the sprung couplings hold the vehicles together. Together they provide a flexible connection that nevertheless enables close coupling of long vehicles.

The main distinction between systems concerns whether the couplers have to be attached and disengaged manually by a *shunter* standing between or next to the vehicles or whether the couplers engage automatically on contact and can be disengaged either with an automatic power system or by a shunter at the side of the train. When automatic couplers are fitted they usually take care of the coupling as well as the buffing functions; they are sprung in two directions. A further distinction is that some types also automatically connect the main reservoir and brake air pipes between the vehicles, and moreover some types additionally connect the hotel electric power, traction control and train management power lines.

Whilst in Europe the international standard is still the ancient screw-link coupler with side-buffers, multiple-units and fixed-formation trainsets are usually fitted with fully automatic Scharfenberg-type couplers of German origin (patented by Karl Scharfenberg from Königsberg in 1903). These are also marketed under various names such as Radenton and Dellner. Strangely enough, despite their common ancestry these are not in fact interchangeable. The German-developed BSI (now Faiveley) couplers have also found their niche in this market but are not compatible with the Scharfenberg types.

The most widely applied automatic coupler in the world is the strong, beautifully simple and versatile AAR or Buckeye knuckle coupler of US origin, derived from a design by Eli Janney (patented in 1873). The less simple Willison coupler (patented in 1920 by John Willison from Derby) has found wide application as the SA3 in Russian and Chinese dominated areas and is also the designated European standard automatic coupler. Although it would be expedient to introduce a standard automatic coupler throughout the European continent, this is unlikely to happen for a number of reasons. One reason is that sprung side-buffers are indispensable for closely coupling the present-day long continental coaches whilst still taking them through sharp curves, but it is also the case that the transport budgets of many European nations do not stretch sufficiently to allow this kind of expenditure. Willison-type couplers, however, are in daily use for heavy continental freight flows. [(B, NL) koppeling en buffers; (D) Kupplung und Puffer; (B, F) attelage et tampons]

Cutting (see also *Embankment)*
Earthwork engineering whereby a hill or low mountain is traversed on the level by digging away material on the line of the route under construction, leaving high sides bordering the railway line. The sides of cuttings often have to be reinforced with retaining walls to keep them from collapsing onto the railway line. If the hill or mountain is high, a tunnel may be a better alternative, in which case the tunnel entrance will be situated in a cutting. [(B, NL) insnijding; (B, F) coupure; (D) Einschnitt]

Dark territory
An originally US term to indicate a remote railway line that is not fitted with stop signals en route. In the days before radio it was difficult if not impossible to instruct drivers to stop in emergency situations (e.g. heading towards each other on a single-track line).

Dead
1. An expression to indicate that no traction electricity is available on electric vehicles, which therefore cannot be moved under their own power. Vehicles that are connected to traction electricity are live. Dead and live similarly apply to all electric systems. It is railway practice to consider any electric system to be live until expressly indicated by the relevant authority that it has been isolated and is dead. Even then, additional measures are taken (e.g. earthing) to ensure that no errors from those isolating the system or from remaining (static) voltage can do harm. (Doing so saved my life once when replacing a popped 750 V dc shoe fuse on the Southern third-rail system during a winter with ice on the conductor rail, inhibiting a steady supply following which the shoe fuse went and had to be replaced en route.)
2. A non-powered traction vehicle included in a train formation as a hauled vehicle is denoted as dead. This is often used to forward traction

vehicles to locations where they are required to pick up a train, which avoids having to use a track slot and traincrew to work it *light*. It might also be a failed traction vehicle being transported to a repair depot.

Degraded working of signalling
A British expression to indicate working signals under conditions of faults, failures or *interlock* disconnections for repairs and signalling engineering work. As a result, the full security given by signalling is no longer available, so signallers and train drivers must adhere to various protocols enshrined in rules and regulations to ensure maximum safety under the circumstances prevailing. Many times per day some form of degraded working will occur and usually no one on the train notices unless they are conversant with the normal way of working or happen to look out of the window and see, for example, that a signal at red is being passed after an unplanned stop. However, there is undeniably an increased potential for things to go wrong under these circumstances.

Detonator (torpedo, fog signal)
A detonator is a capsule with metal straps that is filled with an explosive to provide an audible emergency signal and is fastened across the rail head. It explodes under a train wheel and so warns traincrew about any danger ahead. The train has to be stopped immediately and the signaller or his local representative contacted before the train may be moved again. [(B, NL) knalsein; (B, F) pétard; (D) Knallkapsel]

Detraining
This British expression refers to taking passengers off a train. It often describes a situation in which this happens by force of circumstances, as an evacuation of sorts, and can take place on the open line away from a station.

Distance to go authority
This term was introduced with the advent of in-cab displayed *signalling*, but equally appertains to fixed block signalling with signals along the line or with moving block signalling without any signals at all. It refers to the distance ahead that is cleared for occupation by the approaching train.

In the case of fixed block signalling with signals, the distance to go is ruled by the amount of signalling sections with cleared signals ahead. With the better *ATP* systems (e.g. TBL2 or ETCS level 1) this can be shown in the *ATP* display in the cab. In the case of fixed block in-cab displayed signalling without lineside signals (e.g. LZB, TVM or ETCS level 2) the display will clearly show how many sections ahead, separated visually by trackside section-separation boards or *repères*, are clear to be passed and at what speed. With moving block signalling I have yet to see how the distance to go permit will be displayed in the cab. However, as stated earlier, the system is based largely on *Automatic Train Operation*, so a driver in charge of a train movement then is likely going to be at low speed only following an emergency.

Down, Up
British terms used to identify a particular track of a railway line. A down line in Britain usually leads away from the London terminus and the Up line towards London, except in the case of the former Midland Railway where the Up direction would take you to the city of Derby. Locally additions to this direction may be made, such as fast and slow (with the exception of the former Great Western Railway, which used main and relief for that purpose) and goods or avoiding lines.

Driver (train operator)
The person responsible for safely controlling the speed of a train and keeping a lookout for signals, people and obstructions along the track ahead. [(NL, B, F) machinist, conducteur; (D) Lokführer; (NA) engineer]

DSD Driver Safety Device
A British expression for the dead man's switch. Usually a type of pedal (on the Southern EMUs, for example, the lever of the power controller had to be pushed and kept down) that must be depressed against spring action by the driver to prove that he is functioning as required for safety. If the lever or pedal is released, an emergency brake application is initiated which then automatically cuts out traction power and stops the train. Actually, the system can all too easily be compromised by a body falling onto the pedal or heavy bags being put on the pedal or being hung from the controller lever. On older trains it is also possible for a driver to give his arm or foot a break when the train is *coasting* by putting the direction switch in neutral (locking up) and thereby immobilising the DSD. To avoid these dangerous acts the *vigilance switch* (NA: alerter switch) was introduced. [(B, NL) dodeman; (D) Sicherheits Fahrschalter (SiFa); (B, F) l'homme mort]

ECS Empty Coaching Stock, empties
ECS is a set of empty passenger carriages on their way to a depot or the starting point of a train service. A train of empties may also consist of freight wagons. [(B, NL) Leeg materieel; (B, F) matériel vide; (D) Leerzug]

Electrification
The following should be borne in mind: ac stands for alternating current [(B, F) courant alternatif; (D) Wechselstrom; (B, NL) wisselstroom]; dc stands for direct current (B, F) courant continu; (D) Gleichstrom; (B, NL) gelijkstroom].

Electric traction current is normally fed to traction vehicles through a system of wires suspended above the track, called catenary or Overhead Line Equipment (OLE) [(B, F) caténaire; (D) Oberleitung; (B, NL) bovenleiding]. If this system is used, the traction vehicles pick up the current with a pantograph [(B, F) pantographe; (D) Stromabnehmer; (B, NL) pantograaf, stroomafnemer], a folding frame on the roof that has a pick-up shoe sliding along the underside of the OLE contact wire. If all pantographs have been retracted to the rooftops of the vehicles in a train these are *dead* as far as traction current is concerned and cannot move under their own power. It is important to be aware when interfering with the electrics on a *dead* traction vehicle that battery power (often at 110 V dc) may still be available until cut by the Battery Isolation Switch (BIS).

Another means of traction energy transmission is the third rail [(B, F) raille électrique; (D) Stromschiene; (B, NL) derde rail, stroomrail]. A pick-up shoe fitted to the *bogies* of the traction vehicle slides over, against or under the third rail fitted on the *sleepers* at rail level. This system is normally used only for low-voltage (600/750/1,500 V) dc systems. In some systems (e.g. the London Underground) a fourth rail is fitted for the return current.

The tension, or pressure, of the electric current is expressed in volts (V) and if higher than a thousand volts in kilovolts (kV). In the case of ac, the frequency of pulse changes per second is expressed in hertz (Hz), kilohertz (kHz) or megahertz (mHz). The power to deliver traction of a locomotive is expressed in watts (W), if higher than 1,000 watts in kilowatts (kW) and if higher than a million watts in megawatts (mW). Its pulling power at the drawhook is officially indicated in megapond (Mp). The electric current drawn to do the job is expressed in amperes (amps). At lower tension networks (volts) the current drawn (amps) to supply a particular power (watts) will be higher than doing the same job on a higher voltage ac network, a process that heats up the conductor wires. Hence many dc main-line networks with their lower voltage can be distinguished by two heavy contact wires in the catenary, to reduce the chance of wires overheating and breaking.

In continental Europe there are four main railway traction electricity networks: 1.5 kV dc, 3 kV dc, 15 kV ac at 16.67 Hz and 25 kV ac at 50 Hz. 25 kV ac has *de facto* become the standard the world over, the frequency being 60 Hz in nations that generate their industrial current at that frequency. Because modern electric traction usually is wholly or partly prepared for working off any electrification system, it is no longer the different traction current systems that inhibit electric traction from crossing changeover points but the *ATP* systems applied. It is also for that reason that national borders no longer limit traction energy systems; changeover points may be found anywhere, including within national networks.

Embankment

Earthwork engineering whereby the railway is taken across a low-lying area on top of a longitudinal artificial hill in line with the rail route. An embankment is the opposite of a *cutting*, and in more hilly or mountainous areas embankments and cuttings often follow each other in order to keep the line level or the gradients within specification. It is also a case of carefully calculating the spoil from the tunnels and cuttings to make up the body of the embankments, the so-called balanced construction of a line. If an embankment would be very high, it is usually better practice to put in a *viaduct* to save space, save on construction effort and ensure the stability of the line. [(B, NL) spoordijk; (B, F) remblai, terrassement; (D) Bahndamm, Böschung]

EMU/DMU Electric/Diesel Multiple-Unit (see also *Multiple-unit remote control*)

A passenger or freight trainset with its own traction and with control cabs at either end, for which reason no separate locomotive is needed for traction. In some texts translated from foreign languages into English they are described as motor-trains, self-propelled units and in North American texts as railcars. [(B, NL) treinstel; (B, F) rame; (D) Triebwagenzug]

Experience

Experience is needed to control train traffic safely, whether as a driver, a train manager or a traffic controller. It is only after starting to work trains or signalling equipment on their own, following training, learning the necessary routes and spending considerable time working under both close and remote tuition, that new workers really start to gain experience. Making mistakes is an almost given part of that process, although the level of mistakes made should not be lethal for those who get the basics right as a result of effective training and tuition. On the other hand, for every fatal mistake made by one person, hundreds of others safely work trains with the same training, good or bad.

Fail-safe

This is one of the basic safety principles of rail transport operations and indeed many other types of safety-critical work. If *safety-critical* equipment fails it does not create a dangerous situation. For example, if signals fail they are not allowed to show a less restrictive aspect than they would show if they had not failed. Basic signalling is designed to show a red (danger) aspect on failure, and the power brakes of a train are designed to apply in case of failures, such as accidental division of the train or leaks in the brake system.

Flyover (flying junction)

A flying junction is one that avoids the dangers associated with 'crossing track' conflict by carrying the conflicting tracks on a viaduct or bridge. The only tracks that touch each other are separating or merging routes on which the trains normally travel in the same direction. The opposite type of split-grade junction is the burrowing junction or dive-under, whereby the diverging tracks go down into a short tunnel under the main tracks. As well as eliminating conflict between opposing trains, which additionally cuts out waiting on that side, the replacement of an at-grade or flat junction with a split-level junction is often driven by the fact that permitted speeds through connecting turnouts can usually be increased substantially, which facilitates quick clearance of the junction to allow another service to pass.

Ghosting, swamping

These are British railway jargon terms that denote the detrimental effects of bright, low-angled sunlight on the visibility and discernibility of *signal aspects*. This phenomenon occurs when low-angled sunlight (in the early morning or late afternoon) shines into the coloured and shaped Fresnel or half-spherical glass lenses of the older type of colour-light signals, causing them to light up through sunlight refracting inside the lens glass (ghosting) whilst at the same time making it almost impossible to see the actual signal aspect through the power of the incoming sunlight (swamping). Often the phenomena occur all at the same time, the ghosting non-lit aspects of a signal being equally powerful as the swamped lit aspect. On more than one occasion train drivers passed a ghosting or swamped signal without adherence to the rules about 'signals not or imperfectly shown', guessing the actual signal aspect based on previous aspects passed.

When considering ghosting problems at the root of an incident, do not forget that your perception from looking at a photograph differs greatly from encountering the phenomenon at speed and having to take a *safety-critical* decision based on your perception right there and then. The only correct way for a driver to deal with this situation in case of doubt is to stop and call the traffic controller to ascertain that it is safe to pass the signal, but if working with *ATP* drivers should consult their displays as most systems give an indication as to what the true signal aspect is.

These days, all over the world, the older type of colour-light signals with tungsten light bulbs and coloured lenses are being replaced with optical fibre or LED (Light Emitting Diode) signal illumination, which no longer have shaped coloured glass lenses. If sunlight should cause trouble then the lens lights up white. As these signals also have multi-dot lighting instead of a single light source, there are always good visible light dots in such signal aspects. Furthermore, the normally weaker colour light of a signal – the green aspect – can be tuned up to be as powerful as the other aspects. Whilst some problems with colour perception at short distance have been encountered by drivers wearing varifocal glasses, on the whole the LED signals are a major advance with regard to easy and comfortable reading of signal aspects en route.

Headcode

The headcode is the train identity number, in Britain previously displayed with large digits at the head of the train to enable its identification by the signallers along the line. Headcodes died out with the installation of *train describer* equipment but the term is still in daily use in Britain. [(B, NL) treinnummer; (B, F) numéro du train/service; (D) Zugnummer]

Headlights/tail-lights

These are the white and red lights that indicate the presence of a train on the track and show in which direction it is likely to move. The need for train drivers to use their headlights to see ahead of the train, for example in darkness, is limited, as at speeds higher than 40 km/h (25 mph) they are unlikely to be able to stop for whatever shows up in their headlights, but effective headlights undeniably help them to see signs along the route and, on remote stretches, to see wildlife and blow the whistle in an attempt to get the animals off the track. Animals (with the exception of pheasants, wood pigeons, sheep and cows) usually demonstrate more common sense than humans in that respect.

Virtually all over the world the present standard is to show at least three white headlights in a triangle (to provide instant recognition as a rail vehicle amongst other traffic headlights) and one or two continuous or flashing red tail-lights. Many freight trains in Europe, however, again show the old German tail indication board: a square, diagonally crossed red and white board on one or both corners. This results from the fact that when crossing borders non-stop the expensive portable and rechargeable red flashing tail-lights went missing, whilst some people near yards had a habit of removing these tail-lights from parked wagons for their own purposes. The cheap plastic tailboards make sense in that situation, also because there are no longer many signallers or even manned stations along the lines to notice from tail indications whether or not trains still are complete.

Britain shows one, two or three white lights at the front, not necessarily in a triangle, one of which is brighter than the others and may be on the right side in the daytime and on the left side at night. It is a creative interpretation of the idea of head and marker lights. In Switzerland you may see that the end of the train is indicated by a single white light rather than a red light at the right-hand side in the direction of travel, especially on the many narrow-gauge and rack railways.

Note that head and tail-lights only indicate the front and the rear of the train and not, for example, the position of the locomotive; it may be working on a push-pull basis. When shunting is in progress, many networks show a single white light or a single white and single red light at both ends. Flashing head and tail-lights always are an emergency signal and indicate to approaching trains to slow down and, if possible, stop. If a train is parked there is no front or rear and usually both ends will show red lights to indicate an obstruction to other trains on that track, which is similar to a red light on buffer stops. On many networks the red light or red board on buffer stops in fact constitutes a signal and is properly preceded by a distant signal. If a driver hits the buffer stops in that case, he has not only caused a buffer-stop collision but has also committed a Category A *SPAD*.

In-cab signalling systems
These signalling systems do not require lineside colour-light *signals* to separate signalling *block sections*. The only visual displays separating signalling sections are so-called block section boards or *repères*. The driver obtains all the information he needs about the state of the line ahead from his signalling and speed display in the *cab*, so the only reasons to look out of the windscreen are to check on obstructions on other lines and people or animals on the track, or to come to a stop at the *repère* covering a so-called closed *block section*. The reasoning behind such systems is that drivers can no longer reliably spot and interpret lineside signals under all circumstances at speeds above 200 km/h (125 mph). Full in-cab signalling systems in present use in Europe are the French TVM, the German LZB and the international ETCS level 2.

Interlock
Interlock is one of the basic safety principles of the railway (and exists in other safety-critical environments). It denotes a situation in which a particular action cannot be undertaken unless a demand or a series of demands with safety implications have been satisfied. For example, a signal at the beginning of a required route will not show a proceed aspect if a) the path for that particular train has not been cleared by a previous train, b) no conflicting routes have been set up, c) all protecting signals show red (danger) aspects, and d) all turnouts along the route are in the correct position and locked.

Interlock exists at many levels of train operation. For example, in Britain, with its cramped *loading gauge*, trains with power doors have a traction interlock; they will not release brakes and power up unless all exit doors are closed and locked. Domestic examples of interlock include electric coffee grinders and blenders that will not work until the lid is properly closed and locked, or washing machines and dishwashers that do not start until the door is correctly shut.

Jack-knifing
See *Concertinaing*.

Kinetic envelope (see also *Loading gauge*)
The kinetic envelope is the space above the *track* occupied by passing *rolling stock*. It takes into account the maximum permitted sideways and vertical motions made by these vehicles on their suspension. A passing train will hit anything intruding into the kinetic envelope.

Level crossing

The location where rail and road traffic cross each other's right of way and priority has to be established. Rail traffic has priority over road traffic by law in the vast majority of nations, which is recognition of the fact that a rail vehicle at speed cannot be stopped when the driver sees a road vehicle approaching the crossing.

In most nations level crossings are allowed only on single- or double-track lines with a permitted line speed up to 160 km/h (100 mph) and below certain levels of occupation, measured in train and road vehicle movements per hour. If those values are exceeded, grade-separated crossings (tunnels or bridges) must be installed. In fact, on most networks no new level crossings are allowed except in well-defined circumstances concerning road traffic speed and density versus train speed and frequency. Many nations try to eliminate level crossings from their rail networks due to the impossibility of enforcing safe behaviour by pedestrians and road users. Level crossings therefore carry too much risk for the safety of trains, especially where road freight vehicles are concerned. [(B, NL) spoorwegovergang, overweg; (B, F) PN/passage à niveau; (D) Bü/Bahnübergang]

Light

A British expression for a locomotive without a train that is worked under its own power to a particular destination. This usually concerns a so-called balancing trip to bring the locomotive to a place for its next booked working with a train from that destination. If at all possible, however, such a locomotive will preferably be forwarded *dead* in a train service going to that destination, thus saving on track slot and crewing costs (and possibly in recognition of the fact that light locomotives can be surprisingly difficult to stop in an emergency). In The Netherlands, at the time when locomotives were still used for all types of trains, one or more dead locomotives between the working locomotive and the train were frequently seen for balancing reasons on lines to and from the borders with Germany and Belgium. Now, in the great majority of cases, locomotives cross the border with their train.

Line voltage

A British expression to indicate the electric power for traction on the overhead electric lines or the third rail. If this has to be switched off, isolation of the power is requested by contacting the local traffic controller or the relevant electrical controller directly. In the case of the main-line third-rail direct current (dc) systems south of London and around Liverpool, Paris and New York, and earlier on the Maurienne line in the French Alps, however, there is the possibility of short circuiting this traction power against the nearby running rail with the short-circuiting bar to ensure that no traction current remains available.

Loading gauge

The loading gauge – the space above the *track* exclusively reserved for occupation by passing trains – is usually the same as, or larger than, the *kinetic envelope*, but whilst in the case of the kinetic envelope nothing is allowed to intrude, in the case of the loading gauge nothing on a train is allowed to come outside the gauge. It is the guaranteed edge where a passing train will not hit lineside objects. Nevertheless, if a train needs to run out of gauge, all sorts of special measures are necessary. These include extremely careful plotting of the route that the train will take (e.g. avoiding narrow tunnels), whilst traffic on neighbouring tracks has to be stopped and special routes through stations along tracks away from platforms have to be used. In such cases, *signals* or overhead electrification masts and wires may have to be temporarily removed.

On the European continent the so-called Berne gauge rules the international loading gauge, although many nations internally have substantially larger loading gauges than that. Britain, however, has a considerably smaller loading gauge, and whilst British rolling stock in theory can run freely on continental tracks, only HS1 from London St Pancras to the Channel Tunnel will accept Berne gauge vehicles from the continent without restrictions. On the other hand, the Channel Tunnel Shuttle vehicles between France and Britain run within the largest European loading gauge by far. In North America the loading gauge of many railway lines has been extended to accept rolling stock loaded with double-stacked intermodal containers, but many lines east of Illinois and along the north-eastern seaboard still have an insufficient loading gauge to accept these vehicles when loaded to capacity. [(NL) profiel van vrije ruimte; (F) profil transversal; (D) Lichtraumprofil]

Loop

See *Siding*.

Manually operated level crossings

These are *level crossings* at which the barriers are lowered and raised by a crossing keeper, either at the location of the crossing or remotely controlled and monitored by a CCTV link. As the keeper checks whether the level crossing deck is free of obstacles, manual crossings may still be protected by full-width barriers.

Maximum speed
An expression that causes confusion, as there is more than one definition of maximum speed. First, there is maximum line speed, which may be split as applicable to passenger/postal *bogie* stock and to other/freight stock. Then there is maximum train speed, which is usually based on the type of running gear, braking capabilities and, again, whether it concerns passenger/postal or freight stock limited by the maximum line speed. Along any line there may be limited maximum speed areas, known as permanent speed restrictions, to cover certain locations where the maximum line speed would cause excessive wear or even danger (notably on curves), and temporary speed restrictions at locations where track has become defective and is awaiting repair.

If, for example, during ultrasonic checks serious track faults are identified, urgent emergency speed restrictions may be implemented, the main result of which is that trains must be stopped at covering signals and drivers informed of the location and the applicable speed restriction. If the location is later signed with the relevant boards and, for example, *ATP* is adjusted to enforce adherence to the restricted speed limit, it then becomes a temporary speed restriction.

Multiple-unit remote control
The expression refers to the possibility of automatic remote control of multiple traction vehicles in a trainset by one train driver. This multiple-unit traction facility was used virtually from day one by connecting the electric traction control circuits on self-propelled carriage units. Hence the reason why these trains became known as multiple-unit sets (see *EMU/DMU)*. The opposite method of work to increase traction power is running in tandem, whereby two or more traction vehicles are coupled up at the head of a train (double, triple or quadruple heading), but each traction vehicle has its own crew on board to work it.

Double-headed diesel or electric traction units may be operated as multiple-units via jumpers in their traction control circuits unless their multiple-unit control systems are not compatible. This is an irritating characteristic in Europe where, as a result, in some cases the old US 8-notch standard traction MU system was installed instead. Another possibility is a wireless radio-remote traction control link. Multiple-unit operation enables a locomotive positioned at the opposite end of a train, or mid-train, to be remotely controlled from unpowered driving trailers with driver's cabs (remote control cars) at the far ends of the train or, in the case of *shunting*, by a man with a portable radio-remote control box on the ground. His task is to take care of moving the train as well as the turnouts, automatic loading or unloading, uncoupling, coupling up and brake testing of the train. Incidents have happened, however, when this man with the magic box had failed to ascertain which radio-locomotive he was actually controlling. Whilst working the buttons and staring at the stationary locomotive he thought was defective, a similar loco elsewhere caused havoc.

NX train traffic control

e**N**trance-e**X**it types of rail traffic control systems were a typical offshoot from the rapid development that electro-mechanical relay-based science went through during World War 2. Such systems were available well before that period, but the war brought about widespread know-how in many different control applications, improved reliability and US mass-production capabilities that sought new markets. This was then aided by Marshall Plan facilities to repair the destruction of rail hardware and control equipment in Europe, as well as the need to minimise labour demands due to the shortage of available manpower. Consequently, reconstruction of post-war European rail networks was often carried out with up-to-date and labour-saving equipment. Hence the choice of automatic signalling, automatic level crossings and highly automated traffic control systems on the European continent, and the reason why US-manufactured electro-mechanical traffic control equipment became so dominant until well into the 1970s.

NX has considerably reduced the work of the signaller, from pulling and pushing sequences of many sometimes very heavy lever handles in a mechanical locking frame to simply entering the desired route on a panel showing the entire layout and signals by pushing a start and an end button for the various sections to be traversed by a train. The NX system then visibly checks whether the requested route is available and, if so, sets all the necessary *turnouts*, locks them and indicates this on the panel. It finally indicates the successfully requested and locked route with a line of (usually) white, yellow or green lights along the route on the panel and enables the necessary signals to come off for a train to proceed. No other route that even remotely conflicts with the cleared route can then be accepted. The *train* for which this route was set up, having received proceed *signal aspects*, now moves along this route, which is indicated on the panel by red lights coming on for every *track circuit* occupied, whilst the concurrent (white, yellow or green) light line for that track circuit section extinguishes. It is only when the train has occupied the track circuits in that manner and has cleared them again (the red lights then also extinguish) that another – conflicting – route may be set up. NX is still in use in many traffic control centres worldwide, and it is fascinating to see signallers working with it, especially when *train describer* systems have been added.

Modern traffic control centres work with computer-screens and a mouse or a tracker ball. Click on the start point and then on the end point and, from that moment on, the indications are pretty similar to the situation described above. But instead of clattering rows of relays in the relay room next door or below your feet, a silent and surprisingly small computer does the job for an unexpectedly large area, and far faster and over far greater distances than the NX installations do, using wireless secure telephone links.

On, Off

British expressions indicating whether or not a *signal* shows a proceed aspect. If a signal shows a proceed (safe) aspect it is 'Off'; if it shows a stop (danger) aspect it is 'On'.

OTDR On Train Data Recorder (data logger, event recorder, trip recorder)

On Train Data Recorder is a British expression for an obligatory type of equipment on every train in Europe. It records speed, application of power or brakes and many other types of actions such as the use of the horn and the cancelling of warnings of train protection systems. All *ATP* actions are recorded on the OTDR. This type of equipment is also referred to as the 'black box', data logger or trip recorder.

After an accident the recordings can be downloaded onto a laptop computer and displayed on screen or in print in various formats, one of which is the graphic format that displays speeds and a number of technical parameters as lines on a time/distance graph accurate to a tenth of a second. It is surprising just what this reveals about the way the train has been driven. Unfortunately, unlike an aircraft cockpit voice recorder, it does not record what was said, for which reason the causes of some accidents (e.g. ***Åsta*** in Norway) still remain enigmatic. What happened is recorded in great detail; why it happened is still unknown.

Such recorders have a long history. Even before World War 1 the Swiss firm of Sécheron-Hasler supplied self-registering speedometers. Initially only speed was recorded, but later this was expanded to include things such as brake-pipe and cylinder pressures against a time graph on a waxed paper tape. The accidents at ***St Michel de Maurienne*** in 1917, ***Harmelen*** in 1962, ***Granville*** in 1977 and ***Åsta*** in 2000 were all investigated with speedometer recorder tapes from the locomotives involved.

Many train drivers talk about 'the spy in the cab', but I have seen more drivers exonerated from serious blame by the system than those who were found to be illegally operating a train, notably by speeding. Perhaps the most infamous case in Britain, on 2 November 1994, was that of a *DMU* train that skidded with locked wheels on wet track covered by leaf mulch for a kilometre (three-quarters of a mile) before slamming into the buffer stops at *Slough* in Berkshire at 50 km/h (30 mph) from 93 km/h (58 mph). Without the OTDR to confirm what happened, that driver would have stood little chance of protesting the allegations of late braking. In any event, when drivers are brought to book for mishandling their train they get to see the graph and what it reveals and so are able to give their interpretation of the incident.

Nevertheless, many networks do not use the OTDR to check on driver competence at work, although such things as speeding are also picked up by other monitoring equipment along the track, but there is no denying that the use of the OTDR to monitor train driving competence, as on the lines on which I used to work, has materially improved the quality of train driving. Bad handling is picked up and dealt with before problems are caused, whilst issues like wheelslip and skid sites clearly reveal themselves, as the wheelslip and slide protection systems on the train are recorded as well, so the issue can be accurately targeted and dealt with by the track authority thereby enhancing safety during leaf-fall periods.

Incidentally, it would be very interesting to see what would happen on the roads if motorists were monitored by GPS-connected trip recorders in their vehicles. Issues in the run-up to an accident, such as location, speed, skidding and use of the pedals, gearbox and steering wheel, would be revealed in detail, which would facilitate accurate apportioning of blame (whether on the driver himself or, for example, the highway authorities for not gritting during periods of snow or black ice). I am sure that the roads would quickly become much safer, and insurance companies would no doubt benefit enormously. On the other hand, the system would doubtless be unpopular with car and motorbike drivers of the speed-junkie variety, for which reason it would be important to ensure that the system is tamper-proof.

Peak hour, rush hour, off peak

The rush hour, or peak hour, is the time of the day when many people travel to work and later back home again, creating a concentrated stream of people into cities and towns in the morning and a slightly less concentrated outbound stream in the later afternoon and early evening. This phenomenon causes many problems, especially on urban forms of public transport, but transportation systems in urbanised areas are geared up to deal with peak hours: the maximum number of rolling stock and staff are at work and the timetable shows the densest amount of trains the network can carry to get the crowds to work and back home again. During the off-peak period between the rush hours, however, most trains are parked in storage sidings not earning their keep or are being maintained in workshops. Urban public transport operations are, therefore, often not remunerative and money has to be found in order to provide subsidies. Any operator that succeeds in maximising the off-peak use of its trains will be commercially successful and may even see a decent return on investment.

For public transport operators, problems with peak hours are based mainly around the following issues:

1. Much rolling stock is needed for the few peak hours per day but is not used off-peak. This makes operation difficult, with a lot of empty stock movements feeding into traffic and a couple of hours later out of traffic twice a day. This not only temporarily occupies much of the network capacity around termini but also extends the flood of non-revenue-earning train movements in the peak period to either side of it.
2. Observation of passenger behaviour on trains during the off-peak period is difficult due to fewer staff being on duty. Consequently, in many areas, crime and serious vandalism become a problem in the relatively empty carriages. For this reason, many are taken out of service and kept in sidings, although equally often, to counter vandalism on parked trains, they are kept on the move in service trains. However, large-scale introduction of CCTV on stations, trains and other premises, plus heavy penalties imposed on perpetrators, has mitigated this problem considerably.
3. In city centre stations expensive space has to be created for private vehicles, whether cars or bicycles, that are parked for a large part of the day and have to be protected against crime. This space requirement and the wish to keep road traffic out of city centres has given rise to the development of so called Parkway or Park & Ride (P&R) stations away from built-up areas. These,

however, cause extra stops for trains, resulting in increased energy use, longer track section occupation and an extension of travel time.

4 The fact that during peak hours large amounts of people are on their way to and from work causes a massive headache to passengers as well as operators. The great majority of these people travel on public transport that is not only crush-loaded and uncomfortable but, in Britain in particular, is also expensive. As these rail networks are operated to maximum capacity, any disruption that occurs rapidly causes massive delays. Unsurprisingly, the negative opinion of rail travel held by a vast number of people is shaped by such experiences and often causes behavioural issues on peak-hour trains that in turn fosters a further deterioration in comfort, adds to the delays and impinges on safety, which results in even more inconvenience for the travelling public.

5 As far as accidents are concerned, peak-hour incidents always cause maximum harm to passengers, as the vehicles are heavily loaded (if not overloaded), many people having to stand rather than being seated. On impact, these bodies fly freely through whatever spaces are available and violently impact with each other as well as with seats and luggage racks.

[(B, NL) piekuren, daluren; (D) Spitzenstunden, Talstunden; (B, F) heures de pointe, heures creuses]

Permanent way (P-way)
P-way is the British term for the infrastructure operator's department that maintains the track. The North American equivalent is Maintenance-of-Way.

Rail (see also *Track*)
The equipment, nowadays often manufactured of rolled and heat-treated steel, which carries and steers trains and makes up the *permanent way*. It is made up into *track* by being fastened to sleepers (NA: ties) and embedded in crushed and graded stone ballast or directly fastened onto a concrete base in the case of paved track. The rails transmit the forces of a passing train through the fasteners and baseplates (earlier: chairs) onto the sleepers, which in turn transmit these forces to the underlying ballast bed. The function of the ballast is a) to distribute the static and dynamic forces generated by a passing train into the undersoil, b) to hold the track in place, and c) to disperse rainwater efficiently. For the latter

reason, ballast should be cleaned regularly to get rid of soil incursions, which both hinder the draining function and promote unwanted plant growth that in turn brings only more wind-blown soil deposits into the ballast.

Nowadays rails worldwide are rolled into a profile – the Vignoles profile – with a horizontal flat foot, in the middle of which is the equally flat vertical web, on top of which is the roughly square rail head. On top of the two rail heads roll the two opposite wheel treads per axle of the passing train. Rail is fastened to the sleepers with a slight inward lean, whilst the tapered wheel treads are shaped to fit into this lean. This is what actually keeps the wheels on the track and steers the train. The wheel flanges do not generally touch the rail head; if they do, the rail head quickly wears.

Rail may be manufactured in short lengths, joined together with fishplates, nuts and bolts into what is known as jointed track. Alternatively the sections may be welded together in longer lengths, called Continuous Welded Rail (CWR). A serious problem with steel rail is its reaction to ambient temperature changes. If it is cold, the rail shrinks in length and may snap in extreme temperatures; if it is very hot, rail expands lengthways and the track may buckle. Both snapping during cold spells and buckling or kinking during hot weather have caused accidents and great care goes into ensuring that the track and its bedding can take the stresses that occur due to these climatic influences. [(B, F) rail; (NL) spoor; (D) Schiene]

Railway

A railway is a means of land transport whereby the vehicles used for traction and for transport of passengers and freight (trains) are both carried and steered by track, which may consist of one (monorail) or, usually, two rails. This means that the job of the train driver is only to control the speed and not to steer the train. The person with limited influence on where the train will go is the traffic controller or signaller, who controls *turnouts* and signals. The fact that different people are in charge of speed and direction contributes significantly to the safety of rail operations.

By steering the train, the track guarantees passage through a relatively narrow *kinetic envelope* that permits the making-up of long trains from individual vehicles and so creates the massive land transport potential that distinguishes railway operation. An additional advantage is that the number of people required to achieve this transport potential is very low in comparison with road traffic for example. A high degree of automation of traffic and train control reduces this human employment requirement to an absolute minimum (e.g. with dedicated Australian mining rail operations work with *ARS* and *ATO* controlled trains, there is no human involvement in the operation of the train and signalling during its long trip), the available jobs being of an administrative, supervisory and maintenance nature rather than operator jobs. Public railways are indeed working towards achieving this position.

It should be noted that not all railways use or have used steel wheels on steel rails. In the late Middle Ages wooden track and grooved wheels were used on many central European mining lines. On Dartmoor in Devon are the remains of the Hay Tor granite quarry tramway, which used grooved granite blocks set apart at the required gauge as track, whilst in Paris, for example, Metro lines have tracks shaped to carry and guide rubber-tyred rolling stock. This sort of track bears a close kinship with what has become known as the guided busway, whereby specially adapted street-running buses arguably become rail vehicles as their track carries and steers them. Their drivers only take care of acceleration and stopping, in the same way a train driver does. [(B, NL) spoorweg; (B, F) chemin de fer; (D) Eisenbahn]

Redundancy

Redundancy is the third main safety principle that rules safety-critical rail (and maritime and air) operations. Failure of safety-critical systems components never disables the complete system and thus avoids creating potentially dangerous circumstances. Often redundancy is based on two-out-of-three architecture of computer-based control systems. Normally three computers run the show and as they are all programmed with the same software and receive the same inputs, their control decisions should be the same. If, however, one computer starts to deviate but the other two continue to agree, the deviating computer is locked out from further decision-making and a fault indication is given to the operator who usually has the means to reboot all three computers in order to bring the locked-out one into circuit again.

Another type of redundancy to ensure reliable functioning may be provided by fitting two or three safety-critical operating systems in tandem, whereby one single system still operates the safety-critical equipment if the others should fail. Accurate and clear fault reporting to the service and maintenance department is very important, and is often done automatically through a GSM (Global System for Mobile Communications) network. This also enables decisions of an operational nature, such as replacement.

Repère

A repère is a French/European lineside board with a blue background and a yellow triangle or arrow pointing towards the track it governs. Under *in-cab signalling* systems it replaces a signal as the signal section separating agent. Like signals, repères can be at danger ('closed'), but this is indicated only by the *ATP*/signalling display in the cab. Repères may have a subsidiary signal attached that allows trains to pass them at very restricted speed into a closed section under strict operating rules, comparable to a permitted passing of a signal at red.

Rolling stock

The term rolling stock encompasses all vehicles that run on and are carried and steered by rails. [(B, NL) rollend materieel; (B, F) matériel roulant; (D) Eisenbahnfahrzeuge, Rollmaterial]

Rolling stock classification

The power source is often the first reference when classifying rolling stock, and may be steam, an internal combustion engine or electric. Internal combustion engines are subdivided by transmission: mechanically, electrically, hydraulically or hydrostatically. The most common classification system, however, involves the wheels or axles. There are several of these wheel/axle notations in existence, and I have deliberately selected the full UIC axle notation as the most unambiguous. This system departs from the premise that axles and wheels on a train actually form one unit, referred to as an axle. It then applies a numerical digit to undriven axles and a capital alphabetic character to driven axles, whereby an added (o) identifies whether the driven axles are powered individually and independently from one another or whether they are coupled. It further identifies axles in a separate *bogie* frame by adding a (') to the notation. A coach on two bogies with four axles, then, is a 2'2'. One of the heavy 12-wheelers on two 3-axle bogies is obviously a 3'3', and a 4-wheeled van is a 2.

Diesel and electric locomotives

Most modern locomotives are shown as either Bo'Bo' or Co'Co' types – machines with independently driven axles in either two 2- or two 3-axle bogies. Some 6-axle machines have three 2-axle bogies; these are known as Bo'Bo'Bo' (or Tri-Bo') traction. If a locomotive has bogies with coupled driven axles, such as diesel-hydraulic and certain types of French-built electric machines with monomotor bogies, a 4-axle locomotive is a B'B' and a 6-axle locomotive a C'C'. Older British types of diesel-electric locomotives had two non-driven fixed axles with six driven axles in two bogie frames to spread the weight, 1Co'Co1', but the famous Pennsylvania Railroad GG1 electric locomotives were of the 2'Co'Co'2' type. Japanese railways build locomotives for lightly laid curved track that have an undriven axle or a carrier-bogie amidships, Bo' 1 Bo' and Bo'2'Bo'. Similarly, many twin-bogie 6-axle machines for lighter track are in fact A1A'A1A' or Bo1' 1Bo' types. A simple shunting locomotive with two independently driven axles is a Bo, but on two coupled axles a B, on three coupled axles a C, and on four coupled axles a D. A Norwegian class El 8 electric locomotive is a 1'Do1', a machine with a single undriven axle in a pony-truck, four independently driven axles in the locomotive frame and one undriven axle in another pony truck. The similar-looking class 1000 of Netherlands Railways or BR18 of the German Railways, however, are 1A'BoA1', a machine with an undriven axle and a single driven axle in a bogie, two independently driven axles fitted in the locomotive frame and another bogie with a driven and an undriven axle. Other specials are the British Crossley Co'Bo' and the US Fairbanks-Morse C-liner Bo'A1A' diesel-electric locomotives. Some French Maurienne electric locomotives from 1927 were 1A' Bo'+Bo'A1' double electric locomotives with a non-driven and a driven axle in a bogie and a bogie with two independently driven axles per single locomotive.

Steam locomotives

A 'Pacific', in the Whyte notation a 4-6-2, is usually a 2'C1' in the UIC notation, but may be a 2'C1 with the trailing unpowered axle fixed in the frame with Cortazzi sliding axleboxes rather than in a pony-truck. A British Class 9F and the well-known German Classes 50 and 52 (Decapods or 2-10-0 according to Whyte) are 1'E types. A Union Pacific 'Big Boy' non-compound Mallet machine (Whyte 4-8-8-4) is a 2'D'D2'; note the fixed rear driven axles. There was a World War 2 German 'steam-motor' locomotive with individual drives that under the Whyte notation would have been a 2-8-2, but in the UIC notation is a 1'Do1'.

Route knowledge

In many accidents described in this book route knowledge is something that was often conspicuously lacking. It appertains to traincrew being conversant with the features of a railway line in all sorts of operating conditions to enable a train of any type to be driven with confidence and in the most economical way. In fact, this is not usually an officially defined knowledge on many networks, and it often looks a bit like an impressive piece of wizardry to a visitor in the cab. Yet in many cases it is dependent on the amount of work the driver in question has privately invested over time in order to acquire it. As long as there are humans driving trains along older railway lines with all their historical oddities and imperfections, route knowledge will play an important role, and any driver who drives more than once along a route will of necessity start to acquire it, as do car drivers on their daily commuter run and animals on their pathways through the jungle where they live. Airline pilots will tell you that their first time on a glide-path approaching certain runways during a landing is quite a nerve-tingling experience, even after having been thoroughly trained on the simulator, until they acquire confidence through their particular kind of route knowledge. The same also applies to maritime officers bringing a vessel into port or sailing along a river, especially under difficult conditions.

In Britain traincrew route knowledge is viewed as important and includes awareness of the location of uphill and downhill gradients, curves with particular characteristics, relevant controlled signals and their particulars such as route indications and accompanying speed restrictions and their involvement in more than one *SPAD*, signals on awkward locations, level crossings, tunnels and large bridges, leaf-fall skid and wheelslip sites, permanent and longer-term temporary speed restrictions as well as stations, complete with relevant features of their layouts (shunt moves) and the braking points to stop at them. (In the UK there are no station reconnoitring boards displayed along the track as on other networks.) A British train driver will not take a train along a certain route until he has learned it and signed for it, because when something goes wrong the responsibility rests fully with that driver.

Many signalling engineers rate train driver route knowledge as undesirable because it is too ephemeral and too connected with the drivers and their personal characteristics, making it of variable quality and outside of objective control. Speed indicating and in-cab types of signalling were introduced to enable drivers to operate trains without prior route knowledge, the driver just having to follow the various speed indications at speed signs, the signals or on the display. Unfortunately, this does not take into consideration the varying braking performances of trains, especially freight trains.

In previous years (before October 2009) lack of robust route knowledge led to a number of collisions and derailments on the classic speed signalled networks in Germany, Belgium and The Netherlands. The definition of route knowledge and what it should contain is now under discussion and it would perhaps be desirable if further evaluation was undertaken regarding how the signal/driver interface could be improved in order to ensure better driver compliance with *signal aspects*. Dutch signalling engineer Wim Coenraad has determined that route knowledge, especially in the case of speed-indicating signalling, essentially makes a train driver able to work with the historically determined anomalies found in the signalling and course of the track, a judgement with which I agree.

There is an acknowledged concern that route knowledge may engender patterns of expectations and habits in drivers that detract from alertness and keeping a proper lookout. To my mind, that has little to do with route knowledge *per se* but is more to do with a lack of mental discipline preventing expectations about signalling aspects ahead when negotiating the peculiarities that a classic network may throw at you, which is really a matter of routine and familiarity breeding contempt. [(B, NL) wegkennis; (B, F) familiarité avec une route; (D) Streckenkenntnis]

Rules and regulations

All train operating companies have extensive sets of rules and regulations to ensure that staff at the sharp end are instructed about how things ought to be done. These rules and regulations are mostly based on – and are often directly traceable to – previous accidents. It is curious to note that rail networks with little investment possibilities tend to have the biggest rulebooks, as rather than invest in changing awkward or even dangerous situations they instruct staff on how to deal with them with yet another set of rules. Knowledge of rules and regulations for drivers is now being checked at least bi-annually on every network in western Europe. At present the rail infrastructure network providers in Europe have mostly taken over the provision of the general rulebooks, although individual train operators may issue additional rulebook appendices to meet their specific needs.

S&T (Signalling & Telecommunications)

A British term for the department that maintains signalling and telecommunications, and is nowadays part of the infrastructure.

Safety

Safety can be characterised as a situation in which any human activity (sport, transport, construction, manufacturing, travel, etc) always ends as intended, without danger to life and without accidental damage to goods and equipment arising from such actions. The main detractor from safety is the risk of death, damage and disease, which should therefore be excluded as far as is possible by means of preliminary assessment of such risks. Subsequently those risks should be eliminated to the maximum extent, without endangering the economic viability or profitability of the operations. Bear in mind, however, that the safest house is the one that is not built and the safest railway is the one on which no train ever travels and the track and signalling have been taken away in order to remove the danger of stumbling or walking into the signal posts.

Safety-critical

This is an operations risk assessment term and identifies any equipment that, or any person who, has a direct influence on continued *safety* of the operations. For that reason, changes in operation with safety-critical equipment or in methods of work and the introduction of new equipment or methods of work in a safety-critical environment must be risk-assessed before introduction and monitored afterwards. The railway environment and everything that goes on there is in a daily state of risk assessment, even more so when something goes wrong involving passengers or staff, as then the previous risk assessment appears to have failed to include, or properly value, all the risk parameters.

Not all risk or safety-critical issues, however, are of such major life-threatening importance. For example, many banisters along stairways will continue to maul fingers sliding along them as top floor edges protrude outwards and come too close to the banister for the hand to pass through. Steps in footways and the quality of flooring material are among other things that continue to cause problems, especially in circumstances such as rain or frost. Many safety-critical risk assessments for new developments are done either by physical or computer modelling. In the above example of the banister and the grazed fingers, a life-size model would point out the danger of grazed or broken knuckles before the situation was introduced in real buildings. Computer modelling is also an invaluable tool in achieving the best safety-critical specifications.

Shunter

1. A British expression for the person who couples up and uncouples rail vehicles, checks that they are fit for traffic and also takes care of the *brake testing* together with the driver. Next to *permanent way* and on-track *S&T* workers the shunter has statistically the most dangerous type of work on European railways.

2. A British expression for a locomotive used solely for shunting purposes. A shunting locomotive may also be known as a 'pilot' when it is assigned to a station or local freight yard. [(B, NL) rangeerlokomotief; (D) Verschublok, Rangierlokomotive; (B, F) locomotive de triage/manoeuvres; (NA) switcher]

Shunting

Shunting is the work of assembling or splitting trains with individual items of rolling stock. In the old days it involved various jobs, such as drivers (and firemen on steam locomotives), shunters to work the couplings and carry out the brake tests, technicians to check the wheels, brakes and bearings, and men in the controlling signal box to work the shunting signals and the turnouts. This work has been thoroughly automated and nowadays there is usually one person in the control tower to oversee the process on computers and work the remotely controlled hump locomotive, with one or two people on similarly radio remote controlled donkey locomotives (the driver/shunter works the locomotive remotely from a transmitter box hanging in front of his chest) to gather up the sorted (classified) trains, couple them up and do the checking and brake testing.

Sorting itself is invariably done at highly automated hump-shunt yards where a remotely controlled locomotive pushes the uncoupled vehicles over a hill to let them roll down under gravity at the other side into the sorting (classification) tracks, the *turnouts* of which are worked based on vehicle recognition by computers that set the turnouts, then measure rolling speed downhill and work rail brakes to avoid excessive speeds. [(B, NL) rangeren; (D) Rangieren, Verschieben; (B, F) manoeuvres, triage; (NA) switching]

Siding (loop)

In this book there are several references to tracks away from the main lines, identified as sidings and loops. In British rail parlance a loop is a track branching off and at its far end coming back into a main line. The track has turnouts at the beginning and end and is typically provided in order to allow a train to get out of the way of another so as to let it pass by. A siding, by contrast, is a track that can be entered and left at one end only and is meant for the berthing (parking) of rolling stock. The meaning of this term differs in other countries.

Signalling

Traditionally signalling is lineside safety equipment installed a) to keep trains apart from each other and so avoid collisions, and b) to ensure that junctions are taken at the proper speed for the line set in order to avoid derailments. Signals, therefore, primarily show a train driver whether or not he is permitted to enter the line section ahead, how far it is cleared for his train and at what speed, and whether as a main-line or a *shunting* movement. If not cleared, a red (danger) aspect is shown, on entry at restricted speed a yellow (warning) aspect is usually displayed, and if the line is clear a green (clear) aspect is shown. Along main lines a yellow aspect always precedes a red aspect, as trains at speed cannot be stopped in time if they suddenly encounter a signal at red. Due to the extended braking distance of trains, they may never be driven at sight with high speed, hence the expression sighting speed [(B, NL) rijden op zicht; (B, F) marche à vue; (D) Sichtgeschwindigkeit], usually up to a maximum of 40 km/h (25 mph), such as when a train has to proceed during signalling failures for example. A yellow aspect is (usually) called a distant signal in Britain [(B, NL) feu jaune, voorsein; (F) avertisseur; (D) Vorsignal] and a red aspect is a stop signal [(B, NL) feu rouge, hoofdsein; (F) carré, sémaphore; (D) Hauptsignal]. Other signals, often combinations of white and/or red position lights, are those that permit shunting moves. Position light signals convey an aspect through signal lights (often white) in a particular position in relation to each other, e.g. circles, triangles, verticals, diagonals or horizontals.

Apart from keeping trains away from each other, signals are also important to ensure that trains come through a junction at the proper speed, as junctions are locations with two or more different speed restrictions due to the layout of the turnouts for the main line and the branch(es). There are two main types of signalling indications for this purpose:

1 A speed-indicating signalling system, which has additional parts of the *signal aspect* (often colour-light combinations or an illuminated digit that when multiplied by ten indicates the speed in km/h to be selected) that give the passing driver a speed order.

2 A route-indicating signalling system, which is employed in Britain and on networks signalled under British influence, whereby the driver receives information from route indicators at the signal about the route set-up ahead, the driver needing to know what speed he must select for the route indicated at the signal. With regard to speed, a red signal gives the speed order 0 km/h or mph.

Any signalling system at junctions, therefore, is basically involved with giving speed orders, and the route-indicating systems do it in a somewhat roundabout way, for which *route knowledge* is necessary.

Under the influence of ATP and high-speed signalling the functions of both the traditional lineside signals and ATP have been combined in so-called *in-cab signalling*, whereby the permitted *speed restriction*, *signal aspects* and the *distance to go* before reaching speed restrictions or signals at danger are displayed in the cab and adherence to their indications is monitored by the ATP part of the system. ATP cab display indications are in principle laid out for speed signalling. In Britain, with its route-indicating signalling, ATP does not work to its full benefit for the train driver for that reason, as in the case of restricted signal aspects it will not repeat the target speed orders that lineside speed-indicating signals give in the cab as continental ATP does. Instead it indicates a target speed of 0 km/h or mph for single and double yellow cautionary aspects, despite the fact that the train may have to proceed for some time under such restricted *signal aspects* on its way through a junction or into a terminus.

Signal aspect
The signal aspect is the complete image that a signal displays to a train driver. This consists mostly of one or more coloured lights, but in the case of a junction signal an additional route or speed indication may be given, and on many networks both are shown.

Signal box
The British term for a location where control of the turnouts and signals as well as their interlocking along a section of track is concentrated and worked remotely. Originally the operating connection between control handles, turnouts and signals was fully mechanical with wires and push rods, but very soon electricity, pressurised air and hydraulic equipment were used in order to ease the physical workload on operating staff. Nowadays the equipment is almost without exception electro-mechanical and electronic in nature and completely automated in traffic control centres.

From very early on, significant effort and intelligence have been invested in mitigating the dangers arising from inappropriate actions by signallers. To that end *interlocking* and particular action sequences have been devised to get the proper turnouts in a desired position and then only the appropriate signals freed to be cleared for a particular train, which then has to travel the cleared route first before it is released again for other trains. It is therefore only during *degraded working* resulting from disconnections for engineering work or after failures that serious accidents can occur and under these circumstances drivers are usually blamed, as the box patently cannot make mistakes. The *OTDR trip recorders* and signalling event recorders have shown up some interesting evidence with regard to things that have actually happened during such disputes. [(B, F) poste de commande; (B, NL) seinhuis, verkeersleidingspost; (D) Stellwerk; (NA) interlocking tower]

Simulator

A simulator is a very useful tool for the training of a train driver, traffic controller, airline pilot or skipper and for the regular monitoring of their operational competence. More importantly, it can be used to teach and maintain ready knowledge to deal with those emergency situations that in reality rarely occur and would otherwise require the wrecking of trains, planes or ships (with the accompanying risk to health) in order to create a realistic scenario. Such issues as fires, the effects of major engine and system faults, a wheels-up landing, how to ditch a plane safely or how to get an empty vessel into a waterway with difficult crosswinds and a fierce tidal current can all be taught using a simulator. Similarly, dealing with traction and brake faults, a high-speed derailment or controlling a long freight train during braking (especially when coming down a steep stretch of track into a level yard or station and pulling out uphill again) all benefit greatly from the use of simulators. They also play their role during route training, as trainees can be subjected to various distractions in order to show them what the potential effects are and the unpleasant surprises that may arise at some locations. Unusual shunting moves or routes that may only be used during diversions, complete with their signal aspects and examples of what to watch out for, may also be easily set up. Finally, there is the benefit that accrues from experiencing emergency routines on a regular basis.

SPAD Signal Passed At Danger

A British expression denoting a red (danger) signal accidentally passed by a train. Category A is an inadvertent SPAD without the permission of the traffic controller, the most dangerous type. Category B is a SPAD arising from a signal failing and coming back to red in the face of an approaching train. Category C is a SPAD due to the signal being put back to danger by the traffic controller to prevent a potential collision or because of another danger ahead. All signals coming back to danger in the face of a train driver are deeply disturbing experiences, as the reason for the sudden change is unclear to the driver and might mean, for example, that another train is approaching on the same track.

A Category A SPAD for an offending driver is usually the precursor to a lengthy time off the track, doing odd jobs in offices while awaiting investigation results and inquiries, and having to write a number of reports and attend all sorts of courses about signalling and train handling. A driver will be made very aware that he has made a serious mistake and on most networks will be allowed only up to three such mistakes in his entire train driving career. [(NL) STS (stop tonend sein) passage; (F) passage illégal d'un carré/sémaphore; (D) Rotlichtvorbeifahrt]

Speed restriction (see also *Maximum speed*)
A speed restriction occurs at any location where the maximum permitted passing speed is lower than the permitted line speed. Such a speed restriction may be permanent or temporary and may pertain to certain types of trains only. [(B, F) TIV/tableau indicateur de vitesse (+ pancarte); (B, NL) snelheidsbeperking; (D) Langsamfahrstelle]

Splitting (opening) a turnout
This refers to the moment when a train negotiates a turnout in the trailing direction that has not been set for its passage and so forces the blades open. In the case of motorised and locked remotely controlled turnouts this causes significant damage as well as being a potential precursor to derailment, and it is for this reason that it is absolutely forbidden. On many systems, turnouts have been fitted with special turnout position signals, especially when, for example, rack rail has been added as on rack-and-pinion mountain railways. [(B, F) ouvrir une aiguillage; (B, NL) wissel openrijden; (D) Weiche Auffahren]

Station limits
A convenient British term that indicates the span of control along a track by signallers or traffic controllers in their (mechanical) signal boxes. This usually comprises the stretch from the home signal to the starting signal or an intermediate block signal in either direction. Note that in the case of complex layouts there could be several station limits throughout the yard or the station if a number of separate signal boxes control different parts of the layout. The station limits of a modern electronic traffic control centre are enormous – well beyond what a signaller could physically oversee. Nowadays a traffic controller has video displays with the track plans showing the signals, turnout positions and trains with their identities to give the necessary information.

Regarding older signal boxes, fringe boxes were those at the ends of the station limits with which the signaller was in contact to accept or pass on trains.

Swamping
See *Ghosting*.

Tank wagon (tank car)
A freight vehicle used to transport powders or liquids in bulk. Normally such a vehicle carries a tank on its frame, the frame taking all the stresses generated. For a time, however, vehicles were constructed in Europe and the US whereby the tank itself took all transportation stresses and there were only short end-frames that contained the drawgear and buffing gear and rested on the bogies, thus reducing the tare weight. In the US these were known as 'stub-sill' tank cars. Incidents occasionally occurred in which the rupture of the tank led to the release of the contents. [(NL, B) ketelwagen; (B, F) wagon-citerne; (D) Kesselwagen]

Telescoping
Telescoping is a characteristic result of a collision between railway vehicles, especially in earlier times. Under the influence of weight transfer effects during impact the front ends of the carriage frames dip down and the rear ends rise up, thus enabling the front end of the following vehicle to shove under the rear end frame of the vehicle in front. The old-fashioned and rather insubstantial coach bodies on top of the frames were unable to withstand the many tonnes of force working on them and gave way, suffering severe damage. Clearly, any passengers in the way of telescoping carriage frames have very little chance of survival or escaping unharmed, such incidents often causing appalling injuries. [(B, NL) telescoperen; (B, F) télescoper; (D) Teleskopieren]

Termination
A British expression indicating a train service that ends under its current *headcode* or train identity number. When the trainset leaves again it is as a different service under a different train identity number. In this situation passengers are often asked to vacate the trainset, as it will be cleaned or go as *ECS* to a depot.

Track (see also *Rail*)
The assembly of undersoil, ballast, sleepers and fasteners that enables suitable trains to travel upon it (see *track gauge*). It is crucial that the

track is as smooth as possible and shows no deviations from the desired top and line. [(B, N) spoor; (B, F) chemin de der; (D) Eisenbahn]

Undersoil has to be prepared to ensure that all parameters such as curves and inclinations are within specified standards so that the proposed traction is able to run successfully. In waterlogged areas the marshland may have to be removed to a substantial depth and replaced with a sub-level sand embankment to create undersoil capable of carrying the weight and dynamic stresses of a train. In hilly or mountainous areas *embankments*, *cuttings*, *viaducts* and tunnels may be needed to take the railway onwards within the power specifications of the proposed traction.

Ballast must be laid and graded to transmit the forces of the train to the undersoil, to drain the track bed and to receive the track on top, after which the spaces between the sleepers are filled with ballast that will be tamped to ensure a proper hold and smooth alignment of the track. A dynamic stabiliser is then used to fully stabilise the track bed. Some track is of the ballastless (paved) type, usually involving the preparation of a reinforced concrete base that takes the rails directly.

Sleepers transmit the train forces to the ballast, carry the rails and fix the track gauge distance. Old-fashioned sleepers were made of wood or steel; modern heavy-duty sleepers are made of concrete. [(B, N) dwarsligger; (F) traverse; (D) Schwelle; (NA) cross tie]

Fasteners fix the rails onto the sleepers. On lightly laid track these could be simple nails (dog spikes) with a protruding lip that were hammered into the wooden sleepers. Elaborate assemblies of baseplates or chairs with a variety of bolts were used in the past, but nowadays a mix of insulated plastic resilient underlay plates and elastic steel clips that fit in horns precast into the concrete sleeper or paved track are invariably used. The modern fastener systems must also protect the concrete sleepers from the damaging vibrations caused by trains, since these attack the concrete in much the same way that a domestic hammer drill does. Loose fasteners are extremely dangerous as they can cause the track gauge to spread under a passing train, resulting in a derailment.

Track gauge

This term refers to the width between the top inside corners (the gauge corners) of the two rail heads fixed on the *sleepers*. This and the width between the wheels of *rolling stock* that needs to run on it are closely related; vehicles with an axle gauge that differs from the provided track gauge cannot be used on that track.

By convention the Stephenson track gauge of 4 ft 8½ in (nowadays set at 1435 mm) is the standard gauge worldwide. Any gauge wider than that is called broad gauge and any gauge narrower than that is called narrow gauge. Among the well-known broad gauges are the British Great Western Railway's 7-ft gauge designed by Brunel and the approximately 5-ft or 6-ft ones used on the Indian, Australian, Irish, Iberian and Russian/Baltic networks. An extensively used narrow gauge is the 3 ft 6 in (1067 mm) Cape gauge found, for example, in Indonesia, Malaysia, Japan, some central and southern African nations and in parts of Australia and New Zealand. Another well-known narrow gauge is the metre gauge, commonly seen in various African nations as well as in Spain, Germany, Austria, France and Switzerland. There are vehicles (e.g. Talgo passenger trains and Transfesa freight vans) with axles that allow the wheels to automatically slide on transition and lock again so as to be able to run on two different track gauges. [(B, NL) Spoorwijdte; (B, F) écartement; (D) Spurweite; (B, NL) smalspoor, normaalspoor, breedspoor; (B, F) voie d'écartement étroite, normal, large; (D) Schmalspur, Normalspur, Breitspur]

Traction (vehicle)
Any rail vehicle that powers a train is called a traction vehicle. This is normally a locomotive, but on an *EMU/DMU* multiple-unit trainset the vehicles with the traction *bogies* are the traction vehicles, the others being *trailers*.

Trailer (vehicle)
A trailer is any rail vehicle not able to move itself or a train, such as wagons, vans, trucks, cars, coaches, carriages, etc.

Train brake (see also *Brake testing*)
A train brake is the equipment employed in order to stop a train. It is not comparable with the equipment used for the same purpose on road vehicles, although certain components (disc or drum brakes) may be similar. A train brake normally works on every vehicle of the train and is controlled from the brake valve or brake controller on the driver's desk. The medium with which the train brake is actuated is usually compressed air. A traction vehicle therefore carries air compressors and associated storage, distribution and control equipment.

In order to provide emergency brake facilities for passengers and fail-safe conditions in an enforced emergency stop in the event of the division of a train, a system was devised by George Westinghouse whereby a brake pipe that has to be kept at pressure is fitted throughout the train (usually approximately 5 bar). If this pressure falls – for example, by the driver working his brake valve, someone pulling the emergency brake handle, or accidental division of the train – the so-called triple valves or brake distributors open up a connection from pressurised brake air reservoirs to the brake cylinders, the pistons of which then push the brake blocks against the wheel treads, or the brake pads are pressed against the brake discs. If the brake-pipe pressure is restored, the triple valves or brake distributors reset and close the connection from the brake reservoirs to the brake cylinders but open up a port from the brake cylinders to the outside air. The compressed air in the brake cylinders vents and the brake blocks or brake pads release, often assisted by spring action. At the same time the brake reservoir pressure is replenished through feed grooves on the triple valves to allow another application. With the original triple-valve system, too many applications in a short time could 'milk' the brake reservoirs of sufficient pressure to stop the train, a situation that could cause brake failure. The correct use of this brake could be rather tricky, its misuse causing havoc to the train and its contents.

However, things have moved on from this situation. Nowadays trains usually have electro-magnetic brake application and release valves worked from an electric controller in the driver's cab, a system known as the electro-pneumatic or EP brake. Freight trains mostly still work with the old system, however, and other trains have it as a back-up in case of electric failures. With an EP type of electrical brake valve operated train brake, the brake pipe for brake operation and safety may have been replaced with an electric train wire that has to be kept powered up to keep the brake from applying. Switches such as the emergency brake handles, for example, disconnect this system and so stop the train.

The above description relates to the mechanical friction brake that converts rolling energy to heat through rubbing brake blocks or pads against revolving parts of the wheels, but there are dynamic brake systems that do not use friction, which is very important during long descents in mountainous areas, for example, when the friction brakes would easily overheat and fail through excessive wear. In the steam days such systems existed whereby steam or compressed air was let into the locomotive traction cylinders in such a way that instead of propelling it, it slowed the train down. With electric or diesel-electric traction, however, the most widely used system is to switch the electric traction motors to generate electricity and so create braking power. The electricity generated can then either be burned off in resistance grids (notably used on diesel-electric traction) or be fed back into the electricity transmission system to power other trains. For diesel-hydraulic or diesel-mechanical power transmission systems there are other possibilities for dynamic braking in the gearboxes to enable limited use of the friction brakes. This naturally saves substantially on wear of wheels and brake components.

It should be clear that the quick-application and release EP air brakes have removed many difficulties concerning the use of the train brake, but braking a train properly is still a bit of an art. Incorrect handling of the brake can lead to train division through broken couplings as well as smashed or buckled rail vehicles, derailments or damage to goods or injury to people on the train, and, as indicated above, to lack of air in the brake reservoirs, the result being that the train cannot be stopped in time. [(B, NL) rem; (B, F) frein; (D) Bremse]

Train describer

Equipment in the traffic control centres that automatically projects the train numbers in the relevant signal sections on a track image panel or VDU screen, the aim being to show where various trains are located. The train numbers jump from section to section as the trains progress, thus giving a reasonably accurate picture of which trains are where.

Train detection

In order to switch signals automatically, safety and signalling systems need to be aware that a train is near. This is achieved by means of train detection systems, which work with track circuits, treadle switches, axle counters and detection loops.

A track circuit is a long-established device that works with a current on an isolated *signal section* of track that is being short-circuited through the wheels and axles of a passing train and so unpowers an electro-magnetic relay switch. This in turn switches a number of other functions such as signalling, locking/unlocking, level crossings, etc. A treadle switch is a normal mechanical switch that is actuated by the flanges of the passing train wheels. An axle counter is a relatively recent piece of equipment that counts the axles of a train moving into section, stores this number in its memory and puts the signal behind the train to danger. When the train leaves the section again all the axles are counted again and the number is compared with the stored number. If both tally, this confirms that the whole train has left the section and the signal is set at a proceed aspect ready for the next train. If, however, there is a discrepancy in the numbers, the system will not release the signal and it stays at danger to protect the vehicle that has possibly broken away and is still in the section. The detection loop is an electro-magnetic mass detection device as used in roads to detect cars approaching (e.g. a vehicle-presence controlled set of traffic lights).

Train orders

Train orders are a form of train traffic control, also known as the dispatcher system. Trains are not regulated by local signallers with home and section stop signals that protect block sections but by a dispatcher who issues train orders to hold, send off or let trains pass or meet at certain locations. This is currently undertaken with sophisticated radio equipment and remote control of infrastructure, but in earlier days the telegraph was the instrument that enabled such traffic control along single lines from control point to control point. This system was difficult to equip with a form of automatic train protection, but it is now possible with Positive Train Control (PTC), a system that is not dependent on signalling for it to function.

Train separation/time interval signalling

These are the two main traditional methods of train traffic management, time interval being the older version. With time interval signalling a train following another is held for a predetermined time interval before dispatch. Theoretically, given identical train speeds, the trains clear the section without one catching up with the other, but a problem arises if the first train comes to an unplanned stop, in which case traincrew immediately must walk forward and back with equipment to warn any oncoming or following train about the obstacle. In too many cases collisions have occurred due to the problem of human walking speed, it being particularly difficult walking on ballast.

Train separation signalling divides a railway line into block sections, entry to which is protected by stop signals that allow the section to be occupied by one train only. If a train fails in section it is protected by the signals. Due to rising train speeds, a warning (distant) signal had to be introduced to allow the train driver to bring his train comfortably to a stand at the stop signal.

Train wheels

Train wheels and their axles are made into a fixed set. On a wheelset of a train the bearings are between the axle and the *bogie* or vehicle frame; there are no bearings between the wheels and axles as there are in a car.

A railway train is defined by being carried as well as steered by the rails of the track it rolls on, which presupposes a close relationship between the track width (gauge) and the width between the wheels on the axle. In order to achieve stable running, the two train wheels of a wheelset have narrow running surfaces that are slightly tapered in opposite outward directions, the wheel tread, on the inside of which a there is a so-called flange. In combination with the slight inward lean of the rails, this wheel-tread taper ensures that the track keeps the train on top of it and steers it through curves. The flanges are fitted mainly to keep the wheelset on the running surfaces of the rails on very sharp curves or uneven track, which is usually also necessary in yards and on turnouts. Flange squeal is the shrill sound that results when flanges touch the rail head in such areas. To the area infrastructure engineer it denotes increased wear and tear of the track, for which he might consider track or flange greasers. If the wheel treads run off the running surfaces of the rail heads, the train becomes *derailed*, which is a very serious condition in the case of a speeding train as all control over direction and braking is instantly lost. [(B, NL) wiel; (B, F) roue; (D) Rad]

Turnout (swich, a set of points)

This device was originally known as a switchpoint and marks the location where two tracks meet or diverge, allowing a train to be sent either way. This is achieved through making short sections of rail movable (the switch rails or blades) between fixed sections of track to which they connect sideways (the stock rails). The two diverging rail lines then leave each other at the crossing, which in many English-speaking nations is also known as the vee, the nose or the frog. This is a vulnerable piece of track, subject to high rates of wear, especially if it is used mainly in one direction only. Modern high-speed turnouts have a movable frog to create a gapless connection that ensures smooth running and decreased wear rates. Turnouts can be manually operated with levers as well as remotely with pull-wires or rods, by air power or (nowadays) invariably electrically, either through direct-drive electric motors or electro-hydraulically.

A turnout can be used facing (the line ahead splits or diverges) or trailing (the two lines meet ahead or converge). When standing at the toe-end of a turnout and looking at the facing (branching) track ahead, if the track branches to the left it is a left-hand turnout and, conversely, to the right makes it a right-hand turnout. This branching line is often subject to a sometimes severe local permanent speed restriction, especially in the facing direction, where the through line may allow a high line speed. Looking at it that way, a turnout has in fact two different speed restrictions at the same location. This may well confuse a driver and has been the cause of accidents. [(B, NL) wissel; (B, F) aiguillage; (D) Weiche; (NA) switch]

Whilst any turnout can be used in four basic directions, two facing and two trailing, combinations of turnouts at one location have given rise to some very intricate-looking pieces of switch and crossing work in the track. There are the single and double-slip turnouts [(B, NL) half en heel Engels wissel; (B, F) traversée jonction simple/double; (D) Einfach und Doppelkreuzweiche] and the three-way turnout, where left-hand and right-hand turnouts have been interlaced to save space [(B, NL) driewegwissel; (B, F) aiguillage à trois directions; (D) Dreiwegweiche]. Turnouts can be even more intricate when track with more than one gauge is involved. Interesting examples are on display in the Great Western Railway Museum at Didcot in Oxfordshire.

If in British railway parlance a turnout is said to be in the normal position, it often means that the track straight ahead is opened and when in reverse position that the branching track is open, but this is not necessarily always the case. On older track in Britain the normal position of a turnout can only be ascertained *in situ*, when the toe of the switch rail nearest to a metal number on a sleeper at the end of the turnout is closed against its stock rail. In order to be able to operate turnouts reliably in all weather conditions, the stretch where the switch blades touch the stock rails is heated, electrically or with piped or bottled gas, to prevent freezing up. This is often clearly visible on remotely controlled turnouts in station yards during the winter where the snow at these locations has melted away. The only alternative to permanent turnout heating is to send people to clear the turnouts of snow and frost with brooms, spades, blowtorches and chemicals. This, however, is a dangerous job if traffic is still moving, and fatalities have occurred on more than one occasion.

Viaduct (bridge)

Viaducts and bridges are examples of engineering infrastructure that allow a road or a railway line to be taken across a lower-lying area on a construction that does not rest on the undersoil along its entire length and has to be strong enough to carry its own weight and absorb the forces of wind and all the dynamic forces that act on it, such as movement of weight along the structure as well as swaying, acceleration and braking.

A bridge or viaduct connects at its ends with the land section of the infrastructure with abutments and additionally may rest on intermediate piers depending on the type of structure. In order to get derailed vehicles safely across a bridge or viaduct, the track has to be fitted with *check rails* between the running rails to keep the vehicles in line with the track and prevent them from rolling to the side and potentially off the bridge. Probably the worst example of such an incident occurred at *Waterval Boven* in South Africa when six out of fifteen carriages fell off the curved Elands Rivier Bridge at 01:30 on 16 November 1949, leaving 64 people dead and 117 injured. [(B, NL) viaduct, brug; (B, F) viaduc, pont; (D) Viadukt, Brücke]

Vigilance switch (alerter switch)

A vigilance switch is an extra facility attached to the *DSD* (dead man's switch) to check the driver's alertness as well as preventing misbehaviour like placing bags on the *DSD* pedal or hanging them from the power controller lever. About every 50-60 seconds a light comes on at the dashboard or a buzzer goes off, to which the driver has to react by lifting his foot off the DSD board and placing it back again within a few seconds (in the case of a warning light, the buzzer starts after 3-5 seconds of not reacting to the light), otherwise an emergency brake application follows. Some older systems require a separate button to be pressed and released to cancel the warning.

The main safety function of the vigilance switch is to deal with a driver who has collapsed on the DSD pedal and so keeps it depressed, sending the train to its doom, as has happened more than once (e.g. *Waterfall*, Australia). The emergency brake is applied via the vigilance switch no more than 60 seconds later. Another characteristic of such equipment is that it is not possible to put the direction switch back to neutral to bypass the vigilance sequences whilst the train is running. As soon as the train moves, the vigilance switch applies the emergency brake if the direction switch is in neutral.

From a driver's point of view it cannot be denied that, despite its important safety role, the vigilance switch can at times be a bit of a pain. This is particularly true when *shunting* or switching, when all attention is concentrated on watching the *shunters* and keeping everything in one piece, for which reason the cabside window droplights are open to be able to look out and to hear instructions. Unfortunately, this also allows the noise from the diesel (or the wind, rain, trains on other tracks, etc) to waft in and mask the noise from the vigilance buzzer, which can cause quite a few unwarranted interventions. In Britain, therefore, the more user-friendly vigilance systems do not require cancelling if the driver works the power- and brake controllers, direction switch and the *whistle* regularly; the system is programmed to accept that as a sign that the driver is doing his job and ought to be fine. In fact, I noticed that I had a tendency to cancel the more user-unfriendly type of vigilance with a routine lifting of the foot off the pedal and putting it down again to the rhythm of slow music in my head, especially when shunting, to avoid getting stopped every so often. [(B, NL) intermitterende dodeman; (B, F) homme mort intermittent; (D) Intermittierender SiFa]

Whistle (horn)
The purpose of a whistle or horn is to warn bystanders along the track of the approach of a train. The term still used in Britain obviously refers to the whistle on a steam locomotive, but nowadays air-operated horns are in use. Normally a train will have a low and a high tone horn to attract extra attention. High-speed trains additionally have a loud position and a soft position; the loud position is used at high speed to throw the sound ahead and the soft position in depots and at stations so as not to assault the ear drums of bystanders. The use of the whistle/horn is recorded on the *OTDR*, since not using it in the run-up to an accident may worsen the consequences of the incident. [(B, NL) fluit; (B, F) sifflet; (D) Pfeife]

Wrong-side/right-side failures
It has already been explained that in the case of failure, railway equipment is designed to *fail safe* in such a way that the situation is not allowed to further endanger trains and their occupants by showing less restrictive *signal aspects* or failing to brake to a stop. This is called a right-side failure. If, however, a dangerous situation nevertheless develops due to the fail-safe procedure partly or completely failing to work, or being bypassed by something unforeseen, then a wrong-side failure occurs that can often cause serious accidents.

4: The Accidents

Meudon, France, 8 May 1842 at approx 18:00

(Axle failure through metal fatigue causing derailment, followed by train fire)

Introduction

This crash, caused by locomotive axle failure, is regarded as the first major railway accident in France and was probably one of the very first anywhere to result in multiple fatalities. No doubt similar incidents had occurred before, but in all likelihood few or no members of the public were involved, and at that time the death or injury of a member of staff was not considered worth reporting. The crash happened on a public holiday and among those killed was a nationally famous figure, the explorer Admiral Jules Dumont d'Urville, and his family. A simple lesson about basic train safety – never completely lock doors on trains – was learned from this accident and, after a few similar incidents on other networks, is still generally applied worldwide today. Incidentally, this locking of doors had a commercial imperative at that time. Trains often went slowly enough for people to be able to jump on and off en route, thereby discovering the joys of fare evasion. Ticket inspectors did not normally travel on the train, and even if they did were unable to make their way along the train to where the fare evader was seated, so the simple solution of locking all train doors just before departure was adopted on many networks. In itself this was not so much of a safety risk for third class passengers as they usually travelled in open vehicles, but second- and first-class passengers travelled in enclosed vehicles with small windows that normally could not be opened when in transit.

The line

The 19.2-km (12-mile) Paris Montparnasse to Le Pecq line, begun in 1837, was built for passenger rather than freight (coal) traffic to the town of St Germain-en-Laye, and like many others in those days was financed largely by the well-known De Rothschild banking family. A spur to the Versailles Rive Gauche (left bank) station, surveyed and built by the eminent French railway engineer Marc Seguin, was soon added in 1840. The line is interesting from a technical point of view as it was initially part-worked by an atmospheric propulsion system that pre-dates by seven years both an Irish attempt at this type of propulsion and that of British engineer Isambard Kingdom Brunel along the sea wall from Dawlish via Teignmouth to Newton Abbot. In fact, it is highly likely that Brunel's French-orientated engineering education influenced what transpired in Devon.

At the time the Le Pecq line went through pleasantly sylvan surroundings, crossing the Seine twice, and a second track was soon needed as Parisians used the line heavily on their leisure days to escape the city, whilst the better-off started to move out to surrounding villages and commute to Paris to work. A surprising part of the travel experience for these commuters was the use of so-called Berlines, two-and-a-half compartment coach bodies that could be fitted on a four-span horse-drawn road frame as well as on a railway vehicle, using a gantry crane in the railway yard. These Berlines, an early example of what today is called multi-modal traffic, would collect people living away from the railheads, and would then be lifted off their road frames onto rail wagons to speed up the journey to Paris. Within two years the railway annually transported 2 million passengers. Another early French way of transporting the city-bound crowds found on this line was the use of 2-axle (4-wheel) double-deck carriages. Eventually the line would be incorporated in the western Paris commuter operations of the Chemin de Fer de l'Ouest (Western Railway Company) and later still the SNCF, whilst most of it is now part of the RER Transilien suburban rail network around Paris. The location of the accident was the cutting near Meudon Bellevue.

The locomotives involved

A double-header team of locomotives, led by a 'Planet' type 1A (2-2-0) and with a 1B (2-4-0) as the train engine, hauled the train. The 'Planet' was a small engine that came from a series of three built by R. & W. Hawthorn of Newcastle upon Tyne for the New York, Providence & Boston Railroad in the US, who eventually cancelled the order. One subsequently went to the Stockton & Darlington Railway in England and two were purchased by the Chemin de Fer de Paris à Versailles. The second machine was a version of the well-known Stephenson 1B, the enlarged version of both the 'Planet' and the 'Patentee' 1A1 (2-2-2) types that were used by many European railways to start up their operations. Logically both types evolved into a 1B (2-4-0) type with considerably better hauling power and better behaviour on the road. The latter were in fact rather advanced machines for their day.

The accident log

Sunday 8 May 1842 had seen one of King Louis Philippe's royal celebrations at Les Grandes Eaux, Versailles. The railway had enabled many Parisian families of all social backgrounds to make their way out to enjoy the sunshine whilst watching the spectacle, listening to the music and taking a stroll in the park. As several festivities came to an end, a large crowd started to make their way back to Versailles Rive Gauche station to return to Paris Montparnasse on the 17:30 departure. This 120-m (396-ft) long 18-carriage train (all 2-axle/4-wheel rolling stock) was packed with an estimated 770 people when it departed and, as described above, was hauled by two locomotives. With the train doing approximately 40 km/h (25 mph) on the approach to Meudon, the front axle of the leading 'Planet' 1A broke, causing the nose of the locomotive to fall down onto the track and the locomotive to slew to its left. The following 1B and the unbraked train pushed this obstacle aside into the side of the cutting, but the sideways resistance against the cutting wall of the derailed 'Planet' then derailed the second machine, which overturned to its right, swinging its front end away from the track as the following train pushed it around. The first carriage broke away and followed its locomotive, ending up on its right-hand side across the second track, thus escaping all of what followed. The following five coaches, however, climbed on top of the locomotives and each other like mating frogs in a pond. Worse still, the wrecks were exposed to burning coal from the fireboxes of the stricken locomotives on the track below and caught fire, the conflagration spreading very quickly through the wreckage of the front of the train. As all the doors had been locked before departure, no one could get out of the first- and second-class vehicles now on fire, and what ought to have been a comparatively minor accident quickly developed into a serious disaster. The dramatic spectacle arose of people trying to get out of the burning carriages through doors and small windows that could not be opened, panicking and screaming for help, only to succumb to the smoke and heat. It took a considerable time before axes and other tools to break windows and open doors could be brought to the site in sufficient quantities to free people from the remaining vehicles, but by then many had been burned alive or sustained severe injuries.

The accident claimed the lives of 43 people at the scene and nine others elsewhere, although some sources mention figures of around 200 dead. Officially, however, 55 people died, including Jules Dumont d'Urville and his family. Identification of the carbonised bodies was not possible in many cases (dental records that are used now were not kept in those days) but an interesting development was the identification of the body of the famous admiral by a Mr Dumontier. He had been the naval physician on board Dumont d'Urville's ship *Astrolabe*, and apart from being a naval physician he also studied phrenology. Phrenology was a branch of Victorian science that purported to classify people's characteristics by the size and features of their skulls (the length of the chin, the height and angle of the forehead above the nose, etc.). He had carefully measured Dumont d'Urville's skull during a journey on the *Astrolabe* and from his notes was able to identify the admiral's skull accurately from the burned remains that were removed from the scene of the disaster.

The aftermath and the lessons learned

Locking the doors of a train whilst in transit was abolished on most of the railways in France, and the broken cast-iron axle was the subject of much scientific research. The Scottish engineer William John Macquorn Rankine published the results of investigations into broken axles in Britain and showed that they had failed through brittle cracking across their diameter. We now know that this is due to metal fatigue and, as such, this incident is linked not only with the ***Viareggio*** and ***Eschede*** accidents and the *Köln* ICE (InterCity Express) derailment in Germany, but also with the *Comet* aircraft accidents in the 1950s (see Chapter 5). Today, metal fatigue is a well-understood phenomenon that rarely causes accidents because vehicles showing signs of the weakness are taken out of service or their fatigue-prone parts are replaced well before problems arise. Testing by German scientist August Wöhler later in the 19th century revealed much of the theory on which our current understanding of the matter is based. However, an earlier, rather fancy and typically Victorian theory was developed, based on the crystalline appearance of the inner break surfaces, that the iron of which the axle had been cast had 'recrystallised' and in some way had reverted to stone rather than iron. It was some time before it was found that this particular type of cast iron simply looked like that under a microscope or magnifying glass.

A few months after the accident a memorial chapel (Chapelle Notre-Dame-des-Flammes) was constructed close by what is now Rue Henri-Savignac. The chapel, designed by the architect François-Marie Lemarié, who lost a son, a sister-in-law and a cousin in the accident, was added to the list of classified monuments in 1938, but was removed in 1959 and shortly thereafter demolished to make way for the widened Rue Henri-Savignac.

The left-bank railway from Paris Montparnasse to St Germain-en-Laye and Versailles was the scene of another accident, on 6 August 1858 at 22:12, when a clutch of carriages coming downhill without traction from St Germain-en-Laye suffered brake failure and ran at speed into the tender of the locomotive waiting to take them onwards to Paris St Lazare, the north-western terminus for the Versailles Rive Droite (right bank) line. There were three fatalities in the first vehicle: the brakesman and two passengers.

On 22 October 1895, one of the most famous rail accidents of them all (many will have seen the picture of the locomotive leaning diagonally nose-down against the façade of *Gare Montparnasse* in the Place de Rennes at least once in their lifetime!) occurred on this network when a slightly delayed 08:45 train from Granville, consisting of 1B locomotive No 721 and its tender, three luggage vans, one postal vehicle and six carriages, failed to stop in time after entering the station due allegedly to the failure of the Westinghouse air brake. As a result, the locomotive, with the train in its wake, slowly ploughed through the buffer stops, concourse and station frontage before plunging down into the street below. Mme Marie-Augustine Aguillard, a newspaper vendor who had just taken over from her husband who had gone to collect the evening newspapers, died as a result of being hit by falling masonry. Of the 131 people involved in the accident there were two fatalities, one in the station and Mme Aguillard in the street below, and of course there were quite a few injuries. Purely based on my own experience with, and knowledge of, the use of a Westinghouse single-pipe automatic air-pressure train brake system, I would venture a guess that the hurried train driver applied and released his brakes a tad too often without giving his brake reservoirs time to recharge, a process known in British railway parlance as 'milking the brake'. As a result, he had insufficient brake-cylinder air pressure left on his train brake to come to a stop in the platform.

The last, equally serious, rail accident in this area occurred on 5 October 1921, when as a result of bad signalling, a rear-ender occurred between two outbound trains inside the *Batignolles* Tunnel, which was then approximately 1 km (¾ mile) long, just outside Gare St Lazare. The preceding, slightly delayed, train to Versailles was stopped in emergency by the Westinghouse brake because of an accidental division of the train, the rear four coaches having broken away. The following train, allowed to proceed against the rules before the preceding train had cleared section, ran into its rear in the darkness. Again, the combination of steam locomotives, vulnerable wooden passenger vehicles and Pintsch coal-gas carriage lighting proved fatal, the fire in the tunnel killing 28 passengers (more according to some sources) and injuring a great many others. This accident caused gas for train illumination to be abandoned in France and the Batignolles Tunnel to be opened up. The present tunnel outside St Lazare is a third of the original tunnel length, at 321 m (350 yd). With ***Quintinshill***, incidentally, this accident shares the rare characteristic, for western Europe, of signallers being arrested, tried, convicted and consequently sent to prison.

The Montparnasse line to Versailles Rive Gauche, being used mostly for local traffic during rush hours only, lost profitability as it in fact mirrored the second, better aligned and shorter line from Paris St Lazare to Versailles Rive Droite on the other bank of the Seine. It was for this reason that the right-bank line rather than the Paris–St Germain left-bank line was eventually used for the extension to Chartres and become part of the national main-line network. Most of these lines were electrified with a bottom-contact third-rail system during the 1920s but are now standard 1.5 kV dc or 25 kV 50 Hz ac lines with overhead power distribution.

***

Staplehurst, Kent, England, 9 June 1865 at approx 14:15

(Permanent-way error leading to derailment and train falling off bridge)

The accident

This was a typical instance of an accident triggered by faulty preparation and consequent bad execution of engineering work on the railway, although it should be made clear that this happened at the beginning of the steep learning curve that the railway was engaged in at the time, leading to our current levels of safety. But even today such engineering work incidents may show a similarity with regard to a certain lack of thoroughness during preparation and whilst on the job. While many can be likened to a farce, others are serious accidents (***Brühl*** in Germany, ***Carcassonne*** in France, ***Ohinewai*** in New Zealand and ***Stavoren*** in The Netherlands). The three latter incidents took place in the present millennium, yet all bear the same unmistakable hallmark that also colours this very early accident in Kent. In a way it proves that, no matter how many rules an organisation makes, in the course of the daily grind it is down to the individual to understand the situation and muster the insight and discipline to keep everything together rather than to allow indecent haste (often no doubt with the best of intentions), disinterest, boredom or distraction to prevail, resulting in damage, injury or death. This particular incident is a sad example of the latter and has given rise to many of the present rules and regulations concerning the organisation of railway engineering work possessions.

It was because the famous writer Charles Dickens, returning from Paris with his mistress Ellen Ternan and her mother (plus the manuscript of his book *Our Mutual Friend*), was on board this boat train when it crashed that the accident gained such notoriety and was the catalyst for many improvements. Incidentally, Dickens had an uncanny knack of being involved in railway incidents. He escaped death quite a few times, even after this accident, for example in Ireland in 1868, after a driving wheel tyre of a locomotive came apart and parts smashed through his compartment inches from his head, as well as being involved in a train fire. In the Staplehurst accident, after establishing his companions were all right, he got out of the wreckage to try and give succour to those more seriously injured than himself. Due to the wholesale destruction of a number of coaches some of the victims were severely mutilated, two badly affected people dying whilst he attempted to help them, and the shock of the accident and what he witnessed mentally overwhelmed him. He neither fully recovered from the shock nor ever felt truly well enough to begin writing books again. Five years later, at 58 years of age, he died from a stroke. Nowadays some would blame untreated post-traumatic stress disorder (similar to shellshock, with the inevitable mental upset that follows) for his death. Of the 80 first-class and 35 second-class passengers on the train, seven females and three males died and 40 were injured to a greater or lesser extent but survived their ordeal. The driver, fireman and first guard all escaped unharmed, the second guard sustaining slight injuries and the third guard being more seriously injured.

The location

The accident happened on a bridge across a small river, the Beult, between Headcorn and Staplehurst. Whilst the Beult is normally barely more than a weed-strewn rivulet, after heavy rainfall or during snow thaw it floods the adjacent fields and for that reason adequate flood yards (hams) are maintained either side of the river. Consequently, the rail bridge near Staplehurst is quite a bit longer than the width of the river in its normal state would warrant. The bridge at that time had eight openings of 6.37 m (21 ft), its total length being 51 m (168 ft). The abutments and masonry piers carried cast-iron trough girders in which wooden baulks were fitted that carried and held the track as normal with cast-iron chairs and wooden keys, the track being situated only 3.3 m (10 ft) above the normal water level. Outside the track-carrying girders there were lighter cast-iron girders that carried the cess walkways, which consisted of a layer of crushed chalk covering corrugated-iron sheeting riveted on to the girders. Between the Up and Down tracks a similar arrangement was used as a centre walkway. No further guiding features for a derailed train were provided, and crossing the bridge could be hazardous for pedestrians as no railings were fitted. The track here is very straight and as good as level, so even in the 1860s speeds of 80 km/h (50 mph) were an everyday occurrence for express passenger trains at this location. The track maintenance was, therefore, of rather a high standard and the accident report mentions that the track was in a very good state of repair. Given later accidents on the South Eastern system in which the state of the track was given as one of the causes (e.g. *Hither Green*), the conclusion must be that track maintenance in this area had lost some of its initial finesse over the years.

The train involved

The locomotive detailed to haul the 'Tidal' boat train that day was 2-2-2 (1A1) No 199. The machine had two running axles, a leading one with 4 ft 9 in wheels and a trailing one with 4 ft wheels, in between of which there was a single driven axle with wheels of 7 ft diameter driven by two 17-in cylinders with a stroke of 22 in. The locomotive without tender weighed in at approximately 32 tons and had a wheelbase of 16 ft 6 in and adhered to the concept that machines with a single driven axle had the freedom of motion (due to avoiding the inevitable loss of traction energy through friction in the bearings between coupling rods) that made them ideal runners for express work. Even by the time of this accident, however, this notion had already begun to erode in favour of machines with more driven axles. The 2-2-2 (1A1) axle notation in fact makes them so-called 'Patentees', although they were already fitted with a later type of larger boiler that generally became known as the long boiler. The machines were designed by James I'Anson Cudworth, the Locomotive Superintendent of the South Eastern Railway (SER), and 16 of them, known as the 'Mail' singles, were put into service in two batches between 1861 and 1866. Their demise followed uncommonly quickly for machines of that era, between 1882 and 1890, after a working life of only 21-25 years. This indicates that they were not considered an unqualified success, otherwise they would have been modernised (upgraded in present-day terms) and worked for at least another 20 years or more. The locomotive's tender was a 6-wheel (3-axle) unit instead of the more usual 4-wheeler. It had a wheelbase of 14 ft and was allowed to be loaded to approximately 24 tons.

The train that day consisted of 14 vehicles, all of them 4-wheelers on two axles; no 6-wheel or bogie vehicles were included, which indicates that the ride must have been rocky at times, despite the good track, but compared with a stage coach on the road, the ride would have been relatively comfortable and smooth. Behind the locomotive with its tender there was a leading guard's van, one second-class carriage, seven first-class carriages, two second-class carriages and three vans. Note the location of the second-class carriages and the guard's vans; in the event of a collision they would be in the more vulnerable position at the extremities of the train. From the leading guard's van there was a communication cord to signal messages to the driver on the footplate by working the locomotive whistle, the driver also using this whistle to send messages to his guards (e.g. about applying their brakes). Three guards accompanied this train. The man in the leading brake van had both the normal wheel-operated parking brake at his disposal and a Creamer's patent quick-acting spring-actuated emergency brake that operated on his brake van and two following passenger vehicles. The other two guards had the normal horizontal wheel-operated parking brake on their vans. It was quite a job to apply these, as I experienced when regularly berthing locomotive-hauled Mk 1 passenger rolling stock at Stewarts Lane depot in Battersea, London, with exactly such a hand wheel. The guard's vans had glazed observatories (otherwise known as birdcages) with a view of the road ahead where they were required to sit and scan the road for signals and problems. Thus, five of the 14 vehicles were braked.

We now come to the issue of braking power on a mid-19th-century train. During his investigation into this accident Colonel Rich concluded that with five braked vehicles and a tender brake (the locomotive itself had no brake whatever but could have its motion reversed, which applied a sort of dynamic braking power to the train) there was sufficient braking power, and he was probably right given the era in which his words were written. However, locomotives at the time were getting increasingly more powerful and faster (60 mph was becoming common by then), yet no comparable effort went into devising technology to stop the mass of a train travelling at such speeds, especially when things went wrong. Westinghouse, with his automatic air pressure brake, and to a lesser extent Smith, with his automatic vacuum brake, were yet to appear on the scene.

The permanent way work involved

On the stretch of the main line from Folkestone to London east of Staplehurst station all of the longitudinal timber baulks carrying the rails on three bridges had to be renewed in a 10-week period. For that reason a crew of four carpenters, a general labourer and three platelayers (track gangers) had been assembled under foreman John Benge. He is described in Colonel Rich's accident report as an experienced, steady and intelligent man (he and his foreman carpenter were proficient in reading and calculating, which was not that common amongst the labouring population of those days). Benge had been with the South Eastern Railway for 10 years and had acted as foreman for almost three years. Under his management they took out and renewed the rail-carrying timber baulks on the three bridges, one after the other, along this stretch of line. Such timber bridge baulks are nowadays fabricated from pressure-treated tropical hardwood, but were then made from local hardwood that was only surface-treated against rot and was therefore prone to relatively fast decay in Britain's often moist and mildewy climate. Moreover, the troughs did not feature holes to drain rainwater, which collected between the iron and wood and caused the wood to rot even more quickly. These baulks are fastened in various ways onto the steel of the bridge, in line with the rails they are to carry, after which the normal cast-iron rail chairs are fitted with heavy coach bolts onto these baulks and subsequently the rails (of the bullhead type) are placed and secured with wooden or spring-steel 'keys' into these chairs. (This track, and this method of supporting it on bridges, can still be seen in Britain, incidentally. In my early days working as a guard on the tracks of the former South Eastern Railway in and around London I well remember seeing exactly this kind of work on bridges around London Bridge and Charing Cross stations.) An advantage of the bullhead rail type, previously known as double-headed rail, over the presently generally applied flat-bottom or Vignoles rail is that the absence of the flat-bottom flange makes the narrower profile of rail somewhat easier to fit in the sharp curves so common on the Southern Railway as well as on the London Underground network. In France this type of track can be found to this day, whilst it formerly was also used in The Netherlands.

The bridge timber renewal work on the South Eastern main line consisted of the carpenters first carefully measuring the dimensions of the new baulks before making the new wood so as to fit into the place of the worn baulk and drilling the holes for the rail chairs. All that work was done away from the track. Then, when the new baulk was ready to be installed, the platelayers would unfasten and lift the necessary rail lengths (which did not correspond with the baulk lengths, incidentally; usually more than one rail length had to be removed), the rail chairs would then be taken off, after which the carpenters could remove the worn baulk and fasten the new one in its place. Finally, the platelayers quickly refitted the rail chairs on top in the pre-drilled holes and fastened the rail with wooden keys again. This was rather a logical and comparatively fast way of working, eminently useful for short possessions that did not interrupt traffic. During the 10 weeks of replacing the baulks on the three bridges, the work had been so well timed in fact that only on three occasions did an approaching train have to be stopped – a ballast train twice and a light locomotive once – none of which was running at the permitted line speed. At no time had one of the faster-running booked passenger or freight trains been stopped out of course, which is important as this would very probably have shown up the errors in the safety arrangements that were at the root of this accident. Or more to the point; the accident would probably have involved another train.

Preparations and safety

Given that at the time there was no possibility of direct verbal or other communication between the signallers and the remote track gang on the bridges, it will be appreciated that this smart work in between the passing trains was possible only because foreman Benge was thoroughly familiar with the time constraints that the timetable imposed on him. He alone was fully responsible for the safety conditions surrounding the work that his gang did. There was an issue in the timetable of which Benge was only too well aware: due to the shallow harbour at Folkestone, the ferry from France could dock only at certain states of the tide and consequently the connecting boat train to London was timed differently every day. It was therefore of paramount importance that the planner of the engineering work, i.e. Benge, consulted the special timetable to ensure that he knew when that 'Tidal' was due to come through his engineering site. He was no doubt also well aware that his employers would not take kindly to seeing this prestigious train full of first-class travellers and foreigners stopped, especially not for a reason such as an overrunning engineering possession.

On the day in question the track gang were working on the final three baulks out of a total of 32 on this job and before breakfast that morning they had already replaced one baulk, with only two more to go in order to finish the entire job. During that breakfast, in the company of his gang, Benge was seen working things out, busily checking his timetable and calculating when the last two baulks on the Beult bridge could be dealt with. With the 'Tidal' working timetable in his hand, as his gangers told the inquiry, he came to the conclusion that the Up 'Tidal' that day would not be at Headcorn until 17:20 and decided to replace the very last baulk, for which two lengths of rail had to be removed, in the interval between a regular Up train due at Staplehurst at 14:34 and a Down train due at 16:15. So later, during the morning and early afternoon, the penultimate baulk was dealt with before work started at 14:30 on the last one after the Headcorn Up section signal, at danger after the passage of the 14:34 Up service, cleared again (open block working!). As it turned out, Benge had made a very serious, albeit common, mistake; a mistake still made by many who look up timetables all over the world. He had erroneously looked up the timetable page for the 'Tidal' on the 10th, the following day. On the day that the accident happened, the 9th, the train was in fact booked through Headcorn about two hours earlier, at 15:15. Right in the middle of the time that he erroneously had given himself to clear the work on that very last baulk.

Benge's leading carpenter had also obtained a copy of the 'Tidal' boat train working timetable, so perhaps he could have additionally checked the accuracy of Benge's arrangement (in the way that more than one ship's officer is used to shoot the sun and stars to calculate a ship's position). The foreman carpenter, however, had previously left his copy lying on the track, where a passing train ran over and shredded it. As he was presently detailed to work under the foreman platelayer who had his own copy, he had not seen the point of asking for another. That, then, had been the first possible chance to correct Benge's mistake and so avoid the accident altogether. Benge's second and, in fact, principal means of avoiding the disaster consisted of his lookout/flagman, John Wills (see ***Vaughan***). As yet there were no other trackside signs or other indications to advertise the engineering possession ahead to the traincrew (see ***Ohinewai***).

It was Wills's important job to protect the worksite against unexpected approach of trains, as well as to protect those trains against mishap from track under repair. The rulebook stated that this 'signal' man had to position himself 1,000 yards – or more, as necessary, in the case of curves or downhill gradients – away from the worksite towards the oncoming traffic (in this case towards Headcorn station). He had to carry a red flag (or a red-showing illuminated lantern at night) and five detonators, also known as fog signals or torpedoes in North America. These bits of standard railway safety equipment (still in use in Britain) were to be mounted on the rail top with lead straps so as to explode with a very loud report when a locomotive's wheels rolled over them. As soon as the locomotive crew heard the signal they should stop the train and ascertain the reason for being stopped, and should not move the train again until fully informed of the situation and what they were allowed to do. Flagman Wills should have fitted these five detonators on the rail head, one such device every 250 yards and then two of them near his protection position, 10 yards apart, whilst walking the track to the 1,000-yard protection position. His red flag was the main visual warning to stop an approaching train, whilst the exploding detonators added a second tier of warning to the traincrew as well as giving a measure of the distance to the danger site and communicating a clear obligation to the driver and guards to stop. This set of visual and audible warnings, plus the 1,000-yard braking distance allowed for, should have been sufficient to stop any train before it encountered the worksite. However, the accident inquiry heard that on the day of the accident Benge gave his flagman only two detonators and the necessary red flag, with the instruction to use them only when the weather turned foggy. Why he did this has not come down to posterity, but from reading about the way many railway companies worldwide have tried to protect their budgets I can easily see some SER superiors giving Benge a few useful hints about not using expensive materials unnecessarily if he wanted to enhance his career. In any event, as a result, Wills walked to his protection location without pacing out and placing the detonators as required, which in all likelihood was no different from what he had done on every other day of the Beult bridge job. When he walked to the same point as always, ten telegraph poles away from the worksite on the bridge, he was in fact using a well-established if inaccurate way of measuring distance. Apart from the fact that he was unaware that, due to the soft ground, the telegraph poles had been placed at somewhat shorter distances than usual along this stretch of line, he would still not have been 1,000 yards away as even the normal distance between poles was not a full 100 yards. For these two reasons he ended up, according to two external witnesses (farm workers in a neighbouring field), not far past a farmer's occupation level crossing called Slowman's Crossing, 440 yards from the bridge. Colonel Rich estimated that Wills was only 554 yards away from the worksite, which is about half the distance that he should have walked, and thus, in US terminology, any approaching fast trains were short-flagged (see ***Vaughan***). In the previous weeks at Staplehurst viaduct Wills had stopped only three slow-moving non-timetabled trains and in all three cases no mishap had occurred, obviously due to the low speed involved. Whether any detonators actually exploded in those instances has unfortunately not been recorded.

A third issue that could have either prevented the accident or mitigated its severity concerns printed notices or bulletins. All over the world these inform signallers, train drivers, firemen and guards about ongoing engineering possessions and so alert them to potential sources of danger and delay at well-defined locations on the line ahead. These notices are nowadays issued to all frontline operational railway staff, who usually have to sign to confirm receipt (see ***Brühl*** and ***Ohinewai***). They detail alterations to the rulebook, signalling and permanent way and give information about locations, dates and times of engineering work, applicable temporary speed restrictions and any other issues of interest for the maintenance of safe traffic. In Britain they appear on a weekly and monthly basis, and although drivers have been caught out by not reading them on a few occasions, on the whole they take these engineering notices very seriously indeed. The SER did in fact issue printed engineering notices to its frontline staff, but only when there were 'protracted repairs'. These, as it turned out, were repairs where the possessions would occupy the line long enough for the booked train service to have to be altered. Ten weeks of 'blitz' possessions between booked trains, quickly opening up and re-laying the track to carry out major repairs on bridges, did not fall into this definition and therefore the local permanent way inspector did not inform his company manager. That gentleman, therefore, did not require the work to appear in printed notices. In fact, to his amazement, Colonel Rich found that the company manager concerned was wholly unaware that these important track engineering works were going on at all.

The accident log

Although the timetable for this train specified 14:38 as the time of departure, on 9 June the 'Tidal' had left Folkestone Harbour at some time between 14:36 and 14:39, according to whose watch was consulted, to begin the steep climb up to Folkestone East Junction with the main line from Dover. This is where the train reversed and where the Cudworth single driver 'Mail' locomotive took over from the team of locomotives that had pushed and pulled the train up the ascent.

Folkestone Harbour station is now closed, unfortunately, so the sight of a train coming out of the station, crossing the revolving bridge between the inner and outer harbours and then climbing away on the ascent to higher ground is sadly lost to us. Specials were very occasionally hauled by steam traction along this branch line, and a steam locomotive struggling to get a train up to East Junction was certainly a spectacle to behold and hear, but it was always well worth the effort to visit and see even electric or diesel-hauled trains tackling this ascent.

Things went smoothly, as usual. According to Colonel Rich, at 15:11 the speeding 'Tidal' passed Headcorn station, about 2 minutes behind schedule, and another 2 minutes later, at 15:13 (Colonel Rich's report erroneously mentions 2.13 pm here, incidentally) the train was at the bridge over the Beult near Headcorn. Given that Headcorn station – as mentioned in the official report – is 2 miles distant from the bridge, it suggests a speed rather closer to a mile a minute than the 50 mph (80 km/h) speed mentioned elsewhere in the report. The locomotive certainly was capable of such speed. At that moment two sections of the off-side (right-hand, six-foot side) rail in the direction of travel had been taken out and 13 ft (4 m) of the old timber baulk had just been removed. The remainder of that baulk was still fastened in the trough girder, awaiting removal and replacement with the prepared baulk that was waiting nearby. The engineering crew were powerless to intervene and could only watch things go badly wrong. Lookout/flagman John Wills sprang into action on hearing and seeing the fast-approaching train and waved his flag, which was picked up by the driver who immediately shut off steam and signalled the request for brake application to his guards with the whistle. Despite presumably being seated in the rooftop observation cupolas (also known as bird cages) of their vans and looking forward, as the regulations required, the guards, especially the man in the leading brake van, did not notice the clearly visible red-flag-waving lookout for reasons that have not been recorded. As an unfortunate result, the guards set to work applying the normal brakes with the cumbersome big hand wheel as for a regular stop, a job that wastes precious seconds, instead of the first man immediately applying the spring-actuated Creamer's patent emergency brake on the three leading vehicles and then applying the heavy handbrake. The fireman on the locomotive screwed down his handbrake on the tender whilst the driver reversed the steam distribution motion and then applied 'counter steam' (see

Vaughan). The net result was that the braking effort was working at its maximum power only about 250 yards from the bridge and that the train was not going to stop before the worksite.

The speed with which the braking locomotive took the hole in the track at the far end of the bridge – 30 mph (50 km/h) according to Colonel Rich, but quite substantially higher in my humble opinion – must have been great enough for the right-hand wheels to more or less fly the first section of missing track and timber, as they left no marks in the few still available wooden parts. Only past the hole of the already removed part of the baulk did those tell-tale flange scores appear, indicating that the machine had tipped to the left and landed there. On landing on its right-hand side the locomotive pushed its left-hand wheel treads off the rail, which was forced out of the chairs towards the left, the top flanges of the cast-iron girder trough in which the baulks were fitted then probably acting as check rails do nowadays, preventing the locomotive from going off the bridge. Whilst derailed it managed to continue in this way, more or less in a straight line, and safely cross the remainder of the bridge and on to land again, where it came to a stop in the ballast 28 yards past the bridge abutment at the Staplehurst end, its right-hand wheels in the four-foot and its left-hand wheels in the cess. The locomotive was comparatively lightly damaged, the worst being that the steam distribution rods (seized in the reverse position, proving that the driver had done his job as described) and its axles were found to be bent and partly seized, which must have made rather an awkward job of its removal. The tender remained coupled to the locomotive but was standing across the Up line along which it had travelled. Its middle axle was broken following the heavy impact and the coupling and buffing gear as well as the braking gear were seriously damaged – further indication of heavy impact. Colonel Rich posited that the broken middle axle had hit the oncoming rail or remaining baulk, suggesting that the locomotive kept the front of the tender up in a proper aircraft landing attitude whilst the nosediving first van behind the tender pushed the trailing end of the tender down.

The first van had broken away after that impact, another indication of just how heavy the blow suffered by the tender was, but remained within the parameters of the Up line track. It was this heavy blow that the tender dealt to the bridge structure that ultimately resulted in the accident being as terrible as it turned out, because the left-hand cast-iron bridge girder under it snapped and collapsed under the following vehicles, tipping them off the track and causing them to be pulled off the bridge and onto the ground or into the water below. Now only the first van was holding up the capsized first passenger carriage (second class), the leading axle of which was propped up on the embankment with the rear axle suspended over the riverbed. The leading first-class carriage hung, still attached to this second-class carriage but with its trailing end resting on the ground next to the riverbed. This was the vehicle in which Charles Dickens and his companions were sitting. The next first-class vehicle lay upside down next to the river, the following five being in various states of destruction, upside down or on their sides, in the river and on its muddy banks. The next vehicle was a second-class carriage on its side, half in the river. The leading end of the final second-class vehicle was on dry land at river level but, as the coupling had held, it hung suspended from the bridge abutment on the Headcorn side. This is the scene as depicted in the well-known lithograph of this accident, in which many of the lesser-damaged vehicles have already been removed. Of the trailing vans, the first two were on their wheels, derailed in the ballast, whilst the last four vehicles were still standing unscathed on the track. In total, seven carriages were completely destroyed as a result of falling off the bridge into the river or from damage caused by rescuers with axes and jemmies attempting to extricate the passengers. Those vehicles that had not crash-landed below but remained on or suspended from the track suffered minimal damage to their draw and buffing gear and crunched ends and sides, all of which were repairable.

The bridge and the track at the accident site had fared badly. On the Up line the left-hand trough girder under the rail had broken and collapsed into the river, whilst the derailing locomotive and the overturning train had largely taken out the cess girder with its footway as well. As a result, the left-hand rail had been torn away, and the track had been bent and the chairs broken on the stretch of the Up line from the end of the bridge to the place where the locomotive had come to a stop. Apart from the girder directly under the train, the second and third outside girders had been broken and the sixth and eight were found to be cracked.

Charles Dickens's comments

Dickens had lived through a period during which travel in general had increased its speed from the 10 mph (16 km/h) of the horse-drawn stage coach to the 60 mph (100 km/h) of the train in which he experienced this accident. He grasped the exciting fact that he was witness to how the world was shrinking beyond the wildest dreams of his ancestors, that something like a trip from London to Paris, which a mere 10 years earlier would have taken two or three days at best, had now been reduced to no more than a day's travel, independent of the weather, on the co-ordinated train and boat services between France and England, and that departure and arrival could be predicted to the minute. Whilst he truly loved that aspect of the changing world, he was naturally profoundly shocked by the sight of what could happen when things went wrong, and, as mentioned earlier, he had had some gruesome experiences of such things. This is part of what he wrote about the Staplehurst accident:

> *I was in the carriage that didn't go down, but that hung inexplicably suspended in the air over the side of the broken bridge. It was caught upon the turn by some of the ruin of the bridge, and hung suspended and balanced in an apparently impossible manner.*

After Dickens had made his way out of the carriage with difficulty, standing on its footboard and hanging on to handrails (it should be made clear at this stage that the carriage doors were locked as usual at the time, see the ***Meudon*** accident), he continues:

> *Some people in the two other compartments were madly trying to plunge out of a window, and had no idea that there was a swampy field 15 feet down below them, and nothing else. The two guards (one with his face cut) were running up and down ... quite wildly. I called out to them.*

He then set out to attempt to get people out of the overturned carriages in the field and river below, and gave succour to the wounded with his brandy flask and his hat full of water, working for hours among the injured and dying. What was happening to him emotionally shines through when he remarks about the sight of a badly hurt and blood-spattered man:

> *He had such a frightful cut across the skull that I couldn't bear to look at him.*

But Dickens nevertheless forced himself to clean his face and wound and gave him a drink. The man uttered, 'I am gone' and died. Dickens's post-accident comment was:

> *No imagination can conceive the ruin of the carriages, or the extraordinary weights under which the people were lying, or the*

> *complications into which they were twisted up among iron and wood, and mud and water.*

Unfortunately, that is something that is still as true today as it was in the 19th century.

The inquiry

Dickens's comments caused a greater interest in this accident than it might otherwise have attracted. In that light it is interesting to read Colonel Rich's comment in his report as he, rather uncharacteristically for the time, felt reasonably free to not mince his words when blaming the railway company for the damage and the loss of life:

> *In all human possibility this train would have reached London safely (even though the road was broken at the Beult viaduct) had the rules of the South-Eastern Company been adhered to. The provisions in those rules, for always using fog signals when rails are taken up, is an additional precaution not generally adopted by Railway Companies.* [Colonel Rich tells us here that the SER nominally worked under more extensive safety provisions than most other railway operators in Britain.] *It appears, however, that for the last ten weeks these rules have been daily disregarded on the line between Staplehurst and Headcorn Stations, and that the Inspector of the Permanent Way, who is supposed to visit every part of the line several times during the week, if he is unable to do so daily, took no notice of such disobedience of the rules, though it does not appear possible for him to have been ignorant of the fact.*

He goes on to express his disbelief that no one had informed the drivers and guards of trains running through the engineering site of what was going on, in order for them to keep a sharper lookout. How much clearer can one get?

Comment

It is too easy to blame John Benge alone for this accident, for which he received nine months' imprisonment. Colonel Rich clearly points to something that today we would call a corporate culture that failed to foster and enforce safety. However careless Benge was in not properly checking the date of his timetable page, it is a matter that to this day can make the difference between such things as a trip to the seaside going well or ending up in a mess. I experienced on a daily basis how passengers (as well as professionals) made mistakes with timetables. It is so easily done, and particularly so when under the stress to perform that Benge must have experienced from his superiors. Making a good impression was in all likelihood the most important issue to him rather than doggedly going for the safe option. As for flagman Wills, I don't think he actually understood why he had to go 1,000 yards. I believe that he had no inkling about the braking distances of various trains, as so many people to this day. He just chose a spot that looked right and, after all, he had previously stopped three trains safely from there.

Another unanswered issue concerns whether Benge actually told Wills to take only two detonators and not to put them on the track, or did Wills just save himself the trouble of having to carry all that equipment around, of having to put them down and pick them up again? But, naturally, he told a different story at the inquest. Moreover, with this same method of working, everything had gone fine for the past 10 weeks, which probably points to similar practice on the tracks during previous years. One of the characteristics of work on the railway was that rules were there to be (occasionally) bent; the railway would often come to a grinding halt if staff and crews did not resort to unofficial methods. I can recall many instances of that myself and have pictures that show the results of others having done that. Another point is that it is so easy to be lulled into a false sense of security and to start cutting corners to relieve the stress of discipline and the waste of time of always doing it right. Until things stop going right, with absolutely no warning. (Even Proverbs 1.27 in the Bible has something to say about that: '... calamity comes on as a whirlwind'.) If fate is unkind, you will find your peace of mind shattered in the most horrendous way, as illustrated by many other examples in this book. How to tell people to stay vigilant whilst doing those boring jobs? See the comparatively recent accident at ***Holzdorf*** in Germany. Benge never saw his wish for promotion in life come true (compare what happened with fireman Caldarelli at ***East Robinson***). His nine months' imprisonment spelled the end of his railway career, even though he would have never committed this sort of stupidity again and therefore would have probably been the best man for the job.

As a result of this crash and similar subsequent occurrences, rail engineering work is organised in a completely different manner nowadays. Obviously, the fact that there are many more services using the line now, and that they travel at higher speeds, adds to the problems for track engineers tasked with organising the necessary repairs. To start with, the responsibility for deciding when to start dismantling the track has been taken out of the hands of the people actually doing the job. This is nowadays all part of a highly detailed planning process that may take place weeks, months or even years before commencement of the actual job. As a result, there is a strict timetable that tells everyone concerned in great detail what to do and when. Certain flexibility has to be built in, of course; there are machines that break down (***Carcassonne***) or people that go sick. Trains or (notably) road freight vehicles with the necessary material or tools may not show up or show up in the wrong order. Getting everybody to do just what is wanted is never easy (e.g. keeping engineering trains in the right order and direction may turn out surprisingly difficult under certain circumstances), but the usual result is that the railway is safely closed down and taken under local possession and only then may rail be taken out or signalling and structures dismantled. At the proper time everything must be in working order to allow rail traffic to start up rail again. Does it always work out that way? No, unfortunately not (***Ohinewai*** and ***Stavoren***). Individual members of the work gangs are still able to commit serious mistakes, machines occasionally break down and many are the times that Monday morning commuters wait in vain for trains due to overrunning engineering work. But planning and organisational mistakes such as Benge's are in the great majority of cases no longer the dividing line between life and death. Of course, mistakes are still made, as every track worker, signaller and train driver can tell you from experience, but comparatively few deaths now occur, and if they do happen they are more likely to be among track workers than the travelling public.

***

The lethal royal wedding in Torino, Italy, 30 May 1867

(Failure to look out for moving trains when on or near track)

Introduction

I first came across this sorry but rather enjoyably gruesome tale at the fascinating exhibition about European royal railway carriages in The Netherlands Railway Museum at Utrecht in 2010. Some diligent work in books and on the internet did the rest. The tale is simply too good not to be included, but if truth be told, the story, unlike the rest of this book, is in all likelihood part-myth. It is quite certain, however, that on the day of the royal wedding in Torino (Turin) on 30 May 1867, mishaps did occur and that part of the story is undoubtedly true. Over the years, however, it has become much embellished, and I leave it to my readers to separate reality from folklore.

Although the railway played a comparatively small role in this strange tale of death on the day of a royal marriage, it is the railway aspect of the story that has the ring of truth about it to me, absurd though it is. It is technically plausible, and a few of the European crowned heads of the day, including the infamous Austrian Empress Sissi, for example, did in fact demand some rather silly things from their staff, such as never turning their back on them, so forcing departing staff to walk away backwards. This is perhaps acceptable in a palace but is patently dangerous near railway tracks. Consequently, there is an important lesson hidden in this story that too many people overlook to this day: keep your eyes open, listen carefully and look around for approaching trains at all times when on or near the track. You may laugh, but accidents have happened and will happen again for this reason. Rail shows and festivals are well-known potential killers for this reason. In fact, the first widely reported British railway accident, the death of the Member of Parliament William Huskisson at *Parkside* station on the Liverpool & Manchester Railway, falls into this category. He had wanted to shake hands with the Duke of Wellington, who resided in a carriage across the track during the opening of the Liverpool to Manchester railway on 15 September 1830 and was overrun by the locomotive *Rocket* while crossing the track. The issue of not properly looking out for moving trains (despite printed warnings in the coaches) was clearly at the root of this accident.

The incident

The occasion was the wedding day of Amadeus Ferdinand, Duke of Aosta and youngest son of King Victor Emmanuel II, and Maria Vittoria dal Pozzo della Cisterna, daughter of Charles Emmanuel, Prince dal Pozzo della Cisterna, on 30 May 1867. As is usual on such occasions, even the smallest details of the wedding had long been rehearsed and palace staff were under great pressure not to allow the least deviation from the arrangements. Nevertheless, things did not go according to plan on the day.

The tale of woe commenced very early in the morning, when the body of the bride's wardrobe mistress was discovered in a large dress cabinet. She had hanged herself next to the wedding dress, with the flowers for the corsages and bouquets on a shelf close by her face. After the courtiers and servants had been suitably calmed, breakfast went ahead as planned and preparations for the departure of the royal procession into the bright sunlight were finalised. However, when everybody at last was seated in their allocated place in the procession it was discovered that the main castle gates were firmly locked and secured. In the turmoil following the earlier grisly discovery the head gatekeeper had completely forgotten to do his job at the required time and so had caused further delay. Profoundly ashamed, he committed suicide by cutting his throat with his ceremonial poniard in the guard-house as soon as he and his assistants had belatedly done their duty under the furious stare of the king, queen and courtiers. By now everyone had become upset and nervous, which must have communicated itself to the horses involved in the procession, as they became boisterous. The commander at the head of the ceremonial mounted military escort, which had already been waiting for hours in the hot sunshine and must have been less than fully alert, suddenly lost control of his steed. Perhaps the nervous animal was frightened by the waving and shouting of people along the route, but after a few bucking jumps the man in his beautiful uniform with gleaming helmet was thrown, broke his neck and died on the cobblestones. To great further consternation, this regrettable mishap was immediately followed by the expiry of the royal solicitor, who at this tense moment suffered a fatal stroke in one of the following coaches. Then, to top off the morning's misery, the bridegroom's best man, a blue-blooded young gent of exalted character, decided that the best course of action following this string of lethal events was to add his own demise to the misery. He shot himself through the head whilst seated on his horse. In the interests of objectivity I must add that some less sensational variations of the tale intimate that the young man was in fact just being silly. Bored whilst waiting in the sun for the procession to start moving he messed about with his pistol, apparently trying to find out whether it was loaded or not, and blew his head off. The question then, of course, is what was such a tense soul doing with a loaded pistol, ceremonial or not, during a royal marriage?

Nevertheless, despite the carnage and delays the ceremony eventually went ahead and after the vows were exchanged and everything else necessary to conduct a royal marriage had run its tedious course, late in the evening the newlyweds finally proceeded to a nearby station to board a special train, which would take them to their honeymoon destination. They were received at the suitably decorated side entrance to the station and were accompanied to their train by the most senior and grey-whiskered stationmaster that could be found. This stately gentleman initially distinguished himself by confidently exercising the fine art of moving backwards whilst bowing towards the following royals, with his shiny top hat held in the crook of his left arm. But despite the guidance provided by the red carpet along the route, the gent must have had a problem navigating, because he failed to notice that he had slightly missed a turn and had come very near the track leading into the platform at which the royal train was waiting. Here he suddenly stumbled in the ballast and fell backwards onto the track. Unfortunately, it was at this very moment that two highly polished and beautifully decked-out steam locomotives, that due to all the earlier delays had had to return to the depot for a water and fuel top-up, were only now returning. They were backing down to couple up to the royal train and, of course, the two smartly-uniformed crews had been instructed not to look out at the side nearest to the illustrious company. They were, therefore, unaware of what was about to happen and had no chance of taking avoiding action. The stationmaster might not have noticed the locomotives, but the locomotives certainly noticed the stationmaster. Lying on his back across the track, in full view and hearing of all present, he was unceremoniously mauled by the wheels of a fully loaded locomotive tender. King Victor Emmanuel II, deathly pale and with good reason believing that the day was jinxed, had had enough. He ordered his adjutant, Count Castiglione, to obtain coaches for the return to the palace forthwith, and at very short notice the Count organised a number of vehicles of sufficient quality and comfort. The Count then oversaw the boarding of the shocked and silent party as well as supervising the correct order of travel for the coaches, after which he was to direct their departure from the station. He positioned himself on the footboard of the right-hand door of the leading coach (carrying the royal family, including the newlyweds) intending to get off after leaving the forecourt, but as the coach left the station forecourt on a sharp right-hand curve he suffered a fatal heart attack, falling off the footboard in agony and straight under the heavy coach's steel-banded rear wheel. The Count was the seventh victim of that most remarkably lethal wedding day. Many of the world's railway accidents would cause fewer victims!

Epilogue

The duchess Maria Vittoria was not a happy woman. Born on 9 August 1847 in Paris, she was an only surviving child out of three. She was an intelligent girl and received a good education, among others she spoke six languages. The man selected to be her husband, who much against his wishes was made King of Spain, forced her to follow him there as his queen consort in November 1870, but she did not enjoy her time away from Italy. Her husband was vain, a compulsive philanderer, and the Spanish court was less than welcoming. The king abdicated in 1873, the same year in which she gave birth to her third son with great difficulty. The exhaustion, made worse by the long trip back to Italy, caused her to die of tuberculosis at the age of 29 on 8 November 1876 in the Villa Dufour at San Remo. She was buried in the Basilica of Superga in Torino.

***

Vaughan, Mississippi, USA, 30 April 1900 at approx 03:52

(Rear-end collision following a 'sawing' shunt manoeuvre)

Setting the scene

Publications on the early history of US railroads all agree on one thing: that although railroad companies from the start, on 4 July 1828, did their best to get passengers and freight delivered on time and in one piece, the personal safety of anyone travelling on a train in those days was not good. By 1853, already 121 people had died in 11 rail accidents, and in the worst year (1907) no fewer than 12,000 rail accident fatalities occurred in the US. Many of these were staff, trespassers walking the tracks and itinerants (hobos) travelling illegally on trains. It should be borne in mind, however, that for most American citizens at that time rail travel was comparatively safer than travelling on highways and by river transport. In his excellent book *Death Rode the Rails* Mark Aldrich describes how the development of US railroads was characterised by the struggle between the need to invest and the wish to earn large returns. This occasionally angered public opinion and finally to the increasing involvement of the authorities in improving transport safety. Liability issues eventually resulted in spending on safety becoming a cheaper option than allowing accidents to happen, for which large amounts of compensation were payable.

The rail transport plans that were required to open up such a large nation were sponsored by the US Federal Government and the various states to stimulate economic development of the land. But it were business interests, often ruthless and corrupt, that principally propelled the construction as well as operation of rail infrastructure and trains. Lack of state funds on the one hand, and commercial issues and the private wish to earn a quick buck on the other, dictated that the building of railways had to be done fast and on the cheap. It was necessary to get the trains rolling, to keep the competitors at bay and to secure what little transport work was initially available, in order to provide the revenue to keep the undertaking alive and satisfy shareholders. In the difficult transport environment that the vast and largely still unexplored nation presented, accepting constraints in the name of safety was simply anathema. Katie Letcher Lyle puts it succinctly in her book *Scalded to Death by the Steam*, about US railroad ballads and the accidents behind them:

> *In the haste to cover the land with a network of tracks errors were made. Contractors built the railroads at a previously agreed price; the faster they could build and the less they spent, the more money they made.*

Hence issues such as the badly graded and prepared road beds, the cheap but lethally dangerous 'strap' rail (wooden track with nailed-on iron strap as the rail head over which the wheel treads rolled) or the lightly constructed iron or steel rail that easily broke and derailed trains. After these problems had generally been dealt with around the 1870s, speeds began to rise and it were single-line crashes that then became the killers. There was also the extensive use of undeniably inventive and cheap but vulnerable wooden and iron bridge trusses and trestles to span rivers and gorges. The result was that in the year 1887 alone no fewer than 21 bridges collapsed under trains (partly due to the use of wood rather than good coal to fire locomotives, which caused sparks). Between 1888 and 1895 there were 251 bridge, culvert and viaduct collapses under trains – a rate of three per month. The accident rate continued to climb steeply because of hurried and low-quality construction, and maintenance of trains, track and formation often was (and in cases still is) just as poor in the hope of keeping operating costs low. What in the 1840s was constructed from raw timber (available in vast quantities in the US) was too often in a dangerous state of decay by the 1860s (for a modern equivalent see ***Red Deer***). Time-interval train regulation on long single-track lines, coupled with people occasionally failing to exercise necessary caution due to fatigue, hold-ups caused by boiler explosions, broken rails, derailments and collisions all contributed to the poor safety record. However, after the passing of the Railroad Safety Appliance Act of 1893, things finally began to improve as had been the case in previous decades in Europe. This was due to the introduction of better rolling stock with the Janney knuckle coupler, automatic air brakes and operations control by dispatchers with the electric telegraph.

The signalling

Distant and stop signals protecting block sections were not employed on the long lines that ran through barely settled territory in the US. The train dispatchers, who even today act as regional traffic controllers throughout much of the world, had oversight of the timetables and the area of the network under their control, had the telecommunications means (which in 1900 meant telegraph connections) to keep an eye on traffic developments and distributed train orders to regulate traffic to the pick-up points. These pick-up points (often a station or depot) usually had two tracks for passing moves and a 3-aspect semaphore signal (at present often with distant signals) to indicate clear passage, a slow passage to pick up train orders on the hop or the requirement to stop in the station. This, then, in conjunction with the necessary route familiarity of the train drivers (engineers) and train conductors, was what was available in the days of time-interval train regulation for safe running between train order pick-up points from around 1875 until well into the 20th century.

In those early days a misread word, name or number on a hurriedly handwritten telegram train order form could spell disaster and sometimes even result in fatalities. Although later, with the introduction of remotely controlled turnouts and far more sophisticated dispatcher's train radio-telecommunications equipment accidents on such lines became increasingly rare during the course of the 20th century (as evidenced by their increasing newsworthiness), it is clear that at the time of this particular accident at Vaughan in 1900 rail safety still relied to a large extent on the will and aptitude of people to get the job done with accuracy. Constant vigilance, thorough route familiarity, a continuous lookout for (hand) signals and obstructions ahead, obedience to the rulebooks and scrupulous adherence to train orders were demanded from traincrews and traffic controllers. Knowledge of schedules, concentration and an observant attitude, together with quick and analytical thinking and good luck, was what made the difference between a safe arrival and disaster.

The development of rail safety at the time

Rail safety in general had received an immeasurable boost when the electric telegraph was introduced from the 1850s, allowing information about moving and stationary traffic, track occupation and clearance to be sent between traffic control points, regional controllers and traincrews. This initially occurred in the more densely populated areas of the north-eastern US seaboard and in western Europe. This surge of technology allowed a better oversight of activity on the layout and enabled decision-making on traffic matters to be concentrated in interlocking towers (signal boxes) and even more centralised regional traffic control offices. A further improvement to safety was obtained with the advent of electro-mechanical locking frames that controlled the signals and turnouts. These contained increasingly sophisticated mechanical and electro-mechanical safety interlocks that prevented most of the errors made by traffic controllers in previous years. Nevertheless, little of this technical sophistication was available to the people running the remote single-track outback railways with passing loops that were prevalent in vast tracts of the South American, Asian, African, Australasian and central and western North American continents. Whilst the telegraph wires appeared along these lines for the exchange of train-running information and the issue of train orders coupled with such lucrative commercial traffic as public telecommunications, no further signalling with independent safety interlocks was provided at intermediate rail hubs and passing loops (see ***East Robinson***). In the event that a following train had to be stopped short of a failed earlier one, the flagman (in Britain, the guard) had to take the place of the distant signal, and if he failed to get to his flagging position in time, or at all, a head-on or a rear-end collision would result. Deaths, injuries and damage all increased because, despite better brakes, vehicle weight, train weight and train speed also kept increasing. Bigger trains on faster schedules meant better business in the US as it did everywhere else, and mistakes by overworked dispatchers and tired traincrews became more frequent, leading not only to a rise in the number of accidents but most of all to an increase in their severity. Greater train frequencies and higher speed overloaded the throughput capacity of the long single-track rail connections between the passing sidings, which in turn were unable to accommodate the increasing number of trains using them. The problem of enabling trains to pass each other or carry out overtaking manoeuvres under these circumstances was solved by the use of 'sawing' shunt moves.

'Sawing' trains past each other

In order to understand this accident it is necessary to appreciate the nature of the shunt move executed to enable trains to pass at a location where not enough siding length was available for all waiting trains to be clear of the main line. This activity was known (also officially in train orders) as 'sawing' and involved filling the passing track (loop) with stopped trains (even trains running in opposite directions if necessary, as was the case at Vaughan). In order to let the non-stop service past, the facing turnout at the end from which the fast train approached would be cleared to take that train on the main track whilst the other train would slowly proceed towards the still occupied turnout at the far end, the train or convoys of trains in the siding moving in opposite directions to run onto the main line at the end where the non-stop had just arrived and so clear the turnout at the main-line exit end. The driver of that train then opened up his locomotive as soon as the trailing turnout was cleared and set for the main line and the relevant flagman gave the all-clear. Provided everyone was on the ball and did the job properly, this was the only possible way of dealing with the situation. In fact, the convoys on the passing track sometimes had to be 'sawed' forward and back several times in order to allow a number of non-stop trains to run along the main track (some express and important long-distance freight trains were run in two or more parts) before they were allowed to proceed onto the main line again. That in itself could be a major shunting exercise if trains in the siding had to depart in both directions but were facing each other, as at Vaughan. One or more trains then had to set back all the way out of the siding first and come through the main track before the final train in the siding could depart.

Incidentally, if anyone thought that 'sawing' trains was typical only of rough and ready US train practice, here is a little Dutch nugget from the 1883 Nijmegen station passenger complaints book. A passenger from 's-Hertogenbosch (also known as Den Bosch) missed his connection to Kleve in Germany at Nijmegen because his train was delayed at Rosmalen. A freight train in the loop there was too long and a 'shunting manoeuvre' was necessary. Of course, that was a signal-protected move, not a flagman shunt.

If the turnout at the approach side toward the non-stop train became blocked, as in this case, because one of the 'sawing' trains had become disabled and could not move back into the siding, a flagman quickly had to walk out to the approaching non-stop train and stop it with a flag, lantern or torpedoes (detonators) and if necessary in the darkness a fusee (a roman candle firework light signal).

The importance of timekeeping

If there is one major way in which mechanical propulsion changed the world then it is in the form of a timetabled schedule. In order to organise the safe operation of trains along any corridor, previously published time slots with scheduled departure and arrival times at crossing points and stations must be arranged. But this immediately gives rise to a whole host of other issues, such as scheduling delivery of refuelling stores, rolling stock maintenance, availability for duty and crew rosters, etc. It was no different with the advent of steam ships, which suddenly made accurate prediction of arrival times on a particular day or even hour possible, one of the reasons that spelled the end of sailing ships on many routes.

Passengers soon took to this aspect of mechanical transport with gusto and eagerly planned their journeys with the published timetables in hand, clearly preferring the operators who delivered the goods and passengers 'on the advertised' (see, for example, Jules Verne's famous book *Around the World in 80 Days*). As a result, a strong imperative developed that freight and especially mails and passengers had to arrive on time. The consequent potential risk to safety in the event of operational difficulties was a lesson that in some cases was not learned until deep into the 20th century (see ***Amagasaki***). On the other hand, a published timetable and tickets or transport capacity purchased on that basis, constitute a type of contract that includes the duty of on-time delivery on the part of the operator.

Punctuality and reliability were boosted not only by the introduction of ever more powerful traction but also by the air (or vacuum) powered train brake from the 1880s onwards. This feature enabled drastically shortened braking distances that in turn resulted in substantially higher average train speeds and so further stimulated the Victorian era's passion for timekeeping. There is, therefore, an undeniable link between what Verne's book describes as possible (to travel round the world within a predictable period by reading railway and shipping timetables) and the factors that determined the way things went wrong in situations such as the sinking of RMS *Titanic* and several other infamous accidents caused by the same obsession with punctual arrival and its consequent influence on operating decisions. Train driver Jonathan L. (Casey) Jones, who died when his locomotive crashed at Vaughan, Mississippi, displayed pretty much the same attitude to driving locomotives that Commodore Edward J. Smith, who died on the *Titanic* in the icy Atlantic Ocean off Newfoundland in 1912, had to sailing Royal Mail liners across the Atlantic (see Chapter 5).

The Illinois Central Railroad (ICRR)

The ICRR was a Chicago-based railroad company chartered in 1851 and capitalised to the tune of $27 million by mainly British and Dutch-American financial interests. The line was strongly lobbied in the US Congress by the future president Abraham Lincoln, who in those earlier days was a lawyer of repute. The first ever allocation of land grants along the line to settlers, amounting to nearly 2.6 million acres, was the result. This new governmental initiative to bring in settlers to the wide open lands of the Midwest soon brought economic benefits to the region through industrial and agricultural development and boosted the carriage of both goods and passengers on the railroad. Another imperative was the certainty of sufficient winter cargo and passengers from the moment the maritime connections between Chicago and the Atlantic Ocean via the Great Lakes and the St Lawrence River froze up, combined with the many difficulties encountered in navigating the unpredictable Mississippi and its tributaries. The inevitable federal mail contract was a further Government stimulant to furnish the railroad with income – amounting, in fact, to another subsidy. These factors together provided sufficient incentive for financial backers to put up the necessary funds to build the line.

As was the case with so many other North American rail operators, the Illinois Central Railroad became instrumental in populating the area it passed through, and within half a century of its opening dozens of towns were founded. The line under ICRR ownership reached New Orleans in 1882, which was facilitated by the fact that after the civil war in 1865 the dilapidated Southern railroad lines between Memphis and New Orleans could be cheaply added to complete the network. By 1900, the year in which the Vaughan accident happened, the ICRR had thus become a well-established operation that managed 5,000 miles (8,000 km) of track, 800 locomotives, 700 passenger carriages, about 33,000 freight wagons and more than 33,000 employees. As already mentioned, improvements in safety were high on the agenda at the time and the company was engaged in the introduction of equipment such as the semi-automatic Janney knuckle coupler (a major US contribution to world railway practice) and the introduction of the Westinghouse automatic air brake. However, as this story shows, these safety enhancing features also introduced a new category of risk through their potential failure as well.

Throughout much of the 20th century the ICRR remained a profitable operation, merging in 1972 with its competitor Gulf, Mobile and Ohio Railroad to become the Illinois Central Gulf (ICG) for some 20 years. The deregulation resulting from the federal Staggers Rail Act of 1980 enabled the by then struggling ICG to implement a successful programme of drastic back-to-basics cuts to the core north–south train operations, which included singling parts of the main line whilst fitting it with sophisticated Centralized Traffic Control (CTC) equipment, together with the sale of branch and main-line track to specialist operators (coal traffic) and shortline feeder operators. In 1999, the Canadian National Railway came across the border and acquired the once more profitable Illinois Central Railroad for the same reasons behind the initial drive to create the line in the 1850s – a fast gateway from the half-yearly frostbound north to the warm deepwater ports of the Gulf. It is for this reason that Canadian traction and rolling stock can at present be seen intermixed with that of the ICRR all the way from Chicago to New Orleans. Whilst CN/ICRR freight trains still pass through Vaughan along the original main line, regular passenger traffic has come to a stop (apart from the occasional Amtrak service), having been superseded by air routes and the Interstate 55 freeway that now connects New Orleans, Memphis and Chicago. However, extension of the limited passenger services is being looked at again.

From the above it is clear that in 1900 the ICRR New Orleans express trains Nos 1-4, popularly known as 'Cannonball Expresses' (as were most American express trains in those days), were important for what we would now call inter-city traffic between chilly Chicago and balmy New Orleans. Driving the locomotives for these trains was the stuff of boys' dreams, which pretty much characterised engineer Casey Jones's train driving career.

Rolling stock involved

The locomotive involved in the accident was No 382, which was not driver Jones's 'own' locomotive, his regularly worked but shared locomotive being No 384. Unlike during his freight days (when he did have his own freight-traffic locomotive, No 638, which he more or less hijacked from its stand at the Chicago World Fair in 1893 and gave up with difficulty when moving to the passenger side of the job), he did not as yet have sufficient seniority in the passenger loco-running hierarchy to qualify for his own locomotive. It appears from various sources that after moving to the passenger side – with its shorter working day, better pay and considerably more public prestige – one of driver Jones's trip-preparing activities became replacing the fitted locomotive steam whistle with his privately commissioned six-pipe calliope instrument every time he took over a different machine. ICRR No 382 was a product of the well-known Rogers Locomotive Works of Paterson, New Jersey, delivered in 1898. The type was a 10-wheeler, otherwise known as a 4-6-0 or a 2'C, a type of locomotive that evolved both in the US and Europe from the 'American' or 4-4-0 (2'B) type.

No 382 was typical of the breed that around the turn of the century would be used on a multitude of services in both Europe and the US, before the steam superheater as developed by Wilhelm Schmidt in Germany found widespread application from 1900 onwards and stimulated the development of more economical but nevertheless rather larger and more powerful high-pressure machines. It was a saturated (non-superheated) steam locomotive with a total weight of 205,550 lb (93.3 tonnes), an adhesive weight of 100,700 lb (45.7 tonnes) and a maximum axle load of 36,923 lb (16.8 tonnes). The machine had two outside cylinders of 19.5 in x 26 in (495 mm x 660 mm) with slide valves for distribution of steam to the cylinders, actuated as usual in the US by inside Stephenson valve gear, working from a boiler pressure of 180 psi and delivering 21,930 lb (9,755 kN) of tractive effort. The steam pressure was generated from a firebox grate area of 31.5 sq ft (2.9 m^2) serving a heated surface of 1,892 sq ft (176 m^2). The length of the locomotive with tender was 60 ft 3 in (18.4 m). Going by the trip timings of No 382 with express train No 1, speeds of 75 mph (120 km/h) and higher were regularly reached and maintained. Despite the moderate (for the US) axle load, this must have been hard on the track, seeing that the cross-balancing issues between motion and wheels (the effect on the track known as hammer blow, a descriptive name, indeed) would not be satisfactorily dealt with until well into the 1920s. After the accident, ICRR No 382 was speedily repaired in the Water Valley shops and put back in service again, only to experience a number of further accidents (in which six were killed in total) until retirement in 1935. Photos of the locomotive just after the accident most probably show a renumbered No 380, the class leader.

The passenger rolling stock of that period in the US was normally of wooden construction on a steel frame. With little exception this was bogie stock on equaliser bar trucks (bear in mind that in Europe at that time there still were many passenger express vehicles on two or three frame-fixed axles) and for premium inter-city services of this nature US railroads also already employed 6-wheel/3-axle bogies to achieve better directional stability and thus a quieter ride. These features would soon come to Europe as well, but at that time US vehicle technology was measurably ahead of that in Europe.

Freight cars (wagons) in the US without fail were bogie vehicles (trucks) whereby the bogies were mostly of the three-piece arch-bar or diamond types, still occasionally seen beneath freight vehicles in museum operations. In those days precious little expensive steel was used in these numerous vehicles. They were made up mostly of wooden bodies on iron or steel running gear and tended to be vulnerable in collisions, another disadvantage of all-wooden construction being coupler equipment failures.

The location

Virtually on leaving Memphis a train would cross the Tennessee state line into Mississippi and then pass Sardis, Grenada, Winona, Durant, Goodman, Vaughan, Canton and Jackson on its way south. Vaughan was (and still is) a small town located roughly halfway between Memphis and New Orleans on the Big Black River, an eastern tributary of the Mississippi. The nearest large town is Jackson. Vaughan station consisted of a north–south orientated main track and a parallel running passing track (loop) on the east side of this, which could be entered through hand-operated turnouts from either end. There was also a house track (siding) west of the main track in the actual station (depot) area. This house track could be entered through a hand-operated turnout from the north end only and served more or less as a local loading/unloading siding, as storage for rolling stock and, if needed, as an additional track to get trains out of the way of higher classified trains passing along the main track. Driver Jones's express train No 1 was not booked to call at Vaughan, but he received train orders that he would be 'sawed' there, so he would have expected slow running or even to be stopped by flagmen in the station area there.

The accident log

The time-card (schedule) for trains Nos 1-4 allowed five hours (in either direction) between Memphis Poplar Street station and Canton, a 188-mile (300-km) run. With a 380 class engine and six or seven coaches this could be comfortably achieved, but with a heavier train and when late running occurred the job became more difficult. Although the train-running department instructions asked for the trip to be cleared in five hours – i.e. that the train be handed over to the next crew five hours after having been taken over from the previous crew – it is known that Jones received train orders instructing him to make up the 90-minute delay the train had accrued on its way from Chicago to Memphis; in other words, he was expected to make the trip with No 1 in three and a half hours. He was told to be at Grenada 35 late, at Durant 20 late and at Canton on time. Just how far this train order was brought about by Jones's reputation for timekeeping, and would as such perhaps not have been given to other drivers, is not known.

Jones and his fireman Simeon Webb arrived at Memphis Poplar Street station with express No 4 from Canton (destination Chicago) dead on time at 18:25 behind locomotive No 384. Their locomotive was uncoupled and stabled, to be replaced by another locomotive for its onward journey. Jones and Webb were now booked to take express No 1 from Memphis to Canton (destination New Orleans) with this machine on the evening of the following day and consequently they prepared for a night's rest. However, because driver Joe Lewis, who was booked to take No 1 to Canton with his own No 382 that night, called in sick, Jones and Webb agreed to 'double out' (step up) to take the 6-coach train to Canton with Lewis's No 382, which was already prepared for the trip by the shed crew. They knew that No 1 had come in late from Chicago, to which delay the crewing problems at Memphis had added, so that by now No 1 was a full 90 minutes late – and that they were expected to clear that delay.

Jones's private loco whistle was speedily moved over from No 384 to No 382 and instead of their booked evening departure Jones and Webb, with conductor J. C. Turner in charge of the train, left Memphis Poplar Street on Sunday 30 April 1900 at 01:05 to their first stop at Memphis Central station, from which they departed after the minimal dwell time of 5 minutes. It should be remembered that in those days in the US, steps had to be brought in to enable passengers to board a train from the very low platform level, so this short dwell time may be seen as an indication that everyone was geared up towards reducing the delay.

On leaving Memphis, a speedy 20-mile uphill run commenced at Memphis East Junction to the summit at Hernando Hill, followed by an even faster 16-mile descent through the station at Love and the Coldwater River valley, past Coldwater and the Grenada District race track, Senatobia and Como stations. They then reached Sardis station for a booked water stop at a distance of 50 miles (2 hr 4 min) from Poplar Street. The running time of No 1 that night from Memphis Central to Sardis was 43 out of 47 booked minutes, but the actual gain in travel time had been far greater due to the short station stop at Central and faster running between Poplar Street and Central. From Sardis the line became more difficult due to curvature, but No 382 responded very positively to firing and driving. Driver Jones expressed his delight with the response of this machine several times to his fireman, although the resulting ride must have been somewhat less than pleasant for the passengers on the train; a conductor in the luggage van in fact complained after the accident that his carefully sorted and stacked luggage had come down several times. Top speed was made from the summit of Hardy Hill all the way to Memphis Junction, where the train had to be stopped to set the turnout for the Memphis Grenada District line as it was normally set for the Water Valley District line. A second stop had to be made after this junction to let the rear-end man who reset the turnout for the Water Valley Line re-board the train. They then reached Grenada station for another water stop, 50 miles from Sardis.

The delay by now had been brought back to 40 minutes within a running distance of 100 miles. The next section of track was flat and fast again all the way to Eckridge, followed by slightly more twisting track uphill to Sawyer, which was followed by a short stop at Winona station, 23 miles from Grenada. From there the line had no restrictions and therefore allowed very fast running to the next stop at Durant station, 33 miles from Winona. Train No 1 got a red order board, meaning that it had to stop to pick up train orders and await a crossing here. These new train orders informed driver Jones that his train was to meet northbound express No 2 at Goodman instead of Durant. Driver George Barnett of No 2, knowing about the initial delay to No 1 and that Jones was the driver of that train, was recorded whilst waiting at Goodman as saying, 'That Jones boy is showing off again.' As a result, they met the virtually on-time No 2 at Goodman as per the revised train orders. When No 1 cleared Goodman the train was running only 5 minutes late and driver Jones had every reason to expect to be able to pass the town of Way, 6 miles north of Canton, right time. He had received no revised train orders regarding northbound local No 26, so the meet would be dealt with at Vaughan as per the schedule on the time-card. Jones told Webb that he was looking forward to his bed in Canton; he was getting tired by now. On coming through Pickens at speed he was as good as on time but to his serious concern he saw no freight trains waiting for him in the passing track there. He told his fireman that several freights must be clogging up the road ahead at Vaughan and he expected something of a mess, hoping it was not going to delay them again.

Vaughan station, 174 miles (278 km) south of Memphis and 12 miles (19 km) north of Canton, was not a booked stop. Here, freight train No 83 had pulled two coupler shanks whilst moving into the passing track to get out of the way of local passenger train No 25. This disabled the train until repairs could have been effected and therefore the incident had delayed local train No 25. In turn this passenger train delayed northbound freight No 72 waiting at Way, which as a result could go no further north than Vaughan to get out of the way of northbound express No 2 to overtake No 72 there, which in turn would then meet No 1 at Goodman. Freight No 72 was sidetracked at Vaughan to meet No 1. Consequently, two substantial freights, No 72 and double-headed No 83, sat facing each other in the Vaughan passing track and together were ten cars too long to be completely clear of the main track. In order to let two sections of northbound local No 26 from Canton into the Vaughan house track to get out of the way of No 1, Nos 72 and 83 had to 'saw' to the north. After this move both freights prepared to 'saw' south to clear the north turnout for the rapidly approaching No 1, but fate now intervened again as during repressurising the air brake pipe to release the train brakes a brake pipe air hose ruptured on the fourth car from the locomotive on No 72. This stalled both trains and the rear of No 83 was therefore blocking the main line in the face of No 1.

In accordance with the rules, brakeman John Newberry of No 83, who had previously flagged No 25 through, was urgently dispatched northwards along the main line to flag No 1 to a stop whilst fireman Kennedy of No 72 tackled the repair of the air hose. No 1 approached through the murky darkness at a speed of 70-75 mph (110-120 km/h) running about 2 minutes late. The crew, on hearing a torpedo (detonator) and becoming aware of the flagman, started to brake but discovered very soon that they would be unable to stop in time to avoid a collision. According to fireman Webb's testimony, Jones whistled continuously to warn people near the trains ahead and whilst braking in emergency, sanded the track, reversed his valve gear and opened up again to make his driving wheels turn backwards (i.e. he 'bombed' his locomotive). As soon as Webb realised that they would be unable to stop in time he informed Jones, who shouted at Webb to jump off. On impact, No 382 and its train smashed through the caboose (brake van) and two wagons of freight train No 83, following which the locomotive lurched and fell over to the right and whilst sliding turned its front end back towards where it had come from. Driver Jones was found in his damaged cab, having been struck in the throat by either a sheared rivet, a bolt or a piece of wood. He was alive but losing much blood. He was brought on a stretcher taken from the baggage car of No 1 to the station building at Vaughan and there, lying stretchered on a cart, he passed away about half an hour after the accident. Webb had been found unconscious along the track and was taken to the station as well, but he regained consciousness only after Jones had died.

A digression about 'bombing' a steam locomotive

Before sensitive souls complain about driver Jones's rough American practice of making a steam locomotive work reverse whilst moving forward, here is a recollection from Mr J. H. Otto from The Netherlands concerning an occurrence he witnessed as an 11-year-old boy. It is taken from Martin van Oostrom's book about the Netherlands Central

Railways Maffei 'Zeppelin' 4-6-0 (2'C) 10-wheelers from 1910. They were München-built machines with a definite US streak when looking at their bar frames and New York Duplex air-brake equipment:

> *On a beautiful summer's day at Nieuwersluis in 1935 I witnessed a rather extraordinary event. I noticed that the gantry-fitted siding entry signal to the left of the high-fitted main-line signal* [right-hand traffic and drive] *on the Utrecht side, covering the entrance to the single siding that was situated between both main tracks, showed stop. The high-fitted main-line signal to its right showed clear, so a through train was due. The distant signal under that main signal, however, for the main-line starting signal further down the line past the station platform, showed caution as the main-line starting signal showed stop.* [All stop signals in The Netherlands have their own distant signal!] *International D-train 290* [from Germany through Arnhem d 09:25, Utrecht CS d 10:12 and Amsterdam CS a 11:12] *approached at speed, hauled by a Class 3600* [the NS classification for these by then 20-year-old NCS 10-wheelers]. *It was one of the few trains in those days that would not first call at Weesperpoort station in Amsterdam but would continue directly via the connecting curve to Centraal Station.* [In fact, that is the route all trains take today; Weesperpoort station no longer exists. For those in the know, the curve to the former Weesperpoort layout can still be discerned from the way a local housing estate curves away from the main line.] *It was a brilliant spectacle; the highly glossed olive-green machine with the murkier green German carriages and the wine-red Mitropa dining and sleeping cars running fast in the hot sunlight under a cloudless sky through this lovely tree-strewn flat Dutch landscape. The fast-approaching train rattled across the small railway drawbridge spanning the Angstel River, but no braking, as expected from the signal indications, was noticed. The driver apparently had not seen the caution-showing distant signal for the following starting signal. Due to the dead straight track, however, he did then notice the starting signal at danger and immediately, at the start of the station platform, shut off steam and applied the emergency brakes. Blue clouds of hot smoke appeared from under the coaches, exhaust from the machine fell silent and its safety valves started blowing off. The alarm signal was given with the locomotive whistle. This apparently did not satisfy the need for braking power and*

'counter steam' was applied; the six 1.90 m [6 ft 3 in] *driving wheels started to rotate slowly backwards whilst the train continued in the forward direction. A few seconds later the signaller was able to pull off the starting signal, at which the driver whistled once more to indicate that he had noticed the change, released the brakes and somewhat later opened up power again. With a clearly audible bark of her exhaust, the 'Zeppelin' then continued the trip to Amsterdam.*

This story (and also that of the driver's actions at ***Staplehurst***) accurately describes the situation in which driver Jones found himself when he noticed the danger ahead. It demonstrates that any steam locomotive driver would thrash his engine in this way if necessary. A brace of 'Western' movies show this manoeuvre, incidentally, when the train nears the heroine of the story who has just been tied across the track ahead. As a matter of fact, the actions described are impossible with diesel or electric traction as all sorts of protective equipment would pop and disable the traction installation.

Excerpts from the company accident report

The official Illinois Central Railroad report as compiled by the General Superintendent, A. W. Sullivan, for J. T. Hanrahan, Second Vice President of the company, is reproduced below. The report, based mainly on testimony from the express messenger William Miller in the mail vehicle behind the locomotive, who, interestingly, received no less than a $25 settlement for what are described as slight injuries, constitutes a clear attempt to mitigate company liability in the event of possible lawsuits and, with the law on its side, load responsibility onto the dead engineer's shoulders.

Chicago, May 10, 1900
Subject: Collision of trains 1 and 83, Vaughan, 4/30/00.

Referring the 478 report No 26 of the Water Valley District, Mississippi Division, and various telegrams from Asst. Supt. Gilleas covering case of passenger train No 1, engine No 382, Conductor J. C. Turner, Engineer [driver] *J. L. Jones, running into rear of freight train No 83, engine Nos 870 and 871, Engineers J. Markette and C. W. Marchison, Conductor E. Hoke, at Vaughan, Miss., 3:52 AM, April 30, 1900, in which accident Engineer Jones of No 1 was killed and the following persons*

injured. Settlement of various cases having been effected as shown by amounts opposite names:
Simon Webb, Fireman Train No 1, body bruises jumping off Engine 382 – $5.00.
Mrs W. E. Breaux, passenger, 1472 Rocheblabe Street, New Orleans, slight bruises – Not settled.
Mrs Wm. Deto, passenger, No 25 East 33rd Street, Chicago, slight bruises left knee and hand – Not settled.
Wm. Miller, Express Messenger, injuries to back and left side, apparently slight – $25.00.
W. L. Whiteside, Postal Clerk, jarred – $1.00.
R. A. Ford, Postal Clerk, jarred – $1.00.

All of the above persons were on train No 1, crew of No 83 having escaped injury by jumping from their train. Efforts are now being made to effect settlement with Mrs W. E. Breaux, and as her injuries were slight it is believed this can be done at small expense. Mrs W. Deto, with whom settlement has not yet been effected, is the wife of engineer Deto of our suburban service at Chicago, and no trouble is anticipated in effecting settlement with her; in addition to which, she was in all probability traveling upon a pass. [Notice how the two broken ribs and grazed knee of Miller – a white employee and provider of the main body of evidence on which this report is based – earned him $25, whilst fireman Webb, who was heavily bruised and knocked unconscious when jumping off the locomotive prior to impact, got just $5. He was black.]

Estimate of amount of damage to property is given in the following tabulation:
Engine No 382 – $1,396.25
Mail Car No 51 – $610.00
Baggage Car No 217 – $105.00
Caboose No 98119 – $430.00
I.C. Box Car 11380 – $400.00
I.C. Box Car 24116 – $55.00
Total damage to property – $2,996.25.

[A later update mentions a further $327.50 of property damage, split into $102.50 track damage, $100 damage to freight in the damaged box cars and $125 in wreck-clearing expenses. In the same update Mrs Breaux has been paid $1 to settle her claim, but the claim of Mrs Deto had not yet been settled. The gash caused by overturned No 382 in the ground beside the track, showing headlight, boiler, driver's-side cylinders, pilot beam and wheel spokes, was still visible more than a decade later. The corn scattered from the second box car that was smashed by the engine grew in the neighbouring fields for many years after. The total damage of this collision amounted to $3,323.75.]

> [Southbound express] *train No 1 met* [northbound express] *train No 2 at Goodman station, No 1 arriving at Goodman on time and taking the siding; it left there 5 minutes late, and at the time of the collision was 2 minutes late. Trains 1st 72* [72 ran in two parts, 1st and 2nd] *and 83 would not have been at Vaughan station for train No 1 but for the fact that No 83, while pulling into the siding to let* [southbound passenger train] *No 25 pass, pulled out two draw-bars, which resulted in delay and prevented No 83 going beyond Vaughan station for the two sections of No 26 and No 1. Trains 1st 72, 83, 1st and 2nd 26 occupied the house track which was clear* [these trains all cleared the main through line]. *1st 72 and 83 occupied the passing track which lacked about ten car lengths of holding the two trains* [the tail end of 83 still occupied the main line at the north side, whilst the tail end of 1st 72 protruded onto the main line at the south side of Vaughan]. *After sawing the two sections of No 26 in at the south end, 1st 72 and 83, while moving south on passing track to clear No 1 at the north end, stopped before going into the clear on account of an air hose* [brake pipe] *bursting on a car in 1st 72, the rear of No 83 fouling the main track* [causing further delay that would force No 1 to be stopped outside Vaughan station until Nos 72 and 83 could be moved completely onto the passing track again at that end].
>
> *Flagman J. Newberry of No 83, provided with the necessary signals, had gone back to place torpedoes* [detonators], *also to signal Engineer on No 1 to stop, and although he* [Jones] *had an unobstructed view of the flagman for 1½ miles, he failed to heed the signals, and the train was not stopped until the collision occurred.*
>
> *Reports received to date indicate that Engineer Jones of the passenger train, who lost his life in the accident, was alone responsible for the accident as train No 83 which was obstructing the main track at Vaughn while sawing by train No 26 was*

properly protected by flagman Newberry, who was provided with the necessary signals and had gone back a distance of 3,000 feet [910 m], *where he had placed torpedoes on the rail; thence continued north a further distance of 500 to 800 feet* [150-240 m], *where he stood and gave signals to train No 1; which signals, however, were apparently not observed by Engineer Jones who had an unobstructed view of the flagman for 1½ miles; nor is it believed he heard the explosions of the torpedoes, as his train continued toward the station at a high rate of speed, notwithstanding the fact it was moving up grade* [several sources indicate that the view ahead from the locomotive at that time cannot have been good and that driver Jones 'bombed' his locomotive and applied his emergency train brake to materially diminish speed and impact], *collision occurring at a point 210 feet* [64 m] *north of the north passing track switch. It is also stated that Engineer Jones of Train No 1 failed to sound his whistle for the station when passing whistling board* [sources mention that he sounded his whistle to warn people ahead as soon as it was clear that he was going to hit].

The explosions of the torpedoes were heard by the crews of trains at Vaughan station; by fireman Webb (colored) on No 1; and by the postal clerks and baggageman on that train. Fireman Webb states that between Pickens and Vaughan stations, after putting in a fire, he was called to the side of engineer Jones, who lost his life in the accident, and they talked about the new whistle which had been put on the engine at Memphis; Jones stated that going into Canton it would arouse the people of the town. This was the first trip with the new whistle and Jones was much pleased with it. [Jones had had this private whistle for a long time and had gained local renown with it.]

Fireman Webb states that after talking with Jones, he stepped down to the deck [the lowest cab-floor section in front of the firebox door] *to put in a fire; and just as he was in the act of stooping for the shovel, he heard the explosion of a torpedo. He immediately went to the gang-way on the Engineer's side and saw a flagman with red and white lights. Going then to the fireman's side, he saw the markers of caboose No 83. He then called to Engineer Jones that there was a train ahead, and feeling that the engineer would not be able to stop the train in time to prevent an accident, told him he was going to jump off, which he did about 300 feet* [90 m] *from the caboose of No 83.* [Jones told Webb to jump.]

Fireman Webb further states that when the torpedo was exploded, train No 1 was running about 75 miles per hour [approx 120 km/h]*; that Engineer Jones immediately applied the air brakes, and that when he* [Webb] *left the engine, speed had been reduced to 50 miles per hour* [80 km/h, contradicting earlier remarks in the report on the speed of this train and the actions taken]. *He also states that had he or Engineer Jones looked ahead, they could have seen the flagman in ample time to have stopped before striking No 83.* [Webb consistently denied ever having said anything to this effect. There are a few questionable issues in these lines of the report. The relatively restricted damage to freight train No 83, of only three largely wooden freight vehicles, is not consistent with being struck by an express train doing the speed claimed but rather more with being hit at a speed of 30 mph (50 km/h), which other sources claim and which supports Webb's as well as passengers' comments about hard braking. The above even implies that flagman Newberry cannot actually have been at the distance from the obstruction that the report claims he was, as the torpedo was heard before seeing him. Furthermore, it should be kept in mind that Jones's view forward to the left through the curve was obscured by the boiler and chimney stack of his locomotive – he worked on the right-hand side – and that Webb was busy firing on the left-hand side of the cab. Only Webb could have seen Newberry through the curve had he been at his window, never mind his night-vision problems with the glow of his fire against the darkness, rain and fog. The compiler of the report is on a mission, which is not concerned with avoiding similar accidents in the future.]

Train No 25 was also flagged by Flagman Newberry and stopped where he stood, which was the same location from which train No 1 was flagged. [Train 25 was a local passenger train with a time-card indication to call at Vaughan and train orders to be 'sawed' through Vaughan. This train was certainly not doing the speed that No 1 was. Furthermore, if we believe fireman Webb's words, this remark could in fact be interpreted as another admission that Newberry cannot actually have been standing where this report claims, but somewhere between his detonator and the caboose of stalled No 83.]

Jones entered the service of this Company as a fireman in March 1888, was promoted to position of engineer in February 1890, since which date his record has been as follows:
– Suspended 10 days February 14, 1891, for collision at Water Valley Yard.
– Suspended 5 days January 17, 1893, running through switch, Carbondale.
– Suspended 5 days for running through switch at Villa Ridge.
– Suspended 10 days December 6, 1893, for striking flat car in siding.
– Suspended 15 days January 4, 1896, rear end collision Extra North and No 92 at Toone, December 20, 1895.
– Suspended 30 days September 3, 1896, for gross carelessness in handling orders at Jackson, Tenn., train 2/25, June 3rd in violation of rules 509, 509a, and 519.
– Suspended 30 days September 3, 1896, sectional collision near Hickory Valley August 27, 1896, train Extra North, engine 618.
– Suspended 10 days September 22, 1897, for not recognizing flagman who was protecting work train Extra, engine 106, as required by train rules.
– Suspended 30 days for having left switch open at cross-over in North yard, resulting in train No 21 running in on siding May 22, 1899.

Engineer Jones was promoted to position of engineer in February 1890 and had a reasonably good record, not having been disciplined for the past three years. He had been assigned to passenger service between Memphis and Canton about sixty days before collision occurred and at the first opportunity thereafter Supt. King had talked to him about the importance of the trains to which he had been assigned, instructing him to use good judgement, especially in stormy weather; to keep close lookout for signals at all times, particularly in approaching and passing through stations and yards; adding that the train he would handle had been successfully handled by other engineers who are on the runs and that satisfactory time had been made. He particularly instructed Jones not to attempt to do any reckless running with a view of establishing a record of making fast time, or better time than the other men on the runs. Jones's work up to the time of the accident had been satisfactory.

As shown above, Engineer Jones was solely responsible for the collision by reason of having disregarded the signals given by Flagman Newberry. I enclose a profile showing the grade and line of road in the vicinity of point where the accident occurred; also the position of the flagman when flagging No 1.

Railroads and justice

Janie Jones and her three children received $3,000 in payouts from life insurance policies her husband had arranged through his two train driver union memberships. She later obtained a settlement of $2,650 from the ICRR through the courts, with the assistance of lawyer Earl Brewer (a Water Valley attorney who would later serve as Governor of the state of Mississippi). But despite driver Jones's length of service with the company, his popular status and reputation for timekeeping, as well as successfully managing the safety of those on train No 1 in the run-up to this crash (and so potentially saving the ICRR a lot of money), based on the findings in the above report, Janie Jones would not receive a more generous settlement because the ICRR had declared her husband guilty. Moreover, rail staff were employed on the premise that they were responsible for their own safety; the company therefore disclaimed liability, hence the union insurance payouts. In various books, incidentally, one can find quite shocking examples of the hard-nosed treatment meted out to people whose health or property had been devastated by railroad activity. For many years the overriding objective was the protection of the all-important rail transport system against such compensation claims.

The staff involved

Jonathan Luther (Casey) Jones was born on 14 March 1863, whilst the civil war was raging. It is widely assumed that his birthplace was Jackson, Tennessee, where he lived most of his working life at 211 West Chester Street. His nickname Casey, acquired to distinguish him from another train driver by the name of Jones, came from a town in Kentucky with the name Cayce, where he lived for a while during his boyhood.

Jones would become a somewhat larger-than-life figure, a star of his times. He rose very quickly through a score of railway jobs (telegrapher and dispatcher, brakeman, conductor, fireman) on the Gulf, Mobile and

Ohio Railroad until 1887, when a yellow fever epidemic on the neighbouring Illinois Central line (a health risk to anyone near the mosquito-infested swamps in those days) caused that company to look for head-end traincrews in a hurry and he grabbed his chance to become a train driver. He soon became a celebrated railroad personality, probably because of his keen intelligence, pleasant personality and certainly also for his timekeeping-above-all-else ethos. Added to this was his distinctive method of self-advertising along the line by means of his six-pipe calliope locomotive whistle that was carefully tuned to play his signature call. When people along the line started to claim that they set their clock the moment they heard Casey Jones's whistle and his train roaring by, it was clear that he had made a name for himself.

Driver Jones was nevertheless not a raucous man. He was a strict teetotaller and took his civilian duties seriously. He also was a religious family man, who changed from Protestantism to Catholicism to marry his wife Janie. Allegations in one song about this accident, that he was high on cocaine whilst driving his train, are a trifle absurd for that reason alone. He was buried at the Mount Calvary Cemetery in Jackson, where in 1947 a monument was unveiled by Charles Clegg and Lucius Beebe in the company of Janie Jones, former fireman Webb, Casey's son Charles (also an ICRR railroad man) and his granddaughter Barbara. Last but not least, indicative of his strong social conscience is the fact that he was held in esteem by his black railway colleagues, which was by no means common in the southern states in those days. One of the engine cleaners at his home depot, Wallace Saunders, was the man who composed *The Ballad of Casey Jones*, which was later streamlined, buffed up and widely popularised by white country singers.

Jones's fireman Simeon (Sim) Taylor Webb was born at McComb, Mississippi, on 12 May 1874, the son of a railway carpenter of 48 years' employment with the ICRR, John Webb. Sim was schooled in reading, writing and arithmetic (which in those days was quite unusual for a black worker's child) as well as in bricklaying, but he loved the atmosphere of the railroad where his father worked far too much to even consider a job in the building trade. Because of his father's good name within the local ICRR hierarchy, Sim was taken on initially as a call boy but was later allowed to start firing locomotives well before he had reached the required age of 21.

After his recuperation from the Vaughan accident, Webb went back to work with another driver, with whom he survived another accident during which the engine again turned over. Then, 19 years after the accident and having come through the stress that World War 1 put on

US Atlantic seaboard railroad crews, Webb asked for leave of absence from his fireman's duties on health grounds and never returned. He took up the job that he had trained for, bricklaying, but later rued that decision as he missed railway life dearly. Following the accident he increasingly gained national renown and for that reason travelled extensively for many years, being regularly invited to retell the story of that fateful night on radio and television and at personal appearances. He died peacefully on 13 July 1957 in hospital in Memphis.

Flagman John Newberry, born in 1873, is the man who is surrounded by controversy as to what he actually did during the course of this accident. Whilst the official report absolved him of all blame, the fact is that on looking deeper into what is likely to have happened that night his actions become questionable. Those members of staff connected with aspects of the accident and liable to come under close scrutiny had about a month to agree their story before they were questioned for the official investigation and so could avoid contradicting each other. Despite that, the report itself, as indicated above, contains many contradictions. Not that being dishonest is particularly unusual in these sorts of investigations (see ***Amagasaki*** in Japan and ***Åsta*** in Norway, or read the revealing illustrated appendix in *British Tramway Accidents*). In US airline circles the blood of cockpit crew may still boil as a result of the whitewashes that were the Civil Aeronautics Board accident reports on a spate of very similar fatal accidents with Boeing 727 aircraft during the early 1960s (chapter 5). Equally, the official *Titanic* accident inquiry reports are a pretty depressing sham.

In Newberry's case there is widespread agreement that he cannot have been at a proper flagging position to stop a fast-running express service, especially given the bad weather conditions that night, as a consequence of which he is *de facto* the person responsible for the accident. Despite the emerging negative feeling towards him, including among his railroad colleagues, he did himself no favours by making some rather unwise remarks in the media. One such incident was his contribution to an article in the *Coffeeville Courier* as late as 1937, in which he made the following remark about fireman Webb, who by then was a media celebrity of sorts, about his jumping off following driver Jones's clear instructions:

> *That Negro fireman, seeing the inevitable coming, nigger-like leaped from the train, but luckily his head struck the ground leaving a clear imprint in the embankment but not injuring him.*

Furthermore, he later wrote to driver Jones's widow, asking for information about her husband as he had an idea for a motion picture that would make a million dollars. She retorted, 'If you had been at your post of duty the night of that wreck, my husband would be living today.' Soon after the accident, Newberry left the railway to run a restaurant in Water Valley and is said to have later been in the road building trade and living in Shaw, Mississippi. He died in 1939 and was buried at Oak Hill Cemetery in Water Valley between his wife Dora and his sister Sadie.

Comment

The following issues are all relevant when considering the causes of this accident:

1 Driver Jones and fireman Webb did not have their booked rest period before working a return train after coming in on time with northbound No 4 from Canton to Memphis in the evening, although Webb stated that after a couple of hours' rest and having some food and hot coffee they were in fact okay to 'double out' and return on train No 1. This would count against them in a modern accident investigation, which would apportion part of the blame for the accident to them. In fact, this is precisely why (in Britain, for example) traincrew insist on taking their booked breaks and rest time.

2 Going by timings and witness statements, Jones did not appear to work with any greater caution to avert the increased risk generated by fatigue. In fact, close to the end of the trip, he told Webb that he was getting tired. This is even more striking when considering the fact that he had virtually completed the journey to Canton in unpleasant circumstances under the stress of making up a 90-minute delay from Memphis. He was running on time and, going by the statements of Webb, looking forward to a more relaxed remainder of the journey (see also ***Buttevant***).

3 Jones's work record as detailed in the report did not paint a picture of a careful railwayman who considered his options before acting. His timekeeping obsession had previously led him to commit serious breaches of the rules, and this, in combination with the outlined fatigue factors, might well have inhibited his all-round performance. There is a case to be made that on nearing Vaughan, Jones was rather more tired than he imagined, sitting on his perch, watching his controls and staring ahead into the murk for the last few miles, possibly relaxing somewhat and consequently less than fully alert to his immediate surroundings

(see ***Harmelen***). He may have been drifting into the relaxing 'last few miles' mindset that has caused accidents worldwide, which may also have caused him to miss the danger signals given by flagman Newberry and then react too late as a result of excessive tiredness.

4 Webb, however, testified that Jones made remarks that indicated that he was awake and functioning and that he was well aware of the potentially chaotic situation at Vaughan. That inevitably leads to the observation that if Jones did expect to be slowed down and possibly stopped at Vaughan, it would have caused him to look out for a flagman on the approach. No train driver wants a crash.

5 Rail freight rolling stock suffered from a number of irksome problems, among which was the use of very old wooden vehicles mixed with new, much heavier and better constructed wagons, which caused problems such as derailments and drawbar pull-outs on a regular basis. Isolation of the train brakes due to deferred maintenance was normal, as were collisions due to heavy new stock derailing the lighter old stock.

6 Whilst the report states that flagman Newberry walked out 3,000 ft to place the detonator and then walked an additional 500-800 ft to take up his flagging position, no mention is made of whether he placed a second detonator at that forward position as well. How, then, did Webb come to testify that they heard the detonator exploding before seeing the flagman?

7 The report mentions that Jones did not sound his whistle at the whistle board, despite several witnesses claiming that he whistled all the way into the accident. It also says that he did not brake, but rolled at speed – uphill – into the collision, after which it mentions that according to witnesses he did brake hard. The report describes the flagman's position ahead of the detonator but then indicates that the footplate crew reacted to the exploding detonator by spotting the flagman. This is inaccurate, but was accepted by company officials, and it is probably this lack of accuracy that enabled Earl Brewer to obtain compensation for Jones's widow from the ICRR.

8 It is altogether likely that Newberry was standing somewhere between the detonator and the caboose of stalled freight train No 83 blocking the entrance to Vaughan station, but not at the distance from No 83 and ahead of his detonator as indicated in the report. Nonetheless, Newberry did successfully flag No 25

to a stop. No 25, however, was a local train booked to call at Vaughan and therefore slowing down, giving Newberry enough time to walk out to his proper flagging position. But when the faulty brake pipe unexpectedly disabled the 'sawing' 1st No 72, he was back in Vaughan yard, to be urgently dispatched from there again to flag No 1. It may well have been that he did not have sufficient time to reach this proper flagging position. Why is that not mentioned? Because the company, as occurred regularly, was intent on blaming the dead driver in order to lessen its own liability?

9 Given that Webb, shortly after hearing the detonator and spotting Newberry, saw that the tail markers on the caboose were so close that a collision was unavoidable (this man had 10 years' worth of experience to estimate braking distances from the cab), it raises a question over the report's claim that the detonator – and the flagman – were placed at anything like 3,000 ft from the caboose.

10 Jones had immediately put his train brake valve in the emergency position and additionally reversed his valve gear, opened up power again and sanded the track under his reverse-rotating driving wheels, then started whistling. This was the absolute maximum he could do in the circumstances, and it materially slowed his train.

11 The Westinghouse automatic air brake, however, even used normally without sand on the track and counter-rotating locomotive driving wheels, should have been well able to stop 6-coach No 1 in the claimed 3,500 ft (1060.5 m) distance at the speed it was travelling, unless perhaps there were an undue number of brake isolations – not a rare occurrence but nevertheless a company liability issue. Given that neither Webb nor anyone else mentioned anything about it, there is no reason to assume that anything was out of order with the air brakes of No 1.

12 The conclusion must be that express train No 1 had been short-flagged (a problem far too common on US railroads at the time) and was involved in the collision as a result. If this premise is accepted, the responsibility for the accident lies not with driver Jones. Interestingly, there may have been a good reason for Newberry's incorrect flagging position, but that, unfortunately, is not discussed in the report.

The real culprit in this accident was the North American train operating environment at the time. Insufficiently robust operational train safety, signalling integrity, lack of sufficient infrastructure to accommodate all traffic and in cases questionable rolling stock fitness for operation all played a role. Those are commercial matters of choice that, unfortunately, occasionally still cause death and destruction. The fatal *Lac-Mégantic* runaway disaster in Canada is a recent example. Moreover, the speed at which train No 1 was booked to travel in these conditions was another example of commercial need outweighing safety considerations. Nor should the harsh go-getting morality of the Victorian US be overlooked as a driver of these events.

As far as its place in the history of rail safety in the US is concerned, the Vaughan crash turns out to have been rather a run-of-the-mill local accident. Limited damage was caused and no one on the train was seriously hurt, with the notable exception of one member of the footplate crew who chose to stay at the controls of his locomotive at the expense of his own life. Had driver Jones jumped off, as he should have, this accident would in all likelihood not have survived in US rail history, certainly not in folklore. We are lucky, in a manner of speaking, that Jones's death inspired Wallace Saunders to compose his famous ballad, as it has preserved the occasion for posterity. Casey Jones himself would, no doubt, have been seriously chuffed about that!

***

Quintinshill, Dumfriesshire, Scotland, 22 May 1915 at 06:50

(Grave signalling error leading to five-train crash with fire)

Introduction

In terms of loss of life and damage to railway equipment, in this case caused by foreseeable errors on the part of signalling staff, this accident was the worst railway crash in Britain's railway history, and since the belated introduction of TPWS train protection in Britain in 2002/03 it is unlikely that an accident of this magnitude will be repeated for the reasons discussed here. It was a typical rail disaster of the early 1900s and involved five trains. Key to the disaster was the fact that no electrical train detection with signal interlocking facilities (available at that time) had been installed, as well as the involvement of obsolete fully wooden coaches with gas illumination. Added to this, the massive amount of splintered wood from the wrecked carriages, tons of coal in the tenders, and fire in the fireboxes of four crashed steam locomotives turned the site into a monstrous inferno, resulting in an excessive number of deaths and injuries. This accident is a graphic illustration of what can happen when railway staff fail to concentrate on their job.

An intriguing feature of the Quintinshill crash is the fact that the assessment of the disaster changed to some extent after the release of Government documents in 2012 and with the discovery of relevant private documents in Carlisle. Many issues, especially relating to the way in which the Caledonian Railway set about limiting its responsibility for what happened, had to be reconsidered in the light of this new evidence. A book, *The Quintinshill Conspiracy*, was the result, which although a thorough assessment of the new evidence, was also an attempt to criticise British society of the time through a 21st century mindset. A society not only grappling with difficult social change but which was also embroiled in a disastrous war. It is important to appreciate the 1915 mindset to see why people acted in the way they did. Why trains were operated in the way they were, why the inquest into this accident (including the Board of Trade accident report from the hand of Lt-Col Druitt) was less than thorough on certain aspects of staff behaviour and why it was written in the way it was. The authors seem to have missed the point regarding how governments or large companies, potentially accused of serious safety failings, tackled the ensuing legal issues in those days. For examples of other such major cover-ups see, for example, the *Titanic* inquiries after the ship's demise in 1912, or the reports on the *Lusitania* sinking, in respect of whether or not she was carrying war materials when she was torpedoed and sunk with the loss of 1,195 lives on 7 May 1915 (chapter 5). That 'zeitgeist' factor has to be respected as reality; there is no point in criticising it.

The location

The accident happened on what is now known as the West Coast main line between London Euston and Glasgow, just over the Scottish border at a point 1½ miles (2.5 km) north of Gretna Junction. On their way to Scotland, the London & North Western trains from London and the Midlands negotiated Shap summit at 915 ft (280 m) and then descended to the city of Carlisle, virtually at sea level on the River Eden. Here the English LNWR locomotives were exchanged for the Scottish ones belonging to the Caledonian Railway. The trains then continued on their way by crossing the sea marshes of the rivers Eden, Esk and Sark into Scotland at Gretna Junction, where the line to Dumfries via Annan branched off and the climb commenced to Beattock summit at 1,015 ft (310 m) via Kirkpatrick and *Lockerbie* (another location that hardly anyone in Britain will forget, for another reason, see chapter 5). It was always a double-track main line, and is now electrified throughout from London to Glasgow. In this area the line ran through lonely countryside and nowadays the A74(M) motorway dominates the scene.

Quintinshill signal box initially was an intermediate wayside section box that controlled the two main lines. The loops and the crossover turnouts between the main lines were installed in 1913 because, even without wartime traffic, the 8½ miles (14 km) into Carlisle Citadel station were among the busiest in the country, worked by no fewer than six railway companies. Efficiently removing freight and local trains away from this bottleneck was necessary to allow the frequent passenger and express trains a clear road when negotiating the 1 in 200 ascent to Beattock summit. The trailing crossover between the two main lines was located almost next to the signal box halfway along the length of the two loops, and this layout normally sufficed to work all traffic between Carlisle and the Glasgow area.

The prevailing circumstances, however, were anything but normal. World War 1 had started almost a year earlier. The passenger timetable had not been thinned out, in the optimistic expectation that the war would be over by Christmas and to maintain the pretence that despite the war everything was running as normal. Therefore many exported and imported goods had to be handled at the ports of Western Scotland and North-western England, to avoid the risks of the mined and enemy-patrolled North Sea. As a result, many extra freight trains had to transport those goods . There was also a great need to move the military to their various destinations, to provide the two northern Navy fleet units with victuals, fuel coal and ammunition, and the Army with its war equipment, whilst at the time of the accident the Whitsun holiday period had generated additional civilian traffic. In short, the railway was being worked extremely hard. Another consequence of this situation was that obsolete but still serviceable rolling stock was pressed into service to provide extra capacity. Added to this, there were many delays to trains that, together with the many extra freight and military services, made operations even harder for the train traffic controllers. (See ***Genthin*** for a similar situation in Germany at the start of World War 2.)

The run-up to the incident

Quintinshill's location in the middle of nowhere introduced another factor that played a role in this accident. The two booked signallers who worked the signal box did not live near their place of work; they either walked or cycled at least 1½ miles (2.5 km) through open countryside in all kinds of weather from Gretna. George Meakin and James Tinsley worked 10-hour shifts each, whilst a relief signalman kept the box open during the remaining four hours between the day shift and the night shift. The day shift started at 06:00 and since the beginning of the war the opportunity had arisen to use the local train from Carlisle via Gretna Junction to Quintinshill occasionally if the circumstances were right. It was a typical railwayman's arrangement, depending on internal operating knowledge and co-operation from colleagues at Gretna signal box.

The train service concerned was the 06:10 local from Carlisle to Beattock, which was booked to connect at Beattock with the 07:49 semi-fast 'Tinto' service to Glasgow Central and therefore had to be kept on time (in some cases, such as this one, the locomotive would go over to haul the Glasgow service as well). If the Down night expresses to Glasgow were running more than a quarter of an hour late, as they regularly did at that time, then the local train, or 'parley' in railway lingo, was sent off first. ('Parley' was a name given to an all-stations local service, after the 1penny per mile third-class 'parliamentary trains' for workers that the Government, by means of the Railway Regulation Act 1844, had originally imposed on the railway companies and which usually ran at rather unsocial hours). The two regular Quintinshill signallers had developed a habit of using this train if it was sent ahead of the late-running express trains from Carlisle, the crux being that this train would be looped out of the way of the express trains at either Kirkpatrick or Quintinshill, but mostly the latter. The signallers firstly ensured that the Gretna Junction men informed them in time, because if the parley ran first and would be looped at Quintinshill then they could stay at home half an hour longer with the bonus of not having to make their way to work by muscle power. The flip side, however, was that the relieving man would not arrive at Quintinshill signal box until 06:30 (half an hour late). In order to hide this little blemish in the arrangements from the eyes of a supervisor, the night signaller would continue as normal until the arrival of the relief man, but would no longer enter the signalled trains in the train register book but on a scrap of paper. The first job of the relieving signaller would then be to copy the contents of this scrap into the train register book in his own writing (to ensure that even an unexpectedly early supervisor would find everything shipshape) and then take over the job. This illegal set-up had worked without a hitch for quite a few months, which is remarkable because the stationmaster at Gretna Junction, their direct supervisor, who was on the platform to supervise the dispatch of the local, must at some time or other have noticed the Quintinshill signaller on the station to board the 06:10 ex-Carlisle parley well after his 06:00 start time, and must therefore have been aware that he could never be on time this way. If he did notice, he failed to do anything about it. In fact, he did little in the way of unannounced supervisor checks at Quintinshill box either.

The accident log

On the morning of the accident a Down freight train to Glasgow had departed Carlisle Kingmoor yard at 04:50 and was sidetracked into the Down loop at Quintinshill at 06:14 to await the two late-running expresses and the local. The 23:45 Down sleeping car express from Euston to Edinburgh and Aberdeen should have called at Carlisle and departed again at 05:50, but this train was running half an hour late. It was closely followed by the 24:00 Down sleeping car express from Euston to Glasgow, booked to depart Carlisle at 06:05, but obviously that train was also running late. The morning Down local from Carlisle Citadel, the parley, booked to follow these two express trains, was therefore sent on its way at its booked time of 06:10. Signaller James Tinsley illegally boarded locomotive No 907 of this train at the stop at Gretna Junction and at 06:24 it arrived at the Down home signal of Quintinshill signal box, just when an empty naval coal train (unofficially known as a Jellicoe Special) on its way south to Wales for reloading was slowly approaching to enter the Up loop. However, the presence of the local was slightly problematic, as the 04:50 Down freight was already occupying the Down loop. What is not clear is why the 06:10 local had not been taken aside further away at Kirkpatrick without stopping at Quintinshill, especially in view of the powerful locomotive available that morning to keep its light load of three coaches and a milk van out of the way of the following expresses. Is that indicative of the railwaymen's unofficial power of train regulating at work? It would have given away the set-up that Tinsley, Meakin and their colleagues had cobbled together, as Tinsley would have had to make his way back from Kirkpatrick in that case. And the accident would not have happened! Anyway, in order to get the stopped Down local out of the way of the two rapidly approaching Down sleeper expresses, it was shunted past the trailing Down main to Up main crossover and then set back through the crossovers on to the Up main line, the locomotive now facing any traffic along that line. It was certainly a permitted move, provided the necessary safety measures to protect the train were applied as per the rulebook.

This, however, is when things started to go wrong. Signaller Tinsley jumped off the locomotive of the local and walked across the Up loop to the signal box to start his day's work by bringing the train register up to date from the notes that signaller Meakin handed him, just when Meakin finished shifting the Up empty coal train into the Up loop at 06:34. It was looped because Gretna Junction had informed him that Carlisle Kingmoor yard was unable to take this train due to congestion. After its arrival in the Up loop he restored the turnouts for the Up main line and, according to his statement, then propped himself in a chair and started reading the morning papers that Tinsley had brought with him. The Up main block instrument, that showed 'Train on Line' for the Up coal empties, should have been left in that position to cover the shunted local. This registered on the similar instrument in Kirkpatrick signal box and prohibited the signaller there from offering trains along the Up main line. One of the two signallers at Quintinshill, however, must have turned that Up main instrument to 'Line Clear' after the arrival at 06:34 of the coal empties in the Up loop and resetting of the turnouts for the main line, for reasons unknown but indicative of the sloppy routines that characterise this accident.

The regulations prohibited, for good reasons, turning the block instrument back to 'Train on Line' to protect the local, as that should have happened before the train was shunted from the Down main to the Up main when it was in that position for the coal train. The Down local was therefore now sitting on the Up main without 'blocking back' protection to Kirkpatrick signal box. Ultimately, of course, they could have telephoned the Kirkpatrick signaller to inform him about the situation at Quintinshill or even put the Up main block instrument back to 'Train on Line' in collaboration, as long as there was no train in section between Kirkpatrick and Quintinshill, but that unusual get-out-of-jail card was not resorted to either. However, in order to secure the Down parley standing on the Up main line, the following further measures were prescribed by the Caledonian Railway rulebook: a) The *fireman* of the train that is detained at a main line signal without track circuit for detection and locking must go to the signal box and check that the signallers are aware of his train and that relevant measures to protect his train with signals have been taken, then sign the train register for agreement, the later Rule 55; b) The *signaller* to put reminder devices (collars) on the levers of the signals covering that train to prevent the levers inadvertently being worked. Three available layers of safety – blocking back, collars and a checking fireman – not used or applied properly! Neither Meakin, as the man who had created the situation, nor

Tinsley, as the relieving man who should have informed himself about the present state of affairs (see ***Warngau/Schaftlach***), observed these duties. Tinsley went straight to the train register and started to copy Meakin's notes. James Hutchinson, the fireman of the shunted local, was in the signal box to check that his train was protected and to sign for it, but he did no such thing with any thoroughness as he assumed that Tinsley, who by now had finished his copying and had started to work the frame, would not overlook a train he had just arrived on. He signed the book and went back to his locomotive on the Up main line after having briefly discussed the war situation with Meakin. Contrary to established wisdom, at no time were there more than three men in the box; the happy bunch of rabbiting railwaymen that kept the signallers from concentrating on their work according to so many writers about the accident were never there. What, however, might be more important to focus on is the shift change – a known contributor to mishaps.

The first of the two late-running overnight expresses from the south, the one to Edinburgh and Aberdeen, was offered by Gretna Junction and Tinsley immediately accepted it, offered it forward to Kirkpatrick, which also accepted, and then pulled off his Down main line signals before the double-headed express blasted uphill along the Down main at approximately 60 mph (100 km/h). Slightly later, the second overnight express, to Glasgow, was offered and he similarly accepted it and passed it forward at 06:38, again routinely pulling off his Down main line signals in rapid succession. Immediately thereafter, however, Kirkpatrick box, now not stopped by the overlooked blocking back measure, belled 4-4-4 to offer the first of two Up troop specials downhill along the Up main to the south. Without hesitation Tinsley again accepted it and offered it onwards to Gretna Junction, where it was also accepted. He now, equally routinely, pulled off his Up main signals at 06:46 – *the very thing that could have been avoided with blocking back and collars on the levers*. Without considering what he was doing, and in all likelihood mentally working on the 'keep traffic moving' autopilot, he managed to completely overlook the parley standing on that same Up main line facing the approaching troop train. The local that he had travelled on to work was sitting in the morning light some 200 ft (60 m) from his signal box. Additionally, of course, had track circuit train detection with turnout and signal interlock been fitted he could not have done this, but despite its importance, Quintinshill was not at the top of the list to have this expensive equipment fitted. Perhaps, apart from his own lack of attention to the state of matters on the track outside his windows, Tinsley might also have been wrong-footed by the fact that, contrary to normal signal box layouts, the frame at Quintinshill was installed facing the rear wall rather than the front windows overlooking the tracks, so as a minimal last chance of averting disaster, he did not have a train with a quietly smoking steam locomotive somewhere in the corner of his vision that might have stopped him from doing what he did. In any event, the trap for a perfect catastrophe had been set. Two minutes later the troop train came into view.

The driver on the 2'B (4-4-0) McIntosh superheated locomotive No 121 stood no chance with his troop special. His train was made up of 213 yards (195 m) of obsolete rolling stock of wooden construction as the result of the unavailability of more modern steel-framed stock for this kind of job and consisted of four 4-axle Great Central Railway passenger bogie carriages, eleven 3-axle Great Central Railway passenger carriages, five 3-axle vans of the Caledonian Railway and one 4-axle open freight bogie wagon. This consist had been put together to transport the 1st Battalion of the 7th Regiment The Royal Scots from their base at Larbert to the port of Liverpool, to embark for the actions planned at faraway Gallipoli. Travelling along a 1 in 200 falling gradient from Beattock summit under clear signals at 70 mph (115 km/h) – rather fast for these ancient wooden carriages with their gas lighting and the recently replenished coal-gas tanks under them – the train came round a curve into a slight cutting. Its footplate crew were suddenly confronted with the sight of the robust front end of the McIntosh 'Cardean' class inside 2-cylinder 2'C (4-6-0, 10-wheeler) No 907 with its three 4-axle bogie passenger coaches and single milk van on the track ahead, virtually when on top of this train.

The impact of the two locomotives on the Up main line, within earshot and under the eyes of Tinsley and Meakin, was fearsome beyond belief. The 'Cardean' locomotive of the local fell on its side and was pushed back 40 yards (35 m), its tender and train breaking away and moving back a further 136 yards (98 m). The lighter and far more seriously damaged locomotive of the troop train was deflected sideways and catapulted on its side across the Down main line, ending up with its smokebox ramming the standing Down freight in the Down loop. Its tender separated from the locomotive and blocked the Up main line, two of the carriages jumping over the two locomotives and the rest spreading themselves liberally between the Up main and the stopped freight in the Down loop whilst disintegrating in the process, yet still managing to damage several further wagons of that freight train. The troop train was reduced to 65 yards (55 m) of smashed wood. The local, thanks to the protection offered by its big locomotive, was not as badly damaged as might have been expected, and had been abandoned in timely fashion by its crew who ran for cover under the Jellicoe Special's coal empties. In the signal box an alarmed George Meakin jumped off his chair and shouted, 'Whatever have you done, Jimmy?' The stunned James Tinsley answered, 'Good heavens, whatever can be wrong? The frame's all right and the signals are all right!" Meakin, now beginning to see the picture more clearly, bellowed, 'You've got the parley sitting there,' followed by, 'Where's the 06:05?' and started to throw the Down main signal levers in the frame back to danger. That same thought about the second approaching night express had already occurred to various surviving railwaymen along the track, among whom was guard Graham of the parley, who started to run towards the approaching sleeper, waving his red flag at the oncoming express.

But it was all too late. A minute after the first collision, the 620 tons of the double-headed sleeping car express, hauled by 2'B (4-4-0) Dunalastair IV class saturated steam locomotive No 140 and Dunalastair IV class superheated steam locomotive No 48 with ten 4-axle side-corridor seating cars and three 6-axle sleeping cars, appeared on the scene along the Down main line, powering hard uphill at about 60 mph (100 km/h). The express ran headlong into the wreckage of the other two trains, killing many of the dazed and injured soldiers in the wreckage and others on the track who were trying to extricate victims from the already blazing troop special. The leading locomotive tossed the tender of the troop train locomotive across the tracks into the Down freight, which scattered a few wagons like balls on a snooker table. The whole carnage involving five trains had taken place in a matter of minutes, Tinsley and Meakin only coming to their senses when Gretna Junction enquired where the expected troop train was.

Fire now took hold of the wreckage. Fed by escaping lighting gas from underfloor carriage tanks on the troop special, by hot ash from fireboxes, coal from overturned locomotive tenders and masses of splintered wood, it consumed what was left of the trains and their occupants, whether dead or alive. It lasted through the day and night and well into the next morning, consuming the entire troop train except for the uncoupled vans at the end, four coaches of the night express and five goods wagons. One reason why the fire turned into such an inferno is that there were no local fire appliances; they had to come from Carlisle and make the last bit of their three-hour trip through boggy meadows, only to find when they got there that water was not available in sufficient quantities.

When a roll-call of the remaining Royal Scots soldiers was taken the following day, only 53 out of nearly 500 men answered to their name.

The precise numbers could not be established as administrative records had been carried on the train and destroyed in the fire. The remaining able-bodied men were then taken to Liverpool but were ultimately spared boarding the ship bound for Gallipoli, where they would have been subjected to the kind of military action that led to an even greater orgy of death and destruction. Eight deaths and 54 injuries occurred on the express, and two lives were lost on the local train. Of the soldiers on the troop train it is estimated that 215 officers and men were killed, to which the driver and fireman of the locomotive of their train should be added, and 191 were injured, many seriously. Fire had claimed most of the 226 lives that perished that morning. Altogether 230 people died. It is a curious fact that the charred and unidentifiable remains of three young children were retrieved from the remaining wreckage of the troop train. It was deduced that they must have stowed away on the train at Maryhill in Glasgow during a signal stop. For that reason they were buried in a single grave at Maryhill Cemetery. Something similar would happen again during the triple collision and fire under a road bridge at *Charfield* in Gloucestershire on 13 October 1928.

James Tinsley and George Meakin were, of course, arrested and sent for trial, where they were poorly defended by their barrister and where Tinsley said that he had forgotten about the local train. The jury found them guilty after only 8 minutes of deliberation and consequently they were convicted of culpable homicide and gross neglect of duty at Edinburgh High Court. Tinsley received three years' hard labour and Meakin 18 months' imprisonment. However, both men were released after only 12 months due to trade union intervention, good behaviour and the fact that they were suffering from nervous stress. Fireman Hutchinson was acquitted as there was insufficient proof that he had completely failed to carry out his duty under Rule 55. He had been in the signal box and signed the register after all, even if it had been on the wrong page.

After Tinsley and Meakin's release from prison, however, something quite peculiar occurred. Both convicted men returned to service on the Caledonian Railway, which should have been impossible. Tinsley went back as a lampman and porter at Carlisle, where he worked for 30 more years until his retirement, and Meakin initially returned to the railway as a freight guard, until he was made redundant after the war. He then set himself up as a coal merchant, working from a yard at Quintinshill sidings, next to where he had played his part in the crash. During World War 2 he worked in a Gretna ammunitions factory until retirement due to ill health. Their return to the railway undeniably points at a deal having been done: Probably that Tinsley and Meakin accepted all blame, thus sparing the Caledonian Railway (and the Government) any further investigations that might have uncovered potentially greater problems. In return, the Caledonian Railway ensured that the signallers' families did not suffer hardship.

Comment

This accident is typical of the time before electro-magnetic train detection and signal interlocking was introduced throughout the network. Serious accidents caused by staff error would nevertheless continue until well into the 1970s, and unfortunately even the 1999 ***Ladbroke Grove*** accident comes under this heading. However, the second and third worst accidents in Britain that resulted from staff error were caused by SPADs during bouts of fog rather than by blatant signaller's errors. The first of these was the horrifying triple high-speed crash in the Up and Down fast platforms at *Harrow & Wealdstone* station on 8 October 1952, when following a SPAD in foggy circumstances an Up Scottish night express ran into the rear of a platformed semi-fast local after which a Down Scottish express ran into the wreckage in the narrow track well between two platforms, leaving an unbelievable mountain of wrecked rolling stock in which 112 people died. The second was the *Lewisham* crash of 4 December 1957, when following a SPAD in dense fog a steam-hauled train rammed the rear of a multiple-unit train standing at a signal at danger. Ninety people died in the heavily loaded rush-hour trains and over 100 were seriously injured, when as a result of the collision, the viaduct on the Nunhead line collapsed onto the wrecked main-line train. A train coming down the hill towards Lewisham on that line stopped only in the nick of time when the driver spotted the curious state of his track ahead, despite the fog. Whilst fire played no part in either case, the speed involved and lack of crashworthiness of the passenger vehicles, especially the propensity to telescope, resulted in the high casualty rates.

These types of accident can only be avoided by installing signalling with train detection, such as through track circuits or axle counters, which prevent protecting signals from showing a proceed aspect if a train is detected in the protected section, to which a decent automatic train protection system should be added in order to counter drivers' mistakes. These two features are, in fact, the reason why the number of accidents has diminished to the low levels we see now, and why, when an accident does occur, the casualty rate is usually low compared to the number of people on the train. Even the ***Ladbroke Grove*** crash with its

30-plus fatalities and a great many seriously injured passengers should be seen in the light of the very high impact speed and the fact that a diesel fire broke out. Compared with the number of people who walked away from that wreckage, that casualty figure is still many times better than when similar numbers of people are involved in high-speed road accidents followed by fire, never mind in an air crash. Modern railway passenger rolling stock tends to protect its inhabitants much better than the vehicles of the past.

Signaller Meakin, in his own words, acting as he always did (i.e. not using collars to protect a train – and his job), failed to take the simple and effective measures available to him. He had done his night job safely enough, but was it he who, when ending his shift and routinely closing down before diverting his attention to the morning newspapers that Tinsley brought in, unthinkingly put that Up block instrument back to 'Line Clear' after the arrival of the coal empties in the Up loop, thus preventing the safe shunting of the parley? Had signaller Tinsley forgotten about the train that he travelled on because he had not done its shunt and was preoccupied with the train register book instead of taking a moment to settle and concentrate? He began working the frame acting on blind routines, which was bad signalling practice that the insufficient sophistication of his equipment could not prevent from ending in disaster.

The authors of *The Quintinshill Conspiracy* posit the possibility that James Tinsley suffered from epilepsy, which might explain his erratic behaviour that morning. That is possible, but among the close-knit Gretna railway community Tinsley was not known to suffer from seizures and he did not display such problems either before or after the incident. But his wife worked as a pub landlady and Tinsley's was a familiar face in the bar, where he regularly helped out (even fairly soon after the disaster he was seen working there). An equally likely explanation for his forgetfulness in the signal box on the morning in question might be consumption of alcohol the evening before, having to get up early again. That is not something that an under-fire railway company, intent on damage limitation, would want in the public domain either. The unusual leniency displayed by the Caledonian in this case in all likelihood stems from it successfully doing a deal to forestall further investigation by compelling both signallers to accept immediate responsibility.

I believe that both signallers, like many before and since, unthinkingly relied on their good luck. Inexperienced risk-takers still do that every day, time and again causing accidents. The signallers had no awareness of their human vulnerability to the quirks of fate and not a few people throughout transport history have come unstuck for that same reason.

As to whether their travel arrangements were a contributory factor? That is a red herring. Thinking logically in 2016, there should have been a quiet word about slightly altering booking-on times to formalise this convenient arrangement, thereby removing the stress of its illegality, the distraction of having to copy the workings into the train register book and so keeping attention where it belonged. Looking at it like that might even provide an explanation for the curious lack of early-morning signal box checks by the management, or the absence of corrective action when noticing the signallers' late booking on, as must have been the case at Gretna Junction at 06:10 when the parley was once again sent off first. The Victorian mores, particularly in war time, would never officially condone slack discipline to make staff life easier; but a blind eye could be turned to minor transgressions. It was the unthinkingly stupid and risky manner in which Tinsley and Meakin both did their jobs, especially under the stress of heavy war traffic, that inevitably led to an accident like this, no matter what their travel arrangements might have been!

***

Ciurea, Romania, 1 January 1917 at approx 01:00; and St Michel de Maurienne, France, 12 December 1917 at approx 24:00

(Wartime runaway accidents with fire and ammunition explosions)

Introduction

These are two of the most dreadful accidents in the history of world rail travel, both occurring during World War 1 following a downhill runaway and both suffering a fire with explosions from ammunition. They are included in this book as in neither case did actual hostilities play a role. Regardless of belligerent actions, war time can be a killer as a result of its high demand on transport capacity under extremely poor working circumstances and problems with maintaining equipment to a sufficient standard. Another such major wartime disaster without a shot being fired was the gigantic explosion of the French ammunitions vessel SS *Mont Blanc* in the port of *Halifax,* Nova Scotia, on 6 December 1917 (see Chapter 5). That accident happened at the end of the year that began with 1,000-plus fatalities at the Ciurea crash and only six days before the 800-900 fatalities of the St Michel de Maurienne incident. The explosion at Halifax flattened a substantial part of the city and left more than 1,900 people dead. (Halifax, incidentally, was the nearest rescue base where many of *Titanic*'s 1,500-plus victims had been landed and buried five years earlier.)

The accident at Ciurea

For a number of reasons this accident in Romania is far less well known than the French accident described below, but it was just as terrible and happened early in the morning of New Year's Day 1917. It displayed many characteristics similar to that at St Michel de Maurienne, especially the runaway factor, the exploding ammunition and the way in which wartime train operating conditions upset normal operations. The root cause, however, was rather different and points to the unexpected dangers arising from passengers travelling on the outside of trains.

As was usual at the time, the official records were somewhat incomplete due to Government suppression of the bad news to avoid demoralisation of a nation at war. But army sub-lieutenant Dumitru Done Tautu, who was on the train and survived the crash, recorded much of his knowledge of this accident on tape in 1979 after the start of the demise of totalitarian rule in eastern Europe and so *de facto* became the main source of information. In 1996, when he was 102 years of age, the Centre for Oral History of the Romanian Broadcasting Corporation invited him into their studios for a second interview and another taped record was made. Of interest is his comment on how the train had been put together. It was double-headed with two steam locomotives and was made up of a mix of sleeping cars, restaurant cars and carriages reserved for foreign military missions, as well as ordinary Romanian passenger carriages of all types that had been found berthed at the small but regionally important town of Bârlad. From Mr Tautu's comments, incidentally, it appears that the Compagnie Internationale des Wagons Lits et des Grands Express Européens (CIWL) contributed a fair share of its rolling stock to this particular train. In all likelihood this was rolling stock that was stranded in Romania at the outbreak of World War 1 from such international long-distance trains as the 'Orient Express' for example. At the time, this CIWL stock was built of varnished teak on steel frames, and whilst train heating on the ordinary coaches generally already took place with steam piped from the train locomotive, Wagons Lits rolling stock (which ventured to some of the most remote areas in Europe, each vehicle also having its individual attendant) was heated with small coal-fired water-heating furnaces in the individual coaches. As heating obviously was necessary in the cold eastern European winter when this accident took place, it explains the source of the rapidly escalating fire that ensued after the actual crash. A small number of such wooden second-generation CIWL vehicles survive in various museum collections throughout Europe; the famous steel-built blue, blue and cream or Pullman umber and cream painted third-generation coaches would not come in until the 1920s. Strangely enough, it was the CIWL that again contributed another such wooden restaurant car in 1918 to play its famous role in the history of the Great War as well as World War 2 in the woods near the French town of Compiègne. The German occupiers destroyed that vehicle after the French Army had signed the surrender in 1940.

In late 1916, the German Army under August von Mackensen penetrated deep into eastern Europe in an attempt to drive a straight line 'from Berlin to the Bosporus' through Romania, and to destroy the Romanian army in the process, in neither of which they succeeded. However, this initiative caused the civilian population to flee the hostilities in panic, because Von Mackensen's army contained not only German but also substantial Bulgarian and Ottoman Turk elements. The latter, whether rightly or not, historically had a reputation for ugly violence against civilians in the eastern European consciousness. It was under these circumstances that a train of no fewer than 26 carriages was put together at Bârlad station at about midday on 31 December 1916. This train had been on standby for about three hours before being ordered away at around 15:00, making its way fairly slowly along the part-double and part-single line from Bârlad to Iaşi and stopping at all intermediate stations.

The line is well laid out and runs through flat agricultural countryside fringed by wooded hills, but between the stations of Bârnova and Ciurea it descends steeply through dense woodland over a distance of 16 km (10 miles) into the station at Ciurea, which then consisted of a through line and a loop. In places the line along this scenic stretch into Ciurea descends in gradients as steep as 1 in 15 or 6.7%. Many of the occupants of this already well-loaded train on departure from Bârlad were military staff and soldiers on their way to regroup in units as yet not immediately involved in the hostilities, but each of the stops en route added a new contingent of civilian and military passengers that soon crush-loaded even this long and heavy train. By now people were sitting and standing on the roofs, footboards, steps and buffers of all the vehicles, including the locomotives. After passing Bârnova, an hour after midnight into New Year's Day 1917 and having covered about 100 km (60 miles) in 13 hours, the train commenced its steep descent into Ciurea. The double-header driver signalled his colleague on the train locomotive by whistle to shut off power and then opened his brake valve to keep the heavy train in check on the steep descent. However, despite having worked normally during the intermediate stops, now only the locomotives and the first few carriages applied their brakes whilst the rest of the train suffered brake failure. As a result, the train became a runaway and quickly gathered speed down the twisting line. Yet despite the obvious fact that something was desperately wrong with the speeding train, no one in the carriages thought to pull the emergency brake handles (which would have vented the isolated portion of the train brake pipe and so applied the brakes on the uncontrolled portion – a clear difference with the accident in France described below). Many lives would have been saved.

The main line through Ciurea station was already occupied by a long military freight train, loaded mainly with artillery shells and other ammunition, en route to the war zone and awaiting clearance of the single-line section ahead. The route through the station for the refugee train, therefore, was set up through the loop line, which carried a severe speed restriction. On entering the station the locomotives and the first carriage of the careering train somehow made it safely into the loop line, but from the second carriage onwards the train derailed on the loop turnout and the following vehicles began to derail and turn over, partially crashing into the locomotives and the wagons of the waiting ammunition train and so preventing that train from being moved. When everything finally came to a stand, only two of the carriages of the refugee train, one at the front and one at the very rear, had remained on the track. The inevitable fire in the crashed train fairly quickly spread throughout the splintered wreckage, involving both trains, and after about half an hour, when attempts to get survivors out had finally gathered pace after tools to break into and remove the wreckage had been found in the neighbourhood (see ***Meudon***), it set off the explosive cargo of the military train. This added to the carnage of the derailment and no fewer than an estimated 1,000 people lost their lives in the crash and the explosion. About 300 survivors were taken to nearby Iaşi by relief train at daybreak.

The reason why the train became a runaway could never be conclusively established in sufficient detail, but the likely root cause was simple enough. In the steel remains of the burnt-out wreckage, fairly close to the front of the train, investigators found one stopcock in the brake pipe connection between two vehicles in the closed position when it should have been open to provide train brake continuity. The assumption is that someone, very probably when getting on the train at the stop before Ciurea and sitting on a buffer of that vehicle, inadvertently closed this brake pipe cock with his boot when trying to ensure that he would not fall off his uncomfortable perch, or maybe he had shifted his position to relieve the onset of stiffness in the freezing cold and touched that stopcock. Or perhaps it was someone collaborating with the German occupying forces who knew how to sabotage a train and then jumped off. From that moment on, the situation closely resembled what would happen many years later near *Washington DC* in the faraway US on 15 January 1953. The inadvertently closed-off brake pipe of the isolated section of train kept its air pressure long enough for the brakes not to apply, but that isolated section of train could not react when the driver used his brake valve. As a result, only the few vehicles still fully connected to the locomotives via the open valves in the brake pipe would have their brakes applied, the rest being just 'swingers' (to use British rail parlance) and kept rolling unchecked. Whilst the locomotive and braking coaches could not muster sufficient brake power to keep the long and heavy train in check, let alone stop it, like at ***Eschede***, the understandable reluctance of passengers to pull the emergency brake handle did the rest.

The accident at St Michel de Maurienne

This incident took place twelve months later on the French side of the famous Maurienne Alpine line, built by Italian engineers as the earliest transalpine rail connection. In France the line runs from Chambéry via Culoz and St Jean de Maurienne to the border station at Modane. There the line enters Italy through the Fréjus Tunnel high up in the mountains and then descends steeply to Torino and onwards, notably to the port of Genova via Alessandria. Train 612 was made up of ancient rolling stock of Italian origin that, as is usual in wartime conditions, was heavily used yet poorly maintained. The rolling stock was of normal construction for the period, with wooden bodies on wooden or riveted steel frames. The train consisted of luggage vans at the head and tail, the latter being the only French PLM (Paris-Lyon-Méditerranée) vehicle, with 15 bogie vehicles in between transporting 982 French troops (some sources mention as many as 1,025) on their way home for 15 days' Christmas leave after successfully assisting the badly demoralised Italian Army to stabilise their front following their rout at the hands of the Austro-Hungarian and German armies on the battlefields of Caporetto (Kobarid/Karfreit) from 24 October until 19 November 1917. This battle undid the Italian-Russian alliance at the Eastern front and so freed the necessary German troops to be redeployed by train for the Marne offensive in the spring of 1918 along the stagnant Western front. The battle of Caporetto was, incidentally, one of the first major battles in modern warfare to see the successful infiltration of enemy defensive lines by German stormtroopers, deployed before the usual Great War artillery barrage that preceded the main attack, to disorganise the Italian defence by attacks from the rear.

The heavy home-leave train had slowly worked its way through the freezing cold December night up the tortuous eastern Mont Cenis ramp with two Italian crews and their locomotives. It entered the 13.6-km (8½-mile) long Fréjus summit tunnel at Bardonecchia and negotiated the sharp reverse curves outside the west portal to finally roll into the French Haute-Savoie border station of Modane. Here the train was sidetracked to let several other trains pass and to add two extra Italian vehicles of considerably greater age, their axles fitted in their wooden frames instead of in bogies. The 19-vehicle train was now 350 m (1,170 ft) in length and weighed 526 tonnes. A crew with one of the 230C type 2'C (4-6-0 or 10-wheeler) steam locomotives, No 2592, of the PLM railway company, were available to take this load down the steep Mont Cenis western ramp to Chambéry, where the troops were to change into various normal service trains to their final destinations. After coupling up the locomotive at 23:00 the train was about an hour late and several other trains had passed through the station whilst the home-leave special was held up. Unsurprisingly, the troops on board were becoming critical of the proceedings, whilst the great majority of officers had used the delay to escape further onward travel on this uncomfortable, cold, slow and crowded train by using their right to privately book on an express directly from Modane to Paris Gare de Lyon. That train was conspicuously being readied in another platform and would clearly depart ahead of the delayed troop special. This also might explain the difference between the numbers of occupants given in various reports.

There was a problem, however, with the troop special. Due to the bad state of maintenance of the Italian vehicles, only the first three carriages had air-operated train brakes that were in working order and were through-connected to the locomotive brake valve (what a British railwayman of old would have called a fitted head). According to the military rail transport regulator at Modane, Captain Fayolle, that single locomotive and these three air-braked vehicles were to hold the 526 tonnes of this train safely in check down the steep descent. This, according to the railwaymen involved, was impossible, for which reason driver Louis Girard refused to move the train. His locomotive, using its counter-pressure dynamic brake and its steam brake, was allowed to take only 144 tonnes (approximately equivalent to 3½ such passenger vehicles) down the 1 in 33 gradient from Modane at an altitude of 1,040 m (3,432 ft) to St Michel de Maurienne at 710 m (2,343 ft) and he demanded a second, in fact booked, locomotive. Captain Fayolle, however, had previously assigned that machine to assist a battle-bound ammunitions train. He solved this stand-off by giving driver Girard a jolly military earful and threatening him with a court martial for refusing to execute military orders in time of war. In fact, particularly with this sort of situation in mind, key railway staff such as drivers and signallers had been given military ranks, and because Fayolle was a captain and Girard an adjutant, the driver had little choice. He could only demand brakesmen on the train, to work the hand-operated parking brakes on coaches in the time-honoured fashion with whistle signals in addition to the train brake used on the locomotive and the first three vehicles, and then try and make the best of it. Seven competent railwaymen were hurriedly scrambled and posted throughout the train near parking-brake wheels, but the worn or completely missing brake blocks were likely to render their efforts futile.

The army men on the train by now were threatening to become mutinous – the last thing Fayolle wanted in the middle of a freezing December night high up in the Alps, and so at 23:15, train 612 was finally was whistled away and departed from Modane. Initially driver Girard managed to keep the speed down to 9 km/h (5.5 mph), as later evidenced by the speedometer record tape, but near Le Freney the train could not be held any longer and became a runaway. Whilst the permitted downhill speed limit was 40 km/h (25 mph) throughout, with a roaring counter-pressure brake on the locomotive and overheating brake gear showering sparks along the entire train, soon speeds of over 100 km/h (60 mph) were reached, causing the coaches to lean on the sharp curves and forcing the occupants to hold on to seats and luggage racks to avoid being shoved into the sides or aisles by the centrifugal powers working on them. The red-hot spark-showering brake blocks and running gear was but one problem; worse was that many troops had lighted candles throughout the dark train to keep warm in the hard frost and to see while playing cards (the steam heating and the gas or electric carriage lighting installations were also out of order) and these burning candles tipped over, fell off the tables and rolled under seats in the wildly bucking and swaying coaches. The varnished coach interiors and collected dust soon started to catch fire, which the men had to try and extinguish with jackets and boots but could not keep in complete check.

Thus the train sped out of control in a haze of sparks, smoke and fire-glow through the night and the soldiers began to panic. Some started to jump off into the pitch-darkness along this mountain railway through rocky terrain, with a river nearby, at recorded top speeds of up to 135 km/h (84 mph).

The last stage of the drama soon came. At Le Praz, near St Michel de Maurienne, the front bogie of the 2'C (4-6-0) locomotive derailed on the first part of the sharp reverse curves either side of the Saussaz Bridge across the River Arc, causing the locomotive and tender to lurch so violently that the taut coupling between the tender and the first luggage van failed. That vehicle broke away from the locomotive and derailed its front bogie at an estimated 102 km/h (63 mph). In the counter curve past the Saussaz Bridge, against a high retaining wall, the careering train met its end when this derailed front luggage van slewed round, laid itself on its side against the retaining wall and formed a heavy obstacle against which the rest of the train piled up and crashed. The locomotive, however, escaped this destruction despite being partially derailed (in fact the bogie re-railed itself on a trailing turnout) and came to a safe stop at St Michel de Maurienne station. Driver Girard had been so occupied until that moment that he was completely unaware that he had lost his train. As luck would have it, a troop train with British soldiers on their way to Italy to take over the fight against the Austro-Hungarian and German armies was stopped at St Jean de Maurienne station further down the line, having been held at the last moment after a warning about a runaway on fire near St Michel de Maurienne came through by telephone from the Le Praz stationmaster.

Immediately after the devastating crash the already smouldering fires found ample fuel and quickly spread, consuming the splintered and compacted wreckage with its cargo of dead and injured soldiers, and the dreadful situation was soon compounded by massive explosions. Many troops, against explicit orders, had brought ammunition from the battlefield on the train as souvenirs, including enemy artillery shells. Hastily rounded-up local police and army personnel kept civilian volunteer rescuers away, sometimes by force. Only the British soldiers from the stopped troop train were later allowed near the blazing train wreck, probably because they were armed, did not understand French, and would not take no for an answer, although in the prevailing circumstances there was little they could do anyway. The main purpose of this blocking measure was to prevent civilians (and with them potential enemy spies; what a heartening propaganda coup!) from witnessing what had happened. In view of the thunderous detonations ripping through the blazing wreckage, however, perhaps it was not such a bad idea after all. When the fire and explosions had finally died away and the remaining debris of mangled steel frames and bogies had cooled down sufficiently to allow approach on the evening of the next day, 425 bodies were officially identified, 135 victims remained unidentified, and 37 dead were retrieved from along the very short stretch of line between Le Praz and the crash site, making a total of 597 men. Only 183 men answered the roll-call, bringing the tally of people whose whereabouts were known, whether dead or alive, up to 780. Altogether, however, probably 900 people died when including those who were never found along the line, the many who were completely obliterated by the fire and explosions, and those very badly injured who finally succumbed in hospital. Except for driver Girard and his fireman there were very few physically uninjured survivors, most of the survivors being those who had mustered the courage to jump off the wildly speeding train and had the luck not to end up in a ravine or river or against a rock wall in the darkness.

As might be expected, Girard and his fireman were arrested and court-martialled. After no less than eight months of trial they were found not guilty, partly due to the no doubt terse testimonies from railway staff and civilians who had witnessed the proceedings at Modane. Five days after the incident, on 17 December, *Le Figaro* mentioned the disaster in a 21-line article, but the inquiry itself and its result were expediently classed as a military secret in order to silence the press and hide the evidence from public scrutiny. Captain Fayolle escaped retribution, despite being single-handedly responsible for far more French military deaths than any enemy soldier at the time could have hoped to achieve. Successive French governments suppressed the release of information about the accident until well into the 1970s. It was then that details still known to the families of survivors and witnesses started to leak out to railway historians, who started to dig and question and finally forced the opening up of the relevant files. For a long time it seemed that no photographs of the wreckage were available, but one picture of a tangled mass of what were clearly steel carriage frames against a rock wall being cleared by a small crane, with uniformed men standing nearby, has come to light. No doubt more exist; photographs did accompany official reports in those days as well. Even today, it remains the worst railway accident in France, and on a European scale it is very near the top. On a world scale, however, especially in lesser developed countries, accidents with a far higher death toll are known of, but unfortunately reliable reports of these are not available.

Like most of the Alpine railway lines, the Maurienne line was changed to electric operation early on, in this case from 1928 onwards. Although the part from the Culoz triangular junction to Chambéry was electrified at 1.5 kV dc with the usual catenary suspended above the tracks, as a result of the narrow bore of the tunnels in the higher regions of the section from Chambéry to Modane the track was fitted with a bottom-contact third-rail system for 1.5 kV dc that had also been applied in the Paris region. Consequently, the locomotives used had to be fitted with pantographs as well as retractable current pick-up shoe gear. From 1928 onwards the line was operated with a small series of rather unusual electric locomotives from very diverse suppliers, of which 161BE3 (1A' Bo'+Bo'A1', 2,580 hp, 129 tonnes) from 1927 has been preserved since 1973 at the massive Mulhouse railway museum. This motley collection of wonderful machines was retired only after the introduction of 21 suitably modified 6-axle Alsthom-built Class CC 6500 C'C' 5,900 kW standard dc chopper-controlled electric machines between 1970 and 1972 (CC 6539/6559). However, after the line was upgraded and refitted throughout with catenary wires in the 1980s, these machines were changed to their normal specification by removal of their shoe gear, although for a long time thereafter they could still be recognised by their green Maurienne livery. As a result of this change, however, any 1.5 kV dc capable electric locomotive type could now reach Modane and the expensive operating characteristics of the CC 6500s soon became their Achilles' heel. Like their cousins on the fast main-line trains they fell out of favour, especially after the cheaper-to-operate and more track-friendly 4-axle machines acquired the haulage power and traction control characteristics necessary for this steep and tortuous line, and they could be run in multiple if needed anyway. At least two of the indisputably good-looking and impressively powerful CC 6500 machines have been preserved in working order based at Avignon: CC 6558 in the green Maurienne livery and CC 6570 in the silver, orange and red TEE (Trans-Europ-Express) Grand Confort livery.

***

Weesp, The Netherlands, 13 September 1918 at 10:30; and Getå, Sweden, 1 October 1918 at approx 18:55

(Embankment collapses with lethal consequences)

Introduction

In 1918, towards the end of World War 1, two non-combatant nations,

The Netherlands and Sweden, suffered catastrophic railway accidents within a fortnight of each other for the same reason: the collapse of an embankment. The first accident took place at Weesp, near Amsterdam, on Friday 13 September at 10:30; the second occurred about a fortnight later at Getå, near Norrköping, on 1 October at approximately 18:55. The Weesp accident is the second worst railway accident ever to have happened in The Netherlands (the worst being the 1962 ***Harmelen*** collision), and the Getå accident remains the worst Swedish railway accident. Pictures of a similar incident on 19 December 1915 in Britain on the South Eastern Railway near *Folkestone Warren* on the Channel coast between Folkestone and Dover point to the fact that this type of accident does not necessarily have to end in disaster (the onset of the landslip was observed by coastal defence soldiers, who stopped the train and evacuated the passengers, after which the line gave way under the train). However, whilst everyone walked away unscathed from the trapped train between Martello and Abbotscliff tunnels in Britain, in The Netherlands 41 passengers were killed and 42 injured, whilst in Sweden 42 people died and 41 were injured. At Getå many people were killed in the ensuing fire in the wreckage, which was an ever-present risk following rail accidents in those days but was something that was nipped in the bud in the Dutch accident. In both cases, however, the lack of protection given by coaches with a wooden superstructure was clearly evident, as it was at ***St Michel de Maurienne***, ***Quintinshill***, ***Buttevant*** and ***Granville***. What is striking about the era of these accidents is how sophisticated many scientific procedures and some types of technical equipment had already become (the equipment used on the *Titanic* being an example). At the same time, and very much so in the case of the *Titanic*, these accidents show that there still were important lessons to be learned.

The accident at Weesp

The countryside of the western Netherlands is famously flat and waterlogged, where road, rail and waterways meet in many places. In order to cross the main inland maritime shipping channels high bridges had to be built, with long embankments to take the railway from datum level to bridge level. This situation remains today as it was at the time of this accident, even though this particular bridge, which at the time consisted of two paired single-track bridges to take the double-track line across the busy Merwede Canal from Amsterdam to the Rijn river, has since been replaced with two higher and longer double-track bridges. These now carry the four-track line at the requisite higher bridge level after the canal was widened and the European standard free height was set at 4.5 m (15 ft) above high water level.

The line involved is the Gooi line, which at the time was a double-track main line from Amsterdam eastwards to Hilversum and Amersfoort, continuing on towards Zwolle and the northern provinces or to the German border at Bentheim or Enschede via Apeldoorn and Deventer. This line was one of the first in The Netherlands along which significant modern commuter traffic developed when the well-to-do left Amsterdam for the leafy Gooi area and used the train to go to work. As a result, trains were usually heavy and well patronised throughout the day and, from the consist of train 102 with its first- and second-class content, it is clear that the people who travelled here were able to pay for their creature comforts.

Train 102 was made up of a Manchester-built HSM (Holland Railway Company) Sharp Stewart locomotive, inside 2-cylinder 2'B (4-4-0) No 520, with a 3-axle tender (staple Dutch motive power at the time), hauling luggage van D507, third-class C912, second/third-class composite BC418, first/second-class composite AB839, luggage van D1921, Postal van P470, third-class C702, first/second-class composite AB688, first/second-class composite AB840, third-class C809, and first/second-class composite AB683. All passenger stock was of the steel frame on bogies and wooden body compartment type with running boards at solebar level, and any British commuter used to the EPB rolling stock on the Southern until the 1990s would have felt thoroughly at home in it. The train had started its journey at 06:59 from Enschede to Amsterdam, a trip that is still part of what Netherlands Railways offers with its frequent clockface inter-city network, although you will no longer automatically end up at the majestic Amsterdam Centraal station but at the rather nondescript Amsterdam Zuid WTC station, or will find yourself in the large underground Schiphol Amsterdam Airport station if you do not leave the train speedily enough at Amsterdam Zuid.

The train picked up coaches from Oldenzaal at Hengelo and a section from Zwolle in Amersfoort, each main section preceded by the ubiquitous luggage van and as such recognisable in the listing of the consist, only the last two coaches added in Amersfoort not being preceded by such a vehicle. The somewhat delayed train was booked to arrive at Amsterdam Centraal at 10:40 and was literally in the last minutes of its journey. After passing non-stop through Weesp at the requisite 45 km/h (28 mph) with steam shut off, the crew on locomotive No 520 started the drag up the bridge embankment, the fireman having a problem keeping a sufficient head of steam with his powdery, clagging wartime coal in the narrow firebox between the big rear driving wheels. Nevertheless, speed had risen to somewhere around 55-60 km/h (about 35 mph) when, approximately 30 m (100 ft) away from the bridge and in front of the astonished faces of the bridge keeper/signaller and his wife when the locomotive suddenly sank and fell away to the right-hand side just as it reached the bridge abutment. The machine slammed into the right-hand corner of the northern bridge bowstring and pushed that entire corner of the bridge off its footing, the bridge now resting on only three of its four corners. Almost immediately a lighting-gas tank under one of the still moving coaches exploded and started a fire, but with great presence of mind the bridge keeper extinguished it with dry sand in buckets he had at his disposal for just such an event. This prevented the already bad accident from becoming substantially worse, as did his simultaneous action of sending a train conductor and a rail worker with red flags in both directions to the next signal box to raise the alarm and to prevent other trains from running into the wreckage. There was a possibility that the signalling and telecommunications equipment had been damaged, so the signals could no longer be relied on to show danger aspects and stop approaching trains.

Behind the locomotive the first luggage van and three coaches were in line hanging down the misshapen embankment side with the luggage van suspended high above the meadows. It had telescoped backwards into the first four third-class compartments of the following carriage, whilst the following second/third-class combine had damaged this third-class vehicle from the rear. The second/third-class vehicle, however, had lost its rear four compartments to the following first/second-class combine but this vehicle suffered the fate of being broken in two, its front half in line with the first part of the train and its rear half pushed two coach lengths past its front half by the rear part of the train when rolling down what was left of the embankment. From AB688 onwards the carriages were little damaged, but only the rear two carriages (C809 and AB683) were still standing on the track. The injuries to passengers were appalling due to the widespread telescoping of carriages, several publications mentioning the rivers of blood at the site.

Given the dire circumstances, there were nevertheless some lucky breaks. Surgeon Mr. Treub from Apeldoorn, on his way to Amsterdam, was unhurt and immediately set to work together with four hospital nuns. They used linen provided by the bridge keeper's wife. One of the few advantages that wartime provided in this case was that shortly afterwards a Captain van Andel happened to pass by with the 2nd Battalion of the 4th Regiment Infantry on their morning march in full wartime outfit between Muiden and Weesp. He and his men were instrumental in providing field tools, stretchers, manpower and armed security, as well as having the authority to requisition and direct substantial additional manpower from a construction site at the nearby dike of the Zuider Zee. He also stopped shipping on the Merwede Canal (the present-day Amsterdam–Rijn Canal) and so provided the main means of transport to get the many dead and injured to hospitals in Amsterdam with a degree of expediency. Furthermore, the military requisitioned medical material from local doctors, factories and other institutions in the neighbourhood. At 11:40 an emergency train from Naarden arrived as the first relief train, the fourth and final one being a Red Cross train from Amsterdam Centraal with medical staff and nurses that arrived at 12:55.

After the initial clear-up of wrecked rolling stock, the embankment was quickly restored to its full height to allow the remaining undamaged track to be used subject to a 5 km/h (3 mph) speed restriction from Thursday 19 September onwards, and after repairing the slight damage to the middle bowstring the speed limit was increased to 25 km/h (15 mph). The locomotive and its tender were salvaged on 4 October and repair to the northern bridge bowstring and the bridge floor started the same day. A temporary pier was built from timber on a wooden support in the meadow under the bridge, after which the damaged parts of the bowstring, bridge floor and supports could be removed and new parts riveted in from 28 October onwards. The first locomotive crossed the bridge on 25 November, and both tracks were open for normal traffic again a week later.

Incidentally, the war provided another bonus to the travelling public in the aftermath of this accident. The two remaining private national railway operators – SS and HSM – had operationally been merged into NS Netherlands Railways in 1917 under a wartime government directive, and diversions along routes that had previously been competitor lines were now freely possible. Although travel from Amsterdam to and from the north via Utrecht was slow in the months during the repairs of the bridge, the delay was minimal in comparison to what would have been the case had it been left to the managements of the former competing companies, notorious for creating petty inconvenience to the travelling public wherever their lines met. In those circumstances the ferry across the Zuider Zee from Enkhuizen to ***Stavoren*** in Friesland would have been the only HSM route open to the north for passengers and freight. In 2007, the name NS Netherlands Railways had been in operational use for 90 years, and there are not many railway operators in the world that can make that claim.

The investigation started without undue delay, the Dutch Government using its wartime powers to instigate a Special Commission to undertake the work. The chairman was no one less than Dr Cornelis Lely, the engineer who drained much of Holland's watery landscape and was an expert regarding the interrelation of water and soil. The Commission was asked not only to investigate the cause of the accident but also to carry out research into other embankments in similar waterlogged circumstances and advise on risks in those situations. This area of The Netherlands consists of a very thick layer of watery peat on top of sand deposited by the rivers that make up the combined delta of the Rijn (Rhein, Rhine) and the Maas (Meuse). As the embankment at Weesp had been built on top of the peat layer, the initial thought was that the embankment had disappeared straight downwards, displacing the peat, but this incident did not show typical signs of such an occurrence, such as changes in the width and direction of nearby ditches or peat-hills arising elsewhere as a result of displacement of lower layers. Furthermore, exploratory drilling through the damaged embankment into the subsoil revealed that the original grassed meadow top was in fact still at its original level. Bearing in mind the sideways movement of the crashing train, the only viable conclusion was that the embankment aggregates themselves had moved sideways; in other words, there had been a landslip.

Another remarkable issue was that the embankment aggregates, especially the lower level still *in situ*, were thoroughly waterlogged to a level well above the surrounding water table (which itself was substantially higher than normal due to recent excessive rainfall). But what had caused this level of saturation of the embankment aggregates, consisting of clean fine- and medium-grain river sand? The embankment, built up from an artificial ditch 1.3 m (4 ft 3 in) deep at its foot, had been left to settle under its own weight, further river sand having been added wherever subsidence was noticed. This method had worked well elsewhere and was used until well into the second half of the 20th century. Moreover, a water separation wall of heavy clay had been inserted down to 4.7 m (15 ft 6 in) below datum between the canal bank and the trench at the foot of the embankment, the canal bottom being only at a depth of 3.75 m (12 ft 4 in). The conclusion was that the accident had been caused by the wet sand of the embankment fluidising under the rhythmic pulses set up by the passing train – which was in fact a well-known phenomenon thought to have been dealt with by the measures just described.

But what had caused the exorbitant saturation of the embankment?

Deeper drilling brought the answers. Under the weight of the embankment the remaining peat had been compressed into a dense 0.4-1.2 m (1 ft 4 in-4 ft) thick watertight layer that actually reached upwards along the embankment sides from the bottom of the ditch to datum level. This had been aggravated by cladding the embankment sides with extracted and compacted peat and building parallel roadways on side embankments higher than datum level, also made up of compacted peat. This prevented water that had collected in the body of the embankment to drain readily from the sides at or below datum level, whilst peat and the clay dam hindered outflow at the canal edge. Furthermore, the clay dam was by no means sufficient to keep canal water away from the embankment, yet of far more importance was the fact that masses of rainwater had recently entered the embankment aggregates via the track and its ballast. Hence the absurdly high water table inside the embankment. The reason why this problem had not arisen earlier was due to the fact that in the 30-year existence of this embankment, the excessive amount of rainfall recently experienced had occurred only once before. The main lesson learned, therefore, was that low-lying drainage should be provided in order to allow water in the body of the embankment to escape quickly into the surrounding water table.

As an indirect result of this accident, the Institute for Geomechanics was set up at Delft. This would much later lead to quite astonishing feats of underwater soil engineering for the massive waterworks that were erected after the serious floods in mid-February 1953, following which construction of the so-called Delta Works was begun to protect the low-lying western Netherlands from another such devastating inundation.

The accident at Getå

The coastal railway line from Åby to Nyköping had opened exactly five years before the day of the accident, on 1 October 1913. In view of the geologically difficult stretch between Åby and Krokek, quite extensive georesearch with exploratory drilling had taken place to determine whether the embankment undersoil would be up to the task. The line, crossing a fjord-like channel called Bråviken at Norrköping, was built into the hillside along the high northern shore of the Bråviken. It was known from experience that the land was prone to landslides, but the exploratory drilling and geotechnical surveys had indicated that the soil was stable enough to hold a single-track railway embankment. As the spring and summer of 1918 had been exceptionally dry in Sweden, the clay around the shore was parched and cracked, so when in early September the rains began to fall and then persisted, the water quickly filled the cracks and penetrated the clay deeply and so produced thick layers of clay in different states of saturation. (A similar effect can also be seen with different layers of snow on hanging mountainsides or the roofs of buildings, when the wetter top layers slide off and cause a wet-snow avalanche.) In the case of clay, the water-saturated layers may slide over and off the drier and harder layers that still have sufficient adhesion to the surrounding soil.

Train 422 was a long-distance express from Malmö to Stockholm. In complete contrast to the train involved in the Weesp accident, it was composed of bogie open saloon stock hauled by a very able type of 2'C1' (4-6-2) 'Pacific' locomotive. The Swedish Class F numbered eleven machines in the 1200 series which were 4-cylinder compounds built between 1914 and 1916 by Nydqvist och Holm Aktiebolaget (later better known as NOHAB) at Västerås. These locomotives were the top of the range in Sweden and when displaced by the rapidly advancing electrification after World War 1 (as in Switzerland, for example, Sweden had no home-mined coal) they were sold to Denmark (Class E 964-974) around 1935/36, where they were so well liked that Frichs in Århus built another 25 machines with a number of later improvements as late as 1943-47. The Swedish machine involved in this accident, F 1200, can now be seen in the Swedish railway museum in Gävle, whilst more than one example has been preserved in Denmark and abroad (including Peterborough, UK). Behind locomotive F 1200 there was a Post Office and luggage van DFo 1107 (lost), luggage van F1 25591 (lost), third-class passenger carriage C3d 2050 (lost), third-class passenger carriage Co5 2039 (lost), third-class passenger carriage Co1 1235 (lost), third-class passenger carriage Co5 2044 (lost), first- and second-class dining car ABo3 2466 (fully derailed but salvaged and repaired), sleeping car Bo1 1015 (on the track), second- and third-class composite passenger carriage BCo 1429 (on the track) and luggage van G3 19003 (on the track).

The train had left the southern city of Malmö behind class F No 1271 at 07:00 and had experienced a rather uneventful trip through the endlessly undulating landscape all the way to Mjölby, where F 1271 was taken off and a replacement machine, class leader F 1200, took over. After a further stop at Linköping the train arrived 12 minutes late at Norrköping, the last station before the accident. At 18:26 another train had passed the shore-bound stretch of line between Krokek and Åby without incident. Between 18:33 and 18:40, the keeper at Getå halt had the strange experience that the wires on the telegraph poles seemed bewitched, making strange sounds and vibrating, but a milk cart passing along the lakeshore road under the railway embankment at about 18:50 experienced nothing untoward. When train 422 had crossed the Bråviken channel bridge on leaving Norrköping 10 minutes late, the first passing point was Åby halt, where the signaller acted by giving 'Tåg Ut' ('Train on Line') to the next signal box at Krokek at 18:54 against a booked passing time of 18:44. As Krokek did not receive the signal, the signaller there telegraphed Åby to request information on the whereabouts of train 422. Åby did not react to that signal, but at 18:57 tried to signal 'Train on Line' again to Krokek, again without response. The telegraph line was down.

The speed of the train at the time of derailment was estimated to be 65-70 km/h (about 40 mph), indicating that the driver was keen to make up time and that his locomotive was well up to the job. Suddenly, however, the leading bogie wheels of the 'Pacific' no longer found track and the machine nosedived into the sloping side of what had previously been the embankment, falling over onto its right side and careering down until it met with resistance of shingle virtually on the shore of Bråviken fjord. Its train followed it naturally, but the carriages suddenly found the bulk of the overturned machine stopping their progress, and the familiar story of the ripping and smashing of telescoping wooden carriages developed. Only the dining car and the following two coaches did not join the heap of wreckage, the dining car hanging down from the remains of the embankment and the other carriages still on the track above. Again, burning coal from the locomotive firebox under the splintered wreckage started a fire, but this time there were no bystanders to extinguish it quickly and soon the heap of entangled wood and steel was well alight, trapping those who were too badly injured to attempt to make their way out. Driver Wahlström, despite having hit his head hard when the locomotive overturned, and suffering from concussion, was able to free himself from the cab, but fireman Carlsson was buried under the spilled coal from the tender. The driver went back to alert the authorities and to protect the train wreckage. Meanwhile, senior conductor Ström was in the last carriage of the train and made his way forward on the track with the same objective as driver Wahlström. The driver told his conductor that the fireman was buried under coal in the cab and then went back to help his colleague whilst Ström headed to Åby to warn the authorities. Along the way he met trackwalker Andersson, who used the telephone in a nearby house to inform staff at Åby halt of what had happened, and from which moment the rescue operation swung into motion. In the meantime another railwayman, ganger Eriksson, approached the site from his house along the track where his wife was crossing keeper. She had closed the barriers in anticipation of the train but then had the strange experience of suddenly hearing the approaching train no longer. She warned her husband, who went out to check what the matter was and spotted some leaning telegraph posts. Realising that something was drastically wrong, he ran back to Getå halt to notify Krokek station about what had happened, get assistance and stop trains.

Of approximately 170 passengers on the train 41 were injured, 5 were missing and 42 were confirmed dead. As at Weesp, the objective was to get the line open again as soon as possible. Between 1 October and 2 December the line was cleared of wreckage, the investigation into the cause of the accident took place and the construction of a new embankment was begun. It was established from interviews with staff and passengers that the fire had started slowly from burning coal deposited from the ashpan and grate of the firebox of the plunging locomotive and not by escaping acetylene gas from the tanks for carriage lighting. But, as also happened in France and Switzerland, Swedish Railways abolished the use of gas for carriage lighting with consummate speed, helped by the rapid onset of electrification of the network based on German and Swiss 15 kV 16.65 Hz ac technology.

The track was soon open again, the line being constructed deeper into the hill with the use of rock walls and retaining walls to keep the embankment in place. The speed limit was lowered to 30 km/h (20 mph) either side of the collapse site and to 15 km/h (9.5 mph) at the site itself. The Geotechnical Commission of Swedish Railways then instigated an in-depth investigation and found that the embankment had been built on top of several alternating layers of glacial sediment and gravel/rock debris fields deposited in stages during various Ice Ages. More surprisingly, they found evidence of a large landslide from very long ago and found that water from the surface (recent excessive rain is again relevant) made its way into the fjord close to this landslide as a subterranean stream and slowly scoured it away. That loss of material is what had destabilised the rest of the rain-saturated clay and shingle subsoil and brought about the landslide. The heavy, speeding train was all that was needed to set off the collapse. (A later landslide here, in 1923, involved only the road along the shoreline.)

Comment

It is interesting to compare both these long-distance trains in The Netherlands and in Sweden. Despite the fact that the Dutch passengers who travelled all the way from Enschede to Amsterdam sat in their train for 3½ hours, they had no more than their compartments and access to a toilet. Catering had to be found at stations during stops. On the other hand, the number of first- and second-class carriages in that train was substantially higher than in the Swedish train, the few Swedish first-class travellers making their trip in the dining car. The Swedish train was a true 10-carriage long-distance train composed of corridor stock with muscular traction in the shape of a 4-year-old 2'C1' (4-6-2) 'Pacific' type of steam locomotive. The Dutch, admittedly with their fairly flat country, considered a 20-year-old 2'B (4-4-0) enough for the 11 carriages of their train, although since 1908 several types of rather more powerful 2'C (4-6-0) 10-wheeler types were available. However, the Sharp Stewart 'Rhine bogies' and what followed from them were excellent and capable locomotives, even in difficult circumstances. One is preserved in the Utrecht railway museum.

In my days at Bristol intercity depot an instance occurred of a railway line sliding down a hillside quite similar to what happened in Sweden. Obviously, in the case of the line at Sapperton Tunnel north portal between Swindon and Stroud, there was no catastrophic collapse or you would have heard about it. Still, the clear dip in the track was severe enough to make life hard for diesel locomotives with heavy freight trains, especially after a severe speed restriction had been imposed at the top of this steep gradient just before the tunnel. When taking a close look at the rail head this was evident from the burns ground in by slipping driving wheels. This situation, however, has now been thoroughly remedied.

***

Bellinzona, Switzerland, 23 April 1924 at approx 02:25

(Head-on crash with fire following a SPAD as well as a signaller's mistake)

The accident

This accident concerned a head-on collision between trains 51B and 70, two international services, which was then made worse by fire. Although both trains were hauled by double-header teams of electric locomotives, the fire was started by two coal-fired steam-heating vans coupled behind them. Swiss vehicles had been converted to take electric heating and lighting from the locomotives, but it was a German gas-illuminated and steam-heated wooden vehicle in the northbound train that caught fire from the steam-heating van. Despite wooden passenger vehicles being outlawed under European agreements in the late 1920s, clearly influenced by this particular accident, many of the older steel-built international passenger vehicles still had steam heating and gas lighting until World War 2. The introduction of axle-driven dc generators with batteries put paid to the practice of carrying large quantities of lighting gas in tanks under the vehicles when Europe introduced modern fleets on a large scale in the late 1940s to replace wartime losses. After the late 1950s only passenger service vehicles with electric heating and lighting were allowed to cross borders, unless by agreement in special services such as steam- or diesel-hauled museum services. Gas lighting is still specifically excluded. Fifteen people died in this accident and there were a small number of injuries.

Traction and trains involved

Forty Be4/6 electric box-cab locomotives were constructed as first-generation production types from 1921 onwards to replace steam traction on the Gotthard line with its long summit tunnel, until their own replacement began around 1955 with the introduction of the celebrated Ae6/6 full-adhesion electric Co'Co' bogie locomotives. The Be4/6 1'B'B'1' was an SLM Winterthur/Brown-Boveri collaboration for the SBB-CFF state railway. It weighed 110 tonnes; its 1,520 kW of power working the driven axles through jackshaft rods, and it was fitted to work in multiple both within class and with the Be4/7, at a maximum permitted speed of 75 km/h (46 mph). One museum machine still remains in working condition. Train 51B, the delayed international portion of train 51, was hauled by two Class Be4/6 electric locomotives, Nos 12342 (train locomotive) and 12329 (double-header).

Seven Be4/7 1'Bo'1'-Bo'1' locomotives were constructed as a first-generation test series for individually driven axles in 1922. They did not go into series production but were nevertheless kept in working order together with the Be4/6s on similar Gotthard duties until replacement by more modern types. They were a collaboration between SLM Winterthur and SAAS (Sécheron) in Genève for the SBB-CFF state railway. The locomotive weighed 111 tonnes, with 1,790 kW of power. These machines were built with hollow-axle drives, one traction motor per axle, and therefore had no jackshafts as was usual for heavy mountain haulers. They were fitted to work in multiple within class and with the Be4/6, at a maximum permitted speed of 80 km/h (50 mph). One museum machine still remains in working condition. Train 70 was also hauled by two electric locomotives, Class Be4/7 12502 (double-header) and Class Be4/6 12322 (train locomotive).

The location

Bellinzona station is one of the important rail hubs on the Zürich to Milano Gotthard mountain line through the Swiss Alps. The 15-km (9-mile) Gotthard Tunnel with its tortuous approach ramps is situated to the north of Bellinzona and separates German- from Italian-speaking Switzerland. The Gotthard route, opened on 1 January 1882, sees heavy international traffic like any other European transalpine route. Following the Swiss experience with steam traction fuel shortages during World War 1 the line has operated with 15 kV 16.76 Hz ac electric traction since 12 December 1920. The difficult original Gotthard route is now in the course of being replaced by the 57-km (35-mile) high-speed Gotthard base tunnel. The original Gotthard line will, however, be kept in service for local and freight traffic in the future.

The accident log

On the night of 22/23 April 1924, international express train 51 from Milano in Italy via Switzerland to Dortmund in Germany was to run in two portions due to a serious delay to the international section of the train. This was nothing out of the ordinary; I experienced a similar situation more than once at Innsbrück in Austria when the Italian section of the train via the Brenner route to Germany and The Netherlands was running late. The portion of the train that starts at the Austrian/Swiss/Italian border is sent off on time, whilst the delayed Italian portion normally follows later and, due to losing its slot, usually amasses further delay en route to its final destination. In this particular case the Swiss portion from Chiasso to Basel had been dispatched from Chiasso on time at 00:10 as train 51. Train 51B was the delayed international section that would follow it. Train 51B therefore departed Chiasso about an hour late and arrived at Bellinzona at 02:20 before departing 4 minutes later, 46 minutes behind schedule. Between Bellinzona station and the Gotthard summit there were four tracks, the two left-hand tracks of which were the freight lines between Bellinzona San Paolo freight yard that continued further northwards for a distance next to the two Gotthard main-line tracks. The connection between the two freight lines and the main lines at Bellinzona was a single ladder of two normal turnouts at the ends and two double-slip turnouts where the intermediate lines crossed.

Train 51B slowly accelerated away from Bellinzona station on its way to the Gotthard Tunnel, travelling under clear signals along the northbound freight line. This move was due to changes that were being made to the use of the track layout here; the train would be switched to the northbound Gotthard main line at this particular junction. Despite the fact that it was only a portion of the actual train, its make-up was substantial. Behind the two Be4/6s there was the ubiquitous steam-heating van with two fire attendants, followed by an ancient first-class coach from Milano to Dortmund of the German Baden Railway company. This was the wooden type of vehicle with gas lighting that was so often at the root of a serious conflagration following a collision. The next vehicle was a modern, steel-built Italian composite first- and second-class coach running from Milano to Basel. These were the three vehicles burned in the accident. The following five coaches stayed on the track and were removed before further fire damage occurred. On board the three damaged vehicles were passengers of eight nationalities: Swiss, German, Italian, American, Norwegian, Czech, French and British. Two of the VIPs on this train were the German former secretary of state Karl Helffrich and the Italian envoy to Denmark, Count Della Torre, neither of whom survived the accident. At the same time, night express 70 was approaching Bellinzona from the north. This train was a combination of the trunk set (Stammzug) 70 from Basel, which at Arth-Goldau had been combined with the set of so-called Kurswagen, train 170, from Zürich. This train was double-headed also. The locomotives, the leading steam-heating and brake van were damaged in the accident, but the rest of the train was saved.

At the controls of Be4/7 12502, approaching Bellinzona with train 70 on a downhill run at approximately 70 km/h (45 mph) was an ageing train driver who had already been off the main line for about a year doing mostly shunt work but who had been recalled to main-line work due to driver shortages in the busy Easter period. His second man, as they approached the crossovers to the freight tracks, noticed that the signal covering the junction showed a red aspect, drew his attention to this fact and said that they should stop. However, the driver remarked that this signal was for freight trains going into the yard only, the clearly visible next signal along the southbound main line into the station was for them and that one showed a green aspect. Nevertheless, the man at the controls must suddenly have had second thoughts about the signal and applied the emergency brake. They saw the delayed international train turning from the freight tracks on to the turnout ladder when their locomotive made the turn on to the crossovers towards the approaching train 51B, at which point the second man jumped off. The driver stayed at his controls and died in the collision, together with the footplate crew of train 51B, the firemen in the steam-heating van of train 51B and 10 passengers in the burned German coach.

The driver of train 70 in all likelihood had not kept himself up to date with changes at Bellinzona. He thought that freight trains were taken on to the tracks on his right-hand side, whilst his express train would be taken in along the southbound main line on main-line signals. He was, even historically, wrong on all counts, but works at the time were taking place to change the layout into a full-time four-track main line controlled from a power signal box, which would enable signallers to use any of the four available lines. To that end signals and turnouts at the crossovers had been partially uncoupled from the main-line interlocking, but there was still the normal interlock between the position of the hand-worked turnouts and their signals at the crossovers themselves. Consequently, nothing would have gone wrong had the driver of train 70 adhered as normal to the red signal aspect that he encountered. However, the pointsman working the crossovers on the main and freight lines had, contrary to the rules, already set this turnout to reverse for southbound freight 8572 coming off the Gotthard main line into the Bellinzona San Paolo yard and booked in after the departure of delayed express 51. That was the reason why the southbound main-line turnout was lying in reverse; the pointsman was saving himself the job of having to shift one extra turnout after the night express to the Gotthard and beyond had gone. Now, had the driver of train 70 been up to speed with his route knowledge and known about the changed use of these tracks for the extended San Paolo yard since 1 June 1921, he would perhaps have stopped at that red junction signal. What was worse, however, was that he let completely erroneous ideas about what signals applied to what type of train on what track overrule his common sense and went at speed past a signal that was clearly at danger along his track. That was then compounded by the error of the temporary local pointsman during resignalling work, and thus the scene had been set for a serious crash. (See ***Buttevant*** for a repeat in Ireland.)

The four locomotives met head-on on the crossovers, the locomotives of train 70 being pushed off the track but staying upright. The leading heating van of train 51B crushed back into the old wooden German carriage following it, but the Italian coach behind it, propelled by the train at the rear, rammed into it from behind. This not only splintered the body into oblivion but also damaged the 1,200-litre lighting-gas tank under the frame, which exploded and immediately set both front vehicles alight. The passengers and their luggage in the following Italian coach had to be evacuated quickly through the windows (the vehicle was leaning to one side and the exit doors at the collision end were crushed shut, a situation that would be repeated fairly often in similar collisions). Whilst everyone got out of that vehicle safely, its interior soon caught fire and was completely destroyed. Despite the fire brigade arriving quite quickly, there was a delay in extinguishing the flames due to confusion about traction current on the wires, and in any event there was no immediate water point close by.

The investigation

It is of interest to follow the chain of events that led to this accident:

1 The recall to main-line service of a driver who had not done that type of work for a year is questionable. Nowadays, at least in my experience, if such a man were selected at all, he would be required to refresh his route knowledge at least for several days and would also have been told about the important changes such as those at Bellinzona. That did not happen, although perhaps that was why he had a qualified second man with him. But, in that case, why did that man not drive the train?

2 It is incomprehensible why the driver of train 70 thought it acceptable to pass that particular signal at danger 'because it only applied to freights going over to the freight lines'. That goes completely against the principle of safe working with signals on any network: If in doubt, stop! That situation of a junction signal at danger followed by a signal with a clear aspect ahead was no different from what may still be experienced daily on rail networks the world over. The Swiss rulebook makes very clear that such a home signal (a controlled signal in British railway parlance) may not be passed at danger unless under very strict agreements between signaller and driver, and that is no different anywhere else that I know of. It is possible that the driver made a distinction between the functions of the home signal for the freight yard and that for the station, which stood somewhat further south and showed a green aspect, but that was still a lethal error. The signal concerned had nothing to do with the freight lines and yard other than to cover the connection into it, and for the driver concerned it should just have been one of his signals along the southbound Gotthard main line. As all signals have to be reacted to appropriately, in such a case of uncertainty he should have stopped and contacted the traffic controller by signal telephone.

3 The attitude of the second man in the cab, who stood by watching it all go wrong, is also puzzling. Why, then, was he in the cab? Yet this is something perhaps that anyone who has worked ships,

planes or trains will recognise. There is a very strong reluctance to do anything to correct perceived mistakes by a more senior person in control in such a situation. I have experienced that myself a few times on board vessels and trains. It is a well-known phenomenon that has contributed to several accidents.

4 The signaller at Ambri-Piotta, confronted with freight train 8572 running 55 minutes late and the on-time night express 70, decided to let the express go first. Whilst this was intrinsically absolutely justified, he warned only the stations between Ambri-Piotta and Biasca about the change and not those between Biasca and Bellinzona. Hence the mistake of the pointsman working the main to freight line turnouts at Bellinzona. Naturally, this can be put down to the signaller's work pressure versus the time required to make a substantial series of telephone calls, but in that case he should have asked the signaller at Biasca to pass on the message. Which, in actual fact, that person should have done anyway but omitted. With improved technology since the 1920s, the change in train order would now have been visible on the train describers.

5 The pointsman at Bellinzona, temporarily employed and manually shifting the turnouts due to the signalling control and track layout changes being undertaken, had reset the southbound main line turnout of the crossovers to reverse (as he should not have done according to the rules) in the expectation of the arrival of freight train 8572 for San Paolo yard. He was, so to say, making his life a bit easier. This, however, still would not have had the consequences it did if the driver of train 70 had acted correctly on the signals.

6 The 'reading through' of signals, whereby a driver can see and make decisions based on the aspects of signals past the nearest signal also exists in Britain (see also ***Genthin***). It is not in itself a problem. It may even be advantageous if you know the layout well, because it is in the nature of driving trains that timely information about signal aspects ahead makes for more efficient driving and quicker clearing of signal sections. But here we see a clear disadvantage leading to a mistaken interpretation of the situation.

7 The signal covering the crossovers from the north was clearly visible for a long distance and for that reason had no distant signal. Had there been a distant signal at the requisite distance it might have triggered brake action by the driver irrespective of

the aspect of the main signal, which in turn would perhaps have made a stop possible or would have mitigated the impact.

Comment

One of the things this accident illustrates is the ambivalence that may be generated by having a single-line connection that is worked in both directions between parallel running main lines. Despite the fact that this accident happened some 90 years ago, it is very recognisable in the way it developed from a number of errors made by various people along the line. Interestingly, like the accident at ***Buttevant***, it also clearly demonstrates the dangers of degraded working of signalling over a longer period of time. All sorts of undesirable traits arise to make life easier that safety-wise are not always covered by interlocking machinery, which would normally refuse to allow such hazardous things. In track alteration situations these days, the signalling installations are temporarily changed to work the new layout, instead of allowing some local pointsman to muddle away.

In fact, what was happening here was that all control from local signal boxes was being transferred to a new electric power signal box for the whole yard and station area plus sections of the main line. Nowadays that new signal box would be built and connected up while the old boxes were still working as normal. Then, when staff had been fully trained and everything new had been tested and found in order, a weekend would be used to transfer all control from the old signal boxes to the new power box in one move and then be thoroughly tested before release to traffic. Not that such arrangements always worked first time (see ***Clapham Junction***), but as a rule, assuming every person does the right thing, that is the only safe way to instigate the migration of control during upgrading of railway signalling. The accident also shows the importance of a railway operator ensuring that safety-critical staff work closely together in order to make the right decisions based on sound knowledge. Individual anomalies such as shown by the driver in this accident are inexcusable, but it is not possible to eliminate this sort of behaviour completely. I encountered some strange aberrations with staff I instructed, but regular testing of rules, traction and route knowledge plus well-prepared safety meetings goes a long way in preventing the situation that occurred here.

Switzerland instigated the Europe-wide ban on the international use of wooden passenger vehicles late in the 1920s and banned all gas-lit rolling stock from its tracks after this accident. Any train coming across the border had to have electrically lit interiors, either powered from the heating connection through the train or, at the time, more usually from axle-driven dc generators that produced lighting voltage when the coach was rolling along. This electricity was then stored in batteries which powered the coach lighting. There was normally enough power to keep a standing coach lit for between half and three-quarters of an hour or so, after which emergency lighting came on. It was only after World War 2 that heating without the use of steam generator vans finally came about.

Electric lighting from batteries brings back memories of sitting on a train in some big station in Germany, awaiting the arrival of delayed Kurswagen carriages that had to be coupled up for the onward trip. All of a sudden the lights would extinguish and a weak blue night bulb would come on, indicating that the batteries were going down. Nowadays coach lighting, heating and air conditioning are all powered from the locomotive through the hotel-power connection and the batteries are used only for lighting when the locomotive is disconnected. That can be felt at the heating/air-conditioning vents as the cooled or heated air stops flowing out when the fans stop, but the lighting remains on.

***

Lagny-Pomponne, France, 23 December 1933 at approx 19:30

(Extremely destructive rear-end crash in fog)

Introduction

This was the worst peacetime rail accident and the second worst rail accident in France, with 230 fatalities and more than 300 injuries, although some responsibility for the high death toll was down to the poor emergency effort after the accident. This tail-end collision occurred in freezing fog on a Saturday evening at around 19:30, due to a SPAD. As such, this was one of those classic major railway collisions that have now been largely brought under control with the introduction of more sophisticated train protection systems such as ATP or PTC. At the time, France was engaged in the introduction of the unified 'Code Verlant' signalling system, whilst as an offshoot France and Belgium had started to fit the 'Crocodile' AWS system. There is a suspicion that this system failed that night for rather simple technical reasons related to the frost.

The line and company involved

The railway company involved was the Compagnie des Chemins de Fer de l'Est, commonly known in France as the Est. It ran services from the

Gare de l'Est in Paris eastwards towards the Alsace and Lorraine regions and the German border. Important cities served were Nancy, Metz, Belfort, Strasbourg, Colmar and Mulhouse, as well as the cities of Luxembourg and of Basel in Switzerland. On the whole the Est main line was not particularly hilly, but it contained some rather long and curving ascents through winding valleys that nevertheless permitted quite high speeds.

The trains involved

Involved in the accident was an extra express, train 55, from Paris bound for Nancy. Due to the Christmas peak demand for seats, the train was made up of older passenger vehicles constructed with wooden bodies on steel frames and hauled by a 10-wheeler (4-6-0 or 2'C) locomotive.

The offending train was express 25*bis* from Paris to Strasbourg, a normal scheduled service which, to cater for high demand, was made up of 17 of the recently constructed Est-type Forestier Metallique steel coaches of 44-46 tonnes each, built at Romilly works. They were hauled by one of the new Est 'Mountain' 2'D1' (4-8-2) steam locomotives of Class 241, No 017, worked by driver Daubigny and fireman Charpentier.

The traction involved

Unlike some of the other French railway companies, the Est initially was not known for big locomotives, but the introduction of the heavy riveted steel Forestier carriages halfway through the 1920s led to traction requirements that saw some of the heaviest steam locomotives in Europe being put into service. Size- and power-wise, these machines were a quantum leap in comparison to the existing stable of locomotives. Besides these machines, the Est operated a larger stable of single-expansion 2'C (4-6-0) steam locomotives than any other French company, mainly because of the conspicuous amount of former German types that had been received as replacement for traction damaged during World War 1. Unlike other French railway companies, the Est did not go for the ubiquitous Class 231 (2'C1' or 4-6-2) 'Pacific' type of steam locomotives, but like the Great Western in Britain, stuck with the Class 230 (2'C or 4-6-0) 10-wheeler type from 1906 onwards until the massive 'Mountains' appeared in 1925. Apart from their size, the new 'Mountain' types were double-expansion compounds, which for reasons outlined elsewhere in this story took quite a bit of getting used to. However, extra express 55 to Nancy was not booked for one of the Class 241s, and therefore in all likelihood was worked by a Class 230 10-wheeler, which might have been one of the ex-Prussian P8 or S10 2'C (4-6-0) war-reparation machines. If train 55 had a French instead of an ex-German locomotive, then in all likelihood it was an Est 230 (later SNCF 230K), a light but powerful 4-cylinder compound type of locomotive built at Épernay in various series between 1906 and 1926. Unfortunately, not a single locomotive of this type is preserved.

Locomotive No 017 (later SNCF 241 A 017 after nationalisation) of Paris-La Villette shed belonged to a series of 41 steam locomotives designed by Mr. Duchâtel. Built at Épernay works, they were the first of the 'Mountain' type express machines in Europe and at 3,500 hp were capable of working trains in excess of 700 tonnes at 105 km/h (65 mph) – increased to 110 km/h (65 mph) after the frame plates near the cylinders were strengthened – uphill along the Est's curving main lines. The prototype was delivered in 1925 and after thorough testing series production started in 1931/32. Thus, at the time of the accident, No 017 was just about a year old. No report mentions whether this might have had any influence on the occurrence of the accident, but I can well imagine it being pertinent.

Est crews, who at the time were rather more used to smaller locomotives (quite a few of which the uncomplicated Prussian simple-expansion machines), can by no means have been fully conversant with the ins and outs of firing and driving these rather intricate monsters with their size-related peculiarities. French compound steam locomotives, fuel-efficient though they were in comparison with simple-expansion machines of similar size and power elsewhere, never had the reputation of being easy to drive unless crews were thoroughly conversant with them. Strangely enough, confirmation of this can be found in the adventures of 241 A 4 and 241 A 65 in Germany. The former was one pilfered by the German Army and left behind in East Germany after World War 2 when the Iron Curtain came down. It was not repatriated and, modified by the East German Deutsche Reichsbahn to work on powdered coal, was found exorbitantly heavy on both fuel and water and therefore soon dumped and scrapped. When 241 A 65 was fairly recently overhauled in Meiningen, Germany, it was used on a special and had to be retrieved well before the planned end of the trip due to running out of fuel and water. In their heydays in France, however, these machines were routinely used from Paris to Strasbourg and Mulhouse without taking water or refuelling (usually working with coal briquettes rather than decent locomotive coal). The French crews knew how to handle them economically at that time, whereas the German crews clearly did not.

Between 1931 and 1933, 49 more of these locomotives were constructed for the Chemin de Fer de l'État, but were transferred to the Est after World War 2 as Class 241 A 42-90. These were replaced on their original État stamping grounds by the equally massive but rather more modern, more powerful and faster 'Chapelon' Class 241 P 2'D1' (4-8-2) locomotives. These in their turn had been displaced by the 1.5 kV dc electrification of the original PLM Sud-Est main line from Paris Gare de Lyon to Lyon Perrache and Marseille St Charles stations. On the twisting Est lines, the lower top speed of the older 'Mountains' was of little consequence.

The Est Class 241 machines were constructed along the principles of the De Glehn-Du Bousquet 4-cylinder compound steam distribution, using the superheated steam at 17 bar (256 psi) after initial tests with lower pressures, with two high-pressure cylinders between the mainframe plates under the smokebox and then reusing it in two bigger low-pressure cylinders below the smokebox outside the mainframe. Exhaust took place through a large diameter multiple blastpipe chimney. The total weight of the locomotive ready for service was 121.9 tonnes (maximum permitted axle weight 18.5 tonnes). The tender weighed in at 70 tonnes, more than 35 tonnes of which in working order (about the weight of a passenger coach) consisted of boiler water. The firebox grate area was a massive 4.43 m^2 (48 sq ft), the driving wheels measuring 1.97 m (6 ft 6 in) in diameter.

Their end came after a long life from the 1960s onwards, when they started to be ousted by the 25 kV 50 Hz ac electrification of the main line to Strasbourg (in fact, pioneered on their home ground in Alsace-Lorraine) and by the introduction of diesel-electric traction on the other Est lines. It took until the advent of the 2,350 kW (3,000+ hp) Class CC 72000 between 1967 and 1974, however, before a single diesel could replace the multiple-united sets of diesels to replace one of these behemoths at the head of a heavy train. The prototype 241 A 1 from 1924/25 is exhibited in the French national railway museum at Mulhouse, whilst ex-État 241 A 65 has been completely overhauled in Meiningen for the Swiss Eurovapor organisation and is in serviceable condition.

The accident log

The day in question was one of those typical holiday mass-migrations that can still be experienced when travelling through France at the start of national public holidays. The stations were overrun and consequently train operations were disrupted and trains badly overloaded despite several extra services having been inserted into the timetables. At the

Gare de l'Est, as at other stations, the situation was hectic and confused. The mix of heavy weekend evening traffic combined with the Christmas holiday exodus from Paris was not helped by a dense freezing fog that slowed the running of trains. The glut of long-distance trains were caught up behind delayed locals on the slow lines and the tightly timed semi-fast trains on the fast lines that called at the larger intermediate stations such as Le Chénay, Chelles, Vaires and Lagny. As a result, the delays to the closely following trains were measured in hours.

Train 55, the 17:25 to Nancy (put together of obsolete rolling stock of wooden construction due to an acute shortage of more modern equipment) had left Gare de l'Est two hours late at 19:25 and was making its way eastwards, past the murk of Paris-La Villette locomotive depot and accelerating into the further suburbs, when the driver started to pick up adverse signals that caused him to slow down. After crawling along for 20 km (12½ miles) the train finally came to a stop at a signal showing a red aspect at Pomponne station, about 1.5 km (1 mile) before Lagny station where the excessive rush-hour crowds were just leaving another long and heavy semi-fast suburban service standing in the eastbound main-line platform. Adhering to the rulebook, the crew of the Nancy train placed fog-signal detonators on the track at the stop signal to cover their train. Not long after the delayed Nancy service had finally departed, the 17-coach express 25*bis* for Strasbourg got the 'right away', about an hour late, and driver Daubigny together with his fireman Charpentier on the 'Mountain' set to work to shift this load. Once past the station throat they accelerated rather smartly to about 100 km/h (60 mph) and were soon rolling along nicely through the fog, no doubt pleased that their locomotive had no apparent trouble with this heavy train in the way the older and smaller types would have done. Around this time at Lagny station the semi-fast had departed and soon thereafter the signal at Pomponne for train 55 to Nancy came off to a caution aspect. The driver had just released the brakes, opened up the regulator and was on the move again when his train was hit violently from behind. Train 25*bis* for Strasbourg, weighing in at approximately 960 tonnes, had run into the rear of the Nancy train at about 100-105 km/h (60-65 mph). Both trains soon came to a stop due to loss of the brake-pipe pressure and application of the emergency brake.

The consequences of the collision were disastrous on an epic scale. The heavy 'Mountain', propelled by the trailing weight of 17 steel coaches, had run through the last five of the twelve wooden coaches of the overcrowded Nancy train, smashing their bodies, folding up their frames, scattering the remains aside and severely telescoping the sixth and seventh coaches before finally coming to a stop, fully derailed but upright. Wheels and axles were the only remaining recognisable parts of the ancient coaches. The immediate death toll was 204 with hundreds injured, many badly. The final toll would increase to 230 fatalities, with more than 300 injured.

Of train 55 not much was left fully intact except for the locomotive and the first five coaches, but the all-steel coaches of train 25*bis* were all undamaged and the great majority were still on the track. Indeed, they would later be used to bring some of the survivors back to Paris. The slow emergency response was partly explained by the fog and the frost, for which reasons the welter of splintered wood from the destroyed coaches was used to build fires next to the line to provide heat for injured survivors and identify the location for the following salvage crews. Driver Daubigny was arrested and made to stand beside the line for hours watching the procession of emergency workers bringing the mostly badly mutilated remains of the victims out of the wreckage and depositing them for identification in a nearby field, being charged with murder by gross negligence. He was brought to trial in 1934, but this did not conclude with a conviction. Daubigny had undoubtedly run irresponsibly fast in the dense fog and during the disruption of services, but the following issues were taken into account:

1. On post-accident testing he turned out to be colour blind and, as a result, although he could see the semaphore signals with their indications in daylight, he had problems with the often weak semaphore colour signal lights at night and in the fog. The company should have checked on this, and it was immediately made part of the recruitment specifications for a train driver throughout France. Quite a few other drivers who did not make the grade immediately lost their driving jobs.
2. On considering what he must have seen with his condition, it was declared possible that he had interpreted the colour-light indications of semaphore distant and stop signals at danger as clear. As a result, the implementation of the new national 'Code Verlant' electric colour-light signalling system was speeded up.
3. The 'Crocodile' AWS ramps and on-board equipment were not fail-safe. Unlike the Great Western Railway ATC and the Strowger-Hudd inductive AWS system, which would always give a warning from the ramp or the permanent magnet in the track in case of power failure, the 'Crocodile' system continuously relied on powered-up electrics for its functioning. It was not fail-safe, therefore, in the event of power failure or contact problems, in

which circumstances there might possibly be no indication, or a wrong indication, in the cab. Ice or frost on the 'Crocodile' ramp could hinder the copper brushes under the locomotive from making sufficient contact and it was considered likely that this is what had occurred on 241 A 017 that night. Nevertheless, it should be noted that the driver of the Nancy train, in similar circumstances, had stopped as normal for adverse signals and did not report a failure of the warning system on his locomotive.

4 There was doubt that on a locomotive of this size the fog-signal detonators at the stop signal, placed by the crew of the stopped Nancy train and presumably exploded by the first axle of the front bogie passing over them, would have been heard by the crew all the way back in the cab, certainly at the speed the locomotive was doing (neither Daubigny nor Charpentier heard anything). Furthermore, a lineside witness reported that the detonators exploded only when the last coach of the 17-vehicle Strasbourg train rolled over them.

5 There was also doubt that the automatic semaphore signalling installation had been sufficiently frost-proof, certainly in combination with the effects of the freezing fog settling on electro-mechanical signalling equipment. This was a further stimulus for the installation of unified 'Code Verlant' colour-light signalling. In the case of densely trafficked suburban lines, installation of the powerful electric colour-lights would be the natural choice, as it was in Britain.

In short, there was doubt about the power of the signalling system and its peripherals to cope both with the inclement weather and developments in the field of traction, so there was insufficient incontrovertible proof to warrant conviction of Daubigny. The Est railway company itself, however, declared that all equipment had worked perfectly and that the accident was entirely due to the driver's imprudent lack of care in the circumstances. They had a point regarding the manner of his driving, which under the circumstances was incautious to say the least. Locomotive 241 A 017, not too severely damaged considering what had happened, was soon repaired and back at La Villette shed again, where the nickname 'La Charcutière' ('the butcheress') was bestowed on it, with that dark sense of humour that characterises railway people. The locomotive never regained a good reputation because it was involved in other, albeit smaller, incidents throughout its life.

Comment

A few things are worth noting. First of all, there is the eerie comparison with the accident that occurred in similar circumstances under the railway viaduct at *St Johns, Lewisham*, on 4 December 1957. In what was the third worst rail accident in Britain, a Southern Railway Bulleid light 2'C1' (4-6-2) 'Pacific' with driver Trew and fireman Hoare from Ramsgate depot ran into the rear of an electric multiple-unit that was stopped at a colour-light signal, causing telescoping as well as bringing the Nunhead line railway viaduct down on top of the train. The death toll was not quite as high as at Lagny-Pomponne, because speeds were slightly lower and no wooden coaches were in use, but an appalling 90 lives were still lost. Another rear-end collision that bears a number of similarities with the Lagny accident is that at ***Genthin*** in Germany, six years later at the outbreak of World War 2.

Something that, to my mind, played a definite role at Lagny was lack of experience with this type of locomotive. How much experience had Daubigny and his driving colleagues been allowed to gather with such a big locomotive, particularly at night and during bad weather? Furthermore, when steam locomotives get caught up in bad delays, an issue that certainly occupies the mind of the driver and the fireman is that the machine still burns coal and uses water, especially in times of cold weather when steam heat from the locomotive to the train must be provided. This might very well have manifested itself in concerns about the remaining amount of water in the boiler and tender, and what the best place would be to refill en route when the locomotive was supposed to clear the trip without refilling. Worry leading to excessive haste certainly has the power to cause distraction and overlooking of other issues that played a role that night, with both crew members doubtless keeping a close eye on the water level indicators. This is known to have played a role in the case of driver Trew in the *St Johns* accident. What about his French colleague on a relatively new and still not completely familiar locomotive? Furthermore, due to the soft exhaust beat of compound locomotives, there was the issue of drifting steam clinging to the boiler top of the locomotive, especially in the event of fog, which would have made sighting the signals that much harder. It probably also contributed to the ***Genthin*** accident. In Britain the Bulleid 'Pacifics' in their original streamlined guise (nicknamed 'Spam Cans' after the shape of the tins that contained processed meat) had a pretty bad name for this feature. The Est 'Mountains' later received smoke deflectors next to the smokebox in order to cause an updraught to take the exhaust away from the boiler top.

A further important question concerns the extent to which driver Daubigny was aware that his reading of signal aspects, especially at night, was compromised by his colour blindness. Had he already had narrow escapes during night trips before but failed to report these for fear of losing his job? Thus accepting the potential risk of accidents as a result? Did his good luck simply run out that night; was that really the first time he had fallen into this trap? And why did he go so fast? Apart from the obvious possibility of plain stupidity or inexcusably giving in to preoccupation with the delay he had suffered, part of the answer may be found in this following bit of information, which again demonstrates the issue of lack of experience with these machines. I recall several former steam locomotive drivers on more than one European network telling me that on many such locomotives there were no speedometers. The speed was gauged by watching the driving wheels, by the exhaust sound and vibration rhythms as well as the movements of the machine. This was initially compared with visually ascertained or sometimes even timed speed impressions, but with growing experience the movements and sounds sufficed to gauge the speed of the train. Over the years this became the established manner of regulating speed and it continued, from habit, after speedometers were installed. For such knowledge to be accurate, however, it required sufficient experience with each particular type of steam locomotive. It is dependent on the driving wheel circumference and thus the rhythm of the wheels on the track joints, as well as the exhaust blast at a given speed. This differed between locomotive classes and because the driving wheels of these 'Mountains' were rather large they gave fewer blasts at a given speed than most other Est types. It is conceivable that Daubigny tricked himself into assuming that he was travelling more slowly than he actually was. This impression would have been reinforced by the fact that the 'Mountain' accelerated so much more easily with this heavy train than the 10-wheelers would have done, compounded by the fact that the 241 A 'Mountains' also had a rather soft suspension and, with their length and bulk, no doubt behaved quite differently – certainly in a rather less lively manner – compared with the older locomotives he had worked. His visual appreciation of the speed that evening was obviously hindered by the darkness and fog that evening, which had made other drivers proceed more slowly than normal and so compounded the delays.

Finally, reference should be made to the urge commonly exhibited by people after a severe delay to make up time (e.g. the way many motorists accelerate away from a traffic jam; see also the level crossing accident at ***Voorst***). A locomotive driver preoccupied by the delay to his train, and perhaps not fully alert to factors such as fog and heavy delays working on the safety of his train, may well override his normal caution in such circumstances, but what may be safe practice in normal visibility can be potentially lethal in fog.

To complete the picture of the three worst rail accidents in France with ***St Michel de Maurienne*** and Lagny-Pomponne, here the brief details of the disaster at *Soissons*. In the early evening of 16 June 1972, part of the 1,600 m (1 mile), 110-year-old Vierzy Tunnel near *Soissons* in northern France collapsed and caused a large amount of heaped rubble to cover both tracks. As no signalling equipment was damaged, and therefore remained functioning as normal, a 6-car DMU railcar set ran at about 110 km/h (65 mph) into the debris at about 20:30, followed by a 3-car DMU on the opposite track that collided with the wrecked train. The accident claimed the lives of 108 people and 240 were injured.

***

Genthin, Germany, 22 December 1939 at 00:55

(High-casualty rear-end collision following SPADs in fog)

The accident

This is yet another accident that demonstrates that although staff error may be the root cause of an accident, it must always be seen in the context of further issues such as for instance poor traffic management on the rail network. In that respect this accident has more than one link with those at ***Ciurea*** and ***St Michel de Maurienne*** in 1917, with ***Lagny-Pomponne*** in 1933 and ***Pécrot***, ***Clapham Junction*** and ***Ladbroke Grove*** many years later. It clearly shows how a breakdown in railway management creates the circumstances in which hazards arise. It further illustrates how one particular mistake, driving trains on expectations, is about the most dangerous thing a driver can do. As at Lagny six years earlier, the accident at Genthin also involved a rear-end collision in circumstances of impaired visibility. There was an AWS type of warning system available, as at Lagny, but the on-board equipment on 2'C1' (4-6-2) 'Pacific' 01 158 had failed, and whilst the machine would normally have been put aside, heavy traffic demanded every bit of available traction.

The accident occurred at 00:55 on 22 December 1939, when express train D180 from Berlin to Neunkirchen/Saar smashed into the rear of the stationary D10 Berlin to Köln express. The heavy locomotive fell on its side and was badly damaged, but was repaired and remained in service until 1975. Six of its carriages derailed and were wrecked, the rear four carriages of train D10 suffering the same destruction. There were 196 fatalities and many hundreds of injured, despite the widespread use of riveted and welded steel carriages in long-distance express trains in Germany since the 1920s. Like in Britain at ***Quintinshill*** in 1915 and in France at ***Lagny-Pomponne,*** the circumstances at the time had forced the railways to start re-using older wooden-bodied vehicles to, mixed with the steel vehicles, cope with the masses of passengers. As a result, there was telescoping and virtually complete destruction of wooden vehicles, with consequent mutilation and loss of life. It remains one of the worst railway crashes in Germany.

Rail traffic in Germany at that time was heavy due to the recent outbreak of World War 2 and was therefore also subject to traffic alterations at very short notice demanded by the Army. This aggravated the already insufficient train capacity for dealing with the normal pre-Christmas and New Year passenger traffic. Additionally, despite the demanding circumstances of war and the already heavy public demand for travel, the Army had also 'negotiated' cheap Christmas ticket prices for trainee and injured soldiers going on leave to their families. Predictably, the trains and stations were absolutely heaving with passengers on their way home for Christmas.

The accident log

Although train D10 to Köln had left Berlin Potsdamer station as advertised at 23:15, there were delays of 5 minutes at Potsdamer station and of 12 minutes at Brandenburg station due to the milling masses on the platforms and in the trains. At Kade, train D10 had to stop at the home signal and section signal to Belicke at danger to await clearance of the section ahead by a non-booked preceding train, so that when it started moving again the delay had increased to 27 minutes and its maximum possible speed was no more than 85 km/h (53 mph) instead of the booked 105 km/h (65 mph). 'Train out of Section' was given for D10 at Kade, the last signal box before reaching Genthin, at 00:34, and that for the following D180 at 00:48. At Genthin the signaller had received the order to pull off home signal A and starting signal F for D10 only 3 minutes earlier at 00:45 – an indication of just how fast D180 was catching up with D10.

D180 to Neunkirchen, the offending train, had left Berlin Potsdamer station on the dot at 23:45, exactly half an hour after D10. This train was also subject to intermediate station delays, but until it arrived at Gross Wusterwitz was running under clear signals and at timetabled speed.

However, the distant signal at Gross Wusterwitz was at caution and the home signal at danger. Fireman Rudolf Nussmann on D180's locomotive 01 158 noticed the situation first and called out the signal aspects to driver Wedekind, who shut off, gave a blast on the whistle but did not apply the brakes. As luck would have it, the signals came off before he even reached the distant, so he opened up the regulator of his 'Pacific' again, but at a fair distance on the approach to Kade, Wedekind once more ran into restricted signal aspects. Puzzled as to what train could be ahead of him, he remarked on this with some irritation to his fireman. Wedekind then shut off again and so slackened his speed somewhat, but again did not apply the train brakes. Once more the home signal and its distant came off on approach, but the section signal and its distant were still kept at danger. However, even they came off before D180 reached them, again without Wedekind touching the brakes. From Kade onwards to Belicke, fog increasingly started to impair vision and D180 at times ran through some rather dense fog banks interspersed by short runs through clear patches. Later, during the inquest, fireman Nussmann stated that from Potsdam onwards the sighting of signals had occasionally been difficult, even though the distant signals were also indicated by countdown marker boards to make spotting them easier (see also ***Stavoren***). An additional problem in the cab of locomotive 01 158 that night was the weather-induced and rather unusual (for these locomotives) clinging of drifting exhaust from the chimney to the boiler top, which impaired signal sighting even more. Beyond Kade was Belicke signal box with its home and section signals, each with its own distant (as usual on most continental networks). Nussmann was busy with his job tending the fire and driver Wedekind, against the rules, did not call out the Belicke signals to him. He had only occasionally said, 'Rudi, look ahead,' which the fireman interpreted as stop firing, shut the firebox door and look out for signals because Wedekind was going to shut off for adverse signals and he wanted an extra pair of eyes. But the fireman needed time for his eyes to get used to the surrounding darkness again after the intense glow of the firebox and during those few minutes was unable to see anything much. On the approach to Genthin the same request from the driver was repeated.

The badly delayed D10 had by now approached the home signal at Genthin, which presently changed to show a clear aspect. As soon as the last vehicle of D10 had cleared the track switch that released the lock on the clear signal that permitted entry into the station limits, the signaller put the signal and its distant back to danger. Which means that driver Wedekind of the following D180 cannot have seen a cleared distant and home signal at Genthin, as he initially stated during the subsequent inquest. It also shows how closely he was following D10, because what he actually saw were the cleared Genthin home and section signals with their respective distant signals for D10. In other words, he was 'reading through'. But then again, he had passed the section signal at danger at Belicke at danger, missed the distant signal and then passed the home signal at Genthin before they had been replaced to danger and he additionally missed the hand danger signals given by a Genthin station crossing keeper at the trackside and by the signaller at Genthin signal box. For an experienced railway driver Wedekind missed a surprising number of danger signals that night.

On noticing D180 passing his section signal at danger, signaller Jakob at Belicke used the train controller's telephone to warn the Genthin signaller of what had happened and that he should be stopped at Genthin. Crossing keeper Adermann at the level crossing situated 850 m (half a mile) before Genthin signal box listened in and immediately picked up a red lantern, a signal horn and, according to his initial report, a few detonators to walk towards the approaching D180 whilst D10 was passing him, accelerating cautiously away after the slow release of the Genthin signals from covering the preceding non-timetabled military special M176 and a consequently late-running local. Looking past D10 towards its rear he then saw D180 approaching him at full speed; so fast, in fact, that Adermann said he had no chance of putting the detonators on the track without endangering his own life. He could only wave his red lantern in large circles to warn the driver of D180. However, he would later change that story and admit that he had been unable to find any detonators. Nevertheless, Wedekind failed to see Adermann's hand signal, his eyes either glued to the green section signal for D10 ahead of him or otherwise simply through inattention. D180 passed the crossing keeper at speed, closing in on D10. When signaller Seeger in his elevated position on the floor of Genthin signal box spotted the triangle of three white headlights of D180 closing in on D10, he panicked, grabbed the lighted red hand lantern used for danger signals and placed it in his side window facing the approaching D180. The first person to notice this red light, unfortunately, was fireman Sztuka of D10. He informed his driver, who then immediately stopped D10 with an application of the emergency train brakes. Virtually at that same moment D180 appeared out of the darkness and fog behind the stationary train and ran into its rear at approximately 120 km/h (75 mph), with disastrous consequences.

The aftermath

Driver Wedekind's report of the sequence of events on D180 only slowly revealed his actions when driving his train on that day, as initially he lied profusely. In short, it was his *expectation* that all the signals ahead would come off on his approach after his earlier experiences until Kade, because he knew that he was following a booked fast train. Therefore, he felt it was unnecessary to apply the brake when approaching distant signals showing caution, because he expected train D10 in front of him to be moving at the same speed as his train.

Things were to a certain extent acceptable when both men on the footplate were watching the line ahead, but when fireman Nussmann was busy with the firebox, driver Wedekind had to observe the signals on his own and failed. This most probably was also due to his tiredness and to his preconception of what was happening ahead, but no doubt was aggravated by the failure of his locomotive's Indusi AWS as well as the difficulty of spotting signals in the fog. It should in fact have made him more cautious, but from an inspection of his work record during the inquest it turned out that he had been involved in quite a number of other SPADs for the same reason and normally did no longer drive on the main line, also because of his advancing age. He was back on the main line only because of severe driver shortages. Also of great importance was the fact that on top of all that he had virtually a 24-hour working day behind him. Yet, until then he had never been involved in an accident; that night his luck ran out in the chaotic wartime circumstances. He was convicted at the Regional Courts of Justice in Magdeburg and sentenced to three years' imprisonment.

Comment

The reason for the manner in which driver Wedekind tried to handle his train will be recognisable to anyone who has that experience. I have to relate my own experiences in order to try and explain. When running into restricted signal aspects you normally start to brake, bearing in mind that you will encounter a red signal somewhere along the line ahead. However, during congested periods and especially on the approach to a big city terminus or a busy junction, you find that instead of having to stop you continue rolling along on restricted signal aspects that keep coming off on approach, just as Wedekind experienced. He failed, however, with regard to the following issue. If you are not booked to stop anywhere at intermediate stations you will try to regulate your speed in such a way that you keep at least two signalling sections at proceed between your train and the preceding train, using the upgrading of the restricted signal aspects to clear as the indicator of the speed of the train ahead. If you just keep them coming off from restricted to less restricted or clear in your face, whilst at all times being prepared to brake to come to a stop, you are doing fine. This gives you time to react if the train in front has to slow down or comes to a stop, but it also allows you to follow it safely up to the station at the closest possible distance. Speed signalling as installed in Germany has this whole procedure spelled out in speed instructions to the driver anyway, which in a way makes things that much easier. Just do exactly as the signals tell you. I assume that driver Wedekind, no longer used to main-line work, deadly tired and seemingly ignorant of the thorough alterations to the daily operations that were taking place on the German rail network as a result of the outbreak of war (such as not repairing the recently failed Indusi AWS on his locomotive, as the machine was badly needed), was not in the habit of consciously taking signalling aspects at face value. He just kept going (as was apparent from his record), based on the expectation that the train ahead of him was moving at approximately his own speed. In fact, under speed signalling in Germany, you receive a specific speed order from signalling aspects, and he committed serious breaches of the rules under those provisions, which is what landed him in jail in the end.

Military interference

In the first three months of World War 2 there were a number of severe rail accidents in Germany, in which a total of 311 people died. The worst were as follows:

- On 20 November, a collision occurred at Spandau, killing nine passengers.
- On 26 November, 15 people died in a collision near Nieder-Wöllstadt.
- On 12 December, there was a head-on collision near Hagen in Westfalen (Westphalia), again claiming the lives of 15 passengers.
- On 22 December, there was an accident between Markdorf and Kluftern on the single line from Friedrichshafen to Radolfzell along the north bank of the Bodensee (Lake Constance). This head-on collision happened in the evening of the same day that the collision at Genthin occurred. A freight train and a passenger train collided head-on, killing 101 people and injuring 28. That day a total of 297 people had died on the railways in Germany in 24 hours, a period total of nearly 500 dead and badly injured.

Quite out of character with the circumstances of the period, however, was that there were no media claims of sabotage by 'enemies of the Reich' to explain these disasters. The official explanation for these terrible fatalities, after so many years of safe travel on the Deutsche Reichsbahn, was that excessive numbers of trains were using the infrastructure, as well as the use of recommissioned obsolete wooden coaches needed to provide enough capacity to meet the increased demand. Which, of course, only comes into play once accidents happen. Unofficially, it proved that the notionally unified German emergency railway management organisation, set up with state pressure between the military and the Deutsche Reichsbahn, was experiencing fundamental problems with the running of rail services. This was because the military had the upper hand, the state with its various war initiatives being behind them.

This struggle between the railway and the military is described by Christian Wolmar in his excellent book *Blood, Iron & Gold*. The military were not interested in other traffic and not open about their rail transport requirements, no doubt due to their overworked administration departments and to forestall information about troop and equipment movements from reaching foreign intelligence. This caused them to demand an unrestricted run once a military special was on the move in order to avoid countermeasures by the enemy closer to the front. This struggle between the military and the German railway officials, who were also trying to serve their usual seasonal overload of passengers, was a cause of gross mutual irritation. It was because of this lack of basic rail operating knowledge that the military for instance managed to overlook the crucial fact that the track gauge in Russia was different from that in Europe and would present a serious logistical barrier during the planned Blitzkrieg campaign at the Russian front. Undeniably, the railways in Britain at the start of World War 2 were far better organised and operated, but then again, they were operated by state-installed rail professionals, not by military people in a state of war-craze and had no Hitler to deal with in their decision-making. It was this ignorance and bullying by the German military, overriding and obstructing the efforts of the Reichsbahn staff to accommodate the glut of passenger, freight and military traffic at the start of the war, that resulted in this shocking number of fatal crashes. As a result, the railways in Germany were badly overstretched and railway management broke down under the effort of providing rolling stock, crew and track slots for the Army as well as for the large numbers of civilians travelling in those last few months of 1939. Soon, however, the situation would be solved by degrees, with track, crews, traction and rolling stock that were requisitioned – or, in fact, stolen – from networks in countries that the Army had overrun.

***

Rickentunnel, Switzerland, 4 October 1926; and Armi Tunnel, Italy, 2 March 1944

(Heavy casualties due to carbon monoxide poisoning)

Introduction

What no ardent lover of steam locomotives will readily admit is that on a number of scores they can be rather dangerous machines. Apart from the very rare boiler explosions or disastrous mechanical failures involving consequent derailments or, for example, release of superheated steam in the cab, there is an often unforeseen danger with steam traction, as illustrated by the two examples that follow. This risk was heightened when circumstances dictated the use of unsuitable coal to fire locomotives. But even with good coal, as was normally the case in Britain, bringing a heavy and slow freight train through the Severn Tunnel, for instance, was a nasty experience for the locomotive crews due to the liberally exhausted smoke and steam from the hard working machines. Experienced older crew members have told me that they sometimes used rubber hosepipes stuck through the bottom of the footplate to breathe through as the atmosphere higher up was absolutely foul, especially for the crew on the second locomotive if the train was a double-header. The runaway and derailment of a coal train on the Great Central Railway at *Torside*, Derbyshire, in 1913 was considered to have been caused by the locomotive crew being rendered semi-conscious by carbon monoxide whilst traversing the notorious (old) Woodhead Tunnel, and German reports on accidents in which steam locomotives were involved always mentioned whether the locomotive had an open or enclosed cab as in the case of an open cab, poisoning of the crew by carbon monoxide in the exhaust gases might have played a part in the incident. (Many types of steam locomotive in Germany, whether tank or tender locomotives, would have fully enclosed cabs by the 1960s.) Similar cases of poisoning from gas in enclosed spaces have been reported in North America and Australia as well. In Australian railway English, such narrow-bore single-track tunnels, often with curves and a stiff climb within their confines, are rather descriptively known as rat-holes.

Rickentunnel near Wattwil

This accident happened on Monday 4 October 1926 at around 12:00 in the comparatively new (1904-10) 8.6-km (5.3-mile) dead straight and narrow single-track Rickentunnel, ascending at 3.85% (1 in 26) from Kaltbrunn at 483 m (1,585 ft) above sea level to Wattwil station at an altitude of 614 m (2,015 ft) between the networks of the Bodensee–Toggenburg Railway (BT) and the State Railways (SBB-CFF-FFS) lines in eastern Switzerland. In the days of regular Swiss steam locomotive operations, the Rickentunnel was noted for its extremely bad air quality, especially under certain atmospheric circumstances, and had acquired the nickname 'Schiefes Kamin' (the leaning chimney). An additional problem was methane gas escaping from soft coal lodes through which the tunnel had been driven. Traincrew of ascending trains in particular, when the steam locomotive would be working hard, were forced to work their machine with a wet cloth over their mouth and nose.

Freight train 6654A, made up of 19 wagons plus a brake van, weighing 253 tonnes and hauled by SBB steam locomotive of the B3/4 type, 1'C (2-6-0) No 1330, stalled inside the tunnel on the incline in the direction of Wattwil. The alarm went out when the train failed to arrive at around 12:20, and stationmaster Zoborist finally decided to dispatch five railwaymen and himself on a battery-electrical draisine into the tunnel. Concern among those at Wattwil station turned into near panic when the draisine was later heard approaching from inside the tunnel and emerged into daylight apparently with no one on board. When the vehicle was stopped, four men were found unconscious or dead on the cab floor, one of the dead being fireman Frommer from the stalled train. Zoborist, who later was found among the dead in the tunnel, was the person who had got the draisine going. Feeling his powers failing he pushed the reverser lever in the return position, opened up power and released the brake so that the draisine would slowly work its way uphill and out of the tunnel. Zoborist and six members of traincrew – driver J. Kläusle and fireman W. Frommer from Rapperswil depot, senior guard J. Meier, conductors A. Brunner and J. Zehnder plus brakesman K. Küng – together with two additional railwaymen from Wattwil station and an entire trainload of beef cattle, died from inhaling carbon monoxide. Emitted by the stalled steam locomotive that burned poor quality fuel in the shape of pressed coal powder briquettes. (The use being made of briquettes was typical in coal-starved regions, these being made of the tons of coal powder residue in the storage bunkers.)

Characteristically, the stricken traincrew members and those who had arrived to assist them had done what they could to save the train from rolling back down the incline to Kaltbrunn instead of immediately trying to make their escape from the tunnel before losing consciousness. (Most experienced traincrew are only too aware of the dangers of a train rolling back out of control and colliding with a following service, there being many such examples throughout the history of rail operation.) A number of wheels were found scotched and some vehicle parking brakes applied, whilst the locomotive driver was found in the eighth van with the necessary equipment to go back and protect his train. He had obviously wanted to take cover when he felt himself ailing. The conductors and their boss were found in the fifth, seventh and twelfth wagons, while brakesman Küng sat dead at his desk in the brake van, his brake fully applied. At a telephone in one of the nearby tunnel refuges the body of railwayman Bleiker was discovered, having tried to make a call. After all the others had been removed, the body of railwayman Emil Zahner was discovered 22 hours later by a crew patrolling the tunnel for the last time before it was opened for traffic again.

From the moment the fate of the train was known, head office in Zürich had been informed and a more intense rescue and salvage effort was organised involving the fire brigade and medical personnel equipped with gas masks. By then quite a crowd had congregated at Wattwil as well as at Kaltbrunn, where the disaster train arrived at 21:00, hauled by a salvage train. The paper strip from the Hasler recorder on locomotive No 1330 showed that the train had steadily lost speed whilst climbing and had then come to a stand, but checking the locomotive gave no clue as to why the train had stalled as technically it was in good working order. As at ***Harmelen*** in 1962, a train with about the same composition and weight as the accident train was sent into the tunnel as a reconstruction (crewed by staff with gas masks, of course) and stalled at virtually the same place. Analysis proved that the briquettes used did not develop enough heat in the confines of the tunnel, and also generated a substantial amount of carbon monoxide – enough to have started to affect people near the locomotive within 3 minutes of it coming to a stop. Regarding the assisting crew, although their gas protection equipment was as prescribed, it was obsolete and partly defective.

As an immediate mitigating measure, the permitted train weight on the line was reduced from 260 to 220 tonnes, rescue equipment was renewed and improved, and a rule introduced that in the event of stalling, the train had to be immediately worked back downhill to Kaltbrunn. But the main issue, steam traction, was soon to be dealt with by the major electrification of the railways in Switzerland, which had in fact started after World War 1 during a period of acute shortages of good imported locomotive coal to keep the railways working, the nation not having its own coal resources. However, it did have plenty of water power to generate electricity and by 1925 Switzerland was a world leader as far as railway electrification and construction of electric traction were concerned.

The line through the Rickentunnel was wired not too long after this accident, electric traction taking over from steam locomotives on 7 May 1927. The nine Bodensee–Toggenburg railway steam locos were sold to the Swiss State Railways for SFR150,000, where they were used for local and shunting jobs as late as 1965. One of the machines was then saved from the torch by the BT railway company and is still in working order, used for excursion trains such as the 'Amor' express, although no longer through the Rickentunnel.

Armi Tunnel near Balvano

A catastrophic wartime accident occurred in southern Italy during World War 2, at about 01:00 on 2 March 1944. It happened on the important single-track main line through the Apennines from Battipaglia, near the Tyrrhenian coast, via Potenza to Metaponto on the Gulf of Taranto. The Allied landings in the south of the country had taken place in September 1943, and initially the run to drive the occupying German Army north and back into the Austrian Alps had been fast and furious, until the onset of winter when the retreating troops regrouped in the central Italian mountains.

The country below Napoli having been liberated, life was slowly reverting to normal. By then US and British troops were tenaciously fighting their way north against fierce German opposition, but the US Army had ground to a halt under the heavy artillery fire raining down from the slopes of Monte Cassino where the Germans under Albert Kesselring had dug in as the last major stronghold before the Allies would reach Roma. The US Army therefore waited for an improvement in the inclement weather and for the massive bombardment that, although eventually smashing the German defences, would also leave the famous and ancient monastery of Monte Cassino in complete ruins. The comparatively short but intense and cruel run of hostilities in Italy since the Allied landings had wreaked incredible destruction of this historic landscape. But in those precarious circumstances the trains in the liberated south had started to run again, in so far as the damaged infrastructure, signalling and traction caused by the retreating Germans would allow. Several other nations would also become familiar with this particular aspect of German displeasure. The Netherlands, for example, had two-thirds of its rail network and rolling stock looted or destroyed after the failure of operation Market Garden in 1944 and before the war finally ended in 1945. The biggest hindrance to normal traffic in the liberated south, however, was not so much the damage inflicted by the retreating Germans but the ongoing German occupation of the north of the country. Coal for the locomotives had to be imported under perilous conditions across the Adriatic from Yugoslavia and was of inferior quality. Its calorific value (capacity to generate heat per given weight) was well below what was normally required for a strenuous trip such as that undertaken by the train involved in this disaster, and it was full of dust that led to the formation of slag. This settled as clinker on the grate under the firebox, further hindering combustion by reducing the amount of fresh oxygen in the fire. In short, it was dirty locomotive fuel, but the railway had to work with whatever it could get.

The traction and train involved

Train 8017 from Napoli to Taranto was double-headed by two types of 1'D (2-8-0) steam locomotives. The leading locomotive was a Class 740, 470 of which were built between 1911 and 1922. These machines were rated at 720 kW (965 hp) and were allowed a top speed of 65 km/h (40 mph). The second machine was a Class 741, an improved version of the previous machine as reconstructed from 1942 onwards with a Franco-Crosti boiler. This consisted of a smaller boiler with an underslung boiler water pre-heater barrel, using hot exhaust gases and exhausted superheated steam from the locomotive cylinders to pre-heat the boiler feed water to near boiling point and so economise on the fuel needed to turn boiler water into steam. These machines used less coal yet had an improved output of 810 kW (1,085 hp). Similar boilers, incidentally, would be tested in the 1950s in Spain on the spectacularly ugly-looking 140.2438 1'D (2-8-0), on some Class 9Fs in Britain and in Germany on a number of Class 50s (both 1'E, 2-10-0, types). Whilst popular with the coal-starved traction departments in Italy, on none of these other networks was the improvement in fuel economy thought worth the considerably higher maintenance requirement, due to sulphuric acid formed in the mix of exhaust gas and cooling exhausted steam, which caused corrosion in the boiler water pre-heater elements. Together both the Italian locomotives were rated at 1,530 kW (2,050 hp), which under normal circumstances was considered sufficient for 500 tonnes at their drawhook along a difficult stretch of mountain railway such as the one in question. With 511 tonnes of train to move, however, train 8017 was 11 tonnes overweight, certainly with the bad coal used.

Those 511 tonnes of train were composed of 42 freight wagons, four passenger coaches and a brake van with accommodation for the brakesman. This latter vehicle, making the train overweight, was necessary because as a result of bad maintenance not all of the vehicles were air braked, so as at ***St Michel de Maurienne*** a brakesman had to be reinstated. It was certainly needed on the downhill run of one of these mountain trips.

The accident log

The train had come to a stop on time at 00:12 on 3 March in the small mountain station of Balvano. After its arrival, two local railway workers, Messrs Caponegro and Biondi, checked couplers, brake connections and the brake rigging and blocks of the train, as was their normal duty. During the inquest later they would report that at that time they found nothing out of the ordinary. Normally, this nightly freight train would not have had a single passenger on board, but these were exceptional times and any train that moved was used for travel, not only here but all over war-torn continental Europe. Night time additionally had its attractions for those people who travelled without the necessary papers or carried goods that they did not want noticed. Black-marketeering and trafficking between the liberated south and occupied north was rife (with food, pilfered British and US Army equipment and even ancient religious artefacts pillaged from destroyed churches to sell to US servicemen for resale at home). On the other hand, there was also a sizeable group of medical students on board, showing that all sorts of normal peacetime human activities were resuming once more.

Train 8017 officially was said to have had 526 passengers in its coaches and in empty freight wagons, but unofficially there were probably about 600 people on board. Most of the illegal travellers and a large number of regular travellers could be found in the freight wagons, although most passengers were seated or sleeping in the four carriages. The two locomotives stood quietly steaming in the darkness at Balvano, the firemen in the cabs breaking the clinker off the firebox grate bars and spreading new coal, working hard to bring their fires up to scratch again. In the cold and rain the locomotive steam blowers produced their soft roar to create a vacuum in the smokeboxes and boiler tubes that sucked fresh air for better combustion through the grates into the fireboxes. The Westinghouse air pumps on the locomotives had worked noisily for a time to bring the pressure of the brake pipe back to the required value before falling silent apart from the occasional hissing blow. Even on this particularly miserable night (the icy rain turning to sleet and later to wet snow) quite a few people were boarding whilst others were stretching their legs until stationmaster Vincenzo Maglio dispatched the train at 00:50. As it drew out of the station on time, Maglio watched the typical Italian circular tailboard under the right-hand buffer of the brake van disappear into the snowy sleet and darkness, to ensure the train had left complete, before he went back into his office to report the departure by telegraph to the next station, Bella-Muro. It was 6 km (3¾ miles) distant as the crow flies but at a considerably higher altitude than Balvano, and that height had to be gained with twists and turns and three tunnels through the mountains. Stationmaster Maglio then handed over his duties to his relief, Giuseppe Salinia. He would have to deal with one further train that night, a similar freight train with passenger accommodation, running under the number 8025 and booked to arrive at Balvano at 01:40, but known to be running late.

Despite the wet and slippery track, train 8017 reached the first tunnel with a good turn of speed, but the line then started to twist and turn uphill at an average climb of 1 in 26. With the leading locomotive creating almost explosive exhaust sounds, the train worked its way through the second tunnel, the drivers constantly checking wheelspin with their regulators and locomotive steam brakes and by regularly sanding the track. Yet, despite the hard work of the crews, the train started to lose speed on the steep section towards Armi Tunnel, the third tunnel on the line. Of the usual narrow bore, this long tunnel had a reverse curve that was in the steepest section towards the summit of the line, and the train eventually stalled due to lack of steam pressure and started to roll back. The crew quickly checked this with the air brake, so only the rear part of the last passenger coach and the brake van were standing outside the tunnel entrance, and then secured their train with the locomotive and tender parking brakes. They then set to the task of cleaning up their fires again to build up sufficient steam pressure to get the train moving. Driver Gigliano on the leading locomotive probably told his fireman, Rosario Rabato, that he should use some of the small amount of decent coal that they carried in order to deal with exactly this sort of circumstance, in order to obtain sufficient heat to get to the summit. From there onwards the quality of the coal would not matter that much. (This much is assumed because their box with good coal on the tender was half empty.) In the process the steam blowers of the two locomotives were belching out thick clouds of acrid smoke through two chimneys into the narrow tunnel bore. At 02:30, shortly before train 8025 was due to stop at his station, stationmaster Salinia at Balvano was made aware that Bella-Muro had not yet informed him about the clearing of the single-line section ahead by train 8017. This prevented him from sending train 8025 onwards, and he therefore made a telephone call to Bella-Muro and was told that train 8017 had not as yet arrived.

Guard Michele Palo of train 8017 sat smoking near the coal-fired heater in his stationary brake van. Because he had not expected a whistle to call him to take action on the brake during this uphill run, he had no reason to believe anything was out of the ordinary. But, with the train having stood motionless for some time, he opened a window and peered forward through the sleety snow, unable to see anything in the darkness. He decided to go and take a look, lit an oil lamp and climbed down to walk forward along the silent train into the tunnel mouth. Some way inside the foul atmosphere of the tunnel he opened a door of a compartment, as far as was possible in the narrow confines, looked in with his lamp and found everyone apparently asleep. Suddenly he began to feel unsteady and bilious, as if in a state of utter drunkenness, so with the greatest urgency he decided to turn back along the deadly, silent train to the outside world, where the windy and cold night air soon restored him to his wits. Shivering with shock, cold and fear he ran back along the railway line to Balvano, to report what had happened and get help. In the meantime train 8025 had arrived at Balvano. Stationmaster Salinia, from recent experience during the hostilities as well as following set procedures in this situation, ordered Caponegro and Biondi to uncouple the leading locomotive of the double-headed freight train and then sent that machine forward under extreme caution to find out what had happened to train 8017 on its way to Bella-Muro. This light locomotive had barely left the station when the crew spotted guard Palo's oil lamp bobbing along the line towards them.

The man was upset, shouting and crying, 'They're all dead, everybody's dead!' No fewer than 521 occupants of train 8017 had died from that lethal and odourless killer, carbon monoxide gas. Poisoned in the confines of Armi Tunnel within minutes of the train stalling, most were found as if asleep; quite clearly there had been no panic on the train. Only a few people in the rear compartments of the last passenger vehicle that had not yet completely entered the tunnel had survived, although unconscious. One of those was olive merchant Domenico Miele, who was wearing his new British-manufactured white woollen scarf to protect himself against the cold. During the slow trip inside the tunnel he had found himself feeling light-headed and decided to wrap the scarf around his nose and mouth against the terrible stink of the smoke, making his way towards the rear passenger carriage to be as far away from the locomotives as possible. Together with four others, he was still alive when found and recovered reasonably quickly, although his hair had turned completely white. One of the other survivors, however, Luigi Cozzolino, was so badly brain damaged that he never fully recovered. At least three black-marketeers in that last carriage are known to have disappeared from the disaster scene and a few others were also thought to have got away. Guard Palo never revealed who provided him with the cigarettes he was smoking or, for instance, whether any particular passengers had travelled with him in his comfortably heated brake van. Due to the wartime circumstances, the US Army under General Gray took responsibility for the investigation of the accident, and it was US Army staff who entered the tunnel first and were soon able to identify the cause of the disaster – the inferior coal being used to work the train. Tests showed that under circumstances of impaired draughting of the fire, the coal gave off substantial quantities of carbon monoxide. The narrow tunnel bore added to the draughting problem in the fireboxes.

As at ***Ciurea*** and ***St Michel de Maurienne*** in 1917, information about this accident was suppressed for a considerable time. The reason in this case was to conceal the extent of the disaster, were the German Army to claim it as a hugely successful act of sabotage.

***

East Robinson, New Mexico, USA 5 September 1956 at approx 03:00

(Heavy head-end collision in siding caused by crew error)

The accident

At first glance this is a case of straightforward crew error when working the hand-operated turnout at the eastern end of Robinson passing loop by the second man (fireman) of train 8. This caused a head-on collision between the arriving westbound train 19 (the 'Chief') and the standing eastbound train 8 (the 'Fast Mail'). As a result, train 19 veered into the loop instead of going through at speed along the main line. However, on more detailed inspection, a number of quite inexplicable issues concerning traincrew behaviour are revealed. Twenty people died at the scene, with two others passing away later. Among the dead were both drivers and the second man on train 19, one of the drivers being just weeks away from retirement. Due to the wholesale destruction of a luggage/staff vehicle in train 19, all the dead were railway staff.

Some history

The year in which this accident happened was not a particularly good one from a safety point of view for the Atchison, Topeka & Santa Fe Railroad (AT&SF), commonly known as the Santa Fe or jokingly as the Ate Tapas, Spit Fire. In fact, 1956 was not a brilliant year for US passenger railroads in general, as 100 people were killed in 66 accidents. Two of those accidents, including the one described in detail here, concerned incompetently handled turnouts at passing loops on long single-track railway lines. The AT&SF and its subsidiaries were also involved in two further railway accidents that year:

- On 22 January, the train driver of a rather overloaded 'San Diegan' service, a 2-car Budd Rail Diesel Car (a diesel-hydraulic multiple-unit train of stainless steel construction), suffered a seizure and became incapacitated whilst doing 70 mph (115 km/h) towards *Redondo Junction*. As the set was on a slightly downhill gradient and was not fitted with alerter equipment (vigilance), the train became a full-blown runaway and tipped on its side on a sharp curve with a 15 mph (25 km/h) permanent speed limit (see also *Waterfall*). The single-pane hardened glass windows shattered on contact with the ground, allowing many passengers to drop out of the capsized train, as would happen at ***Aitrang*** in Germany and at *Hither Green* in Britain. There were 30 fatalities and many more were badly injured.
- On 16 June, staff of AT&SF subsidiary Gulf, Colorado & Santa Fe at *Gainesville*, were fly-shunting two loaded freight wagons on to a string of 20 empties that were standing on a falling gradient and were held only by a single applied parking brake. Predictably, the convoy ran away downhill after impact and ended up on the Fort Worth to Oklahoma City main line in the path of the approaching Chicago express. As luck would have it, this train, headed by two EMD E8M locomotives, was just pulling away from a red signal under 'Stop & Proceed' conditions, and on seeing the approaching rake of freight wagons in his headlight the driver applied the emergency brake, which brought his train to a stop before impact took place. However, the 22 freight wagons hit the stalled passenger train at approximately 20 mph (30 km/h) and caused injuries to the driver, one member of staff on the train and one passenger.

That same year, a Lockheed Constellation and a Douglas DC-7 collided in mid-air right over the *Grand Canyon*. That crash cost 128 people their lives (see Chapter 5).

Rolling stock and traction involved

The period was the last blooming of privately operated US long-distance passenger rail transport before the car and the plane took over. AT&SF was undoubtedly the champion of passenger transport in the US with its celebrated 'Super Chief', the train of the movie stars on the Chicago to Los Angeles route, much as the New York Central Railroad was the main contender on the New York to Chicago run with its '20th Century Limited'. The ambience on board these trains had to be seen to be believed: a stylish, sumptuously comfortable modern passenger environment with air-conditioning and excellent food from restaurateurs of repute on fast runs during which 160-175 km/h (100-110 mph) were the normal speeds over long distances. This standard of opulence was never quite reached on contemporary European trains in the thrifty days after World War 2, not even on the various Trans-Europe Expresses. On the other hand, if people in war-ravaged and politically divided Europe had wanted to travel, not many would then have spend anything like the 37-45 hour on a *train de luxe* any longer, particularly after the demise of such prestigious long-distance trains as the various Orient Expresses.

At the time of the accident the 'Chief' was one of the three AT&SF direct long-distance passenger trains and operated between Chicago in Illinois and Los Angeles in California via Kansas City, La Junta and Las Vegas, New Mexico (not to be confused with the famous tourist destination in Nevada). The 'Chief' was the longest established of these AT&SF trains, having been instigated shortly after the Santa Fe had put down its track all the way westwards across the North American continent. In 1935, however, this historical connection was under threat of relegation by the new and much improved 'City of Los Angeles' via the competing Chicago, Burlington & Quincy Railroad. In order to meet this challenge, the AT&SF decided to give a complete overhaul to the image of its 'Chief' services and put a modern, fully diesel-hauled, air-conditioned, streamlined and lightweight 'Super Chief' service into operation on a 37-hour schedule. This provided substantially more power at the head of a train of lightweight coaches and no longer required stops for refuelling or changing of steam locomotives. This train royally eclipsed the existing 'Chief' (and, more importantly, the 'City of Los Angeles') with regard to comfort and speed. From that moment on, the original 'Chief', adhering to its existing 45-hour schedule, functioned more or less as the semi-fast connection on this route, calling at more stations than the 'Super Chief'. A third premium AT&SF train on this route was 'El Capitan', which was combined with the 'Super Chief' in 1957 (in the declining years of rail travel in the US) but which to a great extent kept its reputation for opulence and luxury intact until its handover to Amtrak during the 1970s. The pride that former AT&SF people (nowadays the Burlington Northern Santa Fe, or BNSF) had in the 'Chiefs' and 'El Capitan' is still evident when discussing operations with them.

Despite the fact that the 'Chief' was ranked lower in the listing of through services on one-and-a-half-day-plus schedules between Chicago and Los Angeles, by 1956 this train had nevertheless lost its heavy pre-war riveted steel coaches and gained the new air-conditioned stainless-steel Budd type of lightweight rolling stock with their ribbed sides, including the restaurant and bar-car combination and a full-length vista-dome car from which the landscape could be viewed at rooftop level through large observation windows. In fact, the fifth coach in the consist of 10 vehicles of the 'Chief' on the night of the accident was just such a vehicle. Unlike the 'Super Chief' make-up, however, the first coach behind the four locomotives was a so-called baggage/dorm, a part-luggage van with overnight bunk accommodation in which railway staff travelled on their way to and from duties ('on the cushions' in British railway parlance) and at least 16 staff were in this vehicle on the night in question. Four 'warbonnet'-liveried GM EMD F3 locomotives hauled this train, forming ABBA set No 30 (a multiple-united combination of a leading locomotive with forward-facing cab, two intermediate cabless units and a trailing locomotive with rearward-facing cab), which together generated 9,000 hp (6,710 kW) from their four 16-cylinder GM 567B diesel prime movers. Driver 'Babe' Foster of Las Vegas depot worked the train from Raton, New Mexico, back to his home depot, together with second man Walt Adams.

Driver Leo Rush was in charge of train 8, the 'Fast Mail', his second man being Pete Caldarelli. They picked up the train at their home depot in Las Vegas to work it to Raton. This was a scheduled service for the US Post Office as well as for the internal railway parcels service, on which a number of facilities for shippers were offered. It consisted of 11 bagged mail vehicles (a rolling Post Office vehicle with sorting facilities had been dropped off at Las Vegas earlier during the trip), one express refrigerator vehicle and two sleepers (one empty and one the vice-president's private business vehicle occupied by a family of two adults and two children). The traction was delivered by another full four-vehicle set of GM EMD F7 locomotives in ABBA set No 41 of 9,000 hp, also in the extremely pleasing red, silver and yellow AT&SF 'warbonnet' livery. Some of the vehicles transported in this train, however, were still of the old 6-axle riveted steel heavyweight stock. Incidentally, it was generally known that drivers Foster and Rush, both on the railway for forty-odd years, had a long-standing dislike for each other. Whether this had a bearing on what happened is uncertain – nothing was proved in that respect – but it crops up as a matter for consideration in the literature dealing with the accident.

The correct procedure for handling passing trains

In the US there were, and still are, long stretches of unsignalled single-track main line that require passing loops at regular intervals. Based on the timetable and on any delays, trains would be regulated from one station via the passing loops to the next station by so-called train dispatchers who telegraphed, telephoned or teleprinted their train orders to local operatives at pick-up points. These train order issuing locations were indicated along the track by the characteristic three-position semaphore signals that informed the footplate crews whether a train was required to stop for a crossing at that point, to slow down to pick up train orders on the hop, or to pass through at speed, whether scheduled or out of schedule. The dispatcher staff at these locations would hand out the teleprinted or handwritten train orders to the train drivers and train conductor (often with characteristic implements bearing a passing resemblance to the on-the-hop manual token exchange tools used in Britain), who were then expected to do precisely what the train orders instructed them.

This was a traffic control system with a long-standing tradition that worked well if all those involved were rigorously on the ball but was nevertheless the cause of a number of serious accidents both in the Americas and the rest of the world. These occurred following what were often quite simple oversights or mistakes due to misreading the handwriting of the man issuing the train orders. In those days before radio telephones became standard equipment on trains, the main problem was that the traincrew were beyond the reach of control staff as soon as they were on the move; they were in so-called 'dark territory' and could no longer be warned or stopped. So after a mistake had been made, nothing could be done other than to wait for the news of a head-on crash to come through, rather like more recent incidents near ***Warngau*** in Germany and at ***Åsta*** in Norway. By 1956, most of the passing loops had been equipped with self-restoring sprung turnouts instead of hand-operated turnouts. Departing trains opened sprung turnouts and after passage these then self-restored to their normal position to ensure that arriving trains (on the AT&SF) would be led into the right-hand receiving track. However, at the east end of Robinson passing loop, at milepost 705, a hand-operated turnout that traincrew had to open and reposition to main-line running was maintained for a number of reasons. None of these turnouts were interlocked with any train detection or signalling equipment, incidentally; they could be moved at any time.

The rules governing operation of this type of equipment demanded that the person responsible for changing the position of the turnout had to wait at the opposite side of the track clearing point for the arrival of the oncoming train and then observe the condition of this train as it passed (Rule 104A), whilst his driver in his seat on the waiting locomotive scanned the other side of the arriving train. Only after that train had passed the clearing point, and was thus proven to be clear of the track of the waiting train, was the turnout operator allowed to walk to the turnout to unlock and reverse it and then walk back to the locomotive and depart (if permitted by the train order). Someone from the train would have walked forward in the meantime to restore and lock the turnout to main-line running, after which he would board the train that had stopped on the main line to await him. On quite a few networks in the world, trains are operated like this even today.

The accident log

Both train 19, the 'Chief', at Raton, and train 8, the 'Fast Mail' at Las Vegas, picked up the following train order: 'No 8 Eng. 41 MEET No 19 Eng. 30 AT ROBINSON. No 8 TAKE SIDING'. This translates as follows: Train 8 worked by driver on locomotive roster set 41 meets train 19 worked by driver on locomotive roster set 30 at Robinson. Train 8 stops in the diverting track. This implies that train 19 can pass at speed through the main track. On that particular night the 'Chief' was running a few minutes late, whilst the 'Fast Mail' was about 30 minutes late. For that reason the meet had been shifted from Gato, 12 miles to the east, to Robinson. Gato had all sprung turnouts where the 'Fast Mail' would have stopped on the main track and the 'Chief' would have rolled past at slow speed through the siding. However, under the revised train order, the 'Fast Mail' had entered the loop at Robinson and come to a stop. Driver Rush switched off his powerful headlights whilst his second man Caldarelli, who had promised trainman Brown, the traincrew member who should have done that job, that he would open the turnout, undid the footlock and then walked back to the clearing point. Switching off such powerful headlights when approaching another train was a matter of showing courtesy to the other driver. (The same principle can be seen in Britain today, after the introduction of similarly powerful 'evil-eye' headlights of rather questionable use.) Driver Rush saw Caldarelli pass the clearing point, walk to the location of the turnout and undo the foot lock before crossing the track and waiting for the rapidly approaching 'Chief', the glow of its headlight having already been visible for some time. With the 'Chief' now close by, and for reasons unknown, driver Rush on the stationary 'Fast Mail' suddenly switched on his headlights to full power again and blew his horn twice, completely confusing Caldarelli near the turnout. Caldarelli must have thought, wrongly, that his driver was trying to tell him something or perhaps was about to depart. Completely losing the plot, he ran across the track back to the turnout handle which he fully unlocked and then pulled the turnout to reverse in the face of the approaching 'Chief', which was doing about 60 mph (100 km/h). Driver Foster on the 'Chief', probably partially blinded by the blazing headlights of train 8 in the siding, nevertheless saw what was happening and applied the emergency brake, but at the speed he was doing there was no chance of stopping. The 'Chief' swung left into the siding and slammed head-on into the standing 'Fast Mail' before the brake had been able to take effect.

Following impact, train 8 was pushed back by a carriage length and the leading locomotives of both trains fell over to the cess side of the loop. The 'Fast Mail' stayed in line on the track behind the three remaining locomotives, seemingly little damaged, whilst the 'Chief' performed the usual North American zigzag concertina pile-up with its three remaining locomotives and the first five passenger vehicles, including the vista-dome car that had its leading bogie derailed. When the emergency crews removed that part of the train that was still on the track and started to clear the first of the derailed wreckage they realised that an entire carriage appeared to be missing as the vehicle numbers for the 'Chief' did not add up. The carriage-end wedged between the fourth locomotive and the first (overturned) sleeper belonged to an unexplained sidewall panel that was visible on top of the upper side of this sleeper. It was the baggage/dorm vehicle crushed between the heavier sleepers and the locomotives. All 16 of the staff who had travelled in it would be dead when the last surviving man passed away in hospital a few weeks later.

Baggageman White of the 'Fast Mail' was the first member of traincrew to get his wits together and take action to get the emergency services on the scene. He got a pair of pliers from the luggage van toolbox, climbed down from his train, crossed the main track and cut his way through the fence to the outside world. He managed to get to the nearby Interstate 25, which despite the time of night was still busy, where he stopped a car and asked its driver to notify the police at nearby Springer to organise assistance. The era of the mobile phone was still 40 years away.

Comment

Contrary to the usual way in which the rail fraternity learns lessons from accidents, the situation with train orders and non-interlocked turnouts in sidings remains to this day, although procedural changes have been implemented to diminish the risks stemming from human error. Following the very similar collision on 18 August 1999 at *Zanthus* in Western Australia, the investigating Australian Transport Safety Board had a very relevant thing to say on page 28 of its report:

> *It is clearly undesirable that a single error should be able to result in a major accident of this nature. While flawless human performance is a worthy aim, it is in reality rarely achievable. It is apparent that the system in operation at Zanthus relied on perfect human performance to ensure safety.*

Substitute any remote single-track accident site throughout the history of rail traffic anywhere in the world and this remark is still utterly valid. It is, in fact, the endorsement of the application of external and driver-independent monitoring systems such as PTC and ATP. And the words are almost literally those used in The Netherlands to describe the reliance on train drivers sighting yellow signal aspects after the ***Harmelen*** crash, after which the fitting of ATP to track and traction was the predictable outcome. However, that was on a very busy network with far more trains per kilometre of track than in many significantly less densely populated parts of the world, which is where the Positive Train Control (PTC) system now being developed in the US comes in. This does not rely on the availability of a signalling system in the way that ATP does, and is being developed precisely for this type of lightly used and remote stretch of single-line operation. The US being the US, however – and rather like the situation regarding the perception of the cost of a similar train protection system in Britain, for example – the massive investment in such a system was not seen as justified in view of what many considered to be the little measurable gain in safety, particularly as the number of railway passengers in the US is comparatively negligible, albeit increasing. It took ***Chatsworth*** to change that perception.

Anyone who has experience of driving trains at night will know how painful other train headlights and the lights of signals can be to approaching train drivers. It is for that reason that signal lights on the rail network in The Netherlands can be switched to half power on clear nights (***Harmelen***). I keep wondering about the reason why driver Rush of the stationary 'Fast Mail', having switched off his headlights as was customary, suddenly and for no apparent reason switched them to full power in the face of the approaching driver. Even more puzzling is the blasting of his horn twice. That single act strikes me as thoroughly odd in the run-up to this accident, as there was no apparent emergency to draw to the attention of either Foster or Caldarelli. Nor could he have been about to depart right at that moment, so why did he do it? With that incomprehensible and unnecessary action he wrong-footed his second man who then, completely unforeseen by any of the people at the scene, set up the accident with an inexplicable error working the hand-turnout.

Now for some Sherlock Holmes-like conjecture: if the known facts, not distorted to fit a theory, point to a certain outcome, then that outcome, however strange, must be what has happened. It is entirely my own opinion, but I cannot escape the impression that the long-running feud with the soon-to-retire Foster provided Rush with the impetus for what was in fact a small-minded prank against a disliked colleague. From the depot crew rosters, as well as from the dispatcher's instructions, Foster would certainly have known that it was Rush who played the headlight trick at the east end of Robinson siding, and that the two blasts of the horn, with no known purpose according to the rules, were tantamount to a one-fingered salute. Had there not been that disastrous intervention of confused second man Caldarelli at the turnout, Foster would have passed by at speed and back to his home depot at Las Vegas with stars in his eyes, a growing headache and a bigger lump of frustration against Rush in his craw. Since both men died in the accident, nothing can be established with any certainty, but in my opinion it is the only plausible reason to explain what happened. Train and other vehicle operators the world over have been known to play similar stupid pranks on each other that have backfired, although not necessarily to the extent as in this case.

Caldarelli had to take the entire blame for the accident and his beloved railway career ended miserably, although from the point of view of rail safety perhaps justifiably (he had passed out as a driver and expected to begin driving soon) in tatters that night. He showed great courage in visiting the grieving families of colleagues who died as a result of his mistake, and found that most were forgiving. To me that is yet another indication that many at least suspected that there was rather more to it than just that bout of inexplicable stupidity on the part of Caldarelli. He spent the rest of his working days, together with his wife Esther and two other musicians, doing dance hall gigs with his accordion band, supplementing his income by parking cars for customers at the local stadium during events. It has some resonance in my own experience that after having to give up train driving for whatever reason, it is difficult to find a similarly rewarding alternative. Likewise, truckers, mariners and pilots also experience the same problem of adjusting to the life that 'normal' people lead.

***

Harmelen (1), The Netherlands, 8 January 1962 at 09:19
(Heavy head-on crash following SPAD)

The accident
This, the most serious Dutch railway accident up to now, was a classic high-speed train crash in the days before ATP in The Netherlands. The type of occurrence that is gradually fading from memory with the ongoing installation of train protection systems in Europe. It was a near head-on collision at a combined impact speed of 162 km/h (100 mph), caused by a high-speed (125 km/h or 75 mph) SPAD at Woerden signal 8 that covered Harmelen junction against express 164 to permit branching-off local 464 to occupy the junction. This happened in medium-to-dense ground fog conditions as a result of winter temperatures in this waterlogged area, witnesses mentioning visibility of 30-150 m (100-495 ft). As previously stated, a feature of colour-light signals in The Netherlands is that they can be dimmed during clear nights. In this case the signal lights were at full power; the switch in the signal box was found in that position and the signaller involved stated it expressly in his post-accident report. No reports from other crews were received that would cast doubt on this point. There were 91 fatalities (89 at the site and two later in hospital) and 54 injuries (32 serious) out of some 1,080 passengers on both trains, and the damage caused to infrastructure and rolling stock was enormous.

The location
Harmelen Junction is located near the town of Woerden on the double-track electrified main line from Utrecht to Gouda and Rotterdam/Den Haag. The line converging at Harmelen Junction is the double-track

electrified branch line from Breukelen Junction on the main line from Amsterdam to Utrecht. In effect, the three junction locations of Utrecht, Harmelen and Breukelen make up a triangle connecting Amsterdam, Utrecht and Gouda with Rotterdam and Den Haag. (As an aside, Breukelen gave its name to Brooklyn in New York, which was then called New Amsterdam, as did the city of Haarlem to Harlem, New York.) Woerden itself, incidentally, was the scene of a crash with one of the British Army of the Rhine home leave troop specials from Germany on its way to Hoek van Holland on 21 November 1960, when a Netherlands Railways driver failed to slow down to a 40 km/h (25 mph) temporary speed restriction to negotiate a set of reversed turnouts for a diversion due to resignalling work, the result of which work we find in this account of the Harmelen crash. The infamous immediate post-war use of obsolete LNER stock for these trains had already ceased; the train was made up of a mix of older stock and more recent West German passenger rolling stock of welded all-steel construction, which contributed to the low casualty toll. Two of the occupants of the military train were killed and 10 were injured in this violent derailment. At Harmelen Junction itself a collision between two freight trains took place about two years before the described accident. On that occasion both train drivers survived the impact.

Services, traincrew and rolling stock involved

Train 164, an express from Leeuwarden via Zwolle, Amersfoort, Utrecht and Gouda to Rotterdam, consisted of 11 all-steel coaches, seven of those being the Plan E-series vehicles constructed from 1954 onwards. The first vehicle and the two last vehicles were from converted 1928 EMU riveted-steel stock, the third last welded-steel vehicle belonging to a series of pre-war former international types. The train weight was approximately 450 tonnes. From Utrecht onwards this train was hauled by Alsthom-built 2,580 hp (1,925 kW) electric Bo'Bo' locomotive No 1131, weighing 80 tonnes, and was estimated to be carrying 900 passengers. The driver was 40-year-old Mr P. van der Leer of Rotterdam Feijenoord depot. The other train involved, train 464, was an all-stations slow from Rotterdam via Gouda, Woerden and Breukelen to Amsterdam. It was a 6-coach EMU Mat '46 (known colloquially to some as 'cucumber' units due to their streamlined shape and green livery), consisting of leading 4-car unit 700 Bo'2'Bo'-Bo'2'Bo' and trailing 2-car unit 297 Bo'2'Bo'. It was estimated to be carrying 180 passengers, and the driver was 41-year-old P. Fictoor of Rotterdam Centraal depot.

The accident log

Train 164 entered Utrecht Centraal station a few minutes behind schedule, hauled by electric locomotive No 1104. Due to the necessary reversal here, a change of traction took place and after splitting off No 1104 from the south end of the train, No 1131 came in from a north-end stabling track and was attached. Neither the shunt driver bringing No 1131 from the depot nor the shunter responsible for coupling it to train 164 noticed anything worth reporting with regard to driver Van der Leer's demeanour. Nor did senior conductor J. Boon from Leeuwarden depot notice anything untoward with Van der Leer when exchanging the particulars of the train. And so, following successful brake-continuity testing, the train departed Utrecht westbound in the direction of Harmelen Junction at 09:11, 6 minutes late.

Train 464 departed Woerden station eastbound in the direction of Harmelen at about 09:15, running on time according to the Woerden station clock as senior conductor P. Verheijden from Eindhoven depot later testified. He stated that this train approached Harmelen Junction at moderate speed and did not brake before the collision. Woerden signaller J. Kwakkenbos, however, reported train 464 as having departed about a minute late at 09:16 and that he expected both trains to be at the junction (very accurately, as it turned out) at 09:19. The scheduled moves involved in the accident at Harmelen Junction were based on a closely following and hourly repeated pattern for the three services, two being at line speed, cutting across each other's pathways within 7 minutes. The westbound express from Leeuwarden via Utrecht and Gouda to Rotterdam would be the first to pass through at xx:14, after which the east-to-north local from Rotterdam via Gouda and Woerden to Amsterdam would cross this line at the junction 4 minutes later at xx:18, another westbound express, from Groningen via Utrecht and Gouda to The Hague, finishing the sequence 3 minutes later at xx:21. These connections all still exist, albeit as part of a rather more frequent and complicated set of north-to-west and Rotterdam via Gouda to Amsterdam 'long way round' service patterns. However, all are now invariably worked with EMUs and push-pull operated trainsets; time and track-space consuming locomotive changes at Utrecht having long since passed into history.

When signaller Kwakkenbos at Woerden received the signal from Utrecht that express train 164 had departed at 09:11, its late running created a problem of conflicting arrival times at Harmelen Junction with local train 464. He was therefore required to decide which train would get preference based on the actual situation at the junction. In the situation as it was, as he stated in his report following the accident, on-time local train 464 would have had preference. But as he took account of the possibility that the driver of express train 164 would attempt to make up lost time, he expected train 164 to be at the junction at 09:19 or 09:20. He decided to give local 464 the road to its junction signal 16 and express 164 the road to junction signal 8. This set-up allowed him to instantly give either train the road as appropriate and was a good example of sensible train traffic control. In fact, the situation is an everyday occurrence in all parts of the world where trains run frequently. Although in those days Kwakkenbos had no train describer equipment to follow the progress of train 164 from Utrecht, he could track local 464 from Woerden station to the junction on his NX control panel by watching the signal-section track circuit occupation indications light up red and then extinguish in sequence. He saw that the Amsterdam-bound local had come up to the distant signal 26 for junction signal 16 before Rotterdam-bound express 164 had reported its presence, and came to the conclusion that, in order not to hold up train 464 and also to allow it to clear the junction in the shortest possible time, the best course of action was to clear the road through the junction on to the Breukelen branch for the local.

Based on the 3-minute travel time from Woerden station, and on the fact that the signaller saw train 464 approaching distant signal 26 showing a yellow aspect, driver Fictoor probably saw this distant signal clear from a yellow to a green aspect. This enabled him to run without stopping past the green junction signal and go through a left-hand set of crossover turnouts 5B to 5A, run 'wrong line' for about 30 m (100 ft), before turning left to gain the right-hand track of the branch to Breukelen through turnout 3 at about 55-60 km/h (35 mph). From my own experience on more than one network, in cases of such last-second clearing of a junction distant signal on approach, most drivers understand that there may be delayed other services in the vicinity and will not hang about. Mr. Fictoor indeed reacted expeditiously, going by the statement of his senior conductor. But Mr. Kwakkenbos had hardly entered this move on the NX traffic control panel than express 164 reported its presence at 2,650 m (about 1¾ mile) from signal 8, showing a red aspect, with a warning buzzer and an indicator light (annunciator) for the track circuit leading up to it. Which then showed occupied a tad quicker than the signaller expected from experience. Kwakkenbos pressed the acceptance button to silence the buzzer, the annunciator light remaining lit as a reminder, but as he had now given local 464 the road the signalling equipment was locked against setting up conflicting moves. As a result he could no longer take back the road to let the express go first; the local had to pass and clear the junction track circuits to release the equipment for other moves. Therefore, driver Van der Leer of express 164 in the fog would find distant signal P713 showing a yellow aspect, ordering him to slow down to 30 km/h (20 mph) and continue braking in order to be able to stop at the red aspect of junction signal 8. (This is something any train driver in this position of running slightly late and trying to make up time dislikes.) Watching the red track-circuit occupation lights for local 464 come up and extinguish in order to quickly reset the junction for the express, Mr. Kwakkenbos suddenly realised that movement had stopped between turnouts 5B and 3, whilst the turnouts-locked indicator lights of these turnouts remained lit instead of extinguishing after the passage of the local, which would have enabled him to reset the road for train 164. The signaller also noticed that the track-circuit occupation light between signal 8 and the junction turnouts was lit – an indication that train 164 had in all likelihood passed signal 8 at danger. Initially, Kwakkenbos suspected a signalling fault and decided to call a man by the name of Van Cleef in the old Harmelen Junction signal box and asked him to check at the junction. This gentleman reported hearing the sound of a crash coming from the junction, but was unable to see anything due to fog and what looked like smoke or dust clouds. The worst had to be expected.

Signaller Kwakkenbos showed exemplary presence of mind. He immediately asked Van Cleef to contact electrical control at Utrecht to switch off traction current from the overhead line equipment and then used the loud-speaker telephones at the junction to warn everyone to stay away from the wreckage and damaged overhead line equipment. He then called the Woerden station manager, Van Zevenberg, to warn him that a collision had very likely occurred at Harmelen Junction. The station manager immediately departed for the junction to take control of the situation there and dispatched his assistant, a Mr Molenaar, to Woerden signal box to assist Kwakkenbos. Molenaar first contacted the local constabulary to warn them of the situation before walking over. These two men together managed to call and mobilise action from Harmelen police to keep open the necessary roads to allow emergency services easy access to the accident site as well as notifying Operations Control at Utrecht to start up the rescue, a local doctor for emergency medical aid, and the Roman Catholic pastor of Harmelen village for spiritual aid. All this was done in the short time between the moment of the collision at 09:19 and 09:23. From that moment Operations Control at Utrecht took over the organisation of the major rescue and salvage effort.

After steel cutting equipment, lifting gear and road cranes had been brought in to clear away the lighter pieces of wreckage and prepare the site for the arrival of the heavy rail cranes, the extent of what had happened gradually started to reveal itself. Locomotive No 1131 had literally flattened the leading carriage of the 4-car leading EMU, torn out the left-hand side of the second and scraped half the third carriage, after which it had rolled upside down off the train it had wrecked, still parallel with the track. It had left its bogies (trucks) back in the massive pile of torn and twisted steel at the impact point. The first carriage behind the locomotive, a riveted-steel standard class coach B5710 of 1928 vintage, in turn had its bogies, underframe, floor and interior ripped away by the curled up nose-end of the underframe of the destroyed leading EMU vehicle and, lying on its left side, draped itself like a limp blanket over the rear of its upturned locomotive. The second carriage, Plan E first class A6544, having had its head-end twisted to the right and partly demolished by the leading coach, ran into the resistance of the pile of mangled steel but, propelled by the mass of the following train, kept moving onwards in a left-hand tangent, veering away from the track it had travelled, past the wreckage of the EMU and finishing its trip by flattening junction-interlocking relay building RH10. Strangely enough, surviving occupants in the rear of this vehicle reported nothing but a few bad shocks and inexplicable movements, words that occur in virtually all the reports from people on the two trains and testimony to the great variety of powers working on the vehicles in the destructive phase of a heavy accident. Three further coaches also derailed leftwards and followed the second coach in line. The last six coaches (removed soon afterwards to Utrecht) remained on the track.

Something fairly similar happened with the EMU. The last carriage of the leading 4-car unit plus the trailing 2-car set were still on the track and were moved away to Woerden. A start could then be made on removing the remaining overhead catenary to allow access to the heavy rail cranes and lift wreckage off the track into the neighbouring field, once all the casualties had been taken away. At 02:00 on the following day the track was cleared and repairs were started. These were finished by 08:00, at which time the railway cranes were withdrawn, enabling wiring trains to be brought in to repair the overhead line equipment, so that at 16:00 that day power could be restored again. Very soon afterwards the first three service trains after the accident passed the site at 30 km/h (20 mph) between walls of destroyed rolling stock under tarpaulins, subsequent trains being subject to a 60 km/h (35 mph) emergency speed restriction until 08:00 on 10 January. From then on, line speed through the junction was resumed.

The inquiry

The tape of the self-recording Hasler speedometer of locomotive No 1131 was retrieved and investigated by Netherlands Railways in conjunction with the Criminal Investigation Laboratories in Rijswijk. It showed that shortly before impact the speed of train 164 had been sharply reduced from 125 km/h (75 mph) to 107 km/h (66 mph), at which speed the collision had occurred. Examination of the controls in the remains of the leading cab in conjunction with power-switching equipment of the locomotive revealed that the power controller and its associated equipment must have been in the 'off' position (meaning that the traction motors could not have been powering) and that the brake valve had been put in a position in or close to emergency. Whilst the brake rigging in the lost bogies was found to be too heavily damaged to conclusively back up such findings, it nevertheless indicated that driver Van der Leer had started an emergency braking manoeuvre, probably at or close to red signal 8 that covered the junction. It is interesting that very few people on his train actually noticed anything at all of this emergency brake application; the actual time between the application and the shocks emanating from smashing into train 464 must have been no more than a matter of seconds.

In order to verify this conclusion, the State Department for Railway Safety proposed to compare the information from this speedometer tape with the braking performance of a test set composed of similar stock to train 164. These tests duly took place on the night of 14 January between Arnhem and Utrecht. The stopping distances whilst braking in emergency of the test train were recorded on its locomotive speedometer and carefully measured *in situ*, these tests indicating that the distance travelled between the start of the deceleration until impact had been 85-105 m (275-345 ft), which indeed equalled the distance from the collision impact site to signal 8, taking into account the reaction time of the driver and the time lag in the application of the pneumatic braking equipment. The inevitable conclusion was that Van der Leer had missed the yellow speed reduction/warning aspect of signal P713 and started emergency braking on seeing the red aspect of signal 8. Permanent way staff that were present at Harmelen Junction and witnessed the collision were also present at the trackside during these tests and commented that, whilst they clearly saw the sparking between the brake blocks and the wheel treads during the tests, no such thing had been noticed during the accident. However, the weather during the tests was dry but at the time of the accident was very moist, which might explain that difference. At the same time, the NX signalling equipment of Woerden signal box and the infrastructure it controlled was rechecked thoroughly against the design diagrams and the reliability of the connections between signaller inputs, indications on the panel and the signal indications and turnout positions on the site was tested several times in varying conditions. Furthermore, the compliance of the installation with the drawings and specifications was checked again, and the light bulbs of distant signal P713 and others were also checked to see whether loosening through vibration generated by passing trains might have caused an intermittent signalling aspect fault that may have caused the yellow aspect to be extinguished. Every single test had a positive outcome. None was found to have played a role in the accident and no other issues were identified that could have caused malfunctions, as the Woerden NX installation had been in service for only half a year, having been commissioned on 28 May 1961 in the course of ongoing post-war repairs, renewals and resignalling.

Actions taken to avoid a repeat

In the course of ongoing rolling stock, infrastructure and signalling renewals and extensions in The Netherlands after the 1970s, all vestiges of what was Harmelen Junction until the 1980s have been obliterated. In stages the connection from Woerden station to Harmelen Junction was extended with extra tracks and a flying junction, whilst control of the junction was moved to the Utrecht regional traffic control centre and Woerden signal box was taken out of use. The main line from Utrecht to Woerden has now been quadrupled and brought on to an embankment (thus also eliminating risks arising from a few earlier AHB level crossings) and Woerden signal box has disappeared following decommissioning and reconstruction of the station layout. It is rather sad, however, that every redundant item was disposed of, thus forgoing the opportunity to convert the redundant signal box into a fitting memorial to the worst accident in Dutch rail history. Possibly as an outbase of the nearby magnificent Utrecht railway museum, it might have been used to explain in an exhibition how signalling works and what was done to prevent such a major disaster from happening again. On 8 January 2012, a small but dignified memorial listing the names of all who died was unveiled to mark the 50th anniversary of the accident. A Dutch TV network aired a programme about the occurrence, from which it was clear that many people still live with barely healed scars from what happened that morning.

After the accident, the Dutch state demanded the expeditious equipping of the network with a system in which the safety of rail passengers was not wholly dependent on a train driver observing the yellow warning aspect of a distant signal under all possible circumstances. The result was the unfortunately very costly and protracted installation of a type of equipment that in Britain would become known as ATP. Due to the many difficulties associated with installing a complex new safety system on a heavily used network such as that of The Netherlands, it took more than 40 years for this situation to be rectified. Equally unfortunate is the fact that many people died in further accidents during the long process of development, testing and equipping, which perhaps could have been avoided had a simpler, cheaper, tried and tested existing system been selected, such as the German Indusi. In fact, the installation of ATP (ATBEG, first-generation ATP) took so long that in the end it had itself become obsolete in a number of aspects and a more modern intermittent system was selected (ATBNG, new-generation ATP) to finalise coverage of the network. For stock using lines equipped with both types, it meant a further amount of equipment on board. On top of all that, owing to the fact that the first-generation Dutch ATP allowed a train to pass a signal at red at speeds of 40 km/h (25 mph) or lower, as with the old US system from which it was developed, many collisions continued to occur until recently. Due to the reduced speed with which the accidents occurred, however, passenger deaths were avoided until a head-on collision occurred west of Amsterdam Centraal station in 2013, which took the life of one passenger. This loophole has now been largely plugged with additional speed-check and stop loops at selected signals (ATB-vv, improved-version ATP) at the cost of several millions of euros. In July 2012, the minister in charge of transport announced that the Dutch Government had committed itself to re-equipping the network with ATP of the ETCS European standard type plan, to be undertaken over ten years. To the Dutch taxpayers the financial burden of fitment of ATP to their rail network has been (and still is) staggering. On the other hand, Europe as a whole has benefited from the consequent gain in train protection expertise and experience. The contribution of this Dutch ATP effort to the development and international know-how of what would eventually become ERTMS/ETCS over the following decades is incontrovertible.

Comment

Few major rail accidents in history had a root cause that was so quickly and conclusively proven as the Harmelen collision. It was driver Van der Leer going through a red light. But was that really all there was to it? As soon as there is a published timetable, timekeeping becomes one of the two ingrained propelling forces in traincrew the world over (as it is with many bus and truck drivers, airline pilots and skippers on all sorts of maritime traffic, including the *Titanic*). Unfortunately, on the railways as elsewhere, it often finds another such driving force – safety-induced caution – in its way. Indeed, weighing up the need to keep time against the need to preserve safety is a daily fact of life for most professional transport operators. The outcome depends largely on company or national habits and on personal choices on the part of the operator in the context of pressure from his employer and people on his train. Which one prevails, which gives way, and to what degree, to the benefit of the other, and for what reason? It should not be overlooked that debatable decisions from the viewpoint of robust safety very rarely lead to accidents and are often to the benefit of many passengers. Such timekeeping-based decisions in the run-up to the Harmelen accident can be demonstrated in two of the contributory issues:

1. Driver Van der Leer was 6 minutes late departing from Utrecht Centraal and decided, despite the fog, to make up time by travelling at the 5 km/h (3 mph) allowed above the timetabled speed, as indeed he was expected to do by signaller Kwakkenbos. This points to a kind of timekeeping ethos in which taking slightly risky measures based on experience and skill is acceptable. This is fine when vision is excellent and the track is dry, but, because it affects power shut-off and start-braking points, good visibility is necessary in order not to overshoot stopping points. Either that or thorough route knowledge allows you to know your position and where the signals are almost blindly. Van der Leer must have had a problem seeing his way in that fog.
2. Had Kwakkenbos kept the fog-bound junction signal for the local train at danger and given priority to the express, the local would have been routed along the main line to Utrecht in case it was involved in a SPAD. Out of the way of the express, this simply being the way signalling control is set up. He after all dealt with a speeding express in dense fog, clearly expecting the driver to be concentrating on making up lost time. Kwakkenbos had, in fact, accurately predicted the arrival time at the junction, but he nevertheless decided that in order to benefit the timekeeping of the on-time local, the turnouts would be set across the path of the express.

There are indications that Van der Leer might have been preoccupied by problems at home, and no doubt his mind was occupied with keeping his train on time. Distraction arising from preoccupation, in my own experience, is a rather sinister issue as the often unnoticed distraction from the current job it causes may be sharply reinforced by fatigue. Indeed, fatigue and absentmindedness at work often go hand in hand. Fog can further increase this negative influence on driver alertness as it may shut out the outside world and allow the driver to sink deeper into his own thoughts. Additionally, there was the sleep-inducing effect of the rhythmically clattering noise caused by the jointed track of the time. If this was perhaps the case with driver Van der Leer, then his ingrained timing and route knowledge might have suddenly set off an alarm in his head that he should by now be approaching Harmelen Junction and that he should be looking out for signals. His distraction and extra speed in the fog, however, had skewed his distance awareness and, instead of the yellow or green distant aspect, he suddenly saw the red aspect of the junction signal come up out of the fog. The degree of shock he must have undergone is hard to imagine.

It is essential for the safe movement of trains that a signaller can trust that trains will stop for his red-aspect signals. After high-speed accidents following a SPAD, discussions often arise internationally about the relative safety given by red signals without the protection of supervisory systems. This is where train protection systems come into play and where the assessment of the merits versus the costs of such systems begins. Any train protection system, whilst expensive, is unlikely to save large numbers of lives and prevent large amounts of damage as the railway is, after all, a very safe means of transport anyway. But it is exactly for that reason that the loss of trust among passengers, the loss of face politically and the amount of damage and death are all considerable when things do finally go wrong. It is only then that a nation decides whether or not to shell out on an effective train protection system. Britain, almost alone in western Europe, delayed introduction of a serious train protection system for a long time before finally installing TPWS in 2002/03 after the ***Ladbroke Grove*** crash. Unfortunately, it is the fate of any train protection system that it will be considered viable only after such major accidents, when politicians feel that the cost will be acceptable to taxpayers.

Conclusion

The impact of this major accident was keenly felt throughout The Netherlands. Although only 9 years of age at the time, I nevertheless remember the rather cold and overcast winter day and particularly the atmosphere of public shock rather well. The radio played classical music as a sign of mourning all day and Queen Juliana and her family cut short their traditional winter holiday in Lech, Austria. On her return she visited the accident site, the Chapelle Ardente with the 91 bodies in the Buurkerk in Utrecht and the casualties in hospital and later attended the various memorial services.

No train protection system will escape criticism after installation. There are degrees of effectiveness and some systems contain weaknesses that still allow accidents to occur. Many such instances can be found in this book. Moreover, many systems took a long time to develop, test and install. The very European notion of national networks developing their own train protection systems was, with hindsight, ridiculous. Had, for instance, Netherlands Railways decided to install the readily available German Indusi system (nowadays called PZB90), then at least two fatal head-on accidents near Arnhem, caused by SPADs at a single-line junction, would in all probability not have happened. One was committed by a German locomotive driver on 31 August 1964, and one by a Dutch driver but involving a Trans-Europ-Express set from Germany on 28 February 1978 on the war-damaged single-track replacement *Westervoort/Arnhem* bridge. Moreover, at that time Indusi was an off-the-shelf intermittent system that could have been installed far more quickly than the Dutch coded track-circuit-based continuous ATBEG system that in many cases required wholesale redesign and reconstruction of track layouts. Finally, Indusi would have provided unbroken coverage for traction from Rotterdam all the way to the top of the Brenner Pass in the Austrian Alps under one warning and protection system. However, the fact that Indusi was an older system (and most of all German, bearing in mind the recent war), whilst ATBEG was a further development of an already more sophisticated US system, no doubt influenced the decision on which system to adopt.

Interestingly, other signalling equipment was available or under development that would probably have prevented the accident from happening in the way that it did. If only Kwakkenbos had had more certainty about the whereabouts of express 164, as at present provided by modern train describer equipment, he would probably have let the express go. And if Van der Leer, on the other side, had had a warning in his cab about the restrictive aspect at distant signal P713, as even the then obsolete German Indusi, British AWS or French/Belgian 'Crocodile' could have given him, he would not have been surprised by that red aspect of signal 8. The benefit of hindsight!

***

Aitrang, Germany, 9 February 1971 at 18:46
(High-speed derailment on a curve)

The accident
One of the Dutch/Swiss 4-car Trans-Europ-Express DEMU sets, at the time working TEE 56 'Bavaria' between München in Germany and Zürich in Switzerland, was involved in a high-speed derailment on a curve with an 80 km/h (50 mph) permanent speed restriction. The train was the forerunner of the service that is encountered later in the accident at ***Lochau-Hörbranz*** in Austria. The casualty toll on the TEE set – 26 dead and 36 severely injured – was excessive for a modern steel-built train of integral construction. Of the 53 passengers and nine staff aboard this luxury train only three survivors escaped without injuries. After the derailment a second train was involved, when a 4-car diesel multiple-unit of the rail bus type set entering Aitrang station ran into the 115-tonne overturned power car lying on its track and collided at a speed of about 40 km/h (25 mph). This secondary collision increased the tally of deaths and injuries, the final toll being 28 fatalities and 42 people injured.

As far as safety on the West German network was concerned, 1971 was not a good year by any means. Following this derailment at Aitrang another high-speed derailment of an express occurred, on 18 May, this time behind one of the fast V200 diesel-hydraulic locomotives on very recently relaid track in hot sunshine at *Kellmünz* in Bayern (not far from the Aitrang accident site) resulting in five fatalities and 46 being injured. Then, on 28 May, the *Dahlerau* single-line collision occurred as a result of gross ignorance on the part of a train-dispatching stationmaster, leaving 46 dead and 25 injured. On 21 July, the infamous *Rheinweiler* high-speed derailment occurred on another sharp curve, resulting from a glitch in the AFB automatic driving and braking (cruise control) installation on the (at the time) top-of-the-range electric high-speed Co'Co' locomotive 103-106-1, with a tally of 23 dead and 142 injured. A week later, the *Mecklar* derailment occurred on 28 July, during which fortunately no one was killed. That accident was in turn followed by the *Hachenburg* collision on 31 December that left seven passengers dead and 34 injured. Together with a number of smaller incidents the West German Deutsche Bundesbahn (DB) left 109 people dead on its tracks that year, many more were injured and the damage ran into many millions of deutsche marks. The loss of prestige for the West German state railway operator, however, just as later happened after the ***Eschede*** and ***Brühl*** accidents, turned out to be the most enduring problem as well as the strongest incentive to set to and improve matters.

The location

The accident occurred at 18:46 in early February at dusk but with clear visibility. The track may have been starting to frost over, which would perhaps have slightly impaired the braking of the train, although the weight of the power car alone makes that questionable. Incidentally, the driving trailer was leading; the heavy power car at the rear of the 4-car train was propelling under remote control from the train driver in the driving trailer. The accident happened at the east end of Aitrang station on the line from München via Buchloe to Kempten and Lindau. TEE 56 was climbing from Kaufbeuren – with its own 80 km/h (50 mph) permanent speed restriction due to curvature – on a 1 in 133 ascent cleared for 125 km/h (75 mph) through Biessenhofen station. This station is situated on a long curve without speed restriction and boasts signals and a junction with the branch line to Marktoberdorf and Füssen. From there it was just a few kilometres until the entry signals into the distinctive station at Aitrang through a left-hand curve were reached, the train slowing down for the 80 km/h permanent speed restriction through the sharper right-hand curve past the eastern station exit, after which a further climb at 1 in 100 to Kempten followed. After a post-war re-laying of curves along this line only Kaufbeuren and the western-end curve at Aitrang had a permanent speed restriction, the rest of the line allowing unhindered running at 120-140 km/h (75-85 mph). In fact, due to the continuous rising gradient, the whole negotiation of the Aitrang speed restriction could have easily been effected by just reducing power in time to let the speed drop as required, before being ready to open up to full power again as soon as line speed allowed. There was no need to touch the brake at all. In any case, in contrast to the ***Berajondo*** derailment in Australia some 30 years later, the line offered many clear reference points for the driver to navigate by.

The Aitrang station layout itself was of an efficient pattern common all over north-western Europe. The two main lines passed through platforms along the outside whilst a single siding or platform line ran through the centre, connected to both running lines at either end. This set-up enabled fast cross-platform connecting manoeuvres between trains in either direction and overtaking manoeuvres from either direction whilst also allowing trains to call in a centre platform away from the main line and then continue or return in either direction (e.g. after termination of a service). At Aitrang, however, the Kempten-bound main and centre lines connected at the western end with a curved turnout, whereby the main line had the smallest radius that ran straight into the curvature of the open line. This may well have had an influence on the way the accident developed. The line on its rather sharp right-hand curve from the eastern exit of Aitrang station disappeared behind a hill. That, together with some other landscape features in these foothills of the Alps, meant that sounds travelled in very unusual ways, as indeed they do in hilly areas. People at and near the station picked up nothing of the noise the accident must have generated and even the wife of the driver of the terminating DMU, who was in a car close to the accident site on her way to the station to pick up her husband, heard no more than an unusual rumble that she paid no attention to.

Derailments caused by excessive speed at well-signed permanent speed restriction sites without aggravating circumstances such as missed temporary speed restrictions, overlooked cautionary orders, using reverse turnouts at too high a speed through misreading signalling etc had become very rare indeed. The inescapable reaction, therefore, was that driver Rahn of TEE 56 was to blame by failing to slow down for the permanent speed restriction on the curve east of Aitrang station. Rahn, however, was a man with 20 years' driving experience who had driven these sets right from the moment of their introduction in 1970 and was very familiar with this route. He was known as a very careful and competent driver, and the records from both the Indusi automatic warning system and the Hasler speed recording system on the train, however much they differed in their readout occasionally, nevertheless showed exemplary driving right up to the moment of the derailment. Driver Rahn died in the accident, so every suggestion regarding the cause of the accident in the official reports and subsequent publications, including this book, is based on surmise.

Rolling stock involved

The train was one of the five 4-car diesel-electric multiple-unit sets, built from 1957 onwards for Dutch (NS) and Swiss (SBB) interests, to operate the international limited-stop first class only Trans-Europ-Express services that were proposed by Netherlands Railways' chairman of the board Mr. Den Hollander. Their numbers were SBB RAm TEE 501 & 502 and NS DE 1001/1003. A female Dutch architect, using the typical 'dog-nose' head-end pattern for Dutch electric multiple-units as the template, cleverly redesigned this Dutch standard type by giving it a higher and more angular profile with flatter side surfaces and with obstacle deflectors under the Scharfenberg automatic couplers. It made for a strikingly recognisable and far more impressive look to the five TEE sets. The power car was a 115-tonne single-ended locomotive with two Werkspoor RUHB 1616 1,000 hp (745 kW) 16-cylinder prime mover diesel engines, driving four electric traction motors (A1A'A1A') through Brown-Boveri dc generators. A separate 300 hp (225 kW) Werkspoor RUB 168 diesel alternator unit was installed for 220 V 50 Hz hotel power on the train. This ample traction power for a relatively lightweight set was needed for good acceleration, to run at sustained high speed in flatter areas and to climb through moderate mountain ranges at good speed. Maximum permitted speed of the set was 140 km/h (85 mph).

The power car and driving trailer cabs were fitted with the 'dead man' driver safety device that, however, had no vigilance (alerter) switch. This was something that would not appear in The Netherlands until the introduction of ATP 20 years later. Much to the surprise of many, initially no automatic warning systems were fitted on cost grounds, not even the Swiss type, although modifications demanded by the various national networks soon cured that surprising omission and apart from the Swiss Signum ZUB system the German Indusi and the French/Belgian 'Crocodile' systems were included.

In comparison with German trainsets used for similar work, the brake system was a somewhat old-fashioned Swiss Oerlikon all-air brake with clasp-type tread brake blocks, rather than electro-pneumatic (EP) equipment with disc brakes and electro-magnetic track brakes. The fitted type of brake had been enhanced with a speed-dependent emergency brake wheelslide protection in view of the high-speed operation intended for these sets. The Oerlikon system allowed a fairly rapid and graduated application as well as release (on a personal note, I had high regard for the Davies & Metcalfe licence-built equipment when I worked the similar brake control system on the Gatwick Express Class 73 ED locomotives). Furthermore, because the sets, even when working two in multiple, were rather short, slow propagation of the brake application and release throughout the train was not normally of concern. In order to work this brake to the best advantage both a brake pipe connection and a main reservoir air train connection were necessary, whereby the main reservoir air came from the two compressors and reservoir on the power car through the main reservoir train pipes only; no auxiliary main reservoir air tanks were fitted on the coaches. Some experts have pointed to the potential for brake problems and the accident report does indeed mention a number of instances where departures were delayed due to brake test faults on these sets, especially non-release of the brakes for which main reservoir air is useful to keep the brake reservoirs on the coaches replenished and to assist release. This was perhaps a case of either cost-cutting or ignorance, similar to the non-fitment of any warning system. Or perhaps through a lack of understanding of certain characteristics of this brake control system. The train was coupled with bar-couplers to the locomotive and among each other, it was a fixed-formation trainset comprising a 38.6-tonne compartment seating car, a 40.6-tonne restaurant car and a 36.8-tonne open saloon seating car/driving trailer (totalling 116 tonnes). These vehicles were very recognisably of the Swiss lightweight type, riding on Swiss SIG torsion bar bogies. These SIG bogies had torsion bars as their secondary suspension rather than the leaf or coil springs that were more usual at the time, something we see more often in road and rail vehicles of the period. To what extent this feature made them ride the curved trailing turnout and eastern-end curve at Aitrang less successfully than other contemporary bogies is unclear. After all, the speed of the train on the curve was substantially over the limit.

As far as fitting out standards were concerned, contrary to German practice in which laminated glass was already used for the inside glazing of double-glazed carriage windows, the windows were fitted with double sheets of toughened glass with passenger-controlled slatted sun blinds in between. Both toughened glass sheets on the TEE set completely shattered on impact, and passengers fell through the large panoramic windows out of the overturned but still sliding coaches. They landed on the track and in the ballast under the overturned vehicles, which was the main reason for the excessive death and mutilation toll. I remember seeing pictures of this accident with body parts visible in the ballast. Another reason for the high casualty list was the shattering of interior coach mirrors, manufactured of mirror glass instead of polished metal. The carnage was compounded by the fact that the seats in the restaurant car were loose, and the tables, in line with contemporary Dutch practice, were hooked into hinges against the bodysides to enable lifting for ease of cleaning the vehicle interior. These flew around the carriage interior, causing further serious injuries. These features rendered the scene in the restaurant car somewhat like that of a battlefield.

Much is made in German publications dealing with the accident of the comparative lack of modern features that other German DMU rolling stock for similar premium services had already been fitted with, such as more stable power cars at either end, EP brake controls acting on disc brakes, electro-magnetic track brakes that automatically activated if the brake controller was moved to emergency, a driver safety device system with vigilance, and fast-acting automatic wheelslide protection on the braking system. On the German coaches mirrors were of the non-glass types, all furniture was solidly fixed and the windows had double-glazing with laminated inside glass panes. Any of these features indeed would have mitigated the outcome of this accident or even prevented it from happening. Arguably the braking equipment on the Dutch/Swiss DEMUs could be described as lacking a certain up-to-date kind of sophistication given the frontline job they had to do, although similar equipment worked fine on the very busy tracks in both The Netherlands and in Switzerland well into the 1990s. Additionally, the 2-car Italian and French TEE diesel sets were in a technical sense certainly not better equipped. Why, then, the German railway authorities accepted the sets for this kind of traffic on their tracks – after only requiring fitment of the Indusi automatic warning system – is far from clear, but the need to upgrade the 'Bavaria' TEE services quickly and at low cost probably played an important role. In that case, however, it was unreasonable to expect cutting-edge technology with old kit. This also explains why the Dutch/Swiss consortium that built the trains was not interested in the advanced safety requirements for passenger vehicles already in force for home-built vehicles in Germany, before putting these trainsets in service there. In fact, both the Swiss and Netherlands Railways administrations were already losing interest in these by then ageing and increasingly difficult to employ trainsets and would not have been averse to getting rid of them. Passengers for their part, however, were happy with these old-fashioned but comfortable trains. Due to good patronage a number of issues invariably arose, the most important being the rather limited and rigid availability of seats in either three or six coaches. Furthermore, these costly to operate diesel-electric trains increasingly performed along extended stretches of electrified track. They kept being moved ahead of new electrification and were then replaced with newly constructed, more capacious and better performing EMUs or, most of all, electric locomotive-hauled TEE sets. The German Railways fully air-conditioned loco-hauled TEE stock in particular, which initially even included so-called vista-dome lookout coaches as used in the US, royally eclipsed anything else on European tracks in terms of comfort, as do the third-generation German ICE trainsets these days (when everything, especially the air-conditioning, works as it should).

That, then, was the situation in which the planned upgrade of the non-electrified Allgäu line, connecting München via Lindau to Zürich, caught up with the old TEE DEMUs. Two administrations (West German DB and Swiss SBB) were seeking suitably comfortable, but most of all readily available, diesel-powered prestige trains as a stopgap whilst looking around for new build for the coming TEE 'Bavaria' services. New-built diesel power was out of the question. Unfortunately, despite their 2,000 hp (1,490 kW) prime-mover traction power, the five sets proved not to be up to dealing with the mountainous character of this Alpine route and, despite flat-out driving, often lost time during the long and tortuous climbs that often caused intermittent overheating of coolant. That was a well-known characteristic of diesel traction of the period and the resulting delays then had to be made up by faster driving along the flatter stretches. This, in fact, was considered as one of the possible contributing causes to the Aitrang accident.

The accident log

The stationmaster/traffic controller at Aitrang had set the road for TEE 56 through the eastbound main platform line for its non-stop passage and the road from the opposite direction into the middle platform for the 18:49 arrival of the 4-car DMU railcar from Kempten. That train would terminate at Aitrang and be berthed there for the night. When he saw on his panel that TEE 56 had passed his home signal he walked over to the window of the control room in the station building to observe it passing. At 18:44 he watched the train coming through and noticed that it was travelling rather fast and that sparks were coming from the wheels of the trailing power car, indicating that the brakes had been applied. He judged that the train would slow down sufficiently in time for the speed-restricted curve just out of his sight. On coming back to his panel at 18:46, however, he noticed that the red track-circuit occupation light for the incoming local from Kempten was already illuminated. He was certain that no message with respect to early running of this train had been issued, and a short moment of consideration made him put all signals for the local back to danger on the assumption that something had gone awry with the express after all. However, that was of no effect as the local had passed the home signal already. At a speed of approximately 40 km/h (25 mph) it ran into the possibly still moving wreckage of the overturned power car of the express on its track and short-circuited the track circuit, which surprised the stationmaster. Sadly, the driver of this service, who in all likelihood had seen the derailment of the express ahead, was among the two fatalities on this train.

The investigation

This section of the story draws heavily from the sterling investigation work done by Hans Joachim Ritzau, as the official work is bland and non-committal and fails to draw a number of conclusions that certainly could have been drawn, as shown below. The following issues were recorded:

- The train did not climb the rail head and derailed upright, but as at ***Berajondo***, the train tipped out of the track and did so shortly after coming through the curved turnout connecting the middle road to the main line. The driving trailer tipped onto its left side, slid across the opposite track, distorting it, and then fell into a ditch parallel to the embankment on which the track had been laid. It came to rest in a more upright position, leaning against the opposite embankment side. The restaurant car fell on its left side and stopped in a similar position behind the first vehicle in the ditch after sliding across the opposite line. The compartment car came to rest across and away from the opposite track in an upright position, pointing away from the track. The power car very likely had been pulled out of the track by its train and, lying on its left side, fouled the opposite track; hence the arriving local set colliding fatally with this heavy vehicle. All three bar-couplers had failed.
- From the Indusi record tape the speed of the train on derailment was apparently 125 km/h (75 mph), which would have been sufficient to tip the train out of the track. In this case the train would not have slowed down at all.
- A Swiss expert read the Hasler tape and from its clearer record of events on the rear power car the derailment speed was taken as 100 km/h (60 mph). The reason for the derailment at this speed is somewhat less clear, but it tallies with the observation of the stationmaster who noticed sparks flying due to a hard application of brake blocks on the trailing power car. This hard braking caused wheelslide, however, which was proven and documented by pictures of the resultant wheelflats on the left side only, indicating the start of overturning. This onset of sliding, in my opinion, made the Hasler speed recording on the power car show a lower speed than was the case, the Indusi readout therefore being more correct. Indeed, the train speed had barely diminished despite the hard braking. A German Indusi expert in the investigating team, however, went sick the moment he was asked to give an official interpretation of the record tape, so escaped having to give evidence based on the record of this system. Indusi records had a reputation for being difficult to work with, especially in the situation where two such systems competed. It

led to confusion and hassle, as virtually all interpretations surrounding the Indusi readout differed. The Hasler tape was not consequently admitted as proof in the final report – a rather peculiar decision, and quite incomprehensible.

- From a scientific point of view, the fact that the leading vehicle was a driving trailer and that the heavy power car trailed should not have made much difference, although the general view is that the derailing and overturning coaches with their higher centre of gravity pulled the heavier and more stable power car with its lower centre of gravity out of the track. On the other hand, there is no guarantee that the coaches would not have derailed even if the power car in the leading position had negotiated the trackwork safely. The performance of the SIG bogies with their torsion bar secondary suspension is not discussed in any depth, but is not likely to have differed much from the then available other types of high-performance bogies with coil-sprung secondary suspension, such as the German MD and the Swiss/Austrian Schlieren types. The SIG bogies also did not give cause for concern when later used in Canada on far less carefully maintained track, albeit that they rode at considerably lower speeds in that role.
- During post-accident tests the braking system of the set functioned as required from both ends, and on the wheelsets of the trailing power car severe corresponding wheel tread flats at both sides were photographed. This indicates that on one side the brakes did apply along the entire train and also that a full-service or, more likely, an emergency application caused wheelslide on the power car before the vehicle started to lean prior to derailment. It also backs up the observation of the stationmaster and the record on the Hasler tape. However, a second rather peculiar development in the judicial process was that no train brake expert was invited to the inquiry to give a viable explanation for the observations and propose a possible brake fault that would cover what was found as evidence.
- At about this time the German press got hold of the opinion of an independent expert, who suggested that an ice obstruction had formed in the train brake pipe somewhere between the leading driving trailer and the rest of the train, which in turn had hindered air pressure reduction in part of the brake pipe and thus application of the brake throughout the entire train. It is, therefore, possible that the driver, noticing brake failure, used the

intercom telephone to request the travelling fitter on the trailing power car to apply the train brakes in emergency from his position. These words came from a German train brake expert, Dr Thoma, who did not contact the inquiry committee first but published his ideas in a special interest magazine, where the public press picked it up and published without asking further questions. This interesting line of inquiry fizzled out in the rumpus surrounding this breach of protocol.

- The initial argument against Dr Thoma's theory is that the train braked as required for the permanent speed restriction at Kaufbeuren and that it is somewhat unlikely that sufficient ice blockage developed from nothing in the brake pipe in less than half an hour in the rather mild frost. But Ritzau points out that, according to record tapes, the train did have a similar problem at Kaufbeuren and started to brake late for the permanent speed restriction there. Consequently, the ice prop in the brake pipe theory cannot be discounted. Ice has substantially more volume than the water it is formed from and thus may well block the brake pipe air-tight. Moreover, such ice blockages will form at the lowest point of the brake pipe train line because that is where moisture collects and where the efficient heat-conducting steel connectors are located in the rubber pipes between vehicles. On the other hand, trains run in much colder climates throughout the world without any such problems at all. The secret, of course, is efficient dehumidification of the compressed air used for brake control (see the brake problems experienced at ***Auckland***). Was the TEE set thus equipped? No!
- The intercom telephone handsets in the leading driving trailer cab and in the travelling train fitter's office in the power car were off their cradles, but the handset in the rear power car cab was still firmly attached. This appears to confirm the suggestion that the driver and the fitter were in contact just before the accident. Based on the theory outlined above, it also suggests that driver Rahn and the Swiss on-board technician, someone who surely had sufficient route knowledge of his own in view of his job, could have been in touch about a brake-functioning emergency that required the Swiss technician to apply the brakes from his position, but that their action came too late to save the train. It supports the ice-blockage theory as well, and explains the fairly violent application of the brake blocks on the trailing power car wheels that the stationmaster observed.

All these issues together officially absolved the dead driver Rahn from responsibility for the accident as it was impossible to produce a case that could apportion blame to him beyond reasonable doubt. However, questions remain as to why Rahn was not more careful after the apparent braking problems at Kaufbeuren, and why his train failed to slow down on the uphill gradient at Aitrang. Perhaps it was a matter of established routine? There is a school of thought that Rahn was chasing minutes because the TEE sets allegedly struggled with timekeeping along the mountain stretches and time had to be made up along the flatter and faster stretches such as where the accident happened. In that case, the possible cause of the accident would be closer to ***Vaughan***, ***Harmelen***, ***Åsta*** and ***Amagasaki***. However, the Hasler tape shows well-considered and steady – in short, exemplary – driving throughout.

Despite the fact that the station was but a few hundred yards from the accident site, the rescue and salvage action was somewhat delayed as the situation was initially unclear for about 10 minutes, but when everything came on stream the rescue was well resourced and well organised. However, it was discovered that handling the wreckage of the Swiss/Dutch trainset posed a problem for the salvage crews as they were unaware of the proper jacking points to lift the vehicles and cutting points to gain entry into the coaches. This accident therefore spawned a number of improvements with regard to emergency equipment and signage of internationally employed coaches. A less savoury feature was that many onlookers made their way to the accident site and clogged up access roads (see ***Granville***), which later led to powers to remove or deny access to people from places and roads near accident sites.

After the accident DB prohibited further use of the sets on German tracks and replaced them with Class 218 diesel-hydraulic locomotives hauling German standard first-class vehicles and later TEE/IC passenger stock. The four remaining sets were rescheduled to TEE 'Edelweiss', soon to be taken over by 5-car, 4-current SBB RAe TEE EMUs before finally being put up for sale. In 1977, they underwent slight external modifications for use in Canada, where the Ontario Northland Railway successfully put them in service on the 'Northlander' services between Toronto and Timmins until 1992. Five passenger vehicles (the four ageing and increasingly less reliable power cars having been scrapped relatively soon after arrival in Canada following replacement with modified GM F7 locomotives) have been repatriated to Europe under the auspices of the TEE-Classics organisation with the idea of using a redundant British Class 50 locomotive to make up a new power car and then work the set in the charter sector. In January 2011, I spotted four repatriated coaches at the Dijksgracht storage facility in Amsterdam. One of the reasons that diesel traction on multiple-unit sets was dropped comparatively early from high-performance premium services like TEE was that it was far from capable of delivering the sort of performance that could match electric traction. However, when hauling a trainset with diesel traction, at least extra locomotives can be placed at the front. The situation at the moment (see ***Lochau-Hörbranz***) is that the line is worked with diesel-hydraulic locomotives with EC coaches from München via Kempten to Lindau and with Swiss electric locomotives from there onwards. The line will be electrified in 2020 and ICE1 or ICE2 electric sets will be allocated to work this connection to Zürich throughout. Aitrang is no longer a staffed station.

Comment

The Aitrang derailment is another example of an accident where the blame appears to sit fairly and squarely on the shoulders of the train driver. That is apart from the fact that non-German rolling stock was in use, which unfortunately triggered a few foolishly bigoted reactions, not dissimilar to what occurs in Britain or France at times. Sadly enough for those commentators, within months of this accident two further fatal high-speed and high-damage derailments. These involved DB rolling stock and also occurred on curves and turnouts, at *Kellmünz* and at *Rheinweiler*, where the causes could be determined with certainty and did not involve the drivers. It did little to convince the press (and thus the public) that there was no underlying cause other than speeding drivers to connect these three apparently similar derailments.

If driver Rahn did have brake problems for the speed restriction at Kaufbeuren, it is difficult to understand why he failed to get the on-board technician to investigate with a view to sorting the problem out. There is nothing worse than driving a train with a serious brake fault, certainly one that might at any moment cause a runaway (which in all likelihood is what happened). Given Rahn's experience and careful style of driving as established from recordings, I wonder just how serious the braking incident at Kaufbeuren was. If, on the other hand, he was completely surprised by a failure of his train brakes at Aitrang, it is a matter of concern why he was still going too fast for the existing speed restriction anyway. Rather than shutting off power to let the rising gradient do the job, was he, after all, trying to make up time? He would probably have had problems later along the route, but not necessarily as devastating as what happened at Aitrang, although coming downhill into the terminus at Lindau on its picturesque island in the Bodensee, connected to the mainland by a dam, might have brought its own problems – such as seeing the train end up in the water.

***

Garden City, Long Island, USA, 8 August 1973 at approx 16:30
(Vehicle intrusion with fire on third-rail electrified line)

The accident
In this incident a car overshot a T-junction and ended up on a third-rail electrified railway line, following an error on the part of the young, inexperienced and unlicensed car driver. No train was involved, however, albeit the train services were cancelled for the rest of that day. This incident exposes the lack of aptitude and proficiency common to many drivers of private vehicles and which too often blights road safety but is regularly shrugged off by motorists as irrelevant.

The location
Garden City is located in the Hempstead area of Long Island, the huge island spit extending east of New York between the East River, Long Island Sound and the Atlantic Ocean. Brooklyn, Queens and Nassau, as well as La Guardia and John F. Kennedy airports, are all situated at the western end of Long Island across the East River. The area is prime New York City commuter country, in much the same way as Kent and Sussex are to London, and public transport is provided by the Metropolitan Transport Authority (MTA) owned Long Island Railroad (LIRR) that operates a diesel and third-rail electric network very much in the mould of the Southern Railway and Merseyrail in Britain or Transilien around Paris. People taking the train from New York Penn station or Brooklyn Flatbush Avenue station to Jamaica Interchange to get to JFK Airport travel the LIRR, or the Toonerville Trolley as it is known to many locals. The accident happened on the short Hempstead branch that diverges from the Port Jefferson-bound main line at Floral Park junction, close to the Belmont Park Racetrack. The names give an indication of the character of the residential neighbourhoods in the area.

The accident log
Late in the afternoon at around 16:30 a two-door saloon car moved off from the Garden City home of one of the five teenage girls travelling in it. A 15 year old, who had not passed her driving test, was at the steering wheel. As the car approached the T-junction at the end of the road, the

driver failed to slow down and drove straight ahead at speed, ending up on the tracks of the LIRR Hempstead branch. The steel car body short-circuited the dc conductor rail against the running rails and set up fierce arcing. However, despite the short circuit, no traction circuit breakers opened due to the way the rail traction supply system was designed. The front of the car quickly caught fire as engine fuel lines overheated and melted, the resulting conflagration spreading explosively through the car. Only the driver and her front passenger were able to vacate the vehicle through the right-hand door, the three panicking girls in the back being unable to fold the front-seat backrests down quickly enough and so perishing in the blaze. The two girls who were able to leave the car vanished from the scene but were later found in a state of severe shock.

Comment

Quite apart from an unlicensed 15 year old driving a car on a public road illegally, there is the matter of the fundamental disregard for safety of having 600-800 V dc in a non-insulated conductor rail located at ground level, as with the ancient top-contact Westinghouse third-rail dc systems found around New York, Boston, London and Liverpool. Every year scores of animals and occasionally trespassers are killed or badly burned through electrocution. Other third-rail systems, such as the German and French bottom-contact types as found on many European RER, U-Bahn and Metro systems, are substantially safer due to the sturdy moulded plastic insulation shrouds that cover the top of the third rail and to the fact that the rail is located at knee-level where it is much more conspicuous and far less easy to step on or stumble over accidentally. Such a system would probably have avoided contact between the car body and the conductor rail and allowed all the girls to leave the car.

The problem of quickly folding the front-seat backrests down to vacate the two-door vehicle in an emergency shows up a distinct but often unappreciated danger of travelling in that type of car. In its report, the US National Transportation Safety Board (NTSB) had a few things to say about a proposed standard location and method of operation of these seatback latches such as standardised indications to rear-seat occupants about how to unlatch the backrests and get out in a hurry. Implementation of this recommendation, however, is not noticeable in such cars to this day and people can still get trapped in a similar fashion during an emergency. The NTSB also noted with concern the lack of a simple traffic sign indicating the necessity to take a left or right turn when approaching a T-Junction in the direction of a railway line, and to provide a crash barrier to protect those who still managed to leave the road and land on the railway line.

Several European nations have experienced similar incidents. For example, in Germany an articulated road freight vehicle ended up on top of an inter-city train passing by below, and in Britain several vehicles have crashed onto railway lines in fairly recent times, one triggering the major fatal head-on collision accident at *Great Heck* in February 2001. Providing decent signage of hazards and protection against vehicles leaving the road in a place where, for example, a railway, a highway or a maritime transport link may be endangered is a typical low-cost measure that could do much to reduce risk to both life and property.

Where the accident in Long Island enters the realm of the unbelievable, however, is that a public railway traction electricity supply system, dating from the very early 1900s, failed to include sufficiently sensitive automatic traction current breakers. These should by design be included in such a system to open automatically in case of overload to cut the electricity, much in the way fuses or miniature circuit breakers do in a family home. This is a basic safety requirement, especially on an already more than usually dangerous third-rail electric system, but was one that inexplicably was not considered necessary on the LIRR in 1973. If it had been in place at the time of this particular accident, the arcing would have stopped and the initial source of the fire would have been removed as a result. With that taken care of, all the girls would have had more time to get out of the car and, whilst on the track in their panic and ignorance, would not have been further endangered by 800 V dc lurking at their feet. Additionally, train traffic would in all probability have been stopped because the traction current had been cut, and railway traffic control would have had an indication of problems on the line, complete with an admittedly very rough location. Last but by no means least, nearby railway staff (a train driver or stationmaster etc) would have been able to remove the traction current by short-circuiting the traction supply with the so-called short circuit bar. This simple piece of equipment is found on all traction (including diesel and steam) operating on the third-rail networks in Britain, and the NTSB report noted that earlier recommendations had already raised this issue. A sign of the times was the recommendation to set up a national universal emergency telephone number (911 in the US). Most other countries followed suit.

***

Warngau/Schaftlach, Germany, 8 JUne 1975 at 18:33
(Head-on collision due to serious mistakes by signallers)

The accident

This was a head-on collision of two diesel locomotive-hauled trains at considerable speed along a single line, following a series of traffic control mistakes. A contributory factor was a feature in the new summer timetable for 1975 (see also ***Schiedam-Nieuwland*** for a similar influence of new timetable features), whereby meetings of opposite train services were scheduled along the single-line section midway between crossing stations Warngau and Schaftlach instead of at one of the stations (somewhat cynically indicated as a 'Luftkreuzung', in the manner of two aircraft crossing at different levels in the sky, like one train jumping over the other). This was done with a view to providing a measure of operational flexibility to ensure enough time would be available for the run-round moves of the locomotives at the Lenggries terminus in the event of delays.

In the two weeks before the introduction of the new timetable, operational staff criticised and protested about these rather theoretical yet dangerous timetabling plans, but had become familiar with them by the time of the accident. Unfortunately, on the day in question, a Sunday, a member of staff, who was recalled from his three-week annual leave to cover for a sick colleague, returned to work without having experienced the rumpus or having been informed about this unusual change to the timetable. Additionally, and more dangerously, he omitted to acquaint himself of the situation by the simple expedient of reading the station working book (as he should have done according to the rules), in which the questionable timetable feature had been clearly marked. The next link in the chain of events were badly handled traffic control exchanges via telephone between Warngau and Schaftlach stations with their passing loops (it concerned the no-signals 'Zugmeldeverfahren' type of train control), which was finally compounded by an accident site where the train drivers could not see each other until within 300 m (990 ft) of impact. This violent head-on collision claimed the lives of 41 people and left 122 people injured.

The line and location

The location was the non-signal and non-interlock block section single-line between Warngau and Schaftlach stations on the line from München

Hauptbahnhof southwards via Holzkirchen to Lenggries. The impact site was at an automatic open level crossing with an Indusi warning from both sides, the system having been fitted to warn train drivers to check the trackside monitoring signals that indicated proper working of the level crossing lights. At this point the railway line ran in a straight course up to the level crossing before making a left-hand turn into a narrow corridor running through a densely wooded area, completely hindering distant sighting of oncoming trains. This explains why the trains collided virtually without braking. The line from München Hauptbahnhof to Holzkirchen via Solln and Deisenhofen is a double-track electrified secondary line. At Holzkirchen the line splits three ways: the electrified line returns to München Ostbahnhof via Kreuzstrasse, where an electrified line branches off towards Rosenheim on the main line from München to Salzburg and Innsbrück in Austria (on which on 9 February 2016 the head-on crash at *Bad Aibling* occurred); the non-electrified line from Kreuzstrasse to Bayrischzell in the German Alps; and the non-electrified single-track line to Lenggries via Schaftlach on which this accident occurred. At Schaftlach a scenic local line to the popular Tegernsee lakeshore branches off and is now regularly operated as a steam-powered tourist attraction.

Owing to the picturesque nature of this sub-alpine landscape close to the major city of München, traffic along all these lines could be quite heavy in the summer season. For that reason, the regular services were supplemented with a number of extras, especially at weekends, and therefore frequent and quite substantial trains operated at appreciable speeds at such times.

Rolling stock involved

Both trains involved in the accident were composed of modern West German locomotive-hauled bogie passenger stock of the 26.4 m (87 ft) UIC-X standard type. E 3594 had six such vehicles and E 3591 had seven. They were of the first-generation so-called Städtewagen (city coaches) type dating from 1952-54, with single slam-door exits at the ends and double slam-door exits in the centre of each vehicle, as well as those of the second-generation unpainted stainless steel variety known as Silberlinge or Silberfische (silverfishes) although they were a dull grey rather than silvery most of the time. They were built from 1960 onwards with their doors in the one-third/two-thirds configuration that was then internationally popular. These coaches had exit doors that had to be opened by hand but were air-powered for automatic closure by the conductor, the first generation of the internationally accepted 'turn and fold' power-closure doors for locomotive-hauled passenger rolling stock that flooded Europe from the early 1960s. The locomotives were both of the Class 218 type, Nos 218 238-4 and 218 243-4, the latest variety of the numerous second generation of post-war West German diesel-hydraulic 4-axle locomotives. They had a 12-cylinder 2,500 hp (1,865 kW) MTU prime mover (in only an 80-tonne locomotive) and, because they had a hydrodynamic brake included in their Voith hydraulic gearboxes in addition to their train air brake system, they were allowed a maximum speed of 140 km/h (85 mph). They are of a type that can still be seen around, albeit in diminishing numbers.

In the light of this accident it is rather sad to realise that the run-round move of the locomotives at Lenggries, which due to time constraints caused by delays triggered the introduction of the 'Luftkreuzung' manoeuvre, was not really necessary had the full technical potential of these locomotives be used. They were enabled to work with driving trailers/remote control vehicles for push-pull operation and these vehicles were indeed being rapidly introduced for just such local out-and-back trips. This line had not received the full necessary allocation as yet, which in summertime was exacerbated by requirements for the many extra services worked.

Some history

Both Erich Preuss and Hans Joachim Ritzau describe the developments leading up to this accident comprehensively, as it badly shook Germany in the way that ***Chatsworth*** did in the US. The background of the 'Luftkreuzung' feature that stood at the root of this accident was the wish to run the necessary extras during the Olympic Games at venues in and around München in 1972. As a sort of theoretical timetabling exercise to make the many single lines in the area work harder by avoiding the sometimes considerable waiting periods at crossing points. To give local staff the freedom to organise a change of crossing points on an *ad hoc* basis depending on the actual situation at the time, a situation was proposed whereby the crossing of two trains on a single line was theoretically planned *between* two crossing stations, leaving it to the operational staff involved to adjust to the prevailing circumstances. There is no doubt that it was one way to mitigate the impact of delays along single lines, which indeed have a tendency to grow quickly and excessively. However, single-line timetable planning based on theoretical head-on collisions has an eyebrow-raising quality to it, even if the single-line section is fully protected with interlocked signals and train protection that would prevent mistakes at either end of the section from turning into catastrophes. But on a non-interlocked, non-signalled single-line stretch, as from Warngau to Schaftlach, where the trains were operated under the 'Zugmeldeverfahren' telephone protocol (in which safe operation was entirely based on the unstinting accuracy with which the two traffic controllers/stationmasters involved did their job) it becomes rather a case of daring the gods. That is fundamentally wrong when the safety of passengers is at stake, and in this case it triggered a grievous crash. As a result, the Deutsche Bundesbahn once more hit the headlines for all the wrong reasons.

When this timetable feature was first introduced, the line from Holzkirchen to Lenggries saw the crossing of two 'Eilzüge' regional fast trains (or in British terminology 'semi-fast' trains) E 3591 and E 3594, on Sundays only, planned on the non-interlocked single-line section between Warngau and Schaftlach. The operations people on the line were astounded by what was being proposed and started to phone around, finally speaking to the timetable compiler at the regional head office in München. Under pressure, the gentleman they spoke to selected Warngau as the crossing station of choice there and then, but he did not officially change the timetabled departure time of E 3591 from Warngau. This is something that would still have been possible and should then have been backed up by issuing notices that detailed the timetable alteration. That is regular practice anywhere where railways operate, but it did not happen in this case. Incidentally, Warngau was not exactly the crossing station of choice for the railway staff working the line. To start with, on Sundays one crossing had already been planned there for E 3591 with E 3590/3592, so now E 3591 had to wait for that train as well as for E 3594. An even more pressing reason was that passengers for E 3591 had to cross tracks that would soon be occupied by a following train.

Incidentally, I once witnessed this type of situation at Ötztal station on the Arlberg route in Austria. Quite a few people, including some elderly and children, were slowly crossing the through tracks on the foot crossing from the station building to the platform for the train to Lindau that I was travelling on, when without any warning a non-stop 'Transalpin' electric push-pull service from Zürich to Innsbrück suddenly came downhill round the curve into the station at speed, bearing straight down on them. Blaring train horns and shouting staff got everyone out of the way in the nick of time, but it was one of those unforgettable moments of deep horror that the railway occasionally serves up. I believe that this foot crossing has now been replaced with a pedestrian tunnel.

Anyway, for the above reasons the request was made to alter the timetable for E 3591 rather than for the trains to cross at Warngau. This was not acted upon, however, and in response a bit of a bodge job was put in place in the local station working books; the timetable for E 3591 and E 3594 received question marks entered with pencil, with the remark that the expected realistic departure time for E 3591 would be 18:*33*, whilst for E 3594 the printed minute figures were removed and replaced with a question mark. In another column the reasons for these alterations were made clear (i.e. the crossing of these two trains at either station as required), so anyone doing their job properly and reading the station working book prior to starting their duties should have been aware. It had worked fine the previous Sunday when E 3594 was a few minutes late and the crossing had indeed been moved to the other station. Fate, however, had yet another trick up its sleeve. On Sunday 8 June, the trains ran virtually according to plan, but because the stationmaster working the afternoon diagram at Warngau station had been away for three weeks and missed the commotion about the new timetable – and had neglected to look in the station working book – he was unaware of the full details of the arrangements. From experience, however, he should have known that such alterations were usually thorough in order to deal with increased loadings, especially at weekends. Obviously, he had a number of other duties to fulfil, such as signalling trains, selling tickets and keeping the station premises in good order, but he would have had time to familiarise himself with the new arrangements.

The accident log

E 3590/3592 departed Schaftlach on its way to München at 18:22 and arrived at Warngau at 18:28, where E 3591, travelling from München to Lenggries, was already waiting for line clearance. E 3594, closely following E 3590/3592 towards München, in its turn arrived at Schaftlach at 18:26. So there was rather dense traffic for a single line that was signalled with a distinct lack of normal technical sophistication. In the passenger timetable E 3594 was booked to depart Schaftlach for Warngau at 18:28, whilst E 3591 was due to leave Warngau for Schaftlach at 18:27. The 'Luftkreuzung' was clearly halfway. The telephone conversation recorder stored the following 'Zugmeldeverfahren' traffic operation exchange regarding these services for the inquiry into the accident. The Warngau traffic controller initiated the call to Schaftlach. (The transcript is in the Bavarian dialect, incidentally.)

1. *Schaftlach*: *'Schaftlach!'*
2. Warngau: [indistinct] then, '*That one is at Warngau.*'
3. *Schaftlach*: *'3594 accepted?'*
4. Warngau: 'From 3, ermm, 29.'
5. *Schaftlach*: *'From 29?'*
6. Warngau: 'Yes!'

Line 1 is clear; the Schaftlach traffic controller answers the Warngau call by reporting with the name of his control station, as he should. No problem there, international good practice is followed.
Line 2 can be interpreted as the Warngau traffic controller reporting back, according to protocol, that the single-line section has been cleared following arrival of E 3590/3592 at his station. In fact, the Schaftlach traffic controller later said that he had asked for a speedy report on the arrival of E 3590/3592. The traffic controller at Warngau probably wanted to offer E 3591 at that moment – 'That one is at Warngau' – but the Schaftlach traffic controller ignores his attempt and breaks into this conversation with his own E 3594, based on the 'Luftkreuzung' alterations.
Line 3, therefore, must be interpreted as the Schaftlach controller assuming that the Warngau controller accepts E 3594 as per the newly agreed local crossing of E 3591 and E 3594 at Warngau, of which the Warngau controller is obviously blissfully ignorant. But now it gets worse.
Line 4 is a departure time the Warngau controller with his E 3591 mentions, newly set at 18:33 as the Schaftlach controller knows but the Warngau controller in all likelihood does not. What did he want to say here? '30, ermm, 29' perhaps? In German, even in the Bavarian dialect, three is *drei* and thirty is *dreissig* (see the underlined 18:*33* altered departure time for train E 3591 earlier), which could then be interpreted by the Schaftlach signaller as '33, ermm, 29 minutes past 6'. Is he mentioning the time at which he intends to let his E 3591 depart to Schaftlach? A lame offering of that train again for acceptance by Schaftlach? Had he actually heard the other man say '3594', or even appreciated the fact that there was a train coming his way?
Line 5 shows how the Schaftlach controller completely ignores this glaring breach of protocol; he clearly interprets the figures mentioned as the agreed departure time for E 3594 in minutes past the hour from Schaftlach to Warngau. This was, in fact, close to the agreed departure time in the station working books for both trains, something both should have realised.

Line 6 is the Warngau controller most probably saying 'Yes' to 29, meaning that that is correct for the train I am offering you, but in fact he uses the final term with which an offered train is accepted. The Schaftlach controller, in a way correctly but in the light of the profound confusion created absurdly ignorantly, interprets this as permission to let E 3594 come to Warngau from 18:29. The trap has been set, and at 18:29/30 both trains are dispatched slightly late towards each other along the single line.

About 4 minutes later the traffic controller at Warngau called his colleague at Schaftlach again, enquiring about a strange, heavy, collision-like sound he had just heard that clearly disturbed him. This was the moment when both men became aware of exactly what they had done. At 18:32 both trains had collided at a level crossing not far from Warngau.

Train E 3591 with locomotive 218 238-4 was climbing quite sharply along a straight line through open agricultural land from Warngau with its seven coaches of about 35 tonnes apiece, doing approximately 75 km/h (46 mph) and despite an emergency brake application it collided at an estimated speed of around 65 km/h (40 mph). Its driver suddenly saw 218 243-4 with its six coaches on E 3594 coming out of the woods round a curve, running downhill at about 92 km/h (57 mph) before colliding at an estimated 88 km/h (55 mph) following its emergency brake application. The theoretical impact speed was 152 km/h (95 mph) but that might have been substantially higher as it is not certain to what extent the trains had actually been slowed down. Both drivers had only about 3 seconds to do something to reduce the impact, and the consequences were exceptionally severe. Both diesel locomotives stayed on the track but each was reduced by a third of its length. The first coach of E 3594 simply jumped over its locomotive to the left in a tangent away from the curve it had just negotiated and landed in a field next to the line. The second coach tried to follow but the energy created by the collision was spent when it had climbed the roof of its locomotive, where it remained. The first mild-steel Städtewagen type coach of E 3591 stayed in line and on the track but telescoped far back into the following stainless-steel Silberling type coach, destroying two-thirds of that body up to the far-end exit door lobby whilst draping the roof of this coach over its own. The damage to the remaining vehicles, however, was comparatively light and repairable. Both train drivers and one of the conductors as well as 35 passengers lost their lives at the accident site, three people died in hospital later and 122 people were injured, some seriously. The damage to railway property was estimated at 4 million deutsche marks, and settlements with parties involved finally ran to another 3.5 million.

The timetable compiler nearly escaped retribution as he had simply used plans that had been aired in a theoretical approach to timetabling problems surrounding train traffic for the Olympic Games in München in 1972. The person who had proposed these plans, however, had changed jobs in 1974 and it could not be proved that he had ever ordered the inclusion of 'Luftkreuzungen' in the 1975 summer timetables for the Lenggries line. Or that he could have foreseen that one of his former underlings would one day use his idea in the way it was used. It was then queried to what extent the timetable compiler was qualified to do this job. This hook had sufficient barbs to get stuck and, as it turned out, a rather undereducated 'school of hard knocks' kind of person ended up with eight months in prison and a 5,000 deutsche mark fine. I wonder about the justification of netting this particular person in this way, as it basically means that he should have seen for himself that this particular work was beyond his abilities and/or mental capabilities. What, then, is the use of managers, who for years were happy enough to let him get on with the job and failed to check and query his work? (See also ***Clapham Junction***.) The traffic controller in Schaftlach received eight months' imprisonment and a fine of 3,000 deutschemarks, while the traffic controller at Warngau got 12 months in prison and was fined 5,000 deutschemarks. His sentence reflected his poor performance on the telephone as well as his failure to read the station working book as he was obliged to.

Comment

During the time that I worked in various positions away from the cab after medical disqualification to drive trains or be near the track (due to the onset of angina pectoris and loss of hearing), I came into contact with roster and timetable compilers on a number of occasions. Virtually never did I have any reason to doubt that these people knew what they were talking about. In fact they had a seriously impressive knowledge of the all-Britain timetables and knew very well how to equip and crew the trains they were responsible for in order to give the service a fair chance of running on time. Yet during the many years I worked on defending my traincrew depot against unjustified attributions of sometimes very costly delays that drivers or conductors were supposed to have caused, I did once run into the sort of theoretically blinkered attitude that was evident to a far greater degree in the Deutsche Bundesbahn people who proposed those 'Luftkreuzungen' for trains on what were rather busy non-signalled single lines. To be fair, however, the sort of thing I ran into by no means even approached the degree of acute timetable-induced violation of basic train safety that preceded the accident at Warngau.

The example I dealt with was a situation in which traincrew were booked to work a train service from their home station in the West Country to a south coast seaside destination in 135 minutes, to change ends in the platform there in 4 minutes and then work the same set back all the way to their home station in another 135 minutes. Immediately after the introduction of this working, regular and substantial delays started to occur that ended up in the lap of the traincrew depot. The train drivers and conductors involved made it clear that after 135 minutes at the controls a driver, simply and very reasonably, needed to visit a toilet. The few minutes thus added to the booked four-minute change of end impacted on other services (yes, long stretches of non-electrified single lines with quite a few request stops), so the combined delay end-figures could be eye-watering when presented to the traincrew depot the next morning, particularly given their impact on the delay-minute budget. Obviously a protest was made, but it took several months of carping before this issue was rectified, the counter-argument being that the timetable as presented was legal within agreed traincrew working-time parameters. That a driver could possibly have a normal person's physical requirements after 2¼ hours of operating a train clearly did not count as a valid argument. The union representatives who supposedly scrutinised the timetables before acceptance had clearly overlooked a thing or two as well. But there was a strong sense of the absurd in the claim, made quite seriously, that a four-minute turnaround time at a midway terminus should suffice within 270 minutes of solid work along a mostly single-line country track with a string of request stops. The dispute was settled in favour of the traincrew depot, mostly on the timetable-managerial argument that at a seaside destination the 4-minute change of end allocation would never do in the event of longer trainsets servicing the bucket-and-spade brigade on nice summer days. In those circumstances the delay minutes would end up in the lap of the timetable compilers, so the change of end time was amended to 10 minutes. Later, I noticed a quiet erosion of that change of end time again, the shortest period implemented being 6 minutes.

Having had that experience, however, my hackles rise somewhat when faced with the idea of not specifying arrival and departure times at a single-line crossing station, however useful that may be in the struggle to create operational flexibility. But, then, to seriously implement a train crossing halfway between stations on a single line with a rather dodgy train control system is close to the limit of credibility. As, for example, was the non-supervision of the operational situation at ***Quintinshill*** in 1915. The traffic controllers' failures in the Warngau disaster fall into the same category of incompetence as that of James Tinsley and George Meakin at ***Quintinshill***, or indeed Bob Sanchez's performance during the ***Chatsworth*** head-on collision in California almost a century later.

Small numbers of such people exist, often undetected for years, and most of the time the railway operations absorb their mistakes or inappropriate work ethic without anything untoward happening – as long as others keep their brain in gear and work around it. Until suddenly things do go wrong, as happened during the *Southall* collision following a 125 mph (200 km/h) SPAD on 19 September 1996. The driver of the offending London-bound express was rummaging around in his bag rather than watching the signals and track ahead and routinely listening out for a warning of the AWS in case of restrictive signal-aspects. Despite knowing that his AWS warning system was switched out his end due to a fault (he had not heard a horn or a bell all the way from Swansea via Cardiff, Bristol Parkway and Swindon to the London suburbs), he was diverting his attention away from driving his train to packing his bag, as he usually did, but now erroneously listening out for an AWS warning to indicate restricted signals ahead. (Other colleagues, as I know from first-hand experience, also did this as a matter of daily routine.) The ATP was technically available, and should have avoided this crash, but had not been in proper use for years due to lack of maintenance to save funds and was therefore also switched out. And so the distracted driver passed two distant signals at caution at full speed until he eventually noticed the red signal and the crossing freight train ahead of him, after which he made an emergency brake application. A major high-speed crash occurred, captured by a nearby CCTV camera, in which seven people died and scores were injured. The offending driver survived as good as physically unscathed; in fact, he was one of the first to report the crash to the traffic controllers, and the transcript of that call tells its own story.

Following such serious lapses in safe working practice, those erring railway people who survive such crashes consequently have to suffer hell on earth in the years that follow. What many of them go through is indescribable. Judge them if you must, but avoid being too harsh. Look honestly at your own blunders in your car and be glad that scores of other people were not dependent on your actions for their survival.

***

Schiedam-Nieuwland, The Netherlands, 4 May 1976 at approx 08:00

(Triple collision due to a 'starting against signal' SPAD)

The accident

A triple collision resulted due to an all-stations slow service to Hoek van Holland departing against a red signal (i.e. a 'starting against signal' SPAD) at Schiedam-Rotterdam West station. It collided head-on with an international express train that was travelling reversible from Vlaardingen Centrum to Schiedam-Rotterdam West to overtake an all-stations slow service from Hoek van Holland to Rotterdam. The accident, which resulted in 24 dead, five seriously injured and scores of walking wounded, was the third worst rail accident in The Netherlands after ***Harmelen*** in 1962 and ***Weesp*** in 1918.

The location

The accident occurred just outside Schiedam-Nieuwland station on the double-track electrified line from Rotterdam to Hoek van Holland between Schiedam-Rotterdam West (a junction station with the Amsterdam CS–Den Haag HS–Rotterdam CS main line, and now known as Schiedam Centrum) and Schiedam-Nieuwland stations. At the time of the accident the line was equipped with reversible colour-light signalling of the NS 1955 type, but not with ATBEG ATP.

The trains and rolling stock involved

Train D215 'Rhein Express' from Hoek van Holland to München, made up of standard first- and second-generation UIC-X types of German (DB) long-distance passenger rolling stock, was hauled by Alsthom-built Co'Co' electric locomotive No 1311 weighing 111 tonnes. These machines were similar to the French SNCF type CC7101-7158, of which CC7107 was involved in the high-speed tests on 28 March 1955 when the world speed record of 340 km/h (210 mph) was attained between Lamothe and Morcenx on the line from Bordeaux to Dax. Only much later would that record be broken by similar test runs with French TGV sets on their high-speed tracks. In fact, No 1311 was specially taken from the production line of the French locomotives and finished to Netherlands Railways specification following a rear-end collision at *Weesp* resulting from a signaller's misuse of interlock override equipment on 19 June 1953, which left brand-spanking-new No 1303 beyond repair. At present these locomotives have all been taken out of service, most having been scrapped.

Train 4125 from Hoek van Holland to Rotterdam comprised two 2-car Plan V-type 2'Bo'+Bo'2' EMUs, built in great numbers from the 1960s until the 1980s following the pattern of the Plan T 4-car 'EMU of the future' type. These rapidly accelerating and fast units were the staple rolling stock for all-stations slow services throughout The Netherlands for a considerable time and the last ones were taken out of use in April 2016.

Train 4116 from Rotterdam to Hoek van Holland was made up of two 2-car SGM 'Sprinter'-type stock Bo'Bo'+Bo'Bo' EMUs. This type of EMU was conceived around 1970, based on the pattern of contemporary all-wheel driven S-Bahn suburban stock around München, as a high-density set for maximum acceleration between closely spaced stations in the Randstad area of The Netherlands, hence the relatively low top speed of 125 km/h (75 mph). Many sets later received an intermediate trailer vehicle, altering the axle-reading to Bo'Bo'+2'2'+Bo'Bo' (which hardly influenced their original performance) and all except those lost in accidents are still in full daily service, having had a thorough overhaul, interior refurbishment and traction equipment upgrade (IGBT dc chopper equipment) in Denmark. Replacement units are coming on stream from 2013 onwards.

Just before the accident the 'Sprinter' set of train 4116 was stopped in the platform at Schiedam-Rotterdam West, its starting signal showing a red aspect. The senior conductor had a problem with a passenger with heavy luggage who got off, on and then off again and then stood in the doorway, clearly unsure as to what to do, holding up the train. The all-stations slow 4125 coming in from Hoek van Holland had passed a yellow aspect and was slowly creeping up to a red signal outside the station along its proper right-hand track, because it was waiting to be overtaken by the delayed D215 'Rhein Express' along the left-hand track before entering the station at Schiedam-Rotterdam West. D215 was a few minutes late and was running reversible as booked along the left-hand track from Vlaardingen Centrum to Schiedam-Rotterdam West, doing 40 km/h (25 mph) as indicated by its signals, to pass train 4125 and then cross over to its proper right-hand track. It was one of the then still existing breed of 'Boat Trains' and connected with ferries from Harwich Parkeston Quay to Hoek van Holland for onward travel into central Europe. As the Hoek van Holland line essentially was an all-EMU all-stations urban line with a half-hourly departure travel pattern throughout the day, it was not really suited to having specials like express or freight trains fitted into the timetable. Reversible signalling, however, enabled the solution of booked overtaking manoeuvres, a situation exploited in the 1976 timetable for the first time (see also ***Warngau***). The delay to D215 explains why train 4125 had to go slow, as the overtaking move should have already taken place a few minutes earlier and its road into Schiedam-Rotterdam West should have been cleared.

The conductor of 'Sprinter' 4116 at Schiedam-Rotterdam West had dealt with the vacillating passenger and jumped back on his train, aware of the few minutes' delay on this busy line as any railwayman in those circumstances would be. In this state of preoccupation he actuated the door-closing switch with his key and the closure button, but overlooked to check both the red-aspect starting signal 60 and the white illuminated 'V' (OFF) platform indications: these were not showing. As a matter of routine he buzzed the departure signal to the driver, who in turn got the green light indicating 'door closure okay' on his desk and reacted to the buzzes by releasing his brakes and opening up. However, he also failed to check the aspect of his starting signal, a common occurrence worldwide, which in Britain is known as a 'ding-ding and away' or more officially as a 'starting against signal' (SAS) SPAD. Train 4116 did what it was built to do, smartly accelerating away in Metro-style from the platform, and when just outside Schiedam-Rotterdam West it split the trailing turnout of the crossover that was lying reversed waiting for D215 to move over to its proper track. Then, when just next to the first coach of the still slowly moving train 4125, at 07:54 train 4116 slammed head-on into slow-running locomotive No 1311 of D215 that was hidden from view behind train 4125 on the left-hand curve. The first carriage of 'Sprinter' set No 2008 wedged itself in between the heavy locomotive and the first carriage of train 4125, taking out the side of that vehicle and virtually demolishing itself, the deaths and serious injuries occurring in this coach. The Plan V set of train 4125 probably prevented the deflected 'Sprinter' from rolling down the embankment there. The driver of locomotive No 1311 managed to jump off in the nick of time before his cab was crushed; the offending driver of train 4116 survived but could only be removed, badly injured, from his crushed cab at around 10:30 after No 1311 had been re-railed and moved away.

One serious issue arising from this accident is that despite the fact that traffic control knew about the collision one minute after it occurred, traction electricity in the overhead line equipment was not isolated for a long time and, with hindsight, endangered the emergency services entering the damaged coaches. In fact, lack of certainty about the state of traction current was common in many more accidents worldwide.

Comment

A number of issues are relevant here, all of them to do with expectations and with routine, which has such a profound influence on the safety of day-to-day running of trains. Routines should really not be broken if not strictly necessary, as it can be a recipe for disaster, despite the fact that drivers and conductors must obey their timetables and signals. As an example, I remember trains from London Victoria to East Grinstead that, in 1992, all ran fast from East Croydon to Sanderstead, except for one service at around 15:45 that was booked to call additionally at South Croydon for home-going schoolchildren. I am not sure how many times those poor children had the pleasure of travelling on that train, but it cannot have been often! The routine of departure from East Croydon and the concentration on managing the train speed by applying the right amount of power at South Croydon so as not to tear up the junction but also to climb the steep incline after it without loss of speed, somehow made it devilishly hard to remember that single booked stop per day at South Croydon for virtually every driver.

In the case of the accident at Schiedam-Nieuwland similar issues apply. Note that none of what follows is meant to defend the actions of the traincrews, but merely to explain them:

1 The line was an EMU-operated stretch with a clock-face timetable. Intermediate red signal aspects were rare (except in times of trouble) and, as a result, it appears that the serious checking of signal aspects on departure was not routinely given much attention. Traincrew normally expected clear signals and based their routines on that assumption.

2 The situation of a hesitating passenger on the platform, holding up the train, occurs regularly and demands tactful but decisive handling by traincrew to avoid unnecessary delay. After such an event traincrew can be irritated, probably thinking about reasons to go in the delay incident report following a complaint from the passenger about lack of helpfulness; it is very easy to be preoccupied with such trivial things. It takes an intuitive alertness to the fact that, above all else, the job requires constant vigilance with regard to things like signals.

3 The way the working of the few remaining international expresses from Hoek van Holland to various destinations in Europe (including TEE 'Rheingold' and D237, the Nord-West Express to Berlin, Warszawa, København and Moskva at one time) and a few daytime local freight services had to be massaged in between the frequent local EMUs simply required extra

alertness from the traincrews, which not everyone can be trusted to have every day. This is where ATP comes in as the last line of defence, being completely automatic and beyond the control of the traincrew. Incidentally, from 1992, Hoek van Holland lost all its direct international daytime 'Boat Train' connections as a result of declining patronage, which was another way of solving the problem.

4 The issue of the traction electricity not having been isolated was due to someone in the stress of the situation overlooking this aspect of the set procedures and no one at the site actually checking that traction current was switched off. Everybody else assumed that the equipment was dead as the pantographs were down, but nevertheless some very dangerous situations occurred when emergency rescue people had to use ladders to enter the vehicles.

Concerning the lessons learned and actions taken, after the 1962 ***Harmelen*** accident ATP was being fitted slowly throughout The Netherlands (by 1975 only 25% of the network had been equipped). Following a few similar accidents, notably the head-on collision at the *Spangen* triangle at Blijdorp Zoo near Rotterdam on 27 December 1982 and the rear-ender at *'s-Heer Arendskerke* on 27 October 1976, both resulting from SPADs, the Railway Accident Council proposed to increase the pace at which the network was equipped with ATP, 60% of the network having been fitted by 1986. However, at Schiedam-Nieuwland, first-generation ATP (ATBEG) as originally specified would possibly not have prevented the accident from happening, as the front of train 4116, where the on-board ATP signal pick-up coils are located, passed starting signal 60 at less than 40 km/h (25 mph), although the brakes were released and therefore the brake criterion was not satisfied. As explained earlier, the system allows passing a signal at danger without intervention in those circumstances, but it would have prevented train 4116 from accelerating beyond 40 km/h, or would have indicated to the driver that he was not permitted to go faster than that speed. This might have made him stop in order to query the situation, so would at least have positively influenced the impact speed of the trains.

There was never any serious doubt in The Netherlands (as there was, and still is, in Britain) about the value of investment in ATP. Netherlands Railways had to defend the slow pace of installation of the system more than once, and after this accident regular reversible travel without ATP, seen as carrying too much risk, was temporarily abandoned until fitment of ATP. Rightly so, given the much later accidents at e.g. ***Pécrot*** and at *Halle-Buizingen in Belgium*. SPADs, however, are still a major problem in The Netherlands. On 24 September 2009 at 22:30, a serious head-on collision occurred between two freight trains in another part of the extensive Rotterdam area rail network at *Barendrecht*, due to a driver overtaken by sickness passing a red signal. ATP is blamed by many to increase unsafe driving by taking away the need to be alert, strangely enough, as they feel that it tends to lull drivers into a false sense of security. I tend to regard that in the same way as the argument that proper cabs on steam locomotives were dangerous due to their making life too comfortable for the crews. Nevertheless, this argument makes the omission of a forced stop on red with ATBEG even more worrying, and selected signals in ATBEG-monitored areas are now being urgently equipped with an additional system (ATB-vv) that enforces a stop on red. Signal 60 at Schiedam-Rotterdam West would be a clear candidate for that additional system. An additional investment in train safety was made through development and fitment of train radio. This gave an obvious additional opportunity to prevent errant trains, with an emergency stop message from traffic control centres. In fact, there are no longer any western European rail operations without any such system, although there is still some way to go to achieve full coverage on many networks.

***

Granville, New South Wales, Australia, 18 January 1977 at 08:27

(Fatal derailment causing bridge collapse)

The accident

An inbound morning commuter express from Mount Victoria in the Blue Mountains to Sydney Central station derailed on the curved Granville Junction under Bold Street overbridge near Granville station. The derailing locomotive and its first coach damaged the bridge supports, as a consequence of which shortly thereafter the steel and concrete bridge deck collapsed onto the stationary train on the track below. The rescue of the survivors and the bodies was very difficult, due to the heat, the threat of further collapse, and LPG gas (for heating on an electrically hauled train!) escaping from damaged tanks in two crushed coaches, threatening an explosion due to sparks from the huge amount of steel cutting equipment that had to be employed to rescue passengers. There were 83 deaths and 213 were injured.

Political influences

The train service operator, New South Wales Government Public Transport Corporation, had recently been subject to savage expenditure cuts that had caused an enormous backlog in the maintenance of track and trains. Cleaning trains, to mention a more ridiculous aspect of this issue, had been stopped altogether, forcing passengers to bring tissues with them to wipe the seats before sitting down. This maintenance backlog was the direct cause of the derailment, but lack of funds was also the reason why in 1977, like at ***Buttevant*** in Ireland in 1980, wooden-bodied coaches were still in daily use and that bottled gas was used to heat them despite the availability of traction electricity.

In fact, the train service involved in this accident had a political background. In order to win further votes by satisfying the obvious need for reasonably fast commuter connections into central Sydney following the run-down of rail services under the previous government, public transport provision by rail had become a hot political issue. Under the new administration, to get quick results where it counted, an express commuter service had been introduced between Sydney and the marginal (but well-off) constituencies in the beautiful and temperate Blue Mountains. Providing rolling stock for this new service had been a matter of stripping out interiors of stored and obsolete wooden main-line sleeper and seating carriages and putting in open saloon interiors. Traction had to be delivered by Class 46 Co'Co' electric freight locomotives. Heating – the Blue Mountains can be chilly – had to be provided by gas heaters on the coaches as neither the electric locomotives nor the coaches were fitted with head-end electric power equipment, fitting of which would have been too costly and time-consuming. Running this express commuter service between the normal all-stations slow services around Sydney was a matter of tight timing (see ***Amagasaki*** and ***Schiedam-Nieuwland***) and timekeeping was important in order for it to be successful. If the train was 3 minutes late entering the Sydney suburban area, stopped local services waiting for it to pass would be allowed in front of it, and that could easily cause a 30-minute late arrival. For that reason the train had received a sort of priority status, and frontline rail staff made a serious effort to ensure that it got through on time. The curve at Granville Junction had a speed limit that was 10 km/h (6 mph) higher than others on the Sydney CityRail network just to keep this train running smartly.

The location

Granville is a western suburb approximately 19 km (12 miles) by train from Sydney. The accident site was at a junction of two main suburban lines, along a four-track railway line passing under Bold Street viaduct in a curve just where the junction turnouts and crossovers are located, followed by a 20 km/h (12.5 mph) speed restriction through Granville station. Signalling consisted of controlled colour-light signals. The bridge was a rather simple but effective construction comprising two abutments at the landward sides and two piers or trellises, consisting of eight steel pillars each, between three pairs of tracks. The bridge deck, constructed of steel with a concrete top surface for the road, was laid in two landward sections from either abutment to the piers and then a central section between the piers, rather in the manner of a Stone Age clapper bridge.

Rolling stock and crew

Traction was delivered by Class 46 Co'Co' dc electric freight locomotive No 4620, a British-built Metropolitan-Vickers product dating from 1957 with a very good pedigree indeed. Similar, albeit 3 kV tension, dc traction equipment had been installed in Swiss-built electric locomotives that went to South Africa. In August 1965, this particular locomotive had been involved in a serious runaway of a freight train followed by derailment, the cause of which was given as brake failure. None of this, however, played a role in this accident. The locomotive was withdrawn from service after the accident and was scrapped in 1979 after only 22 years of service despite not being damaged beyond economical repair. The tape of the Hasler recording speedometer completely exonerated the driver from any allegations of speeding, forcing the investigators to look for another cause.

As mentioned above, the eight coaches were former main-line sleeper and saloon stock that had been converted into medium-density commuter stock with open saloons. They were built with wooden bodies on steel frames, which in this case probably did not materially add to the death toll, as one coach was ripped open by an overhead line stanchion that would have severely damaged a steel coach as well (as confirmed by several examples elsewhere in the world) and the collapsing bridge partially flattened two others. As the ***Eschede*** and *St Johns, Lewisham* accidents demonstrated, steel or modern aluminium coaches will not protect their passengers in that case either.

Driver Edward Olencewicz, aged 52, and co-driver Bill McCrossin, aged 26, both of Sydney Eveleigh depot, escaped serious injury despite their locomotive falling on its right-hand side and the driver having to be freed by bystanders. Unfortunately, rescue proceedings were hindered by thousands of onlookers who had to be removed from the tracks by 250 police officers specially drafted in for the purpose. Some people stole ambulance and rescue equipment, and one was reported to have taken property from the body of a victim. Olencewicz and McCrossin's ordeal was to continue when they and their families began to receive hate mail and offensive telephone calls and the windows at their homes were smashed. This unedifying aspect of part of Australian society was to be repeated with trucker Christiaan Scholl and some of his colleagues after the ***Kerang*** level crossing accident many years later.

The accident log

Train 108 departed Mount Victoria station in the Blue Mountains on the dot at 06:09 for the 126-km (78-mile) journey along a difficult stretch of mountain railway and on into central Sydney, with a booked arrival time of 08:32. The average speed (54.75 km/h or 34 mph) was not particularly high but that was not too important for the many commuters on the train, who were happy enough not to have to be up very early and arrive home very late after negotiating Sydney's peak-hour traffic every working day. That morning, track engineering work caused a signal check at Blacktown that resulted in arrival at Parramatta 3 minutes late, from where the train was to run non-stop to Strathfield. With 25 minutes of travel time left from there, train 108 accelerated to the 80 km/h (50 mph) permitted line speed through the suburbs on its last leg into town and was running almost right time. Approaching Granville station driver Olencewicz routinely shut off power and started to slow down for the 20 km/h (12.5 mph) speed restriction through the curves near the station, when just before Bold Street bridge the right-hand first wheel flange of the locomotive, still doing 78 km/h (48 mph), suddenly climbed the rail head and then toppled and spread the outside rail when poorly maintained fasteners gave way. This was brought about by excessively worn outside curve gauge-corner rail heads in combination with worn flanges and wheel treads on the front axle of the locomotive – a kind of wear that particularly affects locomotives with 3-axle bogies. In German rail parlance 3-axle traction bogies are known as 'Schienenfresser' (track munchers) for that reason.

The locomotive and its first two coaches derailed completely, sending No 4620 and its first coach away at a tangent from the left-curving track to the opening under Bold Street bridge some 40 m (130 ft) ahead. Driver Olencewicz had applied the emergency brake immediately but could not prevent the derailment. The couplers between the first and second coach and the second and third coach broke, whilst coaches three to eight safely came to a stop still on the track with the third and fourth coaches partly under the bridge deck of Bold Street road bridge over which traffic was still passing. The locomotive and its attached first coach, however, had first rammed the intermediate pier of the bridge and then a stanchion supporting the electric overhead line equipment. The coupler between the locomotive and coach one broke, the eight steel supports of the bridge pier failed, and the overhead line stanchion, supported from the overhead line instead of carrying it, sliced into the first coach, which came to rest virtually without its roof and left side. After approximately 10 seconds the now only partly supported bridge deck started to give way, the middle section moving off the damaged pier end and falling onto neighbouring tracks, which in turn caused the part between the damaged pier and the abutment to crash down onto the stationary third and fourth coaches of the train, crushing them down to just centimetres above the ground. It was literally a matter of suddenly seeing a massive concrete beam at your feet where before there was a train floor. A number of road vehicles came down with the collapsing bridge, but luckily none of their occupants was seriously hurt.

Although the locomotive was off the track, 67 m (220 ft) past the bridge, the driver and co-driver were barely hurt. The first coach was sliced open and badly damaged by the destroyed overhead stanchion. Eight of its 73 occupants were dead and 34 injured. The second coach was standing ahead of coach one against a retaining wall, derailed and on its own, well away from the accident site. None of its 64 occupants were killed or seriously hurt, but in coaches three and four, which in the first moments after the derailment appeared to have escaped any harm, the death and injury toll was terrible. Of the 77 occupants of coach three 44 died, and of the 64 occupants of coach four 31 died. The remainder of the coaches, standing undamaged on the track, could be evacuated.

The rescue

During the rescue effort many of the victims with crush injuries were still lucid and talking to those working hard to save them but would die later from what became known as crush syndrome. Myoglobins collected in their injured limbs whilst they awaited their turn to be freed, causing death through kidney failure later in hospital. This led to a thorough review of the way such accident victims were treated; necessary life-saving amputations are performed much sooner now, often at the accident site.

The main problem during the rescue and salvage effort was the fact that the bridge had collapsed in complete steel and concrete slabs, each weighing hundreds of tonnes. This prevented the swift removal of debris from the crushed coaches in order to reach the casualties quickly, the steel and concrete structures having to be laboriously broken up and cut to a manageable size before being lifted out and taken away. Not only did this cause severe delay in reaching the casualties but it also increased the risk of further collapse. Both casualties and the rescuers later described the ominous creaking and groaning of the bridge as it settled. Of inestimable help in preventing such danger was the presence of one volunteering passer-by, structural engineer Ron Scotch, who directed shoring-up measures to protect casualties and rescuers alike. The next problem was an ominous smell of gas that became noticeable to the rescuers with their steel cutting torches and grinders. This came from the gas heating tanks under the crushed coaches, thus bringing a further risk of explosion. The fire brigade brought in gas detectors, water showers to cool the debris, and powerful blowers to disperse the gas and bring more oxygen into the narrow hollows under the collapsed bridge where people worked away in the increasingly stifling heat as the day progressed. Many casualties were subjected to burns from the oxy-acetylene torches cutting through steel very close to them as the rescuers fought their way in to remove the collapsed deck. It was not until 10 hours after the derailment that the last living casualties were taken out, the last of the dead being found in the early afternoon of the following day.

As is usual, unfortunately, the world over, it took an accident of this nature to galvanise the authorities into improving the conditions under which trains were operated. Spending on rail operations increased substantially. The negative aspects of increased car traffic in cities also spurred governments to start spending seriously on the upgrade and renewal of the railways as an alternative means of mass transport. Reorganisations to improve track, signalling and rolling stock were set in motion, similar to what happened much later in Britain after the *Hatfield* derailment following the demise of the privatised and 'commercially' operating rail infrastructure provider Railtrack. Bold Street bridge was rebuilt as a single structure without intermediate piers, as would become the norm for such structures across railway lines. Other viaducts of similar construction to the former Bold Street bridge were thoroughly reinforced. The inherent weakness of such a construction had been known about, but the government of the day had not wanted to pay for the redesign and construction work. Incidentally, had people in Germany known about this part of the investigation, perhaps the ***Eschede*** crash would never have happened in the way it did. The viaduct that was brought down there was a relatively recent structure.

A few weeks before his death from bowel cancer in 1996, the coroner who chaired the inquiry into this accident, Tom Weir, phoned a producer making a docudrama about this accident. In this recorded phone call and in later letters, angry about having been gagged by the NSW government and wanting to tell his side of the story before it was too late, he gave details concerning the key players, their conspiracies and their cover-up attempts to divert political and direct governmental responsibility for the accident during the investigations and the inquiry. He did not live to see the result on screen, but his claims were substantiated in a subsequent government report published decades later (by which time quite a number of those involved had died) and there can be no doubt that the report would not have been as frank as it was had it not been for what Mr Weir brought to light. Once again, it had taken a transport disaster to expose failing and corrupt politicians, for which reason it changed what many Australians felt about the way they were governed and what they wanted from their politicians.

Comment

Like the accidents at ***Brühl*** and ***Eschede*** in Germany, *Hatfield* and ***Clapham Junction*** in Britain and ***Buttevant*** in Ireland, this crash illustrates what a former chairman of British Railways, Sir Peter Parker, meant when he coined that famous expression 'the crumbling edge of quality' in relation to what can be expected from underfunded rail transport. This crumbling edge starts to show and cause mishap when those responsible for funding public transport (such as governments) choose not to provide sufficient finances for maintenance and modernisation, a problem that was common in notably English-speaking nations during the 1970s and 1980s. As a result, serious accidents occurred.

As in the ***Eschede*** and *St Johns* crashes, the Granville derailment caused an overbridge to be brought down on the train, which changed the way bridges across tracks were constructed. Medical research and consequent discoveries concerning injury recognition and treatment led to material improvements in casualty rescue and support. It can, therefore, be said that the victims of the Granville crash did not die for nothing, even if they should not have died at all. Australia did alter its approach to funding public transport for the better after the accident. But as with similar crashes that have political/ideological issues at their root elsewhere in the world, and despite the positive end results, just why was it necessary to have home truths about public transport safety demonstrated at such a cost to those we elect to rule the nations we live in?

***

Wijchen, The Netherlands, 28 August 1979 at approx 18.10
(Head-on crash caused by fitter's errors, equipment failures and driver error)

The accident
This was a head-on crash between two EMU sets, primarily due to one of those erroneously receiving a green signal aspect whilst turnouts ahead were in an incorrect position. It was the result of a chain of events involving a rare combination of two unnoticed insulation faults in signalling equipment, as well as omissions in the working procedures of an electrician. It caused a wrong-side failure of the colour-light signalling. Other contributory factors were a traction fault on a train and ignorance of new signalling practices with consequent procedural mistakes on the part of one of the drivers involved. Six of the occupants of train 4365, including one of the drivers, died at the accident site, whilst one of the seriously injured on this train later passed away in hospital. Less seriously injured were 29 passengers of train 4365, the driver of ECS 74363 and the two senior conductors.

The location
The collision occurred near Nijmegen station on the electrified double-track main line from Nijmegen to Den Bosch in the south-east of The Netherlands. At the time, this line was in transition from being mechanically signalled with Siemens & Halske equipment dating back to 1929 (although this button-controlled system was first introduced in The Netherlands in 1905) to the modern 1954 type colour-light signalling system with CTC and ATP and simplified reversible operation. This equipment was scheduled to be operational in May 1981.

Rolling stock involved
Four-car EMU No 936 of the 1960s-vintage Plan T 4-car and Plan V 2-car types, a 4-car being of 2'2'+Bo'Bo'+Bo'Bo'+2'2' configuration. The set was a representative of the quick accelerating and braking EMUs with electro-pneumatic brakes that dominated all-stations slow operations from the middle 1960s until 2005. The 2-car units were in operation until mid-April 2016. This set had come into Wijchen station from Nijmegen, to reverse there and run back ECS to Nijmegen as train 74363.

Also involved were two Mat '46 2-car units Nos 256 and 244, the infamously lively riding EMUs of post-war Bo'2'Bo' articulated configuration (see ***Harmelen 1***). The driver in these trains sat in the narrow pointed cab in a very vulnerable position in the event of any type of collision, whether with trains or on level crossings. Due to increasing ridership on rail services in The Netherlands they bowed out from service as late as 1993, after a working life of almost half a century. Not because they were unable to fulfil the transport demands put on them through unreliability, but because they were obsolete in terms of comfort and safety.

Causes of the accident

A signalling electrician was working on the equipment at Wijchen in the afternoon of the day in question in preparation for installation of CTC and ATP equipment. He had to add two wires into the electro-mechanical signalling installation, one of them a 34 V feed to the signal relay of home signal D on the Nijmegen side of Wijchen. The job involved altering the configuration of the contacts, for which reason he had to take the relay apart. There were just 20 minutes available between the passing of trains to do this job, so there was clear time pressure and he did not notice that one of the insulator plates between the relay casing and the frame in which it was mounted was absent. Moreover, for the bolts of the contacts he was required to use a special screwdriver, which he had forgotten to bring, and the correct tightening of the contacts was impossible with the normal 3 mm screwdriver he used. He failed to notice that as well. These two omissions meant that the 34 V feed could now intermittently make the steel mounting frame live and so was potentially dangerous.

During the post-accident investigation it was discovered that on only one such relay out of 10,000 investigated examples this same insulation plate was missing. This, however, would not have caused the accident were it not for the fact that a similar insulation was missing in a completely different and unrelated part of the installation. Due to a manufacturing mistake, the core of the proving magnet in the relays for turnouts 19A and 19B (the set of turnouts train 74363 was to use when crossing over to the track on its right-hand side) had not been fitted with triple insulation but with just a single layer. Unnoticed because of maintenance methods employed, this inadequate insulation had also been destroyed through deterioration with age, but it had never caused a malfunction as the steel frame in which it was housed was insulated from the earth as well as the feed, a situation that had now changed due to the missing insulation plate and, most of all, the loose contact at the turnout relay.

As part of the subsequent investigations, Netherlands Railways checked all other such equipment and in 2% of them (about 350 relays) similar omissions were found. Together, both non-related insulation faults – and then only after the interference by the signalling electrician at Wijchen – meant that the designed interlocking fail-safe measures were intermittently nullified. The turnout-position proving magnet is normally powered only when the position of the activating button that works the turnouts and the lie of the actual turnouts are in synch with another. The situation described above, however, enabled that relay to be powered up by a false feed coming through the steel relay-housing frame from another relay, which was completely unrelated and equally poorly manufactured. This in turn allowed the button to be turned but the turnouts to remain as they were, whilst permitting the accompanying signal to come off to a proceed aspect. This was a serious wrong-side failure indeed, but it took a lot more than that to cause an accident.

The accident log

At 16:00 the signalling fitter went home and almost two hours later at 17:55 the Plan T 4-car EMU arrived from Nijmegen as train 4363 at Platform 2, the driver changing ends to work back as ECS to Nijmegen. This was the first occasion after the work of the electrician earlier in the afternoon that turnouts 19A and 19B of the crossover had to be reversed to allow the set to gain the correct right-hand track. Due to the faults described above, the turnouts failed to move, yet the signal came off to green at 18:00 and ECS train 74363 was dispatched and departed. At the same moment, however, the driver noticed that he had a problem with the traction control equipment in this cab, which made it difficult to keep to the indicated 40 km/h (25 mph) when passing through the crossovers. Due to his preoccupation with this he initially did not notice that he had passed the turnouts without being taken onto the right-hand track. He should now immediately have stopped and contacted the Wijchen signaller. He did not, despite knowing that for wrong-line travel a wrong-line running authorisation was required.

This driver had never before worked under the authority of a wrong-line permit, but he had already used the elsewhere increasingly more widely installed reversible signalling that could be used without a wrong-line permit (see ***Pécrot***). He was also aware that this new type of signalling was being installed on this line, although no information to the effect that the new signalling had been brought into use had been published. Implicitly trusting the integrity of the signalling, but acting against several explicit instructions concerning this situation as contained in the rulebook, he continued in the expectation that the traffic controllers at Wijchen and Nijmegen knew that he was travelling wrong line. At an AHB level crossing he brought his train to a stop and let the barriers close, after which he passed it slowly as per the rules (a procedure that Stanley Hall and I experienced in a cab in The Netherlands during a signalling fault, known as 'aanrijden' of an AHB; see Chapter 11 in the book *Level Crossings*). The senior conductor, who was returning to Nijmegen on this ECS, happened to come into the cab and remarked on the fact that they were travelling wrong line. The driver told him that Wijchen must have done this deliberately, in answer to which the conductor shrugged his shoulders and walked back into the train. However, the driver finally started to have doubts. Although the bridge controller's cabin at the drawbridge across the Maas–Waal Canal was unstaffed, at Nijmegen's Dukenburg station he could have stopped and made the call, but would have had to look up the telephone numbers of either Nijmegen or Wijchen signal boxes, so he continued. Then, shortly before reaching Nijmegen station, he saw train 4365 approaching him on the track he occupied and ran back into his train, thereby saving his life. The head-on collision impact speed was 98 km/h (60 mph), the ECS doing 8 km/h (5 mph) and the local passenger service about 90 km/h (55 mph).

The lessons learned

Fail-safe measures usually cover just one equipment fault at a time, but the situation in this case breached that maxim three times over. The faults in the traffic control equipment that caused the signal to come off against fail-safe principles had three causes: two relating to missing insulation in non-related pieces of electro-mechanical equipment, and the third to a rather less than thorough signalling engineer. At Wijchen two undetected faults had been present for a considerable time, so the chances of a third fault breaching the fail-safe wall were high. A signalling engineer, causing an intermittent false feed through the steel relay-housing frame, was all it took. Following the post-accident investigations Netherlands Railways (NS) was told to include checking for earth faults and insulation faults in its maintenance procedures of signalling equipment and also to increase the thoroughness of its checking of newly delivered safety-critical signalling equipment with a view to detecting missing insulation and earth faults. As a result, work on changes to relays was transferred to central workshops in Utrecht, and local electrical engineers were only permitted to take out the old relay and install an already changed and fully tested new one. It was also ruled that if a signalling engineer found that he could not perform his work in the prescribed manner, that he should not start the work, or should reinstate the circumstances he found before starting the work, and then urgently report the situation.

As far as drivers operating wrong line was concerned, a simple, cheap and effective measure took care of that problem. All lines where reversible signalling had not yet been installed and made operational received VS (Verkeerd Spoor, wrong line) signs just past the set of turnouts where the train had to change over to its normal right-hand track. These signs were to be passed only with a specifically written and train-service-related VS wrong-line permit. Of course, these signs would be removed as soon as the line was converted to operation with reversible (bi-directional) signalling, and are indeed rare now. Finally, in order not to have to rely on the availability of lineside telephones when setting up emergency telephone calls between drivers and traffic controllers, a radio telephone system called Telerail was introduced with a modicum of urgency. In fact, testing had started in 1976 but the board of NS had had doubts about the cost of its introduction. The accident at ***Schiedam-Nieuwland***, however, had once again shown that a modern and very busy rail network could no longer do without such a system for reasons of safety, yet in January 1979 introduction had once again been postponed for budget reasons. It was this accident between Wijchen and Nijmegen that finally pushed its introduction through, and from the night of 17/18 September 1988, Telerail radio telephone traffic was available on the entire network in The Netherlands.

Comment

A number of issues played a role in this accident. To start with, a safety-critical signalling engineer who fails to bring all the necessary tools for a job yet nevertheless continues under hurried circumstances is a recipe for disaster. In this case the situation was worsened by two unknown manufacturing errors, aggravated by incomplete maintenance procedures, which together defeated the proven safety inherent in signalling systems. In Britain the implementation of similar renewals at ***Clapham Junction*** nine years later also showed how bad work on the part of signalling engineers under time pressure can lead to an accident.

But the unfortunate chain of events at Wijchen continued. It involved a fairly new driver wrong-footed by both an erroneous green signal aspect and a simultaneous traction fault on his train, compounded by his uncertainty about what to do to with the resulting deviation from routine. Against all basic good practice he continued his journey, drawing dangerously incorrect conclusions due to his lack of essential insight and on his addled assumptions. This latter issue in particular points to the importance of regular (continuous or biannual) re-examinations and regular traincrew safety meetings, during which changes in rules and regulations as well as new traction and infrastructure/signalling work are briefed and discussed. From my experiences during and after my work in the cab I can understand the reasons why this driver dealt with the situation in the way he did, yet cannot see why he failed to adhere scrupulously to safe operating rules and react more positively to his intuitive misgivings. The usual timekeeping issues no doubt went through his head, but that he was doing something out of the ordinary and with potentially dangerous consequences must have been clear to him from the start. Having to stop at an AHB to allow the barriers to close, together with the reaction of the senior conductor, should have made it obvious that he was not under control of a properly working system of bi-directional signalling. Concern with traction control problems on his train prevented him from noticing the turnouts not lying in reverse, a classic example of the negative influence of unreliable equipment and distraction on safety. Had he experienced that situation before passing those turnouts, he would probably have stopped and called Wijchen signalbox.

Finally, the simple but clear expedient of indicating all lines where no signalled wrong-line working was possible with a VS board is to be commended. If all train drivers are made aware by means of the rulebook that a VS wrong-line permit must be carried on the train in order to proceed past such a board, no one has to make any assessment as to whether or not reversible traffic is allowed. Not even with a signal erroneously showing a proceed aspect. That is a useful and cost-effective safety measure indeed.

***

Mississauga, Ontario, Canada, 10 November 1979 at 23:53
(Tank wagon derailment with fire and chemical spill)

Introduction
This infamous derailment, followed by explosions and fire on the train, was due to the disintegration of an axle under a tank wagon in a train that contained 11 wagons with liquid propane gas, three with toluene,

three with styrene and one with chloric acid. The defective wagon derailed as the train was coming into Toronto from Windsor, Ontario (the Canadian city across the Detroit River from Detroit, Michigan, linked by interesting train ferries as well as a rail tunnel). The cause of the axle failure was the seizure of a badly maintained axlebox after running hot due to lack of grease. The direct damage caused amounted to Can$10 million.

The accident

A Canadian Pacific freight train had departed Windsor in the late morning of 10 November 1979 and stopped 1½ hours later in sidings, where tank wagons containing the propane, toluene, chlorine and styrene were added to the consist that from there ran to 106 wagons. Around 18:00 the train departed behind three locomotives, travelling eastwards in the direction of London, Ontario. On passing Milton, about 145 km (90 miles) from London, residents reported seeing sparks and smoke or dust coming from somewhere in the middle of the train and 32 km (20 miles) further down the line, a hotbox on the thirty-third vehicle burned off the axle tap in the bearing at a speed of 80 km/h (50 mph). The now non-supported bogie frame of this wagon was then dragged on a further 2 miles, bouncing on the track, until derailment occurred when slowly running near the junction of Mavis Road and Dundas Street West in Mississauga, a Toronto suburb. At a level crossing the car of a Mr. and Mrs. Dabor was hit by ballast thrown up by the passing train with strangely swaying and rocking cars that started to tilt towards them. Mr Dabor hastily reversed his car away from the crossing, but whilst making a panicked rearward Y-turn manoeuvre he ended up backing his car into a ditch. At that moment one of the derailing tank cars erupted into a fireball and the Dabors got out of their stranded vehicle in panic and ran away from that hell as fast as they could. They had witnessed the dragging and rocking derailed vehicle pulling 24 wagons, among them the tank vehicles loaded with gas and chemicals, out of the track.

After the derailment and fire at 23:53 the alarm was raised immediately. When the fire initially appeared to decrease in intensity the Dabors returned to their car to attempt to get it out of the ditch but were stopped by a police officer who told them to get out of the way. Reluctantly they started to walk away just as all hell again broke loose behind them, a massive blast knocking them into the ditch. At about this time the train driver and his second man, a father and son-in-law team, had discussed how to adhere to the rules about uncoupling the locos. The driver told his son-in-law that if he wanted to attempt to save as much of the train as possible that was fine by him and so, in a fit of supreme heroism or astounding lunacy, the younger man ran back along the train to find the first derailed car, managed to split the brake pipe, release the rear Buckeye coupler of the thirty-second vehicle and signal to his father in law on the locomotives to pull away. The young man rode away with the escaping portion of the train, which included several tank vehicles loaded with propane, when just after midnight another explosion occurred amidst the vehicles left behind. That blast tossed one derailed propane tank wagon 640 m (700 yards) away and opened up the chlorine tank wagon, causing chloric acid to be released.

The fire brigade and emergency services had arrived at the scene within minutes of the derailment and whilst they were attempting to find the train manifest in order to assess exactly what they had on their hands, the second and a third explosions occurred that were loud enough to be heard 48 km (30 miles) away. By around 01:30 the manifest had been found, and on checking the blazing train from some distance, a pool of 90 tonnes of liquid chloric acid was discovered under the burning wreckage of the chlorine tank car. Initially, 3,500 nearby residents were quickly evacuated from their homes, but as the situation continued to deteriorate, the local authorities ordered the evacuation of 218,000 of the 284,000 residents of Mississauga the following day. The fire was eventually brought under control on 13 November, when 144,000 of the evacuated citizens were allowed to go home, but it was not until three days later that the emergency was finally called off. The chlorine had evaporated in the massive updraught of the raging fire and had come down to earth in irritating but not dangerous quantities, people complaining about sore eyes and throats up to a mile away from the scene of the disaster. It remains the biggest evacuation to date in North America, even surpassing the evacuation in New Orleans after the well-publicised devastation caused by hurricane Katrina in 2005.

The lessons learned

Major changes were implemented with regard to the maintenance of railway vehicles and the requirements for detailed documentation. To start with, vehicles on plain oil or grease bearings were phased out with increasing urgency, or had their journals replaced with roller bearings. Dangerous chemicals and explosive cargo were no longer permitted to travel together on one train, and tank vehicles were required to be of sturdier construction capable of withstanding such things as impact from the coupler of a ramming vehicle. Train manifests had to be made much clearer and were to be made available to the appropriate transport safety authorities, not only for accident rescue purposes but also for use in the course of revision of regulations with respect to rail transport of dangerous goods through urban zones. Other measures arising from the accident included the introduction by the American Chemical Council of an emergency service with a 24-hour telephone reporting and support centre, a requirement for a regularly revised and updated emergency response manual to be stowed in each and every emergency vehicle, and railway staff the world over receiving HazChem training.

It might be supposed that, after all this, accidents involving chemical rail transport would be virtually a thing of the past, especially in Europe with its spreading ATP coverage – technical failures excepted. However, a disaster of similar magnitude occurred in July 2013 at *Lac-Mégantic* in Canada, in which a 10,000-tonne crude oil train derailed and caught fire, killing 45 of the inhabitants of Lac-Mégantic village. Another and quite similar derailment, again involving a freight train transporting both chlorine and flammables, had taken place at *Wetteren* in Belgium two months earlier, during which one person died. The Dutch train driver, working for a German operator, was travelling at 80 km/h (50 mph) under signal aspects that restricted the maximum permitted speed to 40 km/h (25 mph) and derailed on a turnout. Explosions, fire and the spread of dangerous chemicals – a mini-replay of the Mississauga accident – were the result. Many people were evacuated for a number of days, but only one fatality occurred as a result of breathing in chlorine gas. The driver missed a restrictive signal before a set of crossovers, but where was the ATP to stop him and his dangerous transport, and why had the findings about combining flammables and toxic chemicals on one train after the Mississauga accident been ignored? On the other hand, the transportation of dangerous goods by rail appears to come under much greater public scrutiny than transport of risky cargoes by road. Of course, such cargoes are good revenue-earners for the manufacturers, shippers and users, and although the idea that they should be banned is preposterous for commercial as well as social reasons (taxes levied!), they are regarded as dangerous for very good reasons. In any event, there appears to be some considerable way to go before we can say that the transport of such dangerous goods by train can be guaranteed as safe. That is a disappointing state of affairs.

***

Winsum, The Netherlands, 25 July 1980 at approx 07:30
(Head-on crash between two diesel units)

The accident
This accident took place under the control of a radio-based traffic control system that used to issue travel permissions. As such it brings non-signal dispatcher and radio-warrant systems as used elsewhere in the world to mind. In the early 1960s the railway in The Netherlands had become loss making due to the steep post-war rise of travel by car and, as a result, economies in the operation of rail traffic had to be found urgently. Use of radio warrants to safely control the comparatively sparse traffic along loss-making rural lines looked promising, the unpalatable alternative being closure as soon as renewal of track and signalling became necessary.

So it was that radio train traffic control was introduced in the north of The Netherlands along this rural line in 1967. Imagine using radio technology to replace signalboxes, manually worked lineside signals and turnouts, many kilometres of cabling plus all the staff needed to work and maintain those items and the scale of potential savings on such operations becomes apparent. The Dutch system was called Vereenvoudigde Centrale Radio Verkeers Leiding (VCRVL) or simplified centralised radio traffic control. As no further signalling safety was provided, the maximum permitted line speed was reduced to 80 km/h (50 mph), the reason for which is not readily obvious as collision at that speed is still is likely to be lethal.

The traffic control system
From a signalling point of view, VCRVL, a typical contemporary radio dispatch system, was pretty similar to the German 'Zugleitverfahren' control, whereby the path through the unsignalled single-track section was reserved and taken in possession for a particular service through telecommunications contact between a traffic controller and the train drivers. It was a radio dispatch system quite similar to that used on many networks with long and comparatively infrequently used single-track sections in the Americas, Australia and New Zealand. The telecommunications side was provided by military and aeronautic radio technology that had made great technical strides (solid-state frequency control technology, the transistor) during World War 2 and the various conflicts that followed it. Its use was based on drivers reporting in on

arrival at certain designated control points with obligatory stops as having cleared the section they had just traversed and then requesting permission from the traffic controller to travel onwards into the next section. In the Dutch and German systems the traffic controller kept a manual record of his dealings on a train traffic graph. He would draw a red line to indicate occupation under the section if a particular service had obtained permission to enter, and when its driver reported he had left the section he would draw a green line under it to indicate clearance. Both the driver and the traffic controller were obliged to keep a time-stamped handwritten record of the radio exchanges concerning the running of the trains.

These radio exchanges normally had a prescribed set of formulations that had to be used during the communication in order to avoid the chance of any misunderstanding. In such systems the controller had no way of seeing where the trains were that he managed, having to rely completely on his graphic record, the word of the drivers, and their adherence to his instructions. This type of control, therefore, was that of a typical protocol system, absolutely devoid of any mechanical or electronic registration, interlock or remote traffic monitoring and therefore relying on exact and unstintingly careful execution of operating procedures by all concerned. Furthermore, it was quite labour-intensive for the few people involved with it, which has been shown to induce tendencies in some to cut corners with regard to exactitude, especially when the job is extremely boring. At the time, however, the latter was not so much a problem in The Netherlands, as all traincrew still worked all types of traffic with the national state railway operator and worked this line fairly irregularly.

All train traffic information available for scrutiny after an accident, therefore, consisted of the train wrecks and the clues they held, the handwritten notes in the book in the cab, and the handwritten books and graphs in the control centre. These were assumed to reflect the actual sequence of events leading up to the incident, something that more than once turned out not to be the case, to be discovered when checking neighbouring books and graphs or later the OTDR downloads. As a touch of modern control technology at the time, however, there was a tape recorder attached to the radio telephone system that recorded the calls but did not time-stamp them. The radio telephone system itself was by design a simple open channel system; the traffic controller and all train drivers could listen to and break in on all exchanges. This was done in order to keep everyone aware of what was being arranged and of what trains were in what section at any time. Technically the single-track railway line had to be adjusted by providing passing loops with sprung, self-restoring turnouts either side at the requisite distances, which would act similarly to the traffic control points for passing trains as in the *Zanthus*, *East Robinson* and ***Chatsworth*** accidents. These sprung turnouts always diverted the arriving trains from the single line into the same loop when negotiated in the facing direction but were opened by the wheels of the departing train from the other track in the trailing direction, after which the turnouts would automatically restore themselves to their design position. The only signal used for such turnouts was an indicator to show the driver of the arriving service that the turnout was properly set. Furthermore, on some networks such as in Germany, there were markers that had to be passed completely by the entire train before arrival was announced, to ensure that no part of that train was fouling the single-line track. Intermediate stations along the line were of no consequence to the traffic controller – a train was in 'dark territory' as soon as it had left a passing loop. Of course, one of the most negative aspects of this low-technology type of railway traffic control was that the controller could not spot illegitimate departures from control points, with their consequent dangerous occupation of a section ahead. He was not enabled to stop trains once they were on a collision course, therefore, which is the main drawback of such an unsophisticated version of this system. For enhanced safety, the control points should at least have train detection equipment with remote track occupation reporting facilities in the control centre to make this traffic control system acceptable. That was the simplification in this Dutch system.

The line

Whilst The Netherlands has a superbly modern and heavily used electrified rail network, there are nevertheless a few pockets of single-track, non-electrified passenger lines. A small network of such minor lines, called the Northern Diesel Lines, exists in a remote corner of the country in the North of the provinces of Friesland and Groningen, the areas bordering the Waddenzee (a large North Sea shoals area) and the estuary of the River Eems (Ems in German). Here, quite uncharacteristically for The Netherlands, silence can still be found if the wind is not blowing too hard. There are small, lonely villages in a wide-open flat and agricultural countryside, where people are unlikely to find work locally other than on farms, in village shops or in the watersports industry. British readers should think of East Anglia, such as the fenlands east of Peterborough and the Norfolk Broads, for a comparison. For that reason, some of the single-track rural heavy rail lines were kept in operation (despite many threats of closure) because they were the only effective connections between this rather remote land and the main regional centres, the cities of Groningen and Leeuwarden in Friesland. These lines enabled commuting to work for villagers under all circumstances. Incidentally, this region also contains the ports of Harlingen, Delfzijl and Eemshaven. The latter occasionally generates a fair amount of rail freight traffic (from an aluminium factory) and are also now the locations where no fewer than five large coal-fired power plants are being planned that will no doubt relieve the quiet of the flat agricultural landscape and will also be followed by manufacturing facilities for plaster and gypsum products made from the burnt fuel waste material from the power plants.

The line involved in this accident runs from Groningen to Roodeschool, the northernmost town in The Netherlands with an operational passenger railway station. This line leaves Groningen westwards together with the line to Leeuwarden, but whilst still within the city boundary it branches off to the North, crosses the Starkenborgh Canal on a drawbridge and then travels virtually straight to Sauwerd Junction where the single-track line to Delfzijl branches off to the right. Our line continues to Winsum and Baflo, from where it turns eastwards and makes its way via Warffum, Usquert, Uithuizen and Uithuizermeeden to Roodeschool. Just before Roodeschool a freight-only line branches off to go a few kilometres further northwards to the deepwater port at Eemshaven.

Rolling stock involved

The incident involved three of the venerable, indestructible, noisy and not very fast Plan X DEII diesel-electric multiple-units from the mid-1950s, built by Allan at Rotterdam. They were Bo'2'Bo' articulated 2-car units with originally two British-manufactured AEC A220 170 kW (230 hp) diesel engines for a maximum permitted speed of 120 km/h (75 mph). Inbound train 8726 from Roodeschool to Groningen consisted of coupled units 87 and 89 and was well patronised with commuters on their way to work in town. Outbound train 8713 from Groningen to Roodeschool consisted of single unit 84 and was lightly loaded. This type of unit is no longer in service.

The accident log

On the morning of the accident the flat countryside was shrouded in dense fog. At 07:00 train 8726, its four carriages well loaded with commuters on their way to Groningen, had received permission to travel from Uithuizermeeden loop to Winsum loop and would call at several halts on its way, Baflo being the last one before reaching Winsum. The driver did not have to call the traffic controller from any of these intermediate stops, his first call being when he reached the Winsum loop and had cleared this section of single line. It was at Winsum that he was scheduled to meet outbound train 8713.

Train 8713 from Groningen, however, was running late. The booked conductor had failed to report for duty and a replacement had to be organised by stepping up a conductor from another service. This had caused a 7-minute delay at Groningen. Having reached the Sauwerd loop, at 07:25 this outbound service 8713 obtained the dispatcher's permission to travel on to the Winsum loop. The driver of inbound 8726 must have heard this radio exchange and by rights should have known that the single line from Winsum to Sauwerd would be occupied until the arrival of train outbound 8713 at Winsum. Its late arrival would delay train 8726 for a few minutes in the fog at Winsum, but that was nothing out of the ordinary and is a typical operating characteristic on a single line.

So, train 8726 with its city-bound commuters arrived at Winsum and reported clearance of the section it had just vacated. The driver then requested permission to depart, but this could not be granted until the arrival of late-running train 8713. Nevertheless, at the booked departure time the train conductor saw the driver gesturing through the fog to him that they were okay to depart and assumed that the driver had received permission to move into the section to Sauwerd. Consequently, the conductor dispatched his train and it noisily departed from Winsum station, its two small red tail-lights above the windscreen of the rear cab disappearing into the fog. It was the equivalent of a 'starting against signal' SPAD had there been a signal.

Given that, as explained earlier, the system was not equipped to report an erroneously departing train to the traffic controller, the radio was not used to stop both trains with an emergency call. As a result, trains 8726 and 8713 met at speed in the dense fog at km 3.050. Witnesses described how both drivers used their horns for a local open level crossing and then slammed head-on into each other at a combined speed of 160 km/h (100 mph). The leading vehicle of unit 84 penetrated deep into the leading vehicle of the opposite train, the two being securely locked into each other, which made rescue efforts very difficult indeed. Nine people died (including both drivers) and 21 people were seriously injured. Unfortunately, having sustained severe injuries during the collision, the surviving conductor was later unable to recall whether he had been aware of not having witnessed the booked arrival of train 8713 on the second track at Winsum.

The following year, after 13 years of accident-free operation of this system prior to the collision, VCRVL was replaced with remotely controlled colour-light signalling, to which ATP in the form of ATBNG was added in the late 1990s. As a consequence, the line speed was raised again to 100 km/h (60 mph). In 2005, a monument was erected near the level crossing where the accident took place.

Comment

The comments after the report on the accident near ***Holzdorf*** in September 2003 are pertinent in this case also. The only remark I would make here is to comment on the suggestion made by some that the driver mistook Winsum for Baflo, the preceding station, rather as happened at ***Berajondo***. I do not think that such navigation errors were the cause at all, for the following reasons:

- The driver reported arrival at Winsum loop to the traffic controller by radio as per the protocol.
- Baflo has a single platform along the single line; there is no loop.
- As a consequence, at Baflo he would not have signalled the conductor that he was ready to leave (and he did not do that at Baflo). That was only necessary at crossing loop stations like Winsum, after receiving the go-ahead from the traffic controller.
- In the dense fog the driver sounded his horn as prescribed for a level crossing, which proves he knew where he was.

My interpretation is that the driver of train 8726 in all probability misconstrued the radio call to train 8713, misunderstanding train orders issued as those pertaining to his train, but then failed to make sure he was safe to depart. It might perhaps also be worth mentioning that in those days the hearing of train drivers was not checked as thoroughly as it is nowadays (compare the vision problems of the driver who caused the crash at ***Lagny-Pomponne***) and perhaps the offending driver had lost some of his hearing, which I know from experience happens almost imperceptibly. Age-related loss of hearing is a major problem with male train drivers of 45 years of age and over, it is what tipped me out of the driver's seat. And, perhaps, the driver of train 8726 had become drowsy (a feature of the job, especially on such wintry early-morning diagrams) or had even gone into a moment of micro-sleep in his seat whilst stopped at Winsum and had missed the call about having to wait, then woke up, took a look at his watch, realised that he had slept past departure time and (routine, autopilot?) gestured receipt of permission to the conductor. Such behaviour, baffling and inexplicable as it is, has internationally been the cause of several such accidents. Had protocol demanded confirmation of receipt of the messages with respect to the change of plans from both drivers involved, it might have prevented this collision, as in the case of sleepiness the confirmation of one driver would not have been obtained. The benefit of hindsight again!

***

Buttevant, County Cork, Ireland, 1 August 1980 at 12:45

(Derailment due to staff error)

The accident

This was a typical accident due to degraded signalling interlocks, with many parallels throughout the history of railways. To my knowledge, however, such an accident has not happened since, and is very unlikely to recur in any developed nation. Someone in charge of a set of disconnected turnouts on a main line assumed that a particular shunt move was about to take place and set the route into a siding accordingly, but did so without the necessary authority from the local controlling signaller. The signaller had the main line set up for an express that was due to come through the station at speed, and as a result of the misunderstanding between these two men the express was diverted into the sidings and derailed catastrophically. The accident left 18 people dead (including two railway staff) and 75 people injured to varying degrees. It also once more confirmed the vulnerability of coaches constructed mainly of wood.

The location

Buttevant was a former station on the 165½-mile (265-km) main line from Dublin to Cork. The station had already been taken out of use, its function having changed to being a signalling block post between Rathluirc (Charleville) and Mallow. Buttevant signal box controlled a distant, a home, a platform starter and an advance starter/section signal for the two tracks in either direction. In British-type mechanical signalling a single distant signal is provided for a maximum of six stop signals per possible route controlled by one signal box. The advantage to a train driver is that if the distant signal is cleared then all the associated stop signals are cleared and passing through at maximum permitted speed is safely allowed. If the distant signal is at caution, however, it cannot be clear to a driver which of the associated stop signals may be at danger, so all stop signals of the station limits have to be approached at a speed slow enough to enable coming to a stop, which may cause an unnecessary loss of time. On the Continent each stop signal has its own distant signal.

The layout

The location comprised a northbound Up line (towards Dublin Heuston station) and a southbound Down line (towards Cork Kent station), with platforms and a footbridge across the main lines still in place. There were sidings on both sides and two trailing crossovers between the main lines. After having been taken out of use as a station and reallocated to engineering for use as a materials depot, changes in the Buttevant track layout had been implemented. The traditional British dislike for facing turnouts on main lines had been set aside to allow more efficient working of ballast trains and on-track engineering equipment into and out of the Buttevant sidings. To that end, two trailing sets of crossovers between the Down main line and the Down sidings had been removed and a facing set of turnouts had been installed, but for reasons of poor communication between various departments these had not been connected with the interlocking of the signal box. Another facing connection between both main lines still existed and was worked from the signal box interlocking but was slated for removal. The reason for this was the delay to rail traffic picking up or setting down ballast trains, as these had to move across the level crossing all the time, for which the gates had to be worked.

Method of working

To work around the disconnection of the facing turnout leading from the Down main line to the Down sidings with a modicum of efficiency a pointsman by the name of William Condon had been appointed to work this turnout under the instruction of the signalman. This entailed unlocking the padlocks of the clips with keys, unclipping the turnout clips and removing the wooden scotches between the turnout blades and the stock rails. Then move the turnout blades into the desired position for the train manouevre with a long iron rod (a pinch bar) and securing the blades in their new position with the scotches, clips and padlocks, before hand-signalling the train movement as required. Every change of position of these turnout-blades required 5-6 minutes of rather hard work. After each movement into or out of the Down sidings the set of turnouts had to be restored and secured for main-line running and the keys were to be handed back at the signal box.

This method of working contravened a number of (somewhat unclear) rulebook instructions that not many people were in fact aware of, but those involved were convinced that it was a safe situation as the signaller presumably was aware of the presence – or not – of the keys, whilst in order to retrieve them, Mr Condon would always have to go up into the box and meet the signaller so that the requirements could be discussed. The signaller, however, was overworked. There were three signallers in Buttevant signal box, who worked seven days a week on 8-hour shifts, i.e. a 56-hour week. Two men would work 12-hour shifts in case of sickness or someone having taken up leave. Not only did the signaller have to work the signalling, he also had to manually open and close a set of gates on the public level crossing next to the signal box whenever an approaching train required his attention upstairs. For this reason the signaller often asked one of the railwaymen present in the signal box to work the level crossing gates for him. These gates were interlocked with the Up and Down home signals and the Up starter signal; these could only be cleared if the gates were open for rail traffic and closed against road traffic. This meant that if the distant signals showed clear, all three stop signals were cleared and the level crossing gates were closed for road traffic. Also of importance is the fact that the turnout padlock keys were not labelled and there were a fair number of similar keys (between 8 and 11 are mentioned), so the signaller was not always aware which key was in and which one was out. Condon had in fact introduced a system with coloured safety pins to allow himself to distinguish the various keys, but did not discuss that with the signallers as the fiefdom system ran deep.

To take in a ballast train from the Down main line the driver would find the Down distant signal at caution prior to being stopped at the Down home signal. Following verbal clarification of requirements, the Down home signal would be cleared and the train would pull into the station area and stop at the new (unconnected) turnout for the siding. This would be set for the siding, after which the train would be hand-signalled into the siding. For a light engine from the Up main line, however, the Up home signal would be at danger and after meeting the signaller the level crossing gates would be opened and the signal cleared for the driver to pull up past the crossover, from where a shunting disc signal would come off to take the movement over to the Down main line to the new turnout. The pointsman would work it as described above from there. In fact, this is what would happen with the light locomotive for the ballast train on the day of the accident.

Other relevant matters

There was a temporary speed restriction in force between mileposts 133.25 and 134 because of track maintenance work, but there was no restriction through the Buttevant station limits. The weather was fair, with good visibility.

The 10:00 Dublin to Cork service was normally classed as a 'Super Express', but on the day in question was running as a 'Special Express No 1' due to the train being overweight by a considerable margin for 'Super Express' timings. It had therefore been allocated an extra 10 minutes to complete the trip. Apart from an extended composition of the train due to heavy demand, an additional ex-maintenance self-service buffet vehicle had been added at Dublin to replace the booked but less-suited old-style buffet vehicle. En route the on-board equipment and commercial stock would be exchanged between the two, after which the older buffet vehicle would be taken out of the set at Cork. A Class 071 diesel-electric locomotive was allowed 323 tonnes on 'Super Express' timings, but for the reasons described above this consist ran at 430 tonnes. Nevertheless, although the train had left Dublin Heuston 3 minutes late, it was a minute early coming through Rathluirc on 'Super Express' timings, which was 8 minutes early on the booked 'Special Express No 1' timings. In fact, the accident report noted that neither the CIE (Córas Iompair Éireann, Irish National Railways) working timetable nor the booked timings for the *ad hoc* running of ballast trains bore much resemblance to the actual working of trains, and commented on that with regard to maintaining an overview of train movements and therewith a number of safety issues. For instance, the ballast train that the light locomotive from Mallow was to pick up was many hours late. This was due to the crew having been told that the booked departure time of 04:40 from Mallow could not be met on the day of the accident because the ballast-cleaning machine had developed a fault. For that reason a revised departure time of 08:00 had been agreed between the supervisor and crew. The traincrew, driver James Mullins and guards Tynan and Kelly, however, had been away from home on this job for the last three days and of their own accord decided to travel back to their depot at Thurles to pick up their wage slips, have a night at home and come back first thing next morning. That was why they and their locomotive appeared at Buttevant around 12:30 when the 10:00 ex-Dublin was about to pass through. Had they been available at 08:00 as agreed, this confusing situation would not have arisen and the accident would not have occurred that day.

Rolling stock involved

The locomotive involved was Class 071 Co'Co' No 075, built by General Motors EMD at La Grange in Illinois, USA. With one GM 12-cylinder diesel engine delivering 2,475 hp (1,846 kW) at 900 rpm, it weighed 99 tons and had a maximum permitted speed of 145 km/h (90 mph). The driver, Bartholomew Walsh, had four years' driving experience. The locomotive was relatively unscathed after its high-speed derailment and was soon back in service. Generator/boiler van No 3191 (for electric and steam train-heating purposes) had been built as a Mk 1 BR vehicle between 1952 and 1956 and was rebuilt with a Lister electric generator set and a Spanner steam boiler in 1971 at the BREL Derby Litchurch Lane facility. This vehicle was severely damaged and not repaired after the accident. Wooden-bodied first-class vehicle 1145, wooden-bodied buffet vehicle 2408, wooden-bodied self-service buffet vehicle 2412 and plywood standard-class vehicle 1491 were all completely destroyed. These were followed by light alloy 'Cravens' standard-class coaches 1529, 1527, 1508, 1542 and 1541. All except coach 1541 received damage but were repaired and brought back into service. Wooden standard-class 1365 and wooden composite standard/brake van 1936 escaped unscathed.

The accident log

Signaller O'Sullivan at Buttevant signal box accepted the express without restrictions from Rathluirc at 12:25 and the light engine from Mallow at 12:29 (but erroneously wrote 12:15 in his book). He received 'Train Entering Section' for the express at 12:39. At 12:30 a ballast cleaner was moved from the Down sidings across the main lines to the Up sidings to enable the ballast wagons in the Down sidings to be picked up and removed by the light engine then arriving from Mallow. Shortly after that ballast cleaner move the siding turnouts worked by Condon were seen by a number of people to be set and secured for main-line through traffic. Soon after, the signaller asked the pointsman to open the level crossing gates, but Condon said he was in a hurry and, unbeknown to signaller O'Sullivan, railwayman Benjamin Stack worked the crossing gates. While Condon was on his way to his set of turnouts, Stack stayed at the level crossing and saw the light engine pass by. The signals were cleared, so he could no longer close the gates against rail traffic. It was then that he heard and saw the express approaching.

Driver Mullins with his light locomotive for the ballast train had arrived at Buttevant 15 minutes after having been accepted from Rathluirc, at around 12:40, guard Tynan with him in the front cab and guard Kelly in the rear cab. They stopped as required at the Up home signal at danger and whilst waiting Mullins noticed that the level crossing gates were being opened and the Down home signal pulled off. His Up home signal then cleared and whilst preparing to pull into the station area he saw someone walking along the opposite platform, but what struck him was that the person then started to run. He halted his locomotive at the Up starting signal at danger where Tynan left the cab to walk over and prepare his ballast train in the Down sidings. O'Sullivan then shouted 'Go ahead' from his signal box, on which instruction Mullins brought his locomotive past the level crossing and stopped past the crossover turnouts to await the turnouts to be reversed with the accompanying clearing of the shunting disc signal. Looking forward, he then saw a train at speed approaching along the Down main line.

Guard Tynan crossed the footbridge between the platforms and saw a man running to the turnout for the Down sidings. At the same time he heard the express approaching and saw that all its signals had been cleared. He ran back to the signal box and told the signaller that there was someone at the turnout to the siding. The signaller looked out, then shouted 'Points, points! Condon, make the points!' and threw back all the cleared Down stop signals to danger, three of four signal levers in the frame. Tynan then went back to the bridge as the express passed at speed.

Driver Walsh on the 10:00 express from Dublin had had an enjoyable ride, having recovered the 3 minutes of lost time with a heavy train. In fact, he had not been told that he had a revised schedule as a result of his train being overweight, and his hard work keeping up 'Super Express' timings after a slightly late start with his 12 coaches had borne fruit. (I know from experience that it is a good feeling to keep time against the odds.) On approaching Buttevant he knew that he was on time and the signals were cleared, so he settled for the last few miles to Cork. Near the Buttevant level crossing, however, he suddenly noticed that the Down home signal was thrown back to danger on him, in response to which he immediately shut off power and applied the vacuum-operated emergency train brake. Approaching the new turnout into the Down sidings he saw a man frantically poking at the track with a long rod and thought he must have killed him when he shot past him into the siding and started to derail. After the locomotive came to rest he left the cab with equipment to protect the Up main line, forgetting to shut down his engine.

At Mallow, signaller John Kelleher had been told by telephone that the light engine for the ballast train would be held on the Up main line to await the passage of the 10:00 and the 10:30 services from Dublin to

Cork when he offered the engine to Buttevant box. He was offered the express with an 'Is Line Clear?' at 12:38, received 'Train Entering Section' at 12:45 and then got 'Obstruction, Danger' at 12:47, as did signaller Thomas O'Doherty at Rathluirc box.

Driver Mullins, standing beyond the level crossing on the main line awaiting the clearing of the shunting signal to move on to the ballast train with his light engine, looked back from his cab and saw the massive cloud of dust thrown up by the derailing express. When it cleared somewhat and he saw the state of the coaches in the siding, he secured his locomotive, climbed down onto the ballast and went across to help. On reaching the derailed locomotive he noticed the engine was still running, so he stopped it. But he did not look at the power controller or the brake valve positions and was therefore unable to answer questions about them during the subsequent investigation.

Post-accident findings

Pointsman Condon had never received an explicit order to set the turnout for the siding; it was all expectation and assumption on his part. He knew that the light engine that had just arrived was waiting to cross over to pick up the ballast train and that it was hours behind schedule, and also knew that the ballast cleaner had recently been moved to allow the locomotive access to its ballast train. He therefore assumed that the level crossing was being opened to get this engine into the sidings quickly as the morning express from Dublin to Cork would be approaching at round about this time. Clues to him were that the main-line signals were all kept at danger but that the level crossing had to be opened suddenly, which indicated a shunt move, and for that reason he was in a hurry in order not to hold up the ballast train moves and thereby the express. Signaller O'Sullivan, however, had earlier mentioned letting the 10:00 and the 10:30 services past first before moving the ballast train, but Condon picked up that tentative information merely as part of the conversation, not as an instruction of any sorts. The signaller had not explicitly said that the light ballast engine would sit waiting past the level crossing behind the shunting signal on the Up main line until after passage of the 10:00. Nor did he ask the pointsman whether the turnouts in the Down main line were locked safe for main-line traffic, and he did not tell Condon that the arrival of the 10:00 was imminent. (See the findings of the French national transport accident investigation organisation after the minor derailment at ***Carcassonne*** in France 27 years later.)

In fact, Condon could have known these things had he had a watch and a better understanding of the local deviations from the working timetable at the time, but that was not the case. And why should an overworked and tired signaller keep informing someone who in effect acted only as a replacement part for the incomplete signalling equipment he worked with, and who ought to do exactly as told, about such details? On the other hand, the signaller was the manager in charge who had to ensure that things went smoothly and safely. The longer the disconnections lasted, the more the people involved developed undesirable practices. These were based on their private understanding of what happened routinely on a day-to-day basis, so errors were increasingly likely to occur with non-routine issues. Acting quickly out of sheer goodwill on his mistaken initiative, Condon hurried to the turnout, undid the padlocks, clips and scotches with a dexterity gained through routine and levered the switch blades set for the siding. He never once looked at the signals again until he saw the engine of the express bearing down on him, which is when he realised that he had made a dreadful mistake. He then tried his hardest to undo it. That way he joined the ranks of railwaymen who caused a bad accident and suffered the hell that follows.

Conclusions

During the inquiry the lack of coherent organisation and the elements of fiercely upheld fiefdom within the Irish railway organisation were scrupulously unravelled and brought to light. It came down to the following. The permanent way people wanted a useful depot for their equipment and stores along this main line in the Cork area and thought up a few logical track layout changes. Being P-Way people, they organised these changes at the disused station of Buttevant expeditiously but without arranging anything else, as that was outside their remit. Signalling & telecommunications people, however, with their own schedules for undertaking jobs, had at best misunderstood the urgency of bringing the track changes into use but, at worst, had actually forgotten about them. Therefore, the three newly installed turnouts in the main lines had remained disconnected from the interlocking in the signal box for no less than four months (and would probably have done so for quite a few months more if the accident had not happened). Therefore it was that the turnouts had to be worked in the way that Condon worked them. Had the relevant S&T manager, a Mr Leahy, realised that the turnouts were to be brought into use expeditiously, then, so he said, he would have put more effort into connecting them to the interlocks and upgrading the track diagram in Buttevant signal box.

But it was not only the signallers who were overworked; the S&T installation people in the area (one crew only) had a backlog of six months' work. In fact, from the early 1970s until well into the 1990s, staff shortages and consequent overtime working just to keep the railway rolling were endemic in the many English-speaking nations of the world. The ***Clapham Junction*** accident was just one example of this situation.

Comment

The log makes everything pretty clear, so there is really little to add. The root cause for the crash centred on the expectations and assumptions of a poorly trained railwayman doing what was an absurdly silly job, but one with major safety implications. I observed the mistakes that could result from such frantic, non-stop working when I started at Slade Green depot near Dartford as a guard and then transferred to the London to Brighton Line at London Victoria as a train driver. The norm was mostly 12-hour days and non-stop 13-day fortnights of work, only slightly tempered by the 'Hidden' rules (framed by Sir Anthony Hidden QC after the ***Clapham Junction*** accident) that marginally reduced the hours of work. However, there were a number of ways of getting round their impact on working schedules, which allowed the trains to keep moving no matter what. Undeniably the pay was very good indeed, but it was not much fun if you also wanted a married life and to spend some of that money on things such as holidays in what spare time was available. In those days at London Victoria there were people with years' worth of unused holiday leave on their files. They used that accumulated time to retire much earlier, or in cases of railway people of foreign extraction, to go back to their home country for extended periods. In my case this situation ended when the infamous privatisation of the railways in Britain started to bite. Gatwick Express was hived off in 1994, and that was where I ended up. The work conditions as well as the basic pay for British footplate crew improved considerably from then on under the relentless pressure exerted by the train driver unions, and the possibilities of a sweeter life outside the cab were rediscovered.

In Ireland (and throughout the rest of the world, if they had not already noticed) the Buttevant accident finally made it clear that wooden-bodied coaches had to go. It resulted in the purchase of a batch of very good and modern British Rail Mk II and later Mk III coaches and thus marked the slow beginnings of what was to become the renaissance of Irish railways, north as well as south of that sad border dividing the island. In fact, the Irish Mk IIIs were better than those in Britain because they had power doors, instead of the jaw-dropping backwardness of the British Mk II and Mk III slam doors without interior door handles. It was only with the later advent of the Class 442 electric 'Intercity' units, based on the Irish Mk III coach body, on the London Waterloo to Bournemouth and Weymouth routes of the Southern that Brits could experience what a really modern Mk III coach was like. And, in Ireland too the use of radio telephone connections between traffic control and train were discovered as a way of improving rail safety.

***

Summit Tunnel, West Yorkshire, England, 20 December 1984 at 05:50

(Tank wagon derailment with ensuing fire in tunnel)

Introduction

This British spectacular involved train 6M08, the 01:40 13-vehicle (100-ton bogie tank wagons) train from Haverton Hill on Teesside to Glazebrook near Warrington. It was loaded with 835 tons of petrol (automobile fuel) and derailed due to a seized axlebox and the consequent disintegration of an axle (similar to the incident at ***Mississauga***) inside the ¾-mile (1,200-m) Summit Tunnel on the Sowerby Bridge line across the Pennines. I have relied on information given by Stanley Hall in his description of this incident, as well as on the official report.

The accident

The train had been checked on departure and checked again after a stop at Healey Mills yard near Wakefield, and even the signaller at Hebden Bridge signal box, just before the tunnel in which the accident happened, noticed nothing untoward when he checked the passing train as prescribed. Indeed, if seized wheels of railway vehicles do not glow, smoke, spark or squeal, it is very hard to pick them up. On reaching the tunnel portal after the long and steep climb, traction power was notched back and the train was rolling easily through the tunnel at 40 mph (65 km/h) when the driver experienced rough snatching movements and then suddenly lost his brake-pipe pressure, following which the train came to a shuddering halt in the darkness. From experience the driver knew that the brake pipe had parted somewhere, which in combination with the jerking movements might point to a divided train. He secured his locomotive and with his British Rail Bardic torch climbed down into the darkness to see what was wrong and to consult with his guard on the best course of action. So far, all his actions had been in accordance with the rulebook, but unfortunately he failed to do the most important thing, which was to place a track circuit clip on the opposite line to put signals for trains on that road back to danger and thus prevent them from running into any obstacles presented by his potentially derailed train. Whilst walking back he noticed a sharp petrol smell and correctly deduced that vehicles were leaking, which in turn pointed to substantial damage to wagons, but he could see very little in the weak light coming from his torch. He was unable to see that some of the vehicles had fallen to their right and were fouling the opposite track. Luckily, they had also short-circuited its track circuit, causing the signal at the tunnel mouth to show a red aspect. However, whilst he and the guard were conferring they suddenly heard a muffled explosion and, fearing that the whole train might explode, ran for their lives.

At the tunnel mouth they used the signal-post telephone to inform the signaller at just after 06:00, some 10 minutes after the accident had occurred. The signaller did his job and by 06:08 Greater Manchester Fire Control had received the message and had passed it on to West Yorkshire Fire Control by 06:13 (Summit Tunnel straddling two fire service areas). Fire engines arrived at the Manchester end of the tunnel 3 minutes later and at the Yorkshire end another 10 minutes later. As luck would have it, British Rail and the fire brigades involved had recently tested response times and the optimal approach roads to use. Despite the serious danger of explosion and a massive fuel fire, the firemen entered the tunnel with torches and started to assess the damage. With hand extinguishers and later piped foam appliances they quelled several localised pockets of fire and decided that the situation allowed removal of the locomotive and the three non-derailed first vehicles from the tunnel. Despite their misgivings, the driver and guard agreed to go back into the tunnel to uncouple the first three undamaged vehicles and pull them away, but that was easier said than done because the fourth vehicle was derailed. Its coupling was stretched hard towards the side and resisted all initial attempts to lift it off the drawhook, and cutting the coupling was not an option due to the fact that it would shower sparks in a very volatile environment. By taking the risk of further derailment by setting back hard against the derailed fourth wagon, however, they finally managed to uncouple and took the locomotive and the first three tank vehicles out of the tunnel.

By 09:45 everything appeared to be under control and railway breakdown gangs were allowed to move in to start clearing up the considerable mess consisting of ten derailed, partially overturned and damaged 100-ton bogie tank wagons, some of them still leaking fuel, before attempting the removal of the damaged vehicles along equally damaged track. However, they had barely started to bring in their tools when a massive fire suddenly erupted, fortunately most of it up two ventilation shafts situated above the train, the flames reaching a height of 150 ft (45 m) above the tunnel. The fire burned for 24 hours, but despite its age, the tunnel, built by George Stephenson in 1840/41 as the first tunnel through the Pennines, stood up well to the inferno.

The cause of the accident

During the investigation, after re-entering the tunnel, it was found that an axle tap of the first axle of the fourth wagon had sheared off in a seized journal. The vehicle had derailed and spread the track, which made the others wagons derail, fall over and divided the train between the sixth and seventh vehicles. The axlebox concerned was a Timken roller bearing type that had been refurbished by the British Steel Corporation, and incorrect re-assembly appears to have been the reason for the failure. Eight months after the accident the tunnel was opened for traffic again.

Comment

To appreciate fully the situation experienced by the traincrew in this incident it would be necessary to find a disused long railway tunnel. Near my home, Shute Shelve tunnel on the former Cheddar Valley railway line from Yatton to Wells was built for Brunel's 7-ft gauge and is consequently rather wider than most, so is not as typical of the accident scene as others might be. By all means take a torch, but nothing more powerful than the old-fashioned British Rail Bardic lamp, and imagine that within that dark and narrow tunnel with a badly damaged train and its highly flammable cargo you are tasked with assessing what has to be done, standing in the foul atmosphere of petrol fumes. And then consider that sudden and ominous 'whoop' sound of fuel catching fire and the consequent risk of explosion with some 900 tons of fuel around you. That sort of raw and immediate fear cannot easily be created in any other way, and you would certainly run like a hare. Still, the men kept their wits and as a result the emergency services were on their way within a quarter of an hour. They cleared what could be cleared quickly and started to work in very dangerous circumstances to remove the wrecked vehicles, displaying dedication and professionalism of the highest degree. In the petrol-saturated air inside the tunnel, with no equipment available to expel the gas from the tunnel, it was sheer luck that when the massive conflagration came everyone was able to get out unscathed. With the force of the fire being directed upwards into the ventilation shafts rather than towards the tunnel portals, the firefighters and railway crews were given that little bit of time needed to leave the tunnel in safety. Pictures of the dramatic sight of the ventilation towers on top of the shafts with flames roaring out of their mouths are truly impressive.

***

Tank wagon failures in the USA

(Several mishaps with tank wagons)

Introduction

These incidents are included not because accidents with chemical transports are any more frequent or dangerous in the US than in the rest of the world, but because the US National Transportation Safety Board (NTSB) described them in a comprehensive yet understandable manner and made them public. As such, they stand for many similar accidents with tank wagon chemical loads that have taken place elsewhere but have not been so clearly documented, even if similar accidents elsewhere can be found in this book (e.g. ***Mississauga***, ***Summit Tunnel***, ***Hamina Port***, ***Viareggio***, and ***Red Deer***). Note that most incidents of this type occur at the premises of shippers and receivers, rather than when in transit.

It is important to bear in mind that development of safer tank wagon transportation is still continuing, driven by governments acting in response to a series of high-casualty accidents such as the *Lac-Mégantic* runaway in 2013. On 5 May 2015, Canada and the US signed agreements to the effect that from the October of that year new equipment must meet DOT-119 specification, and that after May 2017 no DOT-117 tank wagons will be accepted for transport unless upgraded. After January 2021, large consignments of hazardous flammable liquids will have to be transported in tank wagon unit trains fitted with the fast-acting electronically controlled braking (ECB) equipment that is remotely controlled from the front as well as the rear of the train. Better protection, better loading and unloading equipment, much stronger (and therefore heavier) vehicles and improved braking are the way forward according to the NTSB and the TSBC, its Canadian equivalent. The operators, however, are not happy with the front/rear electronic braking system demand and with the requirement that larger shipments have to run as unit trains. Capacity to compete with other modes of transport and vehicle allocation flexibility are at the core of this problem.

Bogalusa, Louisiana, 23 October 1995 at 16:45

At 15:55 on 23 October 1995, staff at the unloading facility of the Gaylord Chemical Company at Bogalusa in Louisiana noticed a 10-15 ft (3-4.5 m) yellowish plume of vapour rising from tank car UTLX 82329 that had arrived loaded with poisonous and corrosive nitrogen tetroxide. The same tank car had been noticed as leaking hydrogen tetroxide vapours the previous month, at which time the wagon had been sprayed with water whilst discharge of the chemical took place and after that fitters from the Union Tank Car Company had replaced four valves. One of those was found to have significant corrosion and showed wear on the valve stem. After the repair the vehicle was brought to the Vicksburg Chemical Company for another cargo of nitrogen tetroxide, after which the wagon was weighed and found to be 9,500 lb (4,310 kg) over the permitted weight but within Gaylord's specification. A Vicksburg employee thought the wagon had been upgraded (with new valves and new bogies) but failed to verify his findings and so the wagon arrived at Gaylord's premises with 10,000 gallons of nitrogen tetroxide, weighing 110,000 lb (49,895 kg). On 12 October, when discharging the cargo into a storage tank and simultaneously into the chemical reactor, the reactor unexpectedly shut down. It was found that the cargo was contaminated with water, and as this could cause corrosion in the plant equipment (nitric acid forms from nitrogen tetroxide when mixed with water) it was decided to unload the remaining contaminated cargo into stainless steel road tank trailers for storage. According to the pump gauges, 10,100 gallons of chemical had been transferred overnight this way, but no further measurements such as weighing the tank car were undertaken. When unloading stopped, a small sample was taken for analysis of the state of the contents from the transfer system and at 16:00 staff decided that the tank car was empty except for some residue. Post-accident weighing, however, showed that only 491 gallons weighing 6,080 lb (2,760 kg) had been transferred. The reason for this was that the carbon steel eduction pipes used during the transfer had been so badly corroded that only a small section of each remained, located high in the tank near the top of the dome and not down at the bottom of the tank from where they would take up the pressurised chemical for transfer into the plant system. Nothing further was done with the wagon whilst Gaylord staff cleaned out the contamination in their plant piping.

Four days later, staff determined that vapours coming off the vehicle were excessive, indicating that the material in the wagon was not diluted enough to permit discharge into the plant's chemical sewer system. The plan was to transfer the remaining waste into another shore-based tank, but this was abandoned after yet another tank that had been filled from this wagon started to give off vapours. It was then determined that the road vehicle tanks used for storage were not fitted with gaskets and valves suitable for nitrogen tetroxide or fuming nitric acid, so these were replaced between 17 and 20 October. On the 19th, samples from the road vehicle tanks and from the tank wagon were taken to Vicksburg for analysis, after which Gaylord staff began pumping material from the tank wagon. By 18:00 the meter showed that 6,700 gallons had been removed to a cargo tank, but post-accident weighing confirmed that only 850 gallons had been shifted. The tank was then purged with water, although this had to be stopped as the pressure inside the tank wagon unexpectedly rose to 60-65 psi. In order to relieve this pressure the wagon was connected to the gas scrubber, and later that day Gaylord staff gravity-discharged the waste from the wagon into the plant's chemical sewerage system. However, this had to be stopped after 2-3 minutes as a large vapour cloud developed.

The next day, a Gaylord employee twice added water to the contents of the tank wagon and in both cases it was noticed, with some disbelief, that the pressure rose significantly as soon as water was pumped in. This pressure was relieved via the gas scrubber connection and two days later, on 23 October, the pressure was noted as a modest 18 psi. Plans were then made to complete the discharge and cleaning out of this unhappy wagon. That day, at 12:54, Gaylord received the test results on the samples they had sent to Vicksburg and to their surprise it turned out to be wet nitrogen tetroxide. The company staff were convinced that this had been completely discharged 10 days earlier and that the result should have been thoroughly diluted nitrogen tetroxide residue. It was decided that the test results were not representative, so between 13:30 and 13:45 more water was pumped into the tank wagon, after which the internal pressure in the tank wagon rose from 18 to 80 psi in 4-5 minutes and at that point the water was turned off. At 14:00 the meter indicated a pressure of 92 psi and then slowly started to go down again after the gas scrubber connection was used to vent the pressure. By 14:30 it had declined to 55 psi and water was turned on again, after which the pressure shot up to the maximum calibrated pressure indication on the meter, 100 psi, at which point water was turned off again. At 15:00 the pressure appeared to be falling again, but by 15:30 it was above measuring level and appeared to be rising again. At 15:55, when a yellowish-brown vapour cloud appeared to leak from the safety valve in the tank wagon dome, the emergency services were informed and staff started spraying water on the tank wagon to avoid the gas spreading. At 16:30 the Bogalusa fire brigade arrived and set up additional hosepipes. At 16:45 the tank-head at the B-end (parking brake end) of the wagon unexpectedly failed and the jacket was blown off, landing 350 ft (106 m) away. The vehicle moved 35 ft (10 m) along its track until hitting derailers welded on at the end and partially jumping off the track. A massive reddish-brown cloud escaped into the air, a situation that continued for 36 hours until it was brought under control through dilution and neutralisation. As a result, 3,000 nearby residents were evacuated and 4,710 people were taken to hospital, 81 of whom were admitted for further treatment. Post-accident investigation revealed that the wagon safety valve had activated at its pre-set value of 375 psi. The same examination revealed massive internal corrosion of the carbon steel tank, the use of which for this sort of transport was discontinued after this accident. New information and materials-handling protocols and instructions were introduced at the same time. The cause of the accident was determined to be Gaylord's inadequate handling of nitrogen tetroxide leading to uncontrolled foaming of the highly corrosive nitric acid.

Sweetwater passing loop, Tennessee, 7 February 1996 at 05:30

On 7 February 1996 at about 05:00, freight service M34T5 sat on the main line at Sweetwater passing loop in Texas awaiting the imminent arrival of a westbound freight train into the loop in order to continue its trip into the single-line section ahead. The westbound train duly arrived, slowly pulling into the loop, and M34T5 received its permission to depart, after which the driver closed his brake valve, selected forward and opened up a few notches. Under low power the head of the train finally started to roll, gingerly accelerating to 2 mph (3 km/h) when after approximately 33 ft (10 m) of movement (the very rear of the train probably still as good as stopped, as on a serious US long-distance freight train this really is just stretching the coupler springs) the train brake pipe suddenly vented and, as a result, the brakes applied hard. The train conductor picked up his radio set and vacated the cab to walk back and see what the matter was. Much to his amazement he found that tank car GATX 92414 had virtually broken in two, discharging its contents of flammable and toxic carbon disulfide into the ballast. The alarm was raised and 500 nearby residents, including the population of a retirement home, were hurriedly evacuated, five of whom were admitted to hospital, one being kept in for further treatment. Two days later, it was decided that the spillage no longer posed a risk beyond the immediate environs of the wreck and the evacuation order was lifted. The wreck was then released for the investigating agencies to start their work at 16:45.

The investigators found that the wagon wreck and the ground around it were now surrounded with polyethylene sheets and wood to cover the spillage and make moving around possible. On noting this, a representative of shipper AkzoNobel expressed concern that pockets of vapour might have been trapped under the sheets that could cause problems when people walked over them. At 01:40 the next morning, when the spill site and wagon were being uncovered to start inspection of the vehicle, a short and low flash fire occurred that momentarily enveloped the people nearby but fortunately caused no injuries. Immediately the fire brigade chief asserted control again and ordered all people off the premises as well as reinstating the evacuation order for nearby residents. After a complete clean-up the evacuation order was eventually lifted on 12 February and the wreck was once again handed over for investigation.

Two general causes for the failure of the 27-year-old wagon were found. The first was the inappropriate ductile-to-brittle transition temperature (30°F) of the selected construction steel for the load and stress bearing parts of the wagon. The ambient temperature at the moment of the failure had been 24°F and during the night the temperature had been considerably lower than that. The second cause was the fact that the tank vehicle was of the stub-sill type, having not originally been constructed with a separate wagon frame with a tank fastened on top, the tank shell itself acting as the frame and taking all stresses transmitted through the stub-sills with the drawgear and the bogies. This had been found to cause buckling and tearing problems where the stub-sills were fastened to the tank, and for that reason the American Association of Railroads (AAR) issued directives for these vehicles to be voluntarily modified from 1975 onwards by welding two beams all the way between the stub-sills on to the underside of the tank plus a separate beam from one side to mid-tank, after which work the vehicles were accepted as complying with the new 1974 directives. In 1990, however, when buckling of stub-sill type tank wagons was still occurring for the reason outlined, voluntary modification became compulsory if the railroads were interested in keeping the scores of these vehicles in traffic.

Owner GATC had sent GATX 92414 to the shop at Hearne, Texas, for modification in 1990. Reinforcement bars had been welded to the bottom of the tank, but the post-accident investigation revealed that this reinforcement had not been executed in accordance with the agreed specifications by welding the bars to the stub-sills and, as a result, the tank still took live traction stresses, especially when accelerating. Because the records of the modification had been discarded after the permitted five years of storage, no conclusions could be drawn as to why the modification had been performed in this unsatisfactory manner, but GATC, to its credit, retired all its remaining 97 stub-sill tank wagons, modified or otherwise, and all tank wagons constructed of the same lower-grade carbon steel were replaced by new ones with fine-grain steel tanks. As a wider national measure, all 18,427 US tank wagons built between 1969 and 1982 were subjected to searching investigations and all emergency staff received new instructions on how to deal with these hazardous chemicals spills.

Conrail, Selkirk Yard, New York, 6 March 1996 at 11:37

Tank wagon UTLX 803627 had been loaded with 31,409 gallons of liquid propane at Sarnia, Ontario (in the Great Lakes country bordering Canada and the US with its most interesting rail and maritime transport history) on 3 March 1996. Three days later, at 09:17, it arrived in the reception sidings at Selkirk yard near New York in the consist of Conrail train COSE5 from Buffalo in upstate New York. Before release for shunting over the hump into the classification yard, where the wagons were sorted into new trains for onward transport, two rolling stock inspectors walked the entire length of the train on either side and did not find anything untoward. So UTLX 803627 was duly pushed over the hump at 11:37 and on its descent the wagon was slowed down with a manual application of the group retarder. The vehicle should have left the retarder at 9.4 mph (15 km/h), but in fact moved substantially faster than that. Yet it was not further slowed by the receiving line retarder and the heavy vehicle must have been a cruiser, seeing that it rolled along its classification (sorting) track at a steady 13 mph (20 km/h). At that speed it slammed with an appreciable bang into a few previously shunted vehicles and its automatic Buckeye coupler engaged. Three minutes later, the tank wagon broke in two and its contents ignited immediately. No doubt in fascinating fashion, the half attached to the other wagons then propelled itself and its comrades for about half a mile (800 m) along its track through the yard. Had Wernher von Braun not already shown the US how to put usable rockets together, this would perhaps have been a defining moment in the space race!

As far as this incident is concerned there is not much more to be said. During an earlier overhaul the vehicle tank steel thickness had been found to be depleted by corrosion in a few places and had been welded up to gain sufficient material. Those welds, however, had not been placed carefully enough and corrosion had been allowed to eat further into the steel covered by the new welds and gradually weakening those as well. On inspection of other members of this batch of vehicles from the Richmond Tank Car Company similar problems were found and all wagons were taken out of service to be sent for scrap.

ICRR Yard, Memphis, Tennessee, 2 April 1997 at 12:05

This incident involves the Illinois Central Railroad that we met with at ***Vaughan***. In this case it related to its typical trade of transporting goods between the Gulf ports and Chicago, and onwards into Canada. At 04:30 on 2 April 1997, tank wagon ACAX 80010 arrived with service GEME01 in the ICRR yard at Memphis. It was loaded with anhydrous hydrogen fluoride on its way from Geismar, Louisiana, to Port Hope, Ontario. A rolling stock inspector detected leakage from this vehicle whilst the train was being shunted later in the day, spotting vapour apparently coming from a crack in the weld of a 66 cm x 99 cm (2 ft x 3 ft) repair patch in the tank. The alarm was raised and 150 residents in nearby dwellings within a half-mile (800-m) radius were evacuated for 17 hours whilst the cargo was transferred into another tank wagon. No injuries were reported.

The incident amounted to no more than that, but it is interesting to read why the leak occurred. The tank wagon, which was 26 years old, was built to US Department of Transportation (DOT) specification for the transport of highly poisonous anhydrous hydrogen fluoride. It had recently been returned from repairs after blistering inside the tank had been found, a problem closely related to transport of that particular commodity in non-protected carbon steel containers. Anhydrous hydrogen fluoride affects normal steel in three ways: it causes hydrogen-assisted stress corrosion cracking; it causes blistering; and it causes stress-oriented hydrogen-induced cracking. What hydrogen does, in short, is react with steel and release atomic hydrogen particles in the process. On release, the hydrogen atoms diffuse into and pass through the steel of its container. If, however, such an atom passes into any minuscule internal void in the steel wall (as can be found in normal carbon steel) it will combine with other hydrogen atoms that happen to make their way into that void, to form molecular hydrogen. These hydrogen molecules are too big to make their way onwards through the steel and start to push the steel molecules apart to form so-called blisters. One of these blisters had been cut out of the tank of this wagon and the hole had been patched with a welded-in piece of sheet metal. It was this that had caused the problems, because this same process caused stress corrosion in heat-hardened areas within steel surfaces such as welds, either during construction or during repairs. This stress corrosion showed up as cracks that propagated within the heat-hardened areas, not into the normal material, and it was through such a crack that the eagle-eyed inspector had seen the vapour escaping. As a result of this incident, steel of a different construction with fewer impurities was prescribed for the transport of anhydrous hydrogen fluoride. This particular vehicle was repaired according to the new regulations and taken back in service.

Pasadena, Texas, 22 November 1997 at 24:00

Once again the sharp eye of someone near a tank wagon was all there was between a bad fright and a disaster. It was noticed that something was out of the ordinary with tank wagon TEAX 3417 being offloaded from 29,054 gallons of liquefied propane/propylene gas (LPG) on the premises of the Georgia Gulf Corporation at Pasadena. A frost ridge had appeared under the tank near the bottom centre. Clearly, the tank was leaking and its contents were escaping.

Despite the leak carrying a realistic risk of fire and explosion in an environment full of static and mobile gas tanks, the discharge of the cargo of liquefied gas from the leaking tank into safe shore storage was completed without mishap. To discharge the residual load the tank wagon was isolated in a less dangerous spot overnight and the job was completed by 07:00 the next day when the last of the load was transferred to a road tank vehicle. The tank wagon was then handed over for investigation. A single circumferential crack in the bottom of the tank shell was found, which turned out to have been caused by a brittle fracture during a single event when the metal of the tank had been exposed to excessive cold, below -50°F. This had occurred when the vehicle had been flushed out with cryogenic nitrogen about a month before the incident. Although this action is very useful in itself, without adequate precautions being taken the tank-shell material can suffer thermal shock failure. This was by no means unheard of, as it was previously recorded in Ontario in 1982, and there was a similar failure in Mississippi two years later. No further damage, injuries or fatalities were recorded in this case.

Clymers, Indiana, 18 February 1999 at 00:05

On 1 March 1993 tank wagon UTLX 643593 had been loaded at the Olin Corporation at Charles, Indiana, with toluene diisocyanate, an intermediate waste product in the manufacturing process of polyurethane foam, well known as a material used in the manufacture of cushions and pillows. Although the Olin Corporation had loaded the tank wagon for temporary storage of this waste product prior to extracting reusable chemical components, another manufacturer, the Essroc Cement Corporation at Clymers, had begun to use such cheap flammable waste material to fire its cement-drying kilns. The vehicle with its contents remained stored at the Olin site for five years until March 1998 when it was finally shipped to Loganport yard. It sat there until December 1999, when it was delivered to the Essroc plant's unloading facility.

On 13 February 1999, Essroc and its chemical experts CPRIN discussed the discharging of this vehicle and decided to connect steam lines to the internal heaters of the wagon to liquefy the contents whilst the discharge lines were attached. This was done by 17:00 and steam was admitted at 18:00, the pressure inside the tank wagon being recorded at 0 psi at 30°F. A sample of the waste drawn for inspection and analysis was described as somewhat thicker than normal. The result of the analysis described the composition but said nothing about its viscosity. At 00:30 the following day the temperature inside the wagon had risen to 80°F and by 02:30 had reached about 120-150°F. Steam heating was stopped when the temperature reached 202°F at 06:00, when unloading of the wagon began. At 09:00 the unloading supervisor reported that he had to stop as the waste was solidifying in the plant transport piping between the unloading manifold and the kiln. The pipes were flushed with toluene to dissolve the waste material and open up the connections again. At 11:00 the pressure inside the wagon tank was recorded as 10 psi and had reached 40 psi two hours later. Discharge restarted and at 16:30 a further report came through that unloading had to be stopped as the pipes were plugged again, and the pipes were flushed with toluene once more. No further unloading was attempted until the following day when the pressure inside the vehicle was reported as 20-25 psi at a temperature of 70°F.

CPRIN was called in again and steam heating was restarted, continuing until 04:30 on 16 February. Samples were drawn again to check the viscosity of the contents, and between 06:30 and 06:45 the samples were noticed to be thin and fluid. The pressure inside the tank was recorded as 45 psi, far higher than the preferred 20-25 psi for unloading as normal. The tank was connected to a gas outlet and scrubber system to relieve this excess of pressure, at which moment the safety valves of the tank wagon were found to be 'whispering' – venting gas under considerable pressure. At 17:30 the relief shift was apprised of the situation, told to monitor the pressure and to connect the nitrogen gas pipe to start unloading as soon as the pressure was at an acceptable level. At 21:15 offloading restarted, but once again the highly viscous waste material clogged the pipes so discharge had to be stopped. The water in the emergency gas scrubber that had been rigged up to clean the escaping gas was boiling and spluttering, and the pressure in the tank wagon had risen to 50 psi. By this time approximately 750 gallons of the waste material had been offloaded. When the supervisor came on duty again the next day he noted that the pressure was down to 40 psi. He opened a vent valve in the discharge line and relieved the pressure to 3-5 psi. The vent valve was then closed and nitrogen gas pumped in, but by 15:00 the pressure had risen to 20-25 psi so steam heating had to be stopped again. The following day, as a new shift were arriving on site at about 11:30, the tank wagon exploded and engulfed the entire discharge area in a fireball. The tank shell itself was blown clean over a number of storage tanks and landed 750 ft (225 m) away from the explosion area, which was completely wrecked. The emergency services were called out, and it was a lucky coincidence that due to the shift change everyone had been indoors, so no one was killed or injured.

The subsequent investigation found that the highly viscous waste matter had blocked vapour and pressure relief valves and that the explosion was due to a chemical auto-reaction at the bottom inside the tank wagon resulting from local overheating of the waste matter. A different procedure with low-pressure heating and a mixing process called sparging would have calmly liquefied the matter and it would have come out easily. Training had been inadequate and there were no comprehensive written instructions from the shipper to deal with this type of matter, whilst no research had been performed as to how to deal with this material at the receiving plant. Moreover, the DOT had issued no guidance on the loading, unloading and transportation of this material in its manuals, and given that the waste was normally solid and stable, it was questionable whether tank wagons were the best means of storing and transporting this material.

Riverview, Michigan, 14 July 2001 at 03:45

An unloading pipe between a tank wagon and shore-based storage facilities failed and disintegrated on the discharge plant of Atofina Chemicals at Riverview, Michigan, at 03:45 on 14 July 2001, leading to the escape of methyl mercaptan (methanethiol) gas. It was not immediately noticed but eventually the emergency services were warned, and shortly after the arrival of the fire brigade commander at 04:09 the gas ignited and created a fireball 200 ft (60 m) high. Burn damage to an unloading connection to a tank wagon on a neighbouring track caused the escape of chlorine, also poisonous and dangerous. The fire was extinguished at 09:30, three employees having been killed in the blast and the fire, several others suffering mainly respiratory problems, and 2,000 nearby residents having been evacuated.

The investigation found that Atofina's unloading procedures and equipment were not sufficiently safe. Inadequate maintenance, storage and procedures for use of unloading equipment all contributed to the corrosion of a metal pipe that connected the top of the vehicle tank with the storage facilities to the extent that it partially disintegrated while gas was flowing through it. The gas-flow control valve of the tank wagon, which should have cut the gas flow when it became too high, was also found to be defective. Nonetheless, despite the deaths and injuries, the worst-case scenario of a major gas fire on a chemical plant, aggravated by the escape of chlorine, had been prevented.

Freeport, Texas, 13 September 2002 at 09:30

Inadequate procedures to control potentially destructive chemical processes in vessels such as road and rail tank vehicles led to excessive pressure in and the consequent explosion of tank wagon DBCX 9804 on the premises of the BASF Corporation at Freeport, Texas, at 09:30 on 13 September 2002. The vehicle had been steam-heated to liquefy 6,500 gallons of chemical waste consisting of cyclohexanone oxime, water and cyclohexanone in preparation for its transfer to a road tank vehicle for ultimate disposal. Following a major explosion the transfer facility with parked rail and road vehicles was comprehensively destroyed, the tank car dome being found in a house approximately a third of a mile (535 m) away. Two shore-based storage tanks were also damaged, which resulted in the escape of fuming sulphuric acid and sulphur trioxide. As usual, the nearby residents were hurriedly evacuated and 28 people were briefly hospitalised with various complaints and minor injuries. Inadequate procedures had allowed the cargo to become overheated, which in turn had set up a runaway exothermic decomposition of the chemical waste, causing the tank of the rail vehicle to become over-pressurised. Once again an incredible stroke of luck had prevented deaths and serious injuries.

***

Clapham Junction, London, England, 12 December 1988 at 08:10
(Triple collision caused by wrong-side signalling failure)

Introduction

This concerns an accident which, like ***Eschede*** in Germany and ***Granville*** in Australia, exposed the extent of the ongoing deterioration in standards of railway operations, fully in line with the warning given by Sir Peter Parker in his final days as British Rail chairman about the 'crumbling edge of quality' resulting from insufficient funding. Another reason for including the Clapham Junction accident is that it is one of several in this book with which I had some form of personal involvement, either directly or through knowing people involved in the accident or its aftermath. In December 1988, I was not living in Britain – that would come five months later – but my friend Jim Vine, a senior British Rail (Southern Region) traction and rolling stock engineer, was appointed to a four-person internal joint inquiry committee on the Clapham Junction accident.

Terry Keating, a former BR South Western train driver, who worked the 06:37 ECS from Wimbledon depot to London Waterloo, was another acquaintance with direct personal experience of this accident. He was one of the five out of 28 drivers who experienced changes to signal aspects that morning when passing signals WF138 and WF47 prior to the collision. The changes that he experienced only later turned out to be possible precursor incidents to the collision. In Terry's case he was slowed down to a red at WF47 by properly restrictive distant signals, but when arriving at WF47 it came off to green and from there he had a clear road all the way to Waterloo. As WF47 was a controlled junction signal, this meant that the road ahead had been reset through an unoccupied line, which in turn made signal WF47 clear to a proceed aspect. It is something that occurs very regularly close to stations, sidings and yards, and under no circumstances needs to be reported. Nevertheless, when Terry stood in the dock he was put through the wringer under intense questioning, which led to physical problems that disqualified him from driving trains. His account of these proceedings during the time we spent together at Stewart's Lane depot graphically revealed the utter horror of being questioned by lawyers with not the least knowledge or understanding of basic railway matters, but whose agenda was to pin the maximum blame on the railway by attempting to wrong-foot witnesses by repeatedly asking similar but slightly different worded questions. They sought to apportion blame for non-existent mistakes and accused witnesses of neglecting to report a signalling failure the consequence of which allegedly contributed to the fatal accident. All five drivers involved – Keating, Mansbridge, Malone, Christy and Priston, in order of the incidents – were eventually cleared as they had breached no rules at all.

A further reason for its inclusion in this book is that, as a result of the accident, the idea of implementing an effective form of Automatic Train Protection (ATP) was seriously considered in Britain for the first time, even though in this particular case such a system would not have prevented the collision. It was a problem with the signalling itself that was at the root of the disaster, not a train handling issue, and ATP would also have allowed the colliding set through as the covering signal showed a green aspect. Nevertheless, the British Government announced two ATP test schemes that, surprisingly, were actually installed. However, when the cost of purchase, development, installation and running of the system was made public, further implementation of ATP, as well as serious maintenance of the test installations, was quietly abandoned. That decision was perhaps defensible in the light of the dire financial situation of British Railways at the time, but it did condemn quite a number of people all over Britain to injuries and deaths in later accidents. The installed but neglected ATP equipment was subsequently revived when the *Southall* and ***Ladbroke Grove*** accidents on the Great Western main line once more exposed the lethal flaws of the obsolete Automatic Warning System (AWS). In fact, these two accidents secured the continued existence – nay, even a piecemeal extension – of the ATP test scheme on the line from London Paddington to Bristol. Fitting of a different (simpler and cheaper) train protection system called Train Protection & Warning System (TPWS) in which the ancient AWS was incorporated finally came about in 2002/03, in the wake of the ***Ladbroke Grove*** accident. By then, true ATP had *de facto* become obsolete, whilst its successor, the European Train Control System (ETCS), was not yet ready for installation. Nevertheless, TPWS, however flawed in comparison with ATP from a signaller's and driver's point of view, did save a number of lives. More importantly, it had proven wrong those bean-counter 'experts', aware of the price of everything but knowing the value of nothing, who advocated that the limited number of lives and injuries saved did not warrant the cost of such systems. Accidents are very expensive.

The location

The collision occurred on the 4-track South Western Up and Down main lines approaching Clapham Junction station, used by services between London Waterloo and the south-west of England. A typically intense Southern commuter railway operation, during rush hour the interval between arrivals at Clapham Junction could be as little as 2 minutes. Slam doors opened by commuters themselves before jumping onto the platforms well before the train came to a stop helped to maintain this sort of throughput. Clapham Junction is one of the busiest stations in the whole of Britain, no fewer than three 4-track main lines (Windsor lines, Portsmouth lines and Brighton lines) coming together for onward connections to both Waterloo and Victoria.

The accident

The triple collision was caused after a train from Basingstoke had stopped at signal WF47 to report a technical SPAD at preceding signal WF138. Due to a fault in the signalling, WF138 showed a green aspect rather than the red aspect necessary to protect this train, as a consequence of which a following train from Poole ran at speed into the rear of the stationary train. Unfortunately, at that moment a third train to Haslemere was passing in the opposite direction and was sideswiped and derailed by the two colliding services. The collision was noticed in the electric power control centre when the whole of the Clapham Junction area went dead, but at the accident site it was unknown for some time whether or not the power had been switched off.

The disaster claimed the lives of 33 passengers and railway staff at the scene, two more passing away later in hospital. More than 500 people were injured, 69 very seriously. It was by no means a good period on the railways in Britain, especially on the Southern. Two weeks after the Clapham Junction investigation hearings had ended, a rear-end collision at *Purley* on the Brighton line occurred following a SPAD at speed on 4 March 1989. Two days later, a head-on collision on a single-lead junction at *Bellgrove* near Glasgow – another money-saving but risk-increasing development of recent years – took place following a 'starting against signal' SPAD.

Rolling stock involved

The Clapham Junction accident involved British Rail Southern Mk I equipment only, built in stages from 1950 until the 1970s to an obsolete carriage construction concept of heavy steel frames with separate, lightly constructed bodies with in many cases steel-plated wooden body frames.

When involved in accidents this rolling stock showed a propensity to telescope when the heavy steel underframes slid over each other, the passenger survival space being ripped apart or compressed. For this type of commuter operation, however, this slam door rolling stock was well-nigh ideal, allowing massive throughputs in the shortest possible station dwell times.

Both the Basingstoke and Haslemere trains were 12-car combinations of three 4-car Class 423 VEP 2'2'+2'2'+Bo'Bo'+2'2' outer-suburban electric multiple-units (EMUs) featuring the successful electro-pneumatic Southern EPB brake system. In 1967, 194 of these units had been introduced to work outer-suburban services from the Southern termini in London into the southern counties. Compared to other types of BR Southern suburban rolling stock they were rather comfortable as they travelled on the British interpretation of the Swiss Schlieren bogies and gave quite a good ride. However, they could be draughty during cold spells. The motor coach had four of the redoubtable 250 hp (186 kW) English Electric EE507 traction motors that enabled a permitted maximum speed of 90 mph (145 km/h). The length of each unit was 80.98 m (266 ft) and the weight 148.30 tonnes. Despite their legendary reliability (they were so basic that nothing much could go wrong with them) they had to go in the late 1990s due to their slam doors and also for crashworthiness reasons.

The Poole train was a 12-car push-pull combination of a 4-car Class 430 REP Bo'Bo'+2'2'+2'2'+Bo'Bo' traction unit with two 4-car Class 491 TC 2'2'+2'2'+2'2'+2'2' remote control trailer units attached, the REP leading the service into London as normal. The REP units, 15 of which were introduced in two batches as late as 1975, had 3,200 hp (2,385 kW) installed traction power – eight EEC 546 400 hp (298 kW) traction motors per unit for 90 mph working when combined with two TC sets – in the two powered end vehicles and had two trailers in between. They were built to work the express services from London Waterloo via Southampton to Bournemouth when the third-rail dc electrification was finally extended to that destination, allowing the last working steam along the line to be retired. One reason why government permission to electrify the line to Bournemouth was given was the fact that no new rolling stock would be built. Existing Mk I locomotive-hauled rolling stock had to be reused after its displacement by the significantly more modern Mk II and Mk III vehicles on other lines through obsolescence. The REPs were specifically built to work together with the unpowered TC sets that had a driving cab at either end of each 4-car set. The REP unit weight was 175.30 tonnes, the TC unit weight being 134.10 tonnes. All were between 80.91 m and 80.98 m long.

The operation of these REP/TC push-pull trainsets was quite innovative. The remotely controlled REP propelled one or two TC sets from Waterloo to Bournemouth, where the REP at the rear would be uncoupled. A modified Class 33/1 1,500 hp (1,120 kW) diesel-electric Bo'Bo' (known colloquially as a 'push-pull Crompton') then coupled up to the front of the TCs and hauled them onwards to Weymouth, from where they would return with the now remotely controlled 33/1 pushing the TCs back into Bournemouth. Here the REP would take over at the front again and haul the TCs onwards to Waterloo. REP and TC units, as well as the push-pull use of the Class 33/1 diesel-electric locomotives, disappeared from the late 1980s onwards, when the powerful traction installations of the scrapped REP units were reused (a long-standing and irrepressible Southern tradition) in the 34 new Class 442 Mk III EMUs. These were able to continue to Weymouth under their own power as the Southern third-rail dc electrification had by then been extended to the resort. These excellent-riding, comfortable and smart-looking 160 km/h (100 mph) units, having bowed out on the South Western, were refurbished and put to work on the altered Gatwick Express and Brighton main-line services from London Victoria, finally bringing serious inter-city comfort and speed as well as good looks to the Brighton route. At the moment they are standing idle again following commissioning of new EMUs for the Gatwick Express.

The accident log

Driver McClymont worked the very busy 07:18 12-car VEP service from Basingstoke to Waterloo with conductor Fritzsche along the South Western Up main line, doing about 60 mph (100 km/h). Fritzsche had decided to stop checking tickets as the train was so overloaded that he could no longer move from coach to coach. He had some 15-20 people in the security cage of his brake van alone, which was by no means unusual at that time of the morning. As they approached Clapham Junction under clear signals, driver McClymont suddenly experienced a technical Category C SPAD at brand-new signal WF138, whose green aspect abruptly changed to red as he approach at speed. Adhering to the relevant rules, he immediately slowed down and managed to stop at the next signal, WF47, the junction signal for the platform loop at Clapham Junction, to report the incident to the controlling signaller. To McClymont's surprise, the signaller professed to have no indication of a signal aspect change on his panel. In no doubt that his train was covered by a red signal aspect, McClymont rang off and went back to his train just as another (a 12-car VEP service ECS Waterloo to Haslemere) passed by at speed on the Down main line. Seconds later, at 08:10, he suddenly experienced his own train moving forward whilst terrible rending sounds and dust or smoke emanated from the rear. He could not see exactly what had happened because of the curved brick retaining wall of the cutting his train was in, but he did not need to guess and immediately phoned back to the signaller to report a heavy rear-end crash. The signaller put all signals back to danger and immediately called the emergency services.

The colliding train from behind was the 06:14 REP/TC 12-car Up express service from Poole to Waterloo, a train not scheduled to stop at Clapham Junction. This train had come from Poole rather than Weymouth because, as a result of vandalism, a derailment had occurred between Branksome and Poole and the line to Weymouth was closed. Due to staff shortages the buffet in the second carriage was not operating but the seating and standing room in this part of the train was rather well patronised. Driver John Rolls had the terrifying experience of coming through a curve in a cutting at a speed of about 60 mph (100 km/h) under clear signals, applying a bit of brake to start slowing down for the permanent speed restriction through Clapham Junction and then suddenly seeing the tail end of another train on his track. Driver Flood, travelling on this train in the rear van, saw a bit of brake going in on the gauge in that compartment before the full emergency brake suddenly came in, after which impact followed. Rolls stood no chance of stopping in time, hitting the stationary Basingstoke train at approximately 35 mph (55 km/h). After the initial impact the train veered away to the right of the Basingstoke service into the kinetic envelope of the Down main line, right at the moment that the 08:03 ECS to Haslemere passed by (the train McClymont had seen coming past when returning from his initial call to report the SPAD).

Driver Alston of the ECS train witnessed the collision between the other two trains and saw the badly damaged coaches of the Poole train coming his way, debris smashing his side windows and entering his cab. Fortunately, it impacted with the second carriage instead of his first vehicle and broke the coupler, which no doubt saved his life. When Alston's vehicle came to a stop he found it was on its own, the rest of his train being well behind him and entangled with the remains of the first two vehicles of the Poole train. McClymont of the Basingstoke train, having climbed back into his cab to get the track circuit clips in order to put signals at neighbouring tracks back to red to protect the wreckage and the people in it, then spoke to an officer of the British Transport Police, Inspector Foster, about what had happened. Inspector Foster went down on to the track and ran back to assess the damage and the need for emergency services. In order to get an overview he climbed on to Battersea Rise road bridge before coming back to the signal-post telephone of WF47 and calling signaller Cotter, advising him that three trains were involved. Then, together with another railwayman, he got a short circuit bar and attempted to discharge the traction electricity on the third rail, but there was no muffled 'thump' as the bar connected because the train wreckage had blown the juice already. The heavy leading motor coach of the Poole REP unit had tossed the rear trailer coach of the Basingstoke train onto its left-hand side on top of the retaining wall and pushed the second vehicle off the track before getting mashed, together with the following buffet trailer, between the coaches of the Basingstoke train and those of the outbound Haslemere train. The first police and fire brigade units arrived on the scene at 08:17 and the first ambulances at 08:21. By 13:04 the last casualty was removed from the trains, the final body being taken out at 15:45.

Shortly after the triple collision a fourth train came hurrying towards Clapham Junction along the Up main line – a Waterloo-to-Waterloo 'rounder' service via Hounslow, Staines and Weybridge. Whilst that train was stopped in the branch platform at Weybridge the Poole train had roared past at line speed of 90 mph (145 km/h) along the Up main line. The 'rounder' then followed it, but lost it after New Malden. Doing about 65 mph (110 km/h), driver Barry Pike suddenly noticed that he had lost traction current close to signal WF152, the distant for WF138. In the circumstances he was sure he could manage to get to Clapham Junction with the train's momentum and decided to roll on. On coming through the curve, however, he suddenly saw the rear coaches of the Poole train 600 yards (540 m) ahead and immediately applied the emergency brake, coming to a stop 187 ft (56.5 m) to the rear of the stricken train. In train terms, that was a close call indeed. Despite the failure of the telecommunications equipment through overload, a typical feature of an accident, driver Pike managed to place a call to Clapham Junction A signal box, to advise that he was looking at what appeared to be the wreckage of three trains ahead of him but that covering signal WF138 on the Up main was nevertheless showing a single yellow proceed aspect. The signaller initially did not believe him, but the driver was indeed correct. The brand-new signalling, commissioned between 04:00 and 05:00 that very morning, had failed wrong side.

The history of the signalling

Despite the enduring scarcity of funds, British Rail (Southern) from 1978 onwards had embarked on a massive resignalling scheme out of Waterloo via Clapham Junction, to be finished in 1988/89, the long gestation period being due to the usual government prevarication with releasing of funding for the work. The main target was to replace old and increasingly failure-prone cabling from the 1920s and from the 1936 resignalling scheme, in places more than 65 years old, with new cables, as well as to modernise the signal installations and concentrate control in a regional traffic control centre rather than in various signal boxes. This ongoing work was known by the acronym WARS, the Waterloo Area Resignalling Scheme. What was unusual about this scheme was that it was handled as piecemeal renewal work under fairly low-placed middle management and not as the single massive resignalling project that it actually was. With a senior project-managing signalling engineer in charge who had overall responsibility for budget, delivery, quality and safety. Resulting from endemic understaffing, this ongoing resignalling work was achieved mainly by available people working long periods without days off, including consecutive weekends, and additionally doing regular daily overtime. This non-stop labour allowed them to earn good money but had a negative impact on the quality of their work.

As a train driver in 1992, four years after this accident and already under post-Clapham Junction 'Hidden' rulings (see ***Buttevant***), but with overtime and weekend work still endemic due to the traincrew depots being desperately short of staff, I was due to earn £10,500 basic per year. In my very first three-quarters of that financial year as a passed out driver, however, I earned more than £28,000. This was wholly due to non-abating 12-hour overtime days and continuously working 13 days every fortnight. For drivers this sort of practice came to an end only after privatisation in 1994, when following an oversight in the rather strange privatisation process, the traincrew unions were able to force the newly created train operating companies into providing decent pay and hours of employment, but for S&T and p-way workers the sharp labour practices continued well into the 1990s.

This situation was far from unique on the railways in Britain at the time but, as this book illustrates, many railway accidents in the UK and elsewhere happened under such tough overtime regimes. Due to these endemic staff shortages, signalling fitters in the WARS project worked unsupervised for long hours. Indeed, managers who should have been supervising and checking the work of the fitters were doing hands-on fitting and wiring work themselves. People were left to their own devices and just had to ensure that they delivered work to the quality required on their own accord and on time. They were, therefore, doing massive amounts of work in great haste, unchecked as to what they did and how they did it. Predictably, corner-cutting, sloppiness and dangerous methods of working developed unnoticed. In fact, many fitters never received serious training for the responsible work they were undertaking. Wire counts – checking the number of wires on the contact terminals of the relays with the design diagrams at hand – were never done. This lapse was at the root of this accident. Moreover, there had been clear precursors to this disastrous event. Between 1985 and 1987 there were no fewer than 15 wrong-side signalling failures on the Southern alone, the most notorious of which (after incidents and accidents) were the *Oxted* incident in 1985 as well as the Northfleet, *South Croydon* and the *Queens Road Peckham* incidents (the latter on 22 May 1988, some six months before the Clapham Junction accident). All were due to faulty wiring work, lack of wire counts, and questionable testing.

Three weeks before the accident one such technician was altering the feed to signalling relays in Clapham Junction A signal box in preparation for the decommissioning of this ageing structure and moving control to the future area signalling centre. This fitter nominally worked under a supervisor. There was also an assistant technician for the fitter and a testing/commissioning engineer. These people never saw much of each other in the course of working on the WARS project, and despite their titles, never saw much of each other's work. The fitter involved was described as a well-liked and hard worker who had been with British Railways since 1972 and was promoted to senior technician in 1981. After a year with the railway he had undergone a preliminary seven-week course. A few short courses later, he was doing wiring but had received no formal training and passed no exams in electrical engineering and electro-technical matters and principles. Yet he became a senior technician simply by staying on in the job, due to the scarcity of people willing to work under the conditions of low basic pay and the very unsocial hours offered by the railway versus the enormous amount of work that needed to be done. Previous to the start of the WARS work, this fitter had worked on the London Bridge and Victoria resignalling and considered himself as someone held in high regard by his superiors – as a reliable worker who could be left on his own and who would finish the jobs he set out to do and who did so on time. He had not taken a day off for a long time.

The work at hand was to replace power feed wires to signal relays in the relay room of the signal box. The old ones were to be cut away and removed, and new ones from a different power source brought in and connected. The old power source, however, had to be kept live to power a number of signals that had not yet been altered, so all connections from this old source were live as well. During this job the fitter had connected the wiring for new signal WF138 and disconnected the wires for old signal WA25. The electro-magnetic relays of WA25, however, were temporarily kept in use for WF138 but were powered from a new feed. This man, never formally trained as a signalling electrician and as such unaware of how to do such things properly, had developed rather unsatisfactory working methods. For instance, he pushed the old wires out of the way instead of cutting them short, rolling them up and then insulating the blank copper ends with insulation tape. And if he already used insulation tape, he would reuse the old ends lying about rather than cut off fresh tape as he should. Whilst finishing his work in the gloomy, dusty and dingy relay room of the signal box that day he was distracted by someone else, forgot what he was doing and accidentally left the very much live old feed wire with its stripped blank copper end hanging, just pushed out of the way. Poor supervision, lack of time and slack safety discipline ensured that no wire counts off the terminals of fuse boxes and relays were done, so the rogue feed wire that for so long had fed relay WA25, now with a new feed used for WF138, was never noticed.

Two weekends later, on 11 December 1988, the fitter was back in the relay room again, working on the replacement of a neighbouring relay, and in the course of that job he disturbed the wires he had previously worked on. At that time the non-insulated and still live old feed wire moved back to its former position, its blank copper end touching its old terminal. This went completely undetected. Again, no wire counts were done that should have revealed this wire. The following morning between 04:00 and 05:00 this resignalling work was commissioned in readiness for the rush hour. What happened next was that this loose-hanging and occasionally moving rogue live wire provided an intermittent feed to the green aspect of new signal WF138 when it made contact with its old terminal. It was only when that wire had temporary lost contact that the signal showed a danger or caution aspect. As soon as it moved back to touch its former terminal, however, the green signal aspect illuminated again. And that was the green instead of the red aspect that John Rolls on his REP/TC combination from Poole saw before he slammed into the end of driver McClymont's VEPs from Basingstoke. It was also the source of the single yellow distant signal that driver Pike of the Waterloo-to-Waterloo 'rounder' via Staines saw when he was looking at the wreckage 66 yards (59.5 m) ahead of him, which should have been covered by a red signal.

After the collision, before photographs had been taken or evidence had been secured, the rogue wire was spotted by an investigating signalling engineer and was partially cut off, rolled up and secured, as it should have been all along. Two other signalling engineers also spotted it and almost cut it off completely but were told to leave it as evidence. They had found it unsightly, offending against good signal wiring practice. The following day yet another signalling engineer cut off another 13 in (33 cm) of the rogue wire, before removing both old and new feed wires and replacing them with another new feed wire. There is a good reason why, after subsequent accidents, railway people in Britain, even those in legitimate investigatory roles, are not allowed anywhere near the wreckage by the railway police unless formally authorised. This did initially lead to a few ridiculous spats between investigating railway people and transport police in those heady days of privatisation with its rapidly developing blame culture, but the issue has now been satisfactorily resolved.

Comment

It is inevitable that rail safety at some time or other will suffer in a situation of severe financial constraints, whether self-imposed or enforced by government, as was the case with British Rail at the time. The business of running trains is expensive and achieving up-to-date railway safety standards unfortunately costs a great deal of money on top of those operational expenses. Such higher levels of safety are too easily put aside as being of little relevance, as was the case in Britain over a long period of time, given the inherent safety of rail operations. An accident such as this was the catalyst that changed opinions (see ***Eschede***, ***Granville*** and ***Chatsworth***).

After Clapham Junction no one accepted the say-so of British Railways with regard to the safety of its operations any longer, and the setting up of an administrative system to implement, prove and check safety measures at all levels was the direct result. This particular accident shook up British Railways quite badly, simply because it exposed its faith in the quality of its basic railway safety as unrealistic, as indeed the ***Eschede*** accident would later do to Deutsche Bahn. More importantly, the accident unsettled the British Government sufficiently to make it provide more funds, together with promises of more to come. It would take two decades, including the privatisation of rail services, before better and contractually fixed rail funding arrangements were provided, but they did eventually come about. Unfortunately, during those years more accidents occurred, before the railways in Britain managed to extricate their way out of the mire in which they were stuck.

There are parallels with such accidents as those at ***Brühl*** in Germany, ***Wijchen*** in The Netherlands and ***Pécrot*** in Belgium. The indecent haste to get people working, with the consequent lack of effective training and supervision, and the leave-them-alone attitude to let them get on with their job, coupled with mental exhaustion due to overwork, were all contributory factors. What the Clapham Junction accident showed is the necessity of formalising safety-critical working methods to prove that they are as safe as possible, of ensuring that standards are adhered to, and of the adequate supervision of workers to ensure that they always carry out their duties in accordance with those standards. The railways in Britain have come a long way since this accident, but sadly it was not until the *Grayrigg* derailment on 23 February 2007 that everyone finally was convinced of the necessity of these often tedious methods to ensure that everything functioned fully as it should. The railways in Britain are indisputably as safe as circumstances and equipment allow, whereas many rail operations elsewhere demonstrate a lack of interest in comparison. Looking at the international accident statistics, the railways in Britain compare favourably, and that is a good thing to know if you have to use them often.

***

Ufa, Russia, 4 June 1989 at 01:15

(Massive natural gas explosion causing train fires)

Some history

At first glance this is not really a railway accident. The railway became involved as a result of what, from a safety point of view, were debatable management decisions by the typical old-style, rigid Soviet bureaucracy. The default position was to stick with the old methods, which led to a disregard for caution, rather than query safety issues and thereby leave oneself vulnerable to the displeasure of superiors, potentially risking one's job. This same mindset had earlier caused the infamous *Chernobyl* nuclear disaster on 26 April 1986 (see Chapter 5) following incompetent handling of the nuclear energy process management during trials. Later, after the demise of the Soviet Union, the same issue would cause the submarine *Kursk* (see Chapter 5) to sink after a torpedo with known unstable hydrogen peroxide propulsion system characteristics exploded and set off others. This torpedo had been seen to drop out of its handling slings during loading and had actually been videoed during that incident, as shown on German TV, but had nevertheless been taken on board instead of removing it for safe disposal. These were exercise torpedoes that, because they did not carry warheads, never had their internal welds in their hydrogen peroxide ducting checked. A faulty or deteriorated weld that ruptured during the fall and caused an unwanted chemical reaction is the probable cause of this disaster. As a result, a series of heavy explosions in the forward torpedo room caused the extremely strongly built submarine to be damaged beyond what its design would allow.

The location

The railway accident happened near the Russian cities of Ufa and Asha in the Ashinsky district of the Bashkir republic, relatively close (in Russian terms) to the foothills of the Ural mountains where the line crosses from Europe into Siberia. Cities within a 200-mile (320-km) range from Ufa are Chelyabinsk and Magnitogorsk.

The cause of the accident

The disaster was caused by a 1.7 m (5 ft 6 in) rupture in the 1,852-km (1,150-mile) long Western Siberia–Ural–Povolzhye gas pipeline that was constructed from 1981 onwards and had originally been intended as a crude oil pipeline of 720-mm (2 ft 4 in) circumference. At some time during its construction, however, natural gas had replaced oil as the great earner of hard foreign currency and a decision had been taken to convert the pipeline for transport of liquefied gas. In fact, this happened close to the date of its originally planned handover for operation in 1984. According to regulations, it was not permitted to pump liquefied petroleum gas and liquefied natural gas through pipelines with a circumference larger than 400 mm (1 ft 4 in), but these were apparently seen as trifling details of little or no consequence under the Soviet regime of the time.

Whilst transmitting a pressurised LPG mix at approximately 22:00 on the day before the accident, compressor operators in the area had noticed that gas pressure fell off beyond their compressor station. But instead of shutting down in order to initiate an urgent investigation (which would delay deliveries and thus lead to dissatisfied customers) they simply speeded up the compressors to maintain throughput. As a result of local geographical conditions and duc to windless and warm weather, the liberally escaping decompressing cool gas stayed close to the ground in the hot and humid night and formed 'gas lakes' of highly flammable propane and butane in two depressions in the landscape close to the pipeline, further gas drifting for about 8 km (5 miles). It was through one of these depressions that the South Uralian branch of the Trans-Siberian railway line ran.

The accident

In the meantime traincrews of passing trains on the Ulu-Telyak to Asha section of the Kuybyshev (South Uralian) Trans-Siberian railway line – 1,710 km (1,060 miles) from Moscow and 11 km (7 miles) from Asha station – had radio-telephoned in to report having travelled through a sort of hazy, strange-smelling cloud. Here another level of officialdom, in this case the railway operations management that dealt with the calls, entered the scene but did not consider that the situation warranted stopping traffic in order to find out what was going on. This southern Trans-Siberian branch was heavily used by international intermodal freight services and delays would create unhappy customers. It basically was the same ex-Soviet mindset that had led to speeding up the compressors rather than investigating the problem earlier. Stopping trains would result in long lines of waiting and diverted services.

A further quirk of fate was that two overnight sleeper trains, Nos 211 (20 carriages) and 212 (18 carriages), running between Novosibirsk and Adler (near Sochi) and back, were timetabled to meet each other in this section and were running on time. These were popular services, used by many for holiday travel between Novosibirsk and the Black Sea resorts. Both trains on the night in question were carrying no fewer than 1,284 passengers (of whom 383 were children who had spent their holidays in youth camps) and 86 traincrew. When the two electrically hauled trains met at 01:15, their wakes through the air violently mixed the dense gas cloud with oxygen, which exploded when the contact point between a pantograph and the 25 kV ac 50 Hz electric wires momentarily sparked. This explosion was a massive thermobaric event equalling the detonation of 300 tonnes of TNT, similar to the nuclear bomb dropped on Hiroshima.

The aftermath

The blast blew both locomotives plus 11 of the 38 coaches off the track, burning virtually the entire lot. No less than 250 hectares of surrounding forest was stripped of its leaves and burned, people 100 km (60 miles) away observed the fireball of 2-km (1¼-mile) circumference light up the night sky, and windows in Asha, more than 10 km (6 miles) from the explosion, were blown out of their frames. Official information stated that 573 of the people on the two trains were killed; other sources give the number as 675, which corresponds with the number of victim's names on the monument erected as a memorial. Of the survivors, 623 were injured to various degrees, the great majority suffering life-changing injuries due to severe burns. Of the 383 children on board 181 were killed and the rest injured, some of those with the worst injuries receiving specialist burn treatment in Britain.

Following the disaster at Ufa, and completely contrary to normal Soviet practice, a visibly furious Mikhail Gorbachev visited the site accompanied by Russian and invited foreign journalists. He was publicly scathing towards the pipeline and railway officials on television, which did not sit well with Soviet hardliners. This unusual response was probably one of the factors that contributed to his demise as president of the Soviet Union. Soon after, he was ousted by the reassuringly macho, old-style communist Boris Yeltsin.

On the other hand, the admittedly late and initially reluctant but later whole-hearted acceptance of Dutch heavy salvage specialists Mammoet to deliver the technology and equipment needed to salvage the wreck of the top-secret submarine *Kursk* showed just what profound political changes could nevertheless take place in that giant nation with its centuries-long history of distrust and suppression of dissent. At present, however, there is an unfortunate retreat into secrecy and obfuscation again after the negative fallout from the war in Ukraine and the destruction there of Malaysia Airlines flight MH17 by a mobile SA-11 BUK surface-to-air missile on 17 July 2014 (resulting in the deaths of 283 passengers and 15 crew). According to a joint international investigating organisation, this missile in all likelihood was erroneously launched by either a Russian or Russian-backed separatist operating crew, mistaking the aircraft for a Ukrainian Antonov An-26 transport. However, the Russian Government and its allies in Ukraine will never admit responsibility as it would cause political destabilisation and thus even bigger problems.

***

Kissimmee, Florida, USA, 30 November 1993 at 12:40

(Heavy transport level crossing accident)

The accident

Many Britons still remember the shocking level crossing accident on the AHB at *Hixon* on 6 January 1968, when a 12-coach electric locomotive-hauled express from Manchester to London Euston smashed at 75 mph (120 km/h) into a Robert Wynn & Sons Ltd operated low-loader (travelling with police escort) carrying a 120-ton transformer. The heavy road transporter was creeping at 3 km/h (2 mph) across the level crossing to allow the crew to check grounding risk and clearance of the 25 kV overhead line equipment when the train struck. Eleven people, including the three railwaymen in the cab of the locomotive, were left dead and 41 were seriously injured. The main contributory cause at Hixon was a complete ignorance of the potential danger posed by an automatically operated level crossing. No one had thought it necessary to contact rail traffic control for trains to be stopped in the event that the crossing deck was not cleared in the time between the start of a closure and the moment a train would be on the crossing. The transporter crew and police escort did not understand that the equipment of an automatic level crossing does not check whether the crossing surface is cleared for the passage of a train after the warning commences, despite literature to that effect having been published.

It might be thought that the report of that particular accident travelled through the world of specialist heavy transport firms, opening their eyes to the risks of automatic level crossings; I remember reading about it in a publication in The Netherlands when I still lived in that country. Nevertheless, an almost carbon copy of that accident happened in the US as described here. In fact, similar accidents have continued to occur, with more recent examples of grounded low-loaders hit by trains at automatic level crossings happening in the US, Canada, France (twice, one of those at ***Tossiat***), Belgium, The Netherlands and Sweden. In virtually all cases, the accidents were triggered by the unsafe practice of stopping on the crossing due to grounding or to check clearances, which could have been avoided if sufficient preparation had gone into choosing a particular route. And most of all: the rail operators should have been notified of the intentions in advance and asked for co-operation.

Fortunately no one was fatally injured in the Kissimmee incident, but six people, including the locomotive driver and second man, were seriously injured, 53 people receiving minor injuries. As luck would have it, the first vehicle behind the locomotive was not in public use when it was crushed under the generator unit and part of the transporter. Had that carriage been occupied there would have been an even greater death toll.

The location

The accident occurred on an AHB level crossing on the access road to the Cane Island power station construction site. This access road branched off Old Tampa Highway and climbed at a gradient of 1 in 26 (3.8%) to the level crossing and slightly past it, reached a summit and then descended at 1 in 23 (4.4%) to the entrance of the Kissimmee Utility Authority (KUA) construction site in Intercession City where a gas turbine electricity-generating module was being installed. The access road complied with the relevant highway design requirements for all road traffic.

The railway line involved, a CSX Transportation line, was single track and non-electrified. It had a 1° curve 0.07% off level with 3 in (7.5 cm) of superelevation (cant) of the outside rail and allowed a line speed of 79 mph (125 km/h). The AHB level crossing (AAR/DOT No 643879-N) had been installed earlier that year and had been in use since 5 August. Whistle boards were provided at the strike-in points for activation of the level crossing equipment. On the day in question the weather was clear and dry.

The train involved

Amtrak (the US National Passenger Railroad Corporation) train 88, the 'Silver Meteor' from Tampa in Florida to New York City, was the train involved in the incident. The locomotive was one of the successful GM EMD F40PH machines, Amtrak No 306, a 3,000 hp (2,238 kW) Bo'Bo' design developed and rebuilt from the first generation of Amtrak passenger diesel locomotives (the GM EMD SDP40F Co'Co' machines) during the later 1970s. This came about because the Co'Co's possessed an unfortunate habit of causing derailments when at speed by compromising track integrity, as did their electric counterpart on the Northeast Corridor, the General Electric E60CP Co'Co'. At the time of this accident the locomotive was carrying virtually its full load of 1,800 US gallons of diesel fuel. The first coach behind the locomotive was a heritage mail van, the second a heritage combined mail and luggage van, the third vehicle a heritage sleeping coach, the fourth a heritage lounge car and the last four were 1970s-vintage Amfleet-type seating vehicles.

Heritage vehicles came from the fleets of the one-time private train operators that Amtrak had taken over and modernised. Amfleet coaches were of a rather distinct stainless-steel Budd/Amtrak design.

The road vehicle involved

The vehicle was a combination of a heavy load transporter and a 1990-built run-of-the-mill Peterbilt 3-axle tractor with a standard cab/sleeper compartment configuration and an 18-speed manual gearbox. The heavy load transporter, however, was an innovative modular vehicle of which the tiller, the drag/steering beam, rested on and was pulled and steered by the 'fifth wheel' of the tractor, the turntable with king-pin assembly that normally takes the trailer. This tiller was attached to two automatically steering 2-axle dollies combined under a bridge frame, on top of which there was a hydraulic tower to raise or lower the load deck. From the top of this hydraulic tower there was a so-called 'gooseneck' bracket attached on its other side to the load deck. The load deck was extended to its maximum length, consisting of three elements bolted together. At its rear end was another 'gooseneck' leading to the second hydraulic tower, resting on a second bridge frame, the ends of which rested on two steering 2-axle dollies. The tiller at this side, however, ran to another steering dolly that could be used to steer the rear section of this rig independently (e.g. to clear obstacles on sharp curves). An interesting feature of this rig was that by lowering the load deck on the ground with the hydraulic towers, both the 'goosenecks' tipped down and opened up spaces behind their hinges between their ends and the load deck. By filling these spaces with so-called shims, pieces of sheet metal, the 'goosenecks' would be restricted in their movement towards the load deck which could then be raised on the fully extended towers to clear 2 ft (60 cm) between the underside of the load deck and the road surface. This feature is relevant to understanding the accident. This whole road rig was approximately 185 ft (55 m) long and required as a minimum a truck driver, a rear tiller steering operator and an equipment supervisor to operate it. On the day of the accident the regional manager of the operator, Rountree Transport & Rigging Inc, accompanied the transport in a separate vehicle, as did two uniformed but off-duty highway patrol police officers in a marked police car. The latter were employed in an approved manner to earn extra overtime. Portable two-way radios were the means of communication between all involved. The load was an 80-ton standby gas turbine electricity-generating unit 14 ft (4.25 m) high, 14 ft wide and approximately 57 ft (18 m) long. The module was covered with blue tarpaulins during the course of the trip.

The run-up to the incident

This was not the first such transport Rountree had carried out for the KUA electricity generating board. During their first trip, on 1 October 1992, the transporter became stuck under a cantilevered bracket-mast that carried the level crossing warning signs for the AHB on Falkenberg Road in Tampa, which should have alerted various bodies that Rountree's scouting missions might have been a tad less than thorough. At that time a CSX local office employee overheard the radio traffic between the transporter crew and police and decided to go and take a look. At the site he asked to see their relevant permit papers (their 'Right of Way: Passage' permit, to be precise) to pass level rail crossings with this oversize load. Neither the Rountree regional manager nor the escorting police officers knew of the requirement to make the railway operator CSX aware of their intentions and to obtain its approval and co-operation.

The cost of the permit at the time was $200, which was mainly to pay for the flagman to secure the level crossings. The rail operator published details of the transport in the crew notices and cab crews then had to radio the flagman to obtain permission to proceed through the area in question. This was the routine way of dealing with temporary obstructions of the track. This same CSX employee then organised the permits and acted as flagman for several such convoys in his control division between October 1992 and March 1993. He later told the inquiry into the accident that he had personally told the Rountree regional manager never to enter a level crossing with such a convoy unless he had the necessary permits and had thereby obtained flagman protection. During this discussion the regional manager had said that it would only take him a minute to clear the crossing, but the CSX man had clearly told him that from the start of the warning he would get only 30 seconds before a train would run into his precious cargo, which meant that he would never be able to clear the track in time if a train approached.

The Rountree regional manager had scouted the route for this particular consignment for a day and a half, planning to use many of the routes they had taken before. He was, however, not present at a meeting between KUA staff and the Rountree vice-president to discuss the particulars of this transport. The vice-president told the KUA delegates that the access road was fine for the move (which in fact it was, subject to some extra work with the rig to clear that hump near the level crossing in the road) and that a flagman would be provided at the crossing. In the meantime, however, the above-mentioned CSX permit issuer and flagman had retired and a new man had taken over his job. This man knew nothing about those 'Right of Way: Passage' permits and how to issue them. For the job discussed in this chapter he eventually handed out only the first page of the two-page form. Unfortunately, it was the second page that contained the necessary names and telephone numbers to call in case of problems.

The Florida Highway Regulations demand that heavy/oversize/low-clearance transport operators notify railway operators of their intended route, and that the notification should include, among other things, full details of all level crossings to be negotiated. The railroad operator is then bound to provide protection with a flagman at the crossings. At the accident inquiry both Rountree and CSX told the authorities that they were not aware of this stipulation. In fact, no other Florida state laws or US federal laws specifically require such special transport carriers and railroads to co-operate with providing protection at level crossings, but by law outsize road transport must obtain permits to operate on public roads by submitting an Overload Permit Request. This includes the dimensions of the vehicle intended to be used and the route written in detail and with a road map. This is checked and if found in order a 'move request number' is issued, which can be converted into a permit. Something like 90,000 such permits are issued annually. Local permits from all the counties that are passed through must also be obtained. Deviation from the route agreed makes any such permit null and void, which financially is a very risky thing if the load crashes through an unforeseen bridge or culvert. At this stage in the Kissimmee incident, however, the safe handling of the dangers of automatic level crossings was not further discussed and agreed upon. Field days for lawyers were therefore in the making.

The accident log

At 11:00 on the previous day, 29 November, the transporter crew oversaw the unloading of the generator module from a barge and its installation on the load deck of their rig. This went smartly enough and at 11:25 the convoy started the trip to the site in Intercession City. There was then a 5½-hour rest period until the early morning of 30 November, which was five hours short of the legal requirement and would lead the NTSB to suggest that fatigue could possibly have played a role in the accident.

That morning the new CSX man took a phone call from the Rountree regional manager. He was ready to move but the requested flagman had not shown up, so he was considering departing without him. The CSX employee urged him not to move and jumped in his car to help out. He guided the convoy across the seven level crossings in his control area and then went back to his office. The rest of the trip, which was booked to cross six more level crossings after the CSX man had gone back to his office in Tampa, went easily enough, despite the fact that no one else took over the job of flagman and there was also an illegal deviation from the booked route to avoid a recently installed replacement bridge. They passed the eighth crossing on this trip, on Pasco County Road 535, without problems, but the ninth crossing required the removal of the cantilevered bracket-mast carrying the level crossing warning lights at Zephyr Hills. Three more crossings were passed without problems, doing about 10-15 mph (15-25 km/h) in urban areas and 22 mph (35 km/h) along more rural stretches of road. At 10:59 the convoy turned off US Route 17/92 on to Old Tampa Highway, about half a mile (800 m) from the KUA plant access road on the left, where a street sign had to be removed to allow the turn to be made. A local KUA man told the Rountree regional manager that a train was due to pass their level crossing in about an hour, and this remark led the manager, who was in charge of this transport, to believe that there was contact between KUA staff and CSX. When they reached the KUA access road at 11:12 and attempted to turn in from Old Tampa Highway they found that the guy wires of a local electricity wire post made the turn impossible. It took 15 minutes to back out on to Old Tampa Highway again and then they went straight ahead for 2½ miles (4 km) where they did a U-turn and came back to the KUA access road to attempt the turn from the opposite direction. Meanwhile, at Tampa station Amtrak train 88 departed at 11:31, a minute behind time. At 12:08 the train called at Lakeland and then departed for Kissimmee, 45 miles (72 km) away.

The generator unit convoy once more reached the KUA access road and started the right turn into it at 12:25. The police escort vehicle had parked on the Old Tampa Highway hard shoulder, where both officers started to calculate the time and mileage entitlement for their escort bill to Rountree, as they had to pay their employer for the hire of the marked police vehicle for this job. (No doubt European eyebrows rise in surprise at this!) On the rising gradient to the level crossing the rear hydraulic tower was slightly extended to keep the gas turbine module level. The tractor passed the AHB level crossing, negotiated the hump in the road and began the descent on the downward incline. The front dollies followed, with the hydraulic towers fully extended to gain ground clearance, but the load deck nevertheless grounded, the generator module straddling the track. In order to increase the bottom clearance to its maximum, at 12:29 it was decided to put the load deck completely down on the road and level crossing surface, tip both 'goosenecks' down, insert shims and then raise the whole construction again. The Rountree regional manager reckoned it would take 6 minutes to do this, but much longer to push the convoy back off the crossing.

At 12:32 the KUA employee who had earlier informed him about the approaching train told him that a train was due in 25 minutes, although both estimates given by this man would later be denied during the inquiry into the accident. The regional manager now tried to contact the CSX train master (controller) at Tampa but got no answer. He attempted to call a CSX toll-free number but got through to an answer-machine with a menu facility. After some pushing of buttons he terminated the call in frustration, but at 12:39 he dialled the number again. After 32 seconds of listening to his mobile phone and pushing buttons he heard the unmistakable whistle of an approaching train. The AHB activated and the descending barriers snapped off on the gas turbine casing. The regional manager ran towards the grounded convoy and yelled through his two-way radio to his crew to get off the vehicle and find a place of safety. The rear tiller steering operator had already vacated his position on the rig, having seen the AHB warning activation. Having checked that his colleagues were clear of the convoy the Rountree regional manager turned and ran.

The driver of Amtrak train 88 spotted a large blue object round the curve through the trees ahead and suspected it was an obstacle on the new level crossing. On getting a clearer view and seeing the stranded vehicle on the crossing, he applied the emergency brake and exited the cab into the locomotive with his second man. At 12:40 the train hit the left flank of the road convoy at an alleged speed of 77 mph (125 km/h) but may in fact have been travelling a tad faster than that. The locomotive cut through the transporter and tossed the gas turbine up into the air before derailing and turning over on its left side, its nose flattened by 1 ft (33 cm) on impact. The gas turbine and most of its transporter deck came to rest on top of the derailed and overturned mail van behind the locomotive, which was largely crushed by this impact. The following three vehicles derailed and jack-knifed slightly. The locomotive and three of the derailed coaches came to rest near two underground high-pressure liquid fuel pipelines either side of and parallel with the railway track, used to transport fuel from an unloading facility in the port of Tampa to destinations inland. At the time of the impact, one was used for unleaded petrol at a pressure of 500 psi and the other for Jet A-1 aviation fuel at 300 psi.

The aftermath
The badly damaged locomotive, given its age and fairly imminent replacement with a newer type, was cut up on site, as also happened with the crushed mail van and the wrecked transporter equipment. As an aside, the NTSB noted that evidence was destroyed in that process. The total material damage was estimated at $14 million.

Rescue vehicles were on site within 11 minutes of a local resident calling the Osceola County emergency telephone exchange at 12:45. The quite badly injured crew had to be freed from inside the locomotive and were evacuated through a removed engine-room roof hatch. All passengers were off the train by 13:37, the crew slightly later, everyone at the site being completely unaware of the pressurised and highly flammable liquids virtually under their feet. Heavy plant and cutting gear were about to be brought in to start clearing up the wreckage when an off-duty pipeline operative who had heard about the accident on the news called his control centre at Tampa and both pipelines were immediately shut down at 13:54. The receiving customers were called to open all their vents in order to get the pressure in the pipelines down to 5 psi, by which time the rescue of passengers from the train was over and the clearing operations were immediately stopped. The fire brigade incident commander at the site said that he had been unaware of the pipelines under their feet, although a junior fire officer had brought back the plans from a training course and put those in the door pocket of his service vehicle, from where they were duly retrieved after the accident. Prior to being reopened, the pipelines at the accident spot were fully renewed, as a pipeline elsewhere had ruptured a few weeks after a similar event, causing a major evacuation of local residents.

Comment
Such are the embarrassments of being caught out by astounding ignorance at all levels, or as the American phrase so descriptively puts it, ‘caught with their pants around their ankles’, following which, naturally, everyone initiates a thorough review and an update of laws, protocols and procedures. Something very similar happened after the ***Clapham Junction*** and ***Eschede*** accidents. In the Kissimmee case, a replacement gas turbine module was duly ordered and delivered to the Tampa quayside, from where a different heavy transport operator took it to the KUA site at Intercession City. This time all the paperwork was in order and the transport went by the book, accompanied by a CSX flagman for the entire trip.

Like the *Hixon* or ***Tossiat*** accidents, this was an event that forced a change in the routine ways of thinking, dispelling much of the previous contempt for what had been perceived as irritating safety measures, and suppressing opposition to sticking strictly to the rulebook that resulted in delays and cost money. What does strike me as odd, however, is the number of official safety rules and regulations that none of the parties involved, including the police, were aware of. These regulations were, however, brought to light again, dusted off and reviewed as to usefulness, which was a good thing. It is nevertheless amazing that even now there is not greater awareness in organisations involved in heavy road transport of the risks and dangers inherent at places where road and rail meet. Surely there are special-interest publishers who highlight such shocking events? But despite initiatives by accident investigation and road safety organisations, heavy loads on level crossings world-wide keep being struck by trains with depressing regularity. The need to contact rail operators before using automatic level crossings with slow-moving special transports somehow does not enter the minds of heavy haul road contractors and operators.

In the case of the Cane Island AHB level crossing: Seven years later, on 17 November 2000 at 16:35, another such outsize load, this time having travelled 2,800 miles (4,480 km) from Salt Lake City, UT, to Intercession City, FL, was struck by Amtrak passenger train 97. Again due to lack of the — by now obligatory — notification of track operator CSXT. The train driver on his GE P-42 'Genesis' diesel-electric locomotive nr. 65, aware of the events in 1993, decided from things he noticed ahead to start braking and reduced speed from 79 mph (126.5 km/h) to 59 mph (100 km/h). The extra 1.7 seconds granted to the convoy made him hit the rear pusher truck only, not the load. It limited the consequences and damage considerably.

***

Cowden, Kent, England, 15 October 1994 at 08:27
(Head-on collision following SPAD)

Introduction

This accident happened at the time I started to drive trains in this particular corner of the south-east of England. I never 'officially' drove the Class 205 and Class 207 2- and 3-car 'Thumper' and 'Tadpole' (nicknames) type of diesel-electric multiple-unit trains that operated on these sylvan, non-electrified Southern lines; before I was due to learn these routes and traction I was moved away to the Gatwick Express train operating company following privatisation in 1994. Nevertheless, I knew the line and its rolling stock well. It was nice and unhurried, running through a very pleasant landscape and therefore a joy to spend time on with experienced tutoring colleagues.

The accident

A head-on collision occurred between two 6-car diesel-electric multiple-units on a stretch of single line in fog on a curved section of track. The cause was a 'starting against signal' SPAD onto the single line by one of the drivers concerned, although it was never positively established whether it was the driver or the guard who was actually driving the offending unit. Both men were found in the wrecked cab, but with the guard in the position closest to the controls. That does not really matter, however, as both men missed a signal displaying a red aspect, albeit that the accident remains the driver's responsibility. There were a number of circumstances pointing to less than satisfactory maintenance of the signal and of the AWS system on the train.

The line

The line is the mainly single-track non-electrified line from Hurst Green Junction to Uckfield. At one time it had been a double-track main line from London Victoria and London Bridge stations via East Croydon and Oxted to Lewes, where the line connected to the route from Eastbourne to Brighton along the coast. As such, it was an alternative to the Brighton main line, but as there no longer appeared to be a need for useful alternative rail connections in the 'efficient' Britain of the 1960s to the 1980s, the line was cut short at Uckfield in 1969 and largely singled, with a few passing loops in the 1980s (which happened to a number of other lines in those days). In fact, it is a wonder that the line still exists as a working railway and not as a cycle path like so many others, however wonderful that alternative is. Much has changed for the better with the appreciation of rail transport now, even if we are still quite some distance away from the world-class railway that various movers and shakers professed they wanted to see.

Despite all this official vandalism, the truncated line survived and, following modernisation of the track and rolling stock after this accident, the numbers of users are growing again, triggering increasing calls to restore double track, reinstate the lifted section and to electrify the entire line via Lewes to Brighton. In that respect this line would follow the

Above: A British HST set doing 125 mph/200 km/h along the Great Western main line towards London Paddington. A moderate fog hides the signals for the Bourton crossovers. For the sake of safety: Welcome Automatic Train Protection!

Below: Berlin technology museum. A German BFE 3rd class carriage from the 1850s. The uncomfortable vehicle is open but allows escape in case of fire. First- and second-class passengers were often not able to do that due to locked doors.

Left: Chillicothe, Ohio, USA summer 1982. A short DT&I freight slowly rolls through the station whilst the conductor on the caboose gets his copy of the train orders. This line has disappeared by now, as did cabooses with their conductors.

Above: Wateringbury on the Medway Valley line in the early 1990s. The level crossing is being opened by the signaller (see Buttevant).

Above: Anti-level crossing misbehaviour measure in the former Soviet Union. *IRSE collection*

Left: Cuijk, The Netherlands. With the kink in the road to slow traffic down, its raised kerbing in the road centre line to discourage zigzag moves, loud bells, 12 quick flashing red LED lights either side and retro-reflective barriers, what could possibly be overlooked about the danger indications of this AHB?

Utrecht Central station, Netherlands. A DD-AR and an ICM unit showing off their retrofitted obstacle deflectors to increase their resistance against derailment after level crossing collisions.
Stanley Hall collection

Burgsteinfurt, Germany: the westbound home signal and distant signal for the starting signal. Just above the disk of the distant is a small black triangle on the mast of the stop signal. That is the Zs1 Ersatz Signal that figured several times in accidents.

GWML BR-ATP of the ACEC TBL2 type, showing space occupancy requirement on board a Class 180 DHMU during the test trips before acceptance in service. This is just one such system, each other system requires about this much again (eg a Eurostar set). An older system such as NS ATB-EG originally required a space about the size of a toilet in an EMU, as Stanley Hall and I once noticed on one of our fact-finding trips.

Above: Stewart's Lane depot, South London. A wooden Great Northern Railway carriage in an Xmas Pullman special. Bar the lack of gas tanks for lighting, this coach goes straight back to the beginning of the 20th century and the time of great train fires.

Below: After the accident but before extensive signalling and signing alterations were implemented: this is the view towards Ladbroke Grove junction from London Paddington along line 2, whilst the normally used line 1 is under possession for engineering work. At the junction this train will be crossed over to the Down Main line at 90 mph/145 km/h. Notice the heavily criticized off-set red aspects of the inverted L signals to the right on the gantry. Line 5 has its signal off for the Relief line.

Ladbroke Grove junction with an inbound HST set doing 90 mph/150 km/h exactly on the location were the inbound front power car hit the outbound local coming toward it. That unit erroneously came off the left track (line 3) merging in front of this train.

Above: The potential danger of level crossings: Langenweddingen station in Germany, 1967, with the destroyed petrol tank vehicle and the completely burnt out remains of the steam hauled double-deck set that smashed into it. A total of 94 people died in the inferno. Cause: equipment failure and staff error. *Erich Preuss collection*

Below: Moulinearn User Worked Crossing, Scotland. The criticised street-side operating buttons are mounted on posts either side of the road. The colliding train came from the right. *Stanley Hall collection*

The famous Flåm mountain railway line, giving a good idea of what a Norwegian train driver encounters en-route. Even at low speed errors may have dire results on lines like these, not least because of deep gorges to the side of the track.

Above: Netherlands Railways, Plan V EMU cab as retrofitted with ATP equipment in the early 1970s. It concerns the panel above the EP brake controller, to the right-hand side of the main dashboard panel in front of the driver.

Left: Rheine, Germany, just before closure of the steam depot in 1978. The driver of an oil-fired Class 041 Mikado (1'D1' or 2-8-2) uses his being detained at a shunt-signal to quickly clean out his fire tubes. Steam locomotives, we all know it, can pollute.

Eschede, Germany: the accident site. The ICE set came at 200 km/h/125 mph towards this site along the middle track, derailed on the just visible crossovers between the main tracks, moved the siding turnout over and partially entered, after which it brought down the bridge from which this illustration was taken.

Above: Weeks before the handover of passenger services to Amtrak in 1971, Atchison, Topeka & Santa Fe train driver David Sell photographed this complete 'warbonnet'-liveried ABBA set for the San Francisco Chief stabled at Richmond, California. *David Sell collection*

Below: Train fires due to using gas for train illumination were finally overcome through the use of electricity. A Dutch carriage in the Utrecht museum shows a British-built, bogie-mounted and axle-driven 'Stone' generator to charge the batteries whilst travelling. Older types were underframe mounted, driven with a belt from an axle.

Right: A 1960s type articulated tram derailed at the top of the Damrak near Centraal station in Amsterdam of the 1980s. Notice the typical scoring of the road surface.

Below: The fixed distant for St Ives, Cornwall in the UK: at braking distance there is a stop signal at danger; in this case the red light at the bufferstops of St Ives station.

Above: A European standard Y28 type of bogie (truck) with wheel-tread problems (look at the just visible bit above the wheelskate), photographed at Gloucester, UK. The axle has been placed on a wheelskate, with which the vehicle may be transported to a place where repairs can be effected. Notice that the brake block of the disabled wheel has been removed. Also notice the yellow solebar brackets to tie down the vehicle on a ferry. On board hydraulic props are used to lift the vehicle weight off the suspension, after which several screw-link chains are fitted to securely immobilize it.

Left: A resilient wheel as used on Üstra, Hannover, Germany, operated trams. DBAG ignored problems that Üstra reported with these wheels; this eventually became an additional cause to the Eschede crash. The rubber inserts between wheel centre and tyre are clearly visible. *Detlef Schneider collection*

Above: A 'just in time' (late 1970s) photograph taken to show the single-track Callender-Hamilton emergency bridge to replace the World War 2-destroyed double-track bridge across the IJssel at Westervoort near Arnhem. Alsthom-built locomotive No 1101 is seen with a brace of DBAG stainless steel 'Silberling' coaches on its way from Köln via Emmerich, Arnhem and Utrecht to Amsterdam. Notice the derailment checkrails fitted inside the track.

Bishop's Lydeard signal box, West Somerset Railway, England. A 'collar' shown fitted for demonstration purposes on shunt-signal lever 20. The blue release lever 21 next to it shows clear signs of regularly having had collars fitted.

example of a lot of other vanished infrastructure that had to be put back again.

The colour-light signalling installed at the time of the singling of the Uckfield line, was the typical low-cost British Rail equipment of the period, a sort of fairly modern but minimal installation. Conceived to remove traditional signal boxes with their workforce and demand for maintenance. In order to save on signalling materials and their control equipment, sometimes absurdly widely spaced 3-aspect signals were installed, as it was done on other resignalling schemes of this period. It was characterised by the long distances between the distant and the stop signals, often rather more than braking distance, which is unpleasant because A) it breaks concentration on the expected red aspect that is generated after passing a caution-showing yellow distant signal, B) drivers look for a braking point not centred on the distant signal but somewhere closer to the red aspect, because of the long time it takes to reach the stop signal. It carries the risk of forgetting. And C) the link between the station stop and the red showing starting signal is lost (see ***Chatsworth***). In this case there was half a mile (800 m) between the offending train at Ashurst station and the a red aspect showing single line junction signal OD58. All things that under normal circumstances would hardly be of consequence, but in the event of problems suddenly may trigger a chain of destructive events.

Despite the signalling being modern and electronically (SSI) controlled, nothing like serious train protection was provided. The obsolete AWS was all there was to draw attention to signals with restricted or danger aspects. There was not even the old-fashioned, but rather effective, British electric token system for single lines with its excellent safety track record over the almost 150 years it was used for often quite frequent services. Nor were there other old-fashioned types of p-way equipment to stop errant trains, such as derailer-turnouts with sand drags. Adrian Vaughan, a former signaller, in his book *Tracks to Disaster* described the situation as putting the best burglar-proof locks on your doors but leaving the doors wide open at night. Quite an apt way to describe the discrepancy between the theoretical safety of a modern signalling system and its actual practical effect, due to absence of essential peripheral safety functions. Having discussed other similar issues with British signalling designers through my membership of IRSE, lack of funds to do the job properly was invariably was at the root of it. Accidents had to happen before lessons were learned.

This modern solid-state signalling was controlled from Oxted (OD) signal box, which was one of the new signalling installations involved in the spate of wrong-side failures due to questionable wiring and testing practices that preceded the ***Clapham Junction*** accident. In fact, when I used to work the double-track electrified branch to East Grinstead (another truncated electrified main line, formerly with through connection to Brighton at Copyhold Junction in that area) on a daily basis, one certainty was that a particular Down starting signal at Woldingham station, before Oxted Tunnel, would never come off during rain. From routine you simply used an alternative stopping point along the platform (much closer to the signal) and called the Oxted signaller straight away to obtain permission to pass the signal at danger, which came without fail. It was just another dubious signalling problem of the period, based on insufficient quality of materials used and neatly handled by the crews to save time.

At the location of this particular accident the 4 mile (6.5 km) single-track section was entered from passing loops either at Hever station or at Ashurst station. Proceeding from Hever, Mark Beech Tunnel came first, followed by Cowden station midway along the single-track line, beyond which the Ashurst loop was reached after 1½ mile (2.5 km). Two-aspect stop signal OD55 controlled access to the single line at the Hever end and OD58 did the same at the Ashurst end. Cowden had no track circuits or signals and was 'dark territory'. Signal OD58 was located about half a mile (800 m) past Ashurst station Up platform stopping point and was invisible from there. As pointed out in the sections about the ***Wolfurt*** and ***Chatsworth*** accidents, this is known as a SPAD trap. Furthermore, although most of the line has none-too-sharp curves and a great deal of straight track that allows rather swift progress, in the area where the collision took place the line twisted rather more than usual, so neither driver could have prevented the accident or mitigated its consequences by seeing the other in time. Moreover, it was foggy. The area, being full of small natural water courses, is known for bouts of occasionally dense fog.

Rolling stock involved

Both trains were made up of two coupled 3-car Class 205 diesel-electric multiple-units (DEMUs). Each therefore contained six vehicles, the Up train with travellers for Croydon and London being well patronised as it was the morning rush hour. Each 3-car DEMU, weighing 122 tonnes, was made up of one traction vehicle and two trailers. The traction vehicle was fitted with an English Electric 4SRKT turbocharged 600 hp (450 kW) 4-cylinder diesel with generator – noisy, dirty, full of vibrations but indestructible. It generated dc for two electric traction motors (the axle chart for the unit being 2'Bo'+2'2'+2'2') and had a driver's cab, a guard's van and a few passenger compartments. It was followed by an intermediate passenger trailer, built in the Southern tradition of slam door compartment stock. That vehicle in turn was followed by a passenger/driving trailer of the same basic layout, but with a driver's cab at the far end to remotely control the traction vehicle when that was pushing in the direction of travel. Needless to say, the vehicles were of the separate frame with tinny body construction that British Rail SR could not shake off until well into the 1970s. The bogies were equally obsolete; their provenance going back to the late 19th century. With worn wheel treads they could give a spectacularly dreadful ride on the rather less than magnificently maintained track of this beautiful line. The brakes, surprisingly when looking at the obsolescence of the train, were neither the slow vacuum nor the somewhat speedier full-air type of brakes, but were the celebrated Southern EP brakes. A nice touch of technical progress.

The rolling stock characteristics mentioned above, together with another unpleasant operating characteristic for the crew – the guard's van being located right behind the diesel motor compartment with its droning noise, vibrations, heat and diesel fumes – ensured that many guards travelled either with the driver up front, in a requisitioned empty passenger compartment in a trailer or in the unused driver's cab at the rear. This was simply to escape headaches after a few return trips between Oxted and Uckfield, never mind after a rush-hour trip into London Bridge or Victoria and slowly back up the hill at continuous full power. The presence of the guard in the cab, however much of a rulebook breach, therefore was nothing out of the ordinary on this line. It was, in a way, more a matter of survival.

The accident log

Down service 2E24, the 08:04 Oxted to Uckfield train, was worked by driver Rees and guard Boyd. Up train 2E27, the 08:00 from Uckfield to Oxted, with driver Barton and guard Brett-Andrews, was much busier and contained many who would change onto the electric service from East Grinstead via Oxted and East Croydon to Victoria or London Bridge. The traincrew had to contend with patchy and sometimes quite thick fog (see ***Harmelen*** and ***Winsum***). Both trains were booked to cross in the loop at Ashurst station and the Oxted signaller had set the routes accordingly. Driver Rees had been shown a proceed aspect at Hever Down starting signal OD55, and had continued through Mark Beech Tunnel, stopped at Cowden station and was now on his way towards

Ashurst along the curves on this single-track section. The recording functions of the electronic equipment proved that the signalling equipment worked correctly and that Ashurst Up starting signal OD58 displayed a red aspect. Despite this, the crew of train 2E27 committed the single-line 'starting against signal' mistake we have seen elsewhere in this book. The train left Ashurst station as it should, but half a mile further on did not stop at signal OD58 with its red aspect at the single-line junction. It cut through the turnout, accelerated and slightly later collided heavily with 2E24 near Cowden station with an approximate impact speed of 45-50 mph (70-80 km/h), about half that at which the ***Winsum*** accident occurred. The front vehicle of 2E24 fell to its left off the track and hung precariously in trees, which prevented it from going down the embankment, whilst its second vehicle was relatively undamaged but had its front bogie derailed. The badly damaged front vehicle of 2E27 came to rest with its front bogie derailed to its left side but its rear bogie still on the track. Both drivers, guard Brett-Andrews and two passengers lost their lives.

The investigation

It was found that the bodies of driver Barton and guard Brett-Andrews occupied the remains of the front cab of 2E27 and that Brett-Andrews was closest to the train controls. The inference was that he had been driving. It was well known that he enjoyed the chance to drive these trains and that he would exchange his main-line duties with other guards to work these branchline duties, just to have a go at the controls. He was also known as a chirpy, cheerful and very talkative sort of person.

The AWS warning system of the leading unit of train 2E27 to Oxted technically could not have been in a working condition due to serious corrosion of the contacts in the cancelling button. Cancelling a warning was impossible as electric current could not flow through the corroded contacts. In turn, that would have inhibited brake release on the train, yet the train was moving as normal. The inescapable inference was that the AWS had been isolated (from my own experience not uncommon on the Southern), but the front of the vehicle was too damaged to enable definitive conclusions with respect to the position of the AWS isolation cock. I personally doubt that repair of the AWS had been requested, or was even considered a priority among the necessary repairs waiting to be done. The fact is that the AWS system was not taken seriously either by traincrews (including myself at the time), fleet or lower operational management. Given that driver Barton had not reported his AWS as defective in the repair book, the accident report came to the conclusion that the AWS must have been working. You make up your own mind. In any case, in the circumstances it is extremely unlikely that an AWS warning did go off and had to be cancelled. Thus attention was not drawn to the red aspect at signal OD58, which was additionally obscured by fog. Such a warning would, in all likelihood, have prevented this accident from happening. The Cowden accident has AWS issues in common with the *Southall* accident three years later.

A so-called banner repeater signal would have had a clear safety-enhancing function with respect to the half mile that separated signal OD58 from the Ashurst station stop and the much longer distance that separated it from its distant signal. A station stop, or any other speed-influencing feature between distant and stop signal, has in the meantime been recognised as a potential trigger for a SPAD (see also ***Wolfurt*** and ***Chatsworth***).

Worse was that signal OD58 that 2E27 passed at danger was found to have a bulb that was wearing out, working at 86.5% of the light power of a comparable bulb in good condition. The inside of its bulb glass was covered with tungsten dust. The inside of the red lens it shone through in turn was covered with excrement from insects trapped in the signal assembly. A signalling technician had exacerbated this unacceptable situation by using an aerosol spray to coat the internal electric installation with a waterproofing agent following electrical problems with moisture ingress during rain (see my Woldingham experiences!). The accident report noted that 'a more than ordinary powerful cleanser was required to clean the back of the lens'. Up to that time cleaning the backs of signal lenses very likely received little attention anyway. Nevertheless, traincrew described OD58 as giving 'reasonably good light', and there is no reason to assume that normally they had difficulty to read this important signal, even with its reduced light power.

Comment

From the above it appears that driver Barton and guard Brett-Andrews, regardless of who was driving, were distracted enough to overlook the insufficiently maintained red-aspect signal OD58 in the fog, despite their combined route and signalling knowledge. This signal was located well away from the end of Ashurst station, preventing them from seeing it when stopped there, and no intermediate repeating signal was provided for an extra indication at this obvious danger point. They were working a train on which what little automatic warning system was provided was in all likelihood not functioning and therefore failed to warn them of the red aspect ahead. And perhaps the two of them were so deeply engrossed in their conversation that they were distracted from their main occupation of watching (fog-bound) signals from a train cab. Who knows? But, bad as this all may be, it is too easy to put the entire blame on the crew. A contributory factor in this case was that BR had grown complacent and lost the edge of its safety-orientated culture at certain operational levels (see, for example, the accidents at ***Clapham Junction***, *Southall* and ***Ladbroke Grove***).

Finally, to my mind the rumpus about Brett-Andrews supposedly driving the offending train is a red herring, despite the clear breach of rules it constitutes. Non-driving staff travelling in cabs happens every day all over the world. Even someone as senior as a former managing director of Netherlands Railways once admitted in a newspaper interview that he used to do it often in order to allow a passenger to occupy his first-class seat in the train and, yes, he even drove the train quite regularly to get a feel of what working trains was like! I admit that I also profited enormously from this practice. This situation, however, was rarely the cause of railway accidents. Driving a train – as in making it go and stop again, especially with control equipment such as the EP brake and the more modern versions of electronically controlled brake and power control equipment that came after it – is not difficult, and the guest is always under tuition. In the end, guard Brett-Andrews was in no different a situation than a learner driver on his first trips on a train with a tutor.

***

Eschede, Germany, 3 June 1998 at approx 10:59

(Derailment followed by overbridge collapse)

The location

The accident occurred on the electrified double-track main line from Hannover via Uelzen to Hamburg, fitted with colour-light signals, PZB and LZB in-cab signalling/ATP. The permitted line speed at the location of the derailment was 200 km/h (125 mph). At Eschede the line has loops at either side of the main lines as refuges to allow slower trains to be overtaken. The turnouts into these loops played a major role in the way this accident developed, very similar to how the incidents at *Ufton Nervet* in Britain escalated into disaster.

The accident and rolling stock involved

The derailment occurred on crossover and loop turnouts at high speed following a wheel tyre failure, the subsequent disastrous crash bringing down an overbridge on the train. The train involved, owned and operated by DBAG Reise & Touristik (Travel & Tourism), was an ICE 1 set, a classic type of high-speed train consisting of two single-ended electric locomotives (power car 401 051 at the front and 401 155 at the rear) with a fixed-formation trainset in between comprising eight second-class coaches, a restaurant car and three first-class coaches.

The disaster left 101 people dead and 105 injured, 88 of whom sustained serious injuries. As this was one of the first ever true high-speed fatal rail accidents, the medical side of the injuries was examined in detail internationally to find out, among other things, whether changes in the design of the interior might help to increase survival rates during such accidents. The fatal (mostly internal) injuries were found to be consistent with the type of impact that led to immediate death. An abrupt stop from about 200 km/h is equivalent to a free fall to earth from a height of 160 m (525 ft).

The run-up to the accident

From 1954 onwards the West German railways had fitted the innovative Minden-Deutz or MD bogie in various sub-types under virtually all their passenger rolling stock until the late 1970s. The final series of this bogie, fitted with disc brakes on the axles and with electro-magnetic track brakes, were passed for travel at 200 km/h. Many networks in Europe and around the world adopted this bogie as well, but around 1980 a new version (MD52), with a number of changes, most of all in the primary suspension to enable a shorter bogie frame, was designed and tested specifically for speeds from 160 km/h (100 mph) up to 310 km/h (190 mph). On the test bench, speeds of 500 km/h (310 mph) were reached, a speed that SNCF French Railways would achieve in reality with tests on the then as yet unopened LGV (Ligne à Grande Vitesse) Est in 2008. In contrast to contemporary bolsterless bogie development for high-speed work with electric multiple-unit trains elsewhere, this bogie still had the classic bolster suspended from four swing links with steel secondary springs at the ends that carried the body, whereas elsewhere in Europe and in Japan flexible secondary air suspension without bolster was used. After testing of various types of bogie with secondary air and coil suspension types this MD52 bogie was selected for the first series of ICE trailer coaches. It is interesting that Japanese National Railways selected the original MD bogie for their first series of Shinkansen high-speed bullet trains in the 1960s, but already then they replaced the originally all-steel bolster assembly with bolsterless air suspension bellows. They ran these bogies at up to 250 km/h (156.25 mph), the speed range of the ICE 1 sets, and their experiences were wholly positive. An example of the first series of Shinkansen EMU can be studied at the National Railway Museum in York.

In the case of the ICE 1 sets, it was found in due course that passenger comfort became seriously impaired at higher speeds when travelling with worn-in wheels. Notably the heavier restaurant vehicles were affected by something colloquially known as 'bistro-brummen', the bistro drone. Cutlery clattered in its baskets, wine glasses moved across tables, coffee cups rattled incessantly in their saucers and reading with elbows on the tables or armrests was rather uncomfortable in the worst cases. Vibrations from imperfections due to wear of the wheel treads were too easily transmitted to the coach body through the steel coil springs of the secondary suspension, setting up reverberations synchronised with the rapidly turning wheel treads on the track. Solutions were looked for, but alternatives such as retrofitting secondary air suspension, or fitting of completely new bogies with air suspension, were ruled out for technical and cost reasons. A different and far less costly solution was found in the application of so-called 'resilient' wheels of the Bochum 84 type (bench tested for 284 km/h or 177.5 mph) to replace the originally one-part monobloc steel wheels. For a number of reasons, not everyone was enthusiastic about this economical initiative.

Resilient wheels

Resilient wheels have a long history. One could argue that the Maunsell wheels of the late 19th century, with their wooden wheel centres surrounded by steel tyres, were resilient wheels. Many examples survive in British and European museum collections. More modern resilient wheels with rubber inserts have been much used for work under light rail vehicles since the 1930s, in Sweden (SAB), the US (Carnegie) and Germany (Duewag). They are an essential part of securing sufficient ride comfort as well as inhibition of interior rolling noise on the famous US-designed PCC type of light-rail vehicles. From personal experience of riding these trams in Den Haag in The Netherlands I can confirm that this aspect worked excellently.

Only in Sweden and Britain was there experience with the application of resilient wheels on heavy rail rolling stock. The Swedish state railway SJ successfully fitted series Da, Dm and Dm3 rod-driven electric freight locomotives (built between 1952 and 1970) with SAB resilient drive and bogie wheels. The Dm3 version, incidentally, was used in the hostile climate above the Arctic Circle on the demanding job of moving heavy ore trains from Kiruna in Sweden to Narvik in Norway, but in the type of traffic that the three types of locomotives worked there was no experience with daily use of these wheels at speeds above 100 km/h (60 mph). The world-famous Swedish Rc-type Bo'Bo' electric locomotives and their Co'Co' derivatives were then fitted with these wheels and in a number of cases operated at speeds up to 200 km/h (125 mph) without much of a problem. These same wheels were then fitted under the AL6 (Class 86) electric Bo'Bo' locomotives in Britain for exactly the same reason that they were selected in Germany – to improve rough riding. In Britain this was due to the debatable decision to fit axle-hung traction motors under high-speed electric locomotives. On discovering that the locomotives damaged the track, like German railways with the ICE 1, they found that they were unable to alter that set-up without incurring huge costs. After the change to resilient wheels, as well as a change in the secondary suspension (sets of flexicoil springs), the results were positive at speeds up to 160 km/h (100 mph).

The German Bochum type of resilient wheels as fitted under the ICE 1 sets were the first ones designated for serious day-to-day high-speed work, however. They were conventionally built up of three parts: the wheel centres, a ring of rubber inserts around that, and then a close-fitting wheel tyre with tread and flange to finish the assembly. Initially only the ICE 1 restaurant vehicles were fitted with the new wheels and the 'bistro drone' disappeared as expected, for which reason fitting the fleet as vehicles came in for wheel-exchange started in a comparatively short time. However, no serious research was carried out as to what the safe wear parameters were and what needed to be done to ensure continued safety of the wheel tyres in the event of fatigue. In Hannover, the local city transport operator Üstra reported fatigue problems with its Bochum-manufactured type of resilient wheels fitted under tramcars in July 1977, flagging up the problem and warning other users of such resilient wheels. DBAG, however, did not consider that they had a problem.

Weeks before the accident the wheelset that caused the crash was reported as having noisy and distorted running due to a wheelflat of 1.1 mm (0.04 in) depth (the norm for action being 0.6 mm) and the wheel treads were no longer round to an extent well outside the approved range. This was entered on the on-board DAVID train diagnostic systems and eight times on the traction diagnostic and reporting ZEUS system from April 1998 onwards. The night before the set was to work ICE 884 *Wilhelm Conrad Röntgen* it was routinely checked with ultrasound. This wheelset gave such unbelievably bad readings that the measuring equipment was faulted after visual inspection of the wheels and the wheelset was left in the bogie. This too had a history; the measuring equipment used was inadequate for the job it had to do, but money for better equipment was not forthcoming.

The accident log

At 05:47 on 3 June 1998, the set departed München Hauptbahnhof on time for Hamburg, well loaded with people going to the North Sea coast for the Whitsun holidays. At Würzburg there was a delay due to awaiting a connection, but the driver worked at making up the excess minutes and, as a result, the train was on time departing from Hannover at 10:33. It then lost a few minutes again as there was a temporary speed restriction of 90 km/h (55 mph) before reaching Celle, after which the train was quickly brought back to the line speed of 200 km/h (125 mph) on its way to Eschede through the agricultural and wooded countryside. At the moment it reached the permitted speed, passengers in the rear of the first coach heard a loud report under the train and something was forced upwards through the floor from below, creating a strange dome-shape to the floor covering of the passenger compartment. The coach also started to ride roughly. One of the passengers took his wife and child to the second coach before going to look for the senior conductor, who was eventually found in the fifth coach. This conductor, not quite able to believe the somewhat incoherent things he was being told, walked back with the passenger to the first coach to check what the matter was before taking action, and so saved his life. On entering the coach he noticed that it was running extremely roughly and making strange noises, but he never had a chance to stop the train as it then fully derailed.

From the moment the noise under the train had been heard and the object had been forced up through the floor into the passenger compartment between two seats, the third axle, which was the first axle of the rear bogie, had lost its right-hand wheel tyre which straightened out slightly and jammed itself in the bogie, partly reaching up through the coach floor and down on to the concrete sleepers. It caused minor damage to these whilst ripping off the zigzag LZB ATP transmitter cable. This went on for 5.5 km (3 miles), the flangeless wheel staying on top of the left-hand rail head.

As already mentioned, there are loops either side of the main lines and a set of crossovers at Eschede. The train first met the trailing turnout of a set of crossovers between the two main lines, where the damaged wheel finally derailed due to the jammed wheel tyre hooking itself in the narrow space between the turnout frog and the check rail. This loosened the 7.2-m (24-ft) long check rail, which was then knocked into the bogie frame, where a metre-long (3-ft) part broke off and jammed itself between the frame and magnetic track brake block. The rest shot through the carriage floor into the first carriage, reminiscent of the 'snakeheads' that early US train travellers were wont to experience on so-called strap rail that disintegrated under their train. The front bogie of the first coach then derailed and moved the blades of the facing turnout to reverse position for the loop on the right side under the train. This suddenly sent the second carriage into the loop, forcing the third and fourth carriages to derail.

It was then that the most unfortunate part of the accident occurred. The rear of the third carriage hit the four concrete centre piers of a rather basic 200-tonne reinforced concrete viaduct carrying Rebberlaher Strasse across the tracks and severely damaged these, the coach breaking itself in half in the process while parting the first three coaches from the front power car and itself from the fourth carriage on impact. Carriage four shot through the collapsing bridge to the right across the track of the loop off an embankment, the viaduct collapsing on the rear half of the fifth coach, breaking it in two. Half of coach five and the whole of coach six were crushed under the collapsed bridge, the rest of the train smashing into it virtually at line speed and concertinaing itself under and against the fallen bridge deck. Only the rear power car remained largely intact. The damage to the welded aluminium coaches was very much as described in other similar incidents in this book, the welds being ripped up cleanly. Two S&T fitters of DBAG Signaltechnik, who had taken shelter under the bridge when the accident began to unfold before their eyes, were among the dead. Their van had been parked on the bridge and was later found in the pile of debris.

The front power car, undamaged and completely on its own, passed through Eschede station, before coming to a stop 2 km (1¼ miles) further on. The Eschede traffic controller immediately blocked all tracks to avoid other trains running into the incident site. The driver had felt a jolt and lost power and then the brakes had come on. He had tried to restore power but was then radioed by the traffic controller to inform him that he had lost his train. He looked back from his cab door, saw the collapsed bridge in a dust cloud in the distance and some derailed coaches of his train and stayed in his cab for two hours until taken out by the emergency services. In fact, he adhered to a ruling that he was not to leave his cab until called out in case of damage to the overhead wires (which there certainly was), but it was the shock of witnessing what had happened that made him stay there.

The first emergency services arrived within minutes, having been warned immediately the accident occurred by members of the public. This efficiency was largely a result of a bad forest fire some years earlier, when the emergency services and their local and regional response co-ordination had failed to stand up to scrutiny. The Eschede area emergency services had learned their lessons, and it showed on this occasion. All casualties retrieved had been transported to various hospitals by helicopter or ambulance by 13:00. Uninjured people had been listed, interviewed and loaded into buses and would reach Hamburg at around 17:00. In that aspect everything ran like clockwork; the main problem was entering the wrecked vehicles through the well-nigh unbreakable windows (see ***Aitrang***). It led to the large-scale fitting of special emergency breakable windows in Europe, although that is something which is now being reconsidered, as during other accidents people have been ejected through them and lost their lives that way. The carriages were soon cleared and transported away, but many clues as to what had happened disappeared from the location with them. One of the problems, for instance, was finding out which wheelset belonged to what carriage as the DBAG administration turned out to be unreliable on that subject. Some carriages even ran on both resilient and monobloc wheels, to mention but one issue.

Nevertheless, amid all this disaster there had also been one massive stroke of luck. ICE 787 *Werdenfelser Land* travelling in the opposite direction was scheduled to meet ICE 884 *Wilhelm Conrad Röntgen* precisely at the Eschede loops. ICE 787 had been running a minute early, whilst ICE 884 had been running about a minute late due to the earlier temporary speed restriction prior to reaching Celle. The situation could, therefore, have been infinitely worse than it already was.

The investigation

As mentioned above, in the debris around the collapsed bridge a van belonging to DBAG Signaltechnik was found. Initially there was talk that this had fallen off the bridge and derailed the train, but no road accident was reported and no damage was found to the front of the leading power car. It was also thought that the train might have been damaged at the engineering site earlier, but no evidence of anything untoward at or past the engineering site could be found until nearer the accident site, where the track appeared damaged. A terrorist attack and other damaging actions were also considered as possible causes.

The day after the accident, however, parts of a train wheel tyre and rubber wheel inserts were found scattered about 5.5 km before the accident site. From that location onwards one side of the sleepers and LZB ATP cable equipment mounted on the sleepers were damaged. It was becoming increasingly apparent that something had happened to a wheel, which was rather unwelcome to DBAG, as by now suspicion began to grow that not everything was as it ought to be with the coveted safety the German railway represented in comparison with other modes of traffic. DBAG made out that this was probably a one-off mishap with a wheelset. At the inquiry Horst Stüchly, the principal inspector with the EBA (German federal rail authority), was unable to give the last date this wheel of the 7-year-old trainset had received ultrasound inspection; neither was DBAG able to provide such information quickly. It was also odd that it was not so much wheel failure as failure of the wheel tyre of one of the resilient wheels. Consequently, all ICE 1 sets had to be immediately checked to discover whether any more potential failures were running around.

Germany's railways were thrown into the sort of disarray that the situation in Britain would later resemble after the violent derailment caused by 85 m (280 ft) of shattered rail at *Hatfield* in October 2000. As more cases of such cracked and broken wheel tyres came to light, the EBA decided on 15 June that the ICE 1 fleet would not be allowed back in traffic with worn wheel tyres. It seemed that DBAG were providing less than full co-operation. When ultrasound tests on wheels were ordered, some DBAG people openly wondered why it was only then that this kind of testing was instigated when it should have been done regularly in the ICE maintenance workshops in München and Hamburg. This point is echoed in Hermann Wolters' book *ICE: Zug der Zukunft* (ICE: Train of the Future). The wheels should have been methodically tested for cracks, wear and roundness every 2,000 km (1,250 miles), but journalists discovered that around 1993/94 these tests had been stopped due to the previously mentioned problems with the test equipment. Rather than modifying and improving the test equipment, checking of ICE wheels from then on had taken place only by hand and eye. The problem was that the resilient wheel tyres cracked from the inside out towards the wheel tread, rather than from the outside to the inside. This was impossible to pick up using the inspection, monitoring and maintenance methods employed.

Conclusions from further research on these wheels, conducted by independent bodies such as the Fraunhofer Institute at Darmstadt and the Üstra Hannover local passenger transport operator, had been ignored in the same way. DBAG research into the proper rate of wear and minimum safe thickness was discontinued without reaching any conclusions that could have provided hard guidance for those who maintained the wheels in everyday traffic. Nor had research been instigated into potentially unexpected types of tyre wear that the wheels might have been subjected to due to the typical stresses working on them in the ICE high-speed/high-mileage traffic environment. Better ultrasound methods for checks were not developed when the existing methods were found to be inadequate. The ICE wheel involved in the accident had been run down to 862 mm from its original 920 mm, the wheel tyre having worn from 60 mm to 31 mm. The minimum circumference for the wheel had been set at 848 mm – on what grounds is not entirely clear – but it probably had much to do with the tyre thickness maintained for traditional steel wheels with shrunk-on wheel tyres. These, however, were not subjected to the types of dynamic stress of the wheel tyres on resilient wheels, which flexed quite considerably on their rubber underlay from round to almost flat like a rubber wheel tyre on the road, thereby stressing the wheel tyre both on the surface of the wheel tread as well as on the inner surface. The thinner the tyres became the more easily they flexed, thus inviting fatigue. In the end, a legal trial against DBAG and three wheel-technology engineers behind the introduction of resilient wheels on the ICE 1 fleet came to nothing. The wheels could have worked fine and individuals could not be held to blame for a company-wide failure to maintain safety. Companies were not the kind of bodies that could be blamed for anything if it was not clear beyond reasonable doubt who in the company had done what to cause the lack of safety or unexpected damage in the first place. Finding an operating culture that generally sacrificed safety in order to lower costs on the way to privatisation was not sufficient for that.

The chief reason for the whole sorry story was the wish to make a quick mark by running prestigious high-speed trains in an environment already dominated by the Japanese with their magnificent Shinkansen, the French with their successful TGV and Italians with their Eurostar Italia rolling stock. But in the DBAG philosophy this entry into high-speed competition had to be done at a price that would not adversely affect the balance sheets the money people in Frankfurt am Main, London and New York would see, as that would stand in the way of the hoped-for success of DBAG's privatisation drive under its chairman, Hartmut Mehdorn. DBAG suffered a massive loss of face, however, both with the travelling public and with those working in the financial industry. Further questionable issues at the time, such as cracking axles and failing air-conditioning on new high-speed sets, as well as surreptitiously withdrawing popular types of discount ticketing, ensured that DBAG acquired an aura of incompetence, which made things decidedly harder for them when another train crash with fatal consequences occurred at ***Brühl*** near Köln on 6 February 2000.

On 16 June 1998, 13 days after the Eschede accident, a Mk IV set on the East Coast main line near *Sandy* in Bedfordshire was brought to an emergency stop after the last coach partially came off the track. One of its monobloc wheels had cracked and lost a significant segment. No one waited unduly before pulling the emergency brake in this case, both crew and customers on the train reacting sharply as a result of what had happened so recently in Germany. Nine people were slightly hurt in this case.

At the end of February 2012, I visited Eschede with friends from Hannover to get a feel for the place, to see the trackwork on which the derailment took place, and visit the impressive monument for the 101 victims and see the replacement bridge in the Rebberlaher Strasse. The location is absolutely unremarkable and is typical of any lineside location where there are loops to take slower trains out of the way of expresses. Traffic was rather dense at the time of my visit, among which was a first-generation ICE set like the one involved in the accident. Nothing other than the monument gave any clue to this terrible accident, with trains mundanely stopping in the loops or just shooting past along the main line. But Eschede has become one of those locations on the railways of Europe that carries the emotional weight that such a heavy accident leaves behind to those in the know. And, of course, the new bridge no longer has supports between the tracks.

***

Ladbroke Grove, London, England, 5 October 1999 at 08:09

(Head-on crash following SPAD)

Introduction

The train services during the early morning of 5 October 1999 were blighted by the fact that it had been the first night with a hard frost. At the First Great Western depot in Bristol several drivers and other traincrew could not make it to work on time due to sickness, late-running trains, dealing with frozen water pipes and cars that would not start. Reallocation of traincrews from their booked diagrams to work on suddenly unstaffed earlier diagrams was the order of the day, a process called stepping-up, which demands the utmost from the traincrew resources people at head office in keeping track of the changes and reviewing their implications later in the day.

As a result of these acute crewing problems, Bristol train driver Brian Rogers voluntarily drove the first morning Up Cheltenham train 1A09, booked empty from Bristol to Cheltenham, into Alstone sidings, changed ends and then moved out again in service via Gloucester, Swindon and Reading to London Paddington. The booked driver for the service was sick and a second driver pencilled in for this job had not materialised in time due to the sudden onset of winter weather, so driver Rogers agreed to do overtime following his night shift. However, he needed to be relieved at Reading in order to avoid going over his maximum of 12 hours per shift, as he had to travel all the way back to Bristol again. Crew resources at Swindon found a Paddington-based driver, Brian Cooper, prepared to volunteer out to Reading on overtime after his night shift and pick up 1A09 there, to take it onwards for the last 30 minutes of its trip into the capital. And so, after the swift handover at Reading's Platform 5, driver Rogers returned to Bristol travelling on an early morning Down service, whilst driver Cooper departed with 1A09 for the non-stop last leg to Paddington. He was on the last 3 miles of this trip, less than 7 minutes away from booking off and going home, when the head-on collision occurred at 08:09.

That morning I did not work my booked second Cheltenham-to-London diagram but was reallocated to a job of running empty from Bristol to Weston-super-Mare in order to start the 06:50 passenger service from there and go via Bristol, Bath, Chippenham, Swindon, Didcot and Reading to Paddington. The trip was uneventful until just before Tilehurst near Reading where the train was slowed on restrictive signal aspects and then stopped at the signal covering Reading West Junction. That in itself was nothing out of the ordinary, if it were not for the odd sight of seeing all Up signals ahead showing a red aspect with a train waiting at each of them whilst there was no Down service clearing platforms anywhere near. Adhering to the rulebook, I donned an orange high-visibility vest and went to the signal-post telephone to report to the signaller, but no one picked up the receiver in Reading signal box. Trying the NRN radio telephone in the cab brought the same result, as did a repeat trip to the signal-post telephone. Clearly a signalling and communications meltdown of sorts had occurred – a new and interesting experience (see ***Clapham Junction***)! I sat down with a book, regularly glancing up to see what was happening, the ATP being a handy tool in these circumstances as it reports a change of signal aspect ahead with a sort of burp sound. But nothing of the kind happened for a considerable time. Eventually the Up main line signal further ahead of me suddenly came off and the train standing at it upped anchor to move into Reading station, the signal returning to red after it. One signal section ahead now being clear, so the signalling appeared to be working after all. A colleague on a local DMU on the Up relief line was busy on his mobile phone and after a while said that something had happened but it was not clear what. Whilst occasionally hanging out of the driver's side cab door to talk to him, I suddenly heard my signal-post telephone ring, jumped off, opened the cover and picked up the handset. Without further introduction the Reading signaller curtly instructed me to get into Reading station, drop off all passengers, change ends and return empty to Bristol, then rang off. The signal at which I was standing came off to two yellows, the one ahead to a single yellow with a No 1 route indicator, and I put my book away to bring the train into Reading's Platform 5.

On arrival the platform was heaving with people, and even more so when those coming off my morning rush-hour train were added, but the situation was oddly quiet, subdued almost, occasional announcements instructing people what to do to get to London but giving no clue as to what had happened. On walking back through the throng along the train to change ends, I met my train manager and asked what the matter was. He was unsure, but a bad accident appeared to have happened closer to London, on hearing which my first thought was something like 'not again' after *Southall* two years earlier. I set up the country-end cab, the signal came off and the train was dispatched back to Bristol without any hindrance by signals along the entire Bristol main line. Passing through stations and junctions at maximum permitted speed was another rather odd sensation – such as rushing through Chippenham at 125 mph (200 km/h) instead of making the routine stop there. At Bristol Temple Meads I was asked to take the train to the city's St Philip's Marsh depot and therefore changed ends again to take the set round via North Somerset Junction. In the shunter's cabin at the Marsh the television finally revealed the truly shocking sight of what had happened earlier that morning. Unable to sleep properly for the following three nights, I wrote down everything I could think of with regard to the possible causes of the accident.

The accident

Recently passed train driver Michael (Mick) Hodder on train 1K20, a semi-fast local service to Reading, Newbury and Bedwyn, passed signal SN109 at danger just before Ladbroke Grove Junction, as had happened to others eight times before. Instead of stopping and reporting as the eight previous drivers had done, however, driver Hodder erroneously accelerated in the Down direction to 51 mph (82 km/h) along a non-reversibly signalled Up connection spur line coming off the Up main line to line 3 that brought him head-on with train 1A09, the inbound express service from Cheltenham running at 84 mph (134.5 km/h). The violent collision took place on the turnout where the spur branched off the Up main line, causing not only virtually the complete destruction of the leading DMU vehicle but also setting off a major fuel fire that obliterated the Up Cheltenham HST power car and the first coach. Thirty-one passengers and traincrew died in the impact and subsequent fire, whilst more than 400 people were injured, some critically and with very severe burn injuries. Train drivers Cooper and Hodder were among the fatalities.

The location

London Paddington rail terminus has 14 platforms for main-line rail traffic. In the station throat these come together into six fully reversibly signalled lines with various intermediate crossover turnouts that run for about 2 miles (3 km) to a western-end throat called Ladbroke Grove Junction where the six reversible lines narrow into four non-reversibly signalled lines to Slough and Reading called the Down and Up main lines and the Down and Up relief lines. Following reconstruction and resignalling in the late 1980s, the layout at Ladbroke Grove Junction had become very fast indeed, the main lines having a permitted line speed of 100 mph (160 km/h) and the turnout Up main/line 2 to the Up spur line/line 3 a permitted speed of 90 mph (145 km/h). It was probably one of the fastest throat layouts in the world so close to a major city terminus. The layout had been conceived with the use of ATP in mind to provide collision-prevention safety, but, as described earlier with the ***Clapham Junction*** accident, the installation of ATP was dropped in Britain whilst no credible train protection substitute had yet been implemented. As a result, the obsolete AWS was all that was available to alert drivers to restrictive and danger signals and to temporary and permanent speed restrictions. The sighting of signals along these six lines, incidentally, had become problematic in places because of low overbridges in conjunction with the dense 25 kV 50 Hz overhead line equipment installed when the Heathrow Express airport rail shuttle plans became reality.

Driver Hodder's insufficient training and route knowledge, one of the main contributing causes to his SPAD at SN109, shows up in his travelling on line 3 Up spur in the Down direction and the fact that he must have missed that SN109 always showed a route indication with any proceed aspect. He should have noticed that and stopped, as prescribed in case of doubt about the signal aspect. The reason why Hodder was wrong-footed at SN109, apart from his insufficient knowledge of the layout, was probably the limited visibility of the signal due to a low overbridge and overhead electrification equipment, and the real possibility that it showed phantom aspects in the bright, low, early morning sun. Before the changeover to fibre-optic and later LED signals after this accident, this area was known for that phenomenon and I experienced it more than once myself. The fact that SN109 and its neighbouring signals on gantry 8 were constructed in a reverse L-shape is often mentioned as a possible reason for the driver's mistake. However, I doubt that, as in my experience all red signal aspects on that gantry were visible at least as easily as normal signals, despite that L-shape. As of a result of the limited visibility mentioned it was necessary to actively look out for them, L-shape or not. It was just that signal SN109 alone had a fat 25 kV overhead line registration arm insulator covering its line of view when approaching along lines 2 and 3 in the Down direction. All traffic control in the area was concentrated at 'Slough New' Integrated Electronic Control Centre (IECC), hence the prefix SN for the controlled signals. Slough New is a relatively up-to-date control centre that is normally run with Automatic Route Setting (ARS) equipment, which records what it does for later perusal. This facility was extensively used for investigation purposes after the accident.

The problem of the warning/protection systems

The lines at the junction were fitted with AWS and one of the ATP test systems (Belgian-manufactured TBL2) that since the *Southall* accident in 1997 had been thoroughly overhauled and reactivated. However, only the Heathrow Express and the First Great Western HST sets were fitted with the necessary ATP on-board equipment; the Class 165/166 'Thames Turbo' local DMUs were not. This was never done following the cancellation of further ATP fitment. Similar Class 165 sets used out of London Marylebone station, however, were fitted with another ATP test system (Swedish-manufactured Selcab) that cannot be used with the lineside equipment of the system fitted near Paddington. It nevertheless means that potential problems with the interface between ATP on-board equipment and both types of trainset are known, which is usually one of the major and costly bugbears when retrofitting existing rolling stock with ATP on-board equipment. Nevertheless, fitting Great Western ATP to the local Class 165/166 DMU sets was rejected on cost grounds, approximately £5 million at the time, which in fact was peanuts. But that decision was fairly typical of the way the railways in Britain were operated. Had the local set been fitted and operated with ATP as originally proposed, this accident would not have occurred. The ATP would have intervened when noticing the local train approaching SN109 too fast and stopped it after going past the signal. Had Hodder noticed this ATP overspeed warning or intervention, it is likely that he would have stopped as normal at SN109 because ATP indicates the signal aspect in its cab display and there would have been no misunderstanding on the matter, not even in the event of phantom signal aspects.

The problem with the AWS system fitted (the functionality of which goes back to the ancient Great Western ATC system dating from 1906) is that it does not distinguish between a danger signal aspect and a restricted signal aspect, thus requiring repetitive cancelling of warnings when running under restrictive aspects (with the odd speed restriction thrown in) as happens frequently near the bigger stations. More importantly, it does not stop a train when the warning has been cancelled. A substantially more advanced system that repeated an image of the signal cancelled in the cab and intervened on red (SR-AWS) was under development at one time but was stopped on the usual cost grounds. At present this layout is fitted with both AWS/TPWS and ATP lineside equipment and both systems are live whenever ATP is fitted on board, which is surely a unique situation in Europe. On the continent, care is taken to avoid having more than one train protection system live in the cab at any one time, and the automatic arm/disarm function, which is also available for AWS/TPWS and ATP, is used to switch between systems on system limits. During an ATP fact-finding trip in The Netherlands, Stanley Hall and I witnessed this automatic arm/disarm feature in action between Nijmegen, Venlo and Roermond on the type of DMU involved in the ***Roermond*** accident.

Rolling stock involved

A class 165 3-car 'Thames Turbo' diesel-hydraulic multiple-unit (DMU) was operating service 1K20 from Paddington to Bedwyn in Wiltshire. The 3-car set weighed 113 tonnes, and the maximum train speed of 90 mph (145 km/h) was delivered from three sets of Perkins diesel engines through hydraulic gearboxes and final drives. These DMUs were built at BREL between 1989 and 1992. The welded aluminium-built unit saw its first coach well-nigh fully destroyed on impact, pretty much in the same way as other trains of similar construction mentioned in this book. The only still recognisable section of that coach was part of one side with a window in it, lying across the roof of one of the coaches of the Up Cheltenham HST, and the rear passenger compartment partly telescoped into the second vehicle. There was slight telescoping between the second and third coaches, not unexpectedly in view of the extreme violence of the collision, but that did not materially add to the deaths and injuries. It was the violence of impact and the destruction of the first coach that did.

Train 1A09 was an HST diesel-electric fixed-formation set, with a maximum train speed of 125 mph (200 km/h) in standard FGW configuration (in the train's direction of travel) of a Class 43 single-ended power car followed by two Mk III type first-class coaches, a buffet/restaurant car with first-class seating and then five standard-class coaches, of which the last one had a small luggage area. The last vehicle was another single-ended Class 43 diesel-electric power car, remotely controlled from the leading one. Each power car had a 2,250 hp (1,680 kW) diesel prime mover (at the time a Paxman Valenta diesel), which in the leading power car was fully used to generate electricity for traction, and in the trailing power car approximately 500 hp (375 kW) was used to generate hotel power for the train to run heating, lighting and air conditioning. When such a set reverses direction the changeover of the engine from traction only to hotel-power configuration can be clearly heard. Apart from the badly damaged front end of the leading power car and the characteristic loss of the BT10 bogies from the Mk III carriages during derailments, this set would have maintained the high passenger safety provided by the Mk III coach design if it had not been for the fuel fire that destroyed the leading power car and first coach. Otherwise loss of life and injuries, certainly the appalling burns suffered by some on this train, would have been comparatively minimal given the impact speed of the accident.

The accident log

As various actions in the IECC traffic control centre were logged and the front and rear on-train trip recorders of 1K20 could be recovered and downloaded, the timelines and the technical information about this trip are known in minute detail. What is not known, however, and is something that is noted more than once in this book, is why the driver acted as he did.

At 08:06 train 1K20 departed from Paddington's Platform 9 with a green aspect at starting signal SN17 for its journey to Reading, Newbury and Bedwyn. Its next signal was a double yellow at SN63, the subsequent signal SN87 being a single yellow with a No 1 position light route indicator informing driver Hodder that he was going to be crossed over to line 3. From Ladbroke Grove Junction the 'Bedwyns' were normally booked to run via the Down main line to Reading, which required crossing over from line 3 via two very fast crossovers (90 mph or 145 km/h) to the Down main/line 1. Unfortunately, the Up main line, leading via Ladbroke Grove Junction into line 2, was set and locked for 1A09, the Up Cheltenham service that was approaching at a speed of about 90-100 mph (145-160 km/h). The pathways of both trains crossed at Up main/line 2, for which reason signal SN109 on line 3 Down direction was displaying a red aspect against 1K20, which would come off after the passage of 1A09, whilst the turnout was set away from line 2 to line 3. At 08:08:15 the AWS warning for SN109 on gantry 8 was cancelled, and at 08:08:25 1K20 passed the signal at 41 mph (66 km/h). As set out earlier, Hodder did not stop his train at SN109 but instead accelerated to 51 mph (82 km/h) against the direction of traffic onto the Up spur line/line 3 where he should have realised that his train ought not to have been.

Train 1A09 had been following another service towards Paddington and was running under cautionary signals that driver Cooper could see clearing to green on his approach. There was no OTDR fitted to this set and its ATP equipment was not operative (which had no bearing on the way the accident developed), but from the signal logs and from surviving witnesses the following can be deduced. At 08:08:50, 10 seconds before impact, the aspect of SN120 at Ladbroke Grove came back to red on Cooper, who immediately cut power and applied emergency brake power. However, it would be to no avail at the speed his train was running at. Hodder powered on until a few seconds before impact. He first pushed his combined power and brake controller to provide braking and then slammed it to emergency braking when it finally dawned on him that his train and the HST were heading for each other. At 08:09 they impacted head-on at a combined speed of 130 mph (210 km/h).

Comment

It was conclusively proved that signal SN109 had shown a red aspect without any route indication. If it is assumed that driver Hodder kept an eye on his signals, then what must have happened at SN109 is that he believed he had cancelled the AWS warning for a signal aspect less restricted than he was travelling under as he accelerated. That less restrictive aspect he *thought* he saw must then have been a double yellow, which would explain his rapid acceleration. This would point to a case of ghosting (phantom aspects), as yellow lenses do stand out under those conditions. He might also have thought that he had cancelled for something he had missed – e.g. a nearby speed board magnet, a failed AWS magnet or portable AWS magnets left behind from engineering work – a situation that is by no means rare. But in that case he should have stopped to report and make sure. On the other hand, perhaps he did not know which signal to look at on gantry 8, although all signals there were recorded as showing red at that time and, in any case, in those circumstances he should have stopped anyway.

It is known from deduction, then, that Hodder cancelled the AWS for a danger aspect at SN109 that he failed to pick up. He then unaccountably accelerated at full power, but that might have been because he had passed Ladbroke Grove Junction and the speed of the line on which this service would normally travel was 100 mph (160 km/h), the little bit of familiarity he had gained with this route kicking in, despite the fact that he had two tracks on his left instead of none. But notice the same confusion about features of the route ahead that were apparent during the accidents at ***Brühl*** in Germany and ***Pécrot*** in Belgium, leading to that inexplicable reaction of speeding away. In all three cases the crucial final mistake was to accelerate, as if to escape further insecurity, whereas stopping would have been the only correct action. Another common aspect of these cases was that driver training was found to be unacceptably flawed.

Driver training

The London area was always short of train drivers. It was for that reason that many Scottish and Welsh railway workers intent on becoming drivers would come to the capital to pass out and then return home at the first available opportunity, which often took many years to materialise. For a similar reason I was accepted for driver training at what was considered to be an 'old' age and despite being overeducated, and I also had the objective of moving away from London as soon as possible. Privatisation helped greatly in achieving that goal; inter-city train operating company depots could easily poach the freshly trained footplate staff from the London area local operators who had forked out for the training. That is how I ended up in Bristol, working for First Great Western. The result of all this, however, was that the local depots in London were perpetually recruiting and training considerable numbers of new traincrew. Mick Hodder entered the ranks of trainee drivers under these circumstances, somewhat surprisingly given some aspects of his background, after he had finished his career in the Royal

Navy. With his Navy background and his pleasant disposition, Hodder was considered good driver material and during his training he did not disappoint. The comments made during the inquest into the accident are clear enough.

It must also be said that training at that time was to a standard that was not materially different from the situation under British Rail, trainees often being judged on rather emotive grounds rather than on predetermined objective grounds. Teaching was often a means of keeping former footplate staff employed, usually being off active service for medical reasons and having received little or no professional teaching training, whilst the organisational back-up and training material provided were far from satisfactory. With the now privatised train operating companies, there was an additional urge to get expensive new staff out onto trains quickly in order to provide passenger miles and reduce the extremely expensive cancellations of services due to staff shortages. Unlike the British Rail of old, however, new footplate staff were no longer sent out for years as second men with experienced drivers, slowly gaining hands-on experience and either getting the hang of the job or falling by the wayside if they failed to come up to scratch with regard to aptitude or ability to deal with the problems caused by the 24-hour nature of the job. These people now had to be trained within a year and had then to gain their route knowledge within an equally short time, after which the settling in (or not, as the case might be) with the 24-hour railway life began. The system required quite different things from instructors and trainees in comparison to what went on before, and it is clear now that neither were really up to meeting the requirements of the situation. The Ladbroke Grove accident showed the limitations of the existing training and provided the harsh incentive to change for the better, which did happen. Nevertheless, it has to be said that even the old way of training produced scores of drivers who were never involved in such accidents.

The layout from Paddington to Ladbroke Grove

At signal SN109 alone eight SPADs had occurred before this one with its fatal consequences. Later, First Great Western operations safety director Alison Forster wrote an urgent letter to Railtrack after a similar incident in which an FGW express train found itself heading towards a Heathrow Express service in exactly the manner that had caught out Hodder, although in this case the driver stopped and reported. There was a definite SPAD gremlin in the area, but the reaction of Railtrack's infrastructure and signalling managers was remarkably lacking in concern, and research to improve the situation was not treated with any urgency.

From a driver's point of view this layout, especially at Ladbroke Grove, was visually cluttered, was intricate route-wise and was also fast. In short, it was rather difficult to grasp. It demanded awareness of the train's position on the track-map at all times, as the view ahead was littered with low overbridges, catenary and massive signal gantries and nothing much stood out as a convenient landmark to navigate by. The ARS often did not send you along the clearest route. Switching track for unfathomable reasons was endemic as each track was reversible and could take you out or in and you could be put over to another track at considerable speed. This made it difficult to get used to anything in detail and at all time it was necessary to know your line number in order to be able to identify which signal on the gantry in the curve ahead pertained to your train. Additionally, at Ladbroke Grove Junction in the Down direction, due to the use of high-speed turnouts in lightly curved main lines, it was not always immediately clear to the poorly initiated whether you would use the crossovers or continue ahead along the line you were on. This, I think, is one matter that shaped Hodder's perception of his whereabouts in the moments before the collision. Having passed the junctions, he must have been shocked to discover where he was.

Once all of the above had been sorted out through increasing experience, gaining familiarity and greater confidence by establishing points along the route at which you intuitively would look out for speed signs and signal indications, the situation became quite enjoyable, oddly enough. Especially because the departures and inbound approaches could be spectacularly fast. If ATP had been used as designed, and all trains fitted with it, the layout would have been world class, which is where the crux of this accident lies. Once it had been decided that ATP would not be fitted to all trains, the risks, merits and demerits of this layout under the reduced circumstances should have been critically reassessed, as indeed happened after the accident. There is no doubt that the decision not to fit at least all regular Paddington trains with ATP was an ill-considered cost-cutting exercise and, if truth be told, Heathrow Express deserves praise for nevertheless fitting full ATP to its brand-new trains.

After the accident many aspects relating to the operation of this layout were changed, as should have happened previously. Flank protection of passing trains was introduced by turning certain turnouts along the path away. The possible variety of allocation of lines to trains was cut back and an Up and Down allocation of lines was instigated whereby reversible use of the signalling was reduced to a minimum, more as a way of working around obstructions and failures. The signal heads and many signal locations were changed, single-lens 'spotlight' 3-aspect signals with optic-fibre lighting taking over from the 4-aspect tungsten light signals with their tendency to start ghosting. The reverse L-shape signals on gantry 8 went. Consequently, the number of SPAD incidents went down and their impact on safety was lessened further with the introduction of TPWS. It is still not brilliant, but no more can be done to improve the situation.

The signallers

There is one issue that still gets signallers throughout the Great Western region angry, in a similar way that driver Terry Keating on the South Western out of London Waterloo never came to terms with his treatment during the hearing on the ***Clapham Junction*** accident. In the case of the Ladbroke Grove accident, it was the allegation that signaller Dave Allen of Slough New IECC could have prevented the accident had he reacted immediately by sending out a 'Stop' message on the cab secure radio. According to the 'experts' who liberally sprouted after the event, he had 20 seconds to do that and so would have prevented the accident. First, it should not be overlooked that seeing two trains on a collision course and then merge on his panel was a terrible experience that signaller Allen had to come to terms with. He then had to deal with this allegation published widely throughout the media. Well, let us have a look at the facts.

Allen's reaction to Hodder's SPAD was primarily based on the experience of the no fewer than eight previous SPADs at signal SN109 and quite a few more at other signals between Paddington and Ladbroke Grove. In virtually all cases before this accident, the drivers had stopped and reported the SPAD, and in only two cases a low-speed collision had occurred closer to the Paddington platforms. So signaller Allen's reaction, based on his expectation from substantial experience, was a hand on the phone to pick up the call from 1K20 as soon as it came through and deal with it expeditiously in order to avoid any unnecessary delay in the morning rush hour. The call never came, however, and by the time this realisation hit, at least 10 seconds if not more had passed (the train having to be brought to a stop and the driver having to initiate a call on his CSR equipment or put on his orange vest, climb down to the ballast, walk over to the nearest phone and hope it works okay). The next step was to call up the CSR radio emergency screen from a menu on a computer and initiate the sending of an emergency 'Stop' message.

That took time, and by the time the message went out the two trains, by then doing 90 mph (145 km/h) and 55 mph (80 km/h) respectively, were almost on top of each other and the collision was inevitable.

Conclusion

The situation was not helped by the fact that the rules and regulations to deal with this situation were to be found in at least three different publications, some of these being no more than photocopied sheaves of stapled pages, and that training in dealing with such situations (which, really, can only be undertaken properly on simulators) had never been carried out in the first place, let alone repeated as it should have been at least once every two years to maintain proficiency. Unsurprisingly, much of this has now changed. I fondly remember doing courses for signallers on faults and failures with AWS, TPWS and ATP. The use of different terminology between signallers and drivers (following privatisation) had created situations in which the two groups no longer understood each other when trying to assess over the telephone what had happened and what was needed, neither having much insight into what went on in the other's work area when these problems occurred. It caused unwanted misunderstandings and consequent delays that these courses set out to eliminate. This sort of quick and decisive *ad hoc* co-operation between infrastructure provider and lead train operator would not have happened in the days before this disaster.

Ladbroke Grove now stands out as the moment after which the 'new' railway in Britain matured further. Different, and in many ways more appropriate, protocols and means of doing things were established as a result of the findings of the inquiry into this accident. It was also the turning point with regard to the making of a decision on some form of ATP. As a result, the implementation of TPWS was speeded up and Britain's railways were substantially safer again by 2003. In the meantime much along the Great Western Main Line has changed considerably. The electrification to Bristol and Swansea is in progress. Crossrail, or to give it its proper name the Elizabeth Line, comes out of the tunnel under Paddington and required a whole new junction layout at Ladbroke Grove and two extra tracks onwards to Stockley Bridge Junction where the line to Heathrow Airport branches off. At the other side of Ladbroke Grove the connections into the Hitachi depot (the former Eurostar depot at North Pole) and extended trackwork to Old Oak Common took care of serious alterations. Reading station is unrecognisable to those who worked it before 2010. Even Swindon has seen improvement with an extra Down platform and the use of the former Up freight spur to bring in passenger trains. ATP of the ETCS variety will be installed. Life goes on.

***

Glenbrook, New South Wales, Australia, 2 December 1999 at 08:22

(Rear-end collision following communication mix-up)

The accident

A double-deck local (inter-urban) EMU collided with the rear end of the slowly moving 'Indian Pacific' long-distance service as a result of incompetent management of a signal failure, incompetent communication between a driver and traffic controllers about handling the signalling failure and a lack of train describer equipment in the controlling signal box to enable the whereabouts of trains to be established. The accident occurred on a double-track electrified line on sharp reverse curves in a fairly narrow rock cutting in daylight on a fine and clear morning. There were seven fatalities and 51 people were injured, the fatalities all occurring on the lower deck of the double-deck front carriage of the EMU. This was caused by severe rearward telescoping of the rear vehicle of the 'Indian Pacific' train, a double-deck Motorail car-carrying wagon.

The location

The collision occurred on the outer-suburban lines from Sydney to the Blue Mountains, on the Main Western line between the stations of Glenbrook and Lapstone. This section of line had previously seen a similar incident, on 16 January 1976, when a freight train had crashed into the rear of a standing passenger train, resulting in 10 people sustaining injuries and the death of one passenger. It is also the line where the ***Granville*** accident occurred on 18 January 1977, and would make headlines again on 31 January 2003 with a fatal crash at *Waterfall*. That accident was caused by a driver who suffered a heart attack on one of the modern double-deck G class 'Tangara' EMUs that, oddly enough, had no vigilance equipment on its 'dead man' safety system. As a result of the obese and desperately ill train driver keeping the footrest of the 'dead man' system down (see ***Roermond***), the train became a runaway on a downward gradient and overturned on a sharp curve. There were seven fatalities including the driver and 41 people were injured.

Glenbrook was the second major crash reported on by Judge Peter

Aloysius McInerney QC, who made a name in the trade of rail crash investigations and the redefining of the basis for safe working systems on railways in Australia. The network is electrified with a 1.5 kV dc OLE system and fitted with colour-light signals that in the Up direction towards Sydney carried the kilometre location with an even decimal digit as the signal number and in the Down direction the same kilometre number but with an odd decimal digit.

The trains and rolling stock involved

The 'Indian Pacific' luxury long-distance train, owned and operated by Great Southern Rail and running between the cities of Perth and Sydney, is arguably one of the better-known trains in the world. This train consisted of 17 stainless steel carriages of the classic ribbed Budd design and was hauled by an NR class diesel-electric locomotive, owned and crewed by the Pacific National Rail Transport Corporation. Locomotive and train turned the scales at 900 tonnes, the overall length being an impressive 426 m (1,400 ft).

The year 1999, incidentally, was not a good one on which to travel on the 'Indian Pacific'. Four months previous to this accident, at 17:06 on 18 August, the train, then carrying headcode 3AP88 and on its way from Adelaide to Perth, collided head-on at slow speed with looped freight train 3PW4N at *Zanthus*. This came about because one of the crew of the freight train operated the turnout to reverse from the main line into the passing loop just when the 'Indian Pacific' was approaching. Luckily, there were no fatalities and the few injuries that were sustained were light, but the damage was serious. This phenomenon of traincrew operating turnouts just when a crossing train is approaching is a disturbingly common feature of operation of long and relatively low-tech sections of single-track lines in sparsely populated areas (see ***East Robinson***, for example).

The 132-tonne NR class Co'Co' is without doubt Australia's most modern and important traction vehicle at the moment. It is based on the US General Electric Dash 9 model and Goninan, the Australian licence-holder of GE traction technology, designed and part-constructed 120 units as its model Cv40-9i with a maximum permitted speed of 115 km/h (70 mph) from 1996 onwards. Their distinguishing feature is the fact that they can be switched to three levels of power output, depending on train weight and the gradient profile of the line ahead. Their V16 4-stroke GE 7FDL-16 engine can operate from 2,125 kW (2,850 hp) via 2,655 kW (3,560 hp) to 3,000 kW (4,020 hp) and that substantially influences their fuel economy. Whilst their staple application is on heavy-haul freight transport, several machines are hired to long-distance passenger operators on 'Hook & Haul' contracts and some of the locomotives in that operating pool appear in rather colourful liveries based on aboriginal designs.

In the Glenbrook accident the EMU set involved was a 4-car double-deck V-class set No V21, working inter-urban service W534 from Lithgow to Sydney. It was 95 m (310 ft) in length and weighed 204 tonnes. Despite the severe damage caused to leading vehicle DIM 8067 (the rear vehicle of the 'Indian Pacific' telescoped 7.5 m or 24.75 ft into this carriage) the set was quickly restored to service as a result of vehicle shortages and the pressing need to provide sufficient rolling stock for the 2000 Olympics in Sydney. The carriage was renumbered to DIM 8020. These sets had as yet no OTDR trip recorders, although this equipment was being installed at the time.

The accident log

The 'Indian Pacific', train identity WL2 on the Sydney CityRail network, was on the last stretch of its long journey from Perth towards Sydney Central (with a booked arrival time of 09:15) when just after 08:01, following the passage of local service W532, the power feed to the track-circuit equipment for signals 40.8 and 41.6 failed. Exactly why this happened was never fully established, the failed components being described during the investigation as normally very reliable. Both signals returned to danger under the fail-safe arrangement and consequently the slightly early-running 'Indian Pacific' came to a halt at signal 41.6. Co-driver Willoughby left the cab to contact the signaller via the signal-post telephone. He and his colleague Marshall had taken over the train at Parkes in the very early hours of that morning and Willoughby had driven the train via Dubbo, Wellington, Orange and Bathurst to Lithgow, Marshall then taking over for the final 156 km (97.5 miles) via Zig Zag, Katoomba, Glenbrook, Penrith, Blacktown and Parramatta to Sydney Central. Willoughby found the telephone cabinet locked, which was not unusual in an urban area, and had to return to the cab to get his key and open it. The Penrith signaller gave him permission to pass signal 41.6 at danger and with a delay of 7 minutes the 'Indian Pacific' started moving again.

The 900-tonne train full of passengers and a few cars was not exactly a wheelbarrow as far as handling on the twisty and hilly Main Western line through the Blue Mountains was concerned, so the crew took the situation seriously, recognising the need to proceed at caution past a signal at danger and creeping past the signal to the next one, automatic

40.8, which was reached in just under 8 minutes and also showed a red aspect. They stopped again and Willoughby jumped down onto the track once more to walk over to the signal-post telephone cabinet, this time with the key at hand, but found that he could not establish contact with the Penrith signaller. One wonders why he did not use the Countrynet radio. In any event, in accordance with the rules, the crew waited a minute and were just starting up their train again at 08:22 to pass this failed automatic signal at danger under 'Stop & Proceed' arrangements when they were shaken up by a collision. Inter-urban service W534, in the charge of driver Kevin Sinnett, a man with six years' experience in the cab of local trains, had crashed into their rear.

Sinnett, also travelling on the Up line from Lithgow, had received the following radio telephone call from the Sydney rail traffic controller, Michael Browne (not to be confused with the local signaller in Penrith signal box by the name of Damian Mulholland), at approximately 08:12. In order to understand the jargon used, some explanation is required:

- An auto is an automatic signal, one automatically operated by passing trains on the associated track circuit or axle-counter section.
- 'Tripping past a signal' describes the action of being stopped by a tripcock application of the emergency brake at a red signal, the way London Underground drivers do in exactly this situation. Some of the red signals on the Sydney network were protected similarly by raised T-bars, tripping a tripcock on the train that passed a red aspect and so stopping it.
- The signals around Sydney are normal red, yellow and green 3-aspect signals, but they additionally have a smaller red aspect, the marker light, outside and below the main cluster, offset towards the track from the signal post. This marker light, irrespective of any other aspect, shows a red aspect if control equipment has failed, forcing the approaching train driver to stop and contact the signaller. Therefore, the remark below about the two reds shows that it was not the signal that had failed but the control equipment that operated the signal.

Mr B: West Control to the driver of W534.
Dvr S: Yes, mate.
Mr B: I've just had a report there from the driver of the Indian ahead of you, 41.6 signal Penrith side of Glenbrook platform is at stop.
Dvr S: Oh, right.

Mr B: Are you around that area yet?
Dvr S: I'm just heading towards Blaxland now, mate.
Mr B: Heading towards Blaxland, all right, okay. Well, I don't know what… he just said it's failed. So I don't know if it's a red marker light or what the story is. It's only an auto.
Dvr S: All right, mate. I'll get on to Penrith anyhow or whatever. Whoever I get there.
Mr B: Yeah, okay.
Dvr S: Thanks a lot!
Mr B: Well, yeah, it's only an auto so just trip past it.
Dvr S: All right, mate.
Mr B: Okay, thanks.

Train service controller Browne, going somewhat beyond his actual authority, thus incited driver Sinnett to pass a failed signal at danger because 'it's only an auto'. Sinnett continued, eventually reaching failed signal 41.6 where he stopped and was contacted by Penrith signaller Mulholland on his Metronet radio telephone. It is the sort of closed channel radio telephone equipment that is in widespread use throughout the world under various names for this sort of suburban and urban telecommunications work. Mulholland answered the request for communication by opening the radio channel to W534.

Mr M: Yeah, 534?
Dvr S: Yeah, who have I got there, matey?
Mr M: Penrith, mate.
Dvr S: Yeah, it is 41, 41.6. I'm right to go past it, am I mate?
Mr M: Yeah mate, you certainly are. Listen, can you get back to us? What was the previous signal showing?
Dvr S: Yellow.
Mr M: Yellow, okay, and what's that signal actually showing, just red or…?
Dvr S: Yeah, two reds, mate.
Mr M: Two reds? No worries. All right, mate, can you just let us know what the signal in advance says when you get to it? Thanks!
Dvr S: Okay, matey.
Mr M: Okay, thanks.

Signaller Mulholland was anxious to know what the signal past this failure was showing as the 'Indian Pacific' did not report from there, but he did not tell driver Sinnett anything about that. So Sinnett hooked up his handset, released his brakes, opened up and departed again. He would later say that from these two conversations he had the idea that the road past this failed signal was clear at least to the next signal. He had, in short, been given expectations about the situation, but in reality signaller Mulholland had not the faintest clue about the state of the track ahead, because he had nothing in the way of track-circuit occupation diagrams or train describer equipment in his signal box to indicate that the 'Indian Pacific' was at least past the failures. He simply assumed that the 'Indian Pacific' had cleared off as enough time had passed and he had heard nothing from them.

As a consequence of the exchanges with Browne and Mulholland, Sinnett's assessment of what was required to travel at caution after having passed this automatic signal at danger was to just be able to stop at the next signal 40.8, of which he knew the location very well. He therefore accelerated to 50 km/h (30 mph) or slightly more, and it was at that speed that he first entered a rather sharp right-hand curve followed by a similarly noticeable reverse left-hand curve on a 1 in 60 downward gradient through a narrow and steep rock cutting leading up to signal 40.8, to eventually reach Glenbrook Tunnel via Glenbrook station. As his train coasted along, ready to start braking for the second failed signal at danger at the end of the reverse curves down the gradient, the tail vehicle of the 'Indian Pacific' was suddenly revealed from behind the rock wall some 100 m (330 ft) ahead. The sheer length of that train, which was just slowly pulling away from signal 40.8, covered more track than the entire braking distance Sinnett needed with his electro-pneumatic brake to stop at signal 40.8 at the speed he was doing. He applied the emergency brake and fled his cab to the lower deck saloon, shouting at people to brace themselves, then left the saloon and so saved his life. Another gent from the lower saloon ran up the stairs and shouted to the people in the upper saloon to hold on. He also saved his life by doing so. The local EMU hit the 'Indian Pacific' rear vehicle, a double-deck Motorail vehicle carrier, violently enough to make the frame of this vehicle telescope into its front carriage. The front lower saloon was badly damaged and all of its seven remaining occupants were dead. Others elsewhere in that carriage suffered grievous injuries. Somehow, however, there was a small stroke of luck. Had it been the next signal, 39.6, that was involved, the accident would have taken place in the dark amid the curved Glenbrook Tunnel between Glenbrook and Lapstone stations.

The investigation

A glut of detailed reports are available on various aspects of this accident. It did send a shockwave through Australia in the run-up to the 2000 Olympics. Judge McInerney found more than 30 unsafe ways of operating that contributed to this accident. The Safeworking Rules, the staff training, the language used during the exchanges between driver and traffic controllers, the lack of recording equipment, obsolete signalling and telecommunications equipment and the lack of train protection on these twisting lines through mountain countryside were all cited as factors.

In reality it came down to two issues. To start with, there was the lack of ATP, as at ***Schiedam-Nieuwland***. Train W534 would still have got past red signal 41.6, with permission in this case, but ATP would have severely reduced its speed and the system would have demanded alertness from the driver. Nevertheless, the impact might still have happened, the situation being made worse by the fact that the rear vehicle of the 'Indian Pacific' was a freight and not a passenger vehicle, but in all likelihood it would not have been as serious. ATP, however, is the last resort, the net that prevents disaster when everything else put in place to maintain safety fails. The antiquity of the signal-post telephone network and the incompatibility of on-board telecommunications systems are the sort of things that we in Britain are only now growing out of. You work with what you have, soldiering on etc.

The second issue was the matter of staff behaviour. Why did the controller at Sydney initiate that Metronet radio telephone call to driver Sinnett on W534? That call would solve or change nothing in the way that things developed after the signalling failure. Neither Sinnett nor any other traincrew could possibly tell him what had failed or why, or what the situation ahead would be like. In short, it was to all intents and purposes a useless call. But it was not entirely without purpose. Controller Browne used this call to prime the driver about just tripping past an automatic signal at danger – the real purpose was timekeeping. As in Japan at ***Amagasaki***, timekeeping was very important on the CityRail network, as Judge McInerney revealed as well, and has already been mentioned in the section concerning the ***Granville*** accident.

In Britain in 1999, after a few serious accidents, the railways were just getting past that stage (it even swung the other way for a while) and this accident – and the contents of Judge McInerney's report, no doubt – caused the same swing in operating priorities in Australia. Like JR West (West Japan Railway Company) at ***Amagasaki***, Sydney CityRail found to its cost that although passengers undoubtedly like to arrive on time, they prefer to do so in one piece and on their own feet. As for signaller Mulholland at Penrith, given the day and age and the sheer length of the lines he had to control, he was forced to work with unacceptably little in the way of being able to see what was going on around him on the network. Close to where I live, on the lovely West Somerset Railway, signallers can see the occupation of their track circuits with red lights on antiquated track diagrams that show their station limits. With the technology available in 1997, I could scan traffic along many European continental corridors on a computer monitor somewhere in the traincrew supervisor's back office at London Victoria station. Staff at the 'Indian Pacific' office at Dulwich near Adelaide could normally have watched the train passing Glenbrook, had the track circuits worked and had they had the inclination and time to do so. It was because of the lack of that readily available but expensive train describer equipment, however, that signaller Mulholland had to guess about trains nearby. And guessing is what he did, which was a dangerous strategy to adopt under the circumstances and predictably that is where things went wrong. He could have solved his problem of lack of information by using the Countrynet radio telephone to try and call the crew on the 'Indian Pacific', as indeed he did after the accident. Or could have called people at Glenbrook and Lapstone to discover whether the 'Indian Pacific' had passed their stations. It was not the kind of train that easily went by unnoticed, even when you happened to visit the toilet. But he did not.

In my opinion, Mulholland had insufficient understanding of what basic signalling is about, such as keeping trains apart etc; and his Safeworking Rulebook was not too helpful on the subject either. Like controller Browne at Sydney, he just wanted trains to roll, be done with this hassle at the onset of the morning rush hour. But he should not have accepted a driver at a failed signal telling him that it was OK to pass it at danger without him having mentioned a word about it. And why did he overlook to mention the possibility of a train in section ahead? Unthinking hurry? It is true that in the signalling lore automatic signals tend to rank lower than controlled signals (for example, in Britain they are the signals that did not receive TPWS protection). This display of haste by the three men involved went beyond acceptable practice and provided the main contributory cause for the accident by generating expectations on the part of driver Sinnett. When an automatic signal remains at red it means that either control equipment has failed, track has snapped or there is a train in the section ahead, otherwise an automatic signal will show a proceed aspect. Therefore, in the event of a failed automatic that trains need to be talked past, if you know that potentially there still is a train in section, the last thing you do is to send another train into section unless it has been established that the previous train has cleared it. Or you inform the driver of the following train to proceed slowly because there might be an obstruction from a train ahead. None of the three people involved appeared to attach much importance to that possibility.

Signaller Mulholland had two options. Either he held W534 at signal 41.6 until he was certain about the location of WL2, the 'Indian Pacific', and whether it was moving. This was the safest option but certain to cause substantial further delay. Or he instructed W534 to pass the signal at danger, to proceed with the utmost caution at such a speed that he would be able to stop for any obstruction ahead and find out what is the matter, as potentially there was another train in section. As for driver Sinnett, he strikes me as a man who did not want to be bothered with problems, but then again, who does? He appears to have been well imbued with the CityRail timekeeping ethos, happy to work his shifts (difficult enough at times, believe me) but had neither the sort of training nor the kind of intuitive insight into signalling and its role in train safety that would enable him to adhere doggedly to the safe option, if he was able to identify which of the two alternatives that was. He took remarks from both the train controller and the signaller at face value and drew understandable but dangerously incorrect conclusions with regard to the safety of his train, when required to pass an automatic signal at danger. However, he belonged to a train-operating culture that primarily concentrated on vigorous timekeeping, and he was sent into the jungle by his traffic control colleagues with the contents of their calls.

The crew on the 'Indian Pacific' were the only professional train operators in this case. Probably their only mistake, if you can call it that, was to not use their Countrynet radio at signal 40.6 where the signal-post telephone had failed. But despite being near the end of their night shift, they accepted that there were problems that would delay them, and from that moment onwards acted to uphold safety, avoid further trouble by being deliberate and giving themselves margins of safety commensurate with the weight and length of their passenger train. Unfortunately, they had to rely for cover on the Penrith signaller, in which task that gent failed. Judge McInerney did have something to say about the apparent lack of urgency of the 'Indian Pacific' footplate crew with their lengthy 900-tonne train, but that was where Judge McInerney merely showed that he was not a train driver but a judge, however good were the accident reports that he otherwise delivered.

***

Åsta, Norway, 4 January 2000 at 13:12

(Head-on collision with fire following a SPAD)

The accident

This was a head-on crash between a DMU railcar set and a locomotive-hauled train on a remote single-track line for reasons that are not fully understood. A devastating fire followed the collision. As far as the circumstances surrounding the accident are concerned, the frost and snow on the ground were not excessive, and low midday sunlight (southern Norway has very short daylight periods in January) that might have caused a phantom aspect (ghosting) of the departure signal at Rustad was ruled out because of the angles of the sunlight.

A digression concerning the Cape gauge

Here we meet the one notable Norwegian contribution in the field of the world's railways, the Cape 3 ft 6 in (1,067 mm) track gauge. After Robert Stephenson had built Norway's first railway line, the Hovedbanen or main line, to British engineering standards, it was felt that railways, although desperately needed to unlock the interior of this mountainous country, were far too expensive for the Norwegian national purse. The principal Norwegian engineer involved, Carl Abraham Pihl, argued for a narrower track gauge to allow sharper curves along mountain contour lines in order to avoid tunnelling wherever possible and to allow rolling stock of only a slightly smaller width and with centre couplers to negotiate those sharper curves. He therefore proposed what came to be known as the Cape gauge. On a world scale this gauge proved ideal for colonial railways through difficult countryside (Norway had no colonies but a lot of difficult countryside) and in a short time the 3 ft 6 in gauge found its way to all continents as a world 'standard' narrow main-line gauge, even if the French and German colonial railways preferred the metre gauge. The reason for the term Cape gauge is a matter of mild controversy; it may relate to the early application of the gauge in the Cape in South Africa or to Pihl's initials, CAP. In South Africa itself, incidentally, the 4 ft 8½ in (1,435 mm) standard gauge was long known as the Cape gauge before its change to the national 3 ft 6 in gauge. Reality (i.e. links with Sweden) soon forced Norway to revert to the 4 ft 8½ in standard gauge, at greater cost than if the lines had been built to that gauge from the outset, and to this day many lines in Norway can take only light axle weights. Which prevents purchase of less expensive off the shelf rolling stock and is a lingering reminder of the lightly built Cape-gauge track of long ago. In many nations with Cape-gauge track, however, the rolling stock is heavier and bigger than the standard-gauge rolling stock as used in Britain for example. In South Africa monstrous coal trains operate on the Cape-gauge line between Sishen and Saldanha, and in Japan and Australia Cape-gauge trainsets reach speeds of 160 km/h (100 mph) in everyday traffic (see ***Berajondo***). Japan even operates double-deck EMUs on its Cape-gauge network.

The location

The community of Åsta (pronounced Ohsta) near Elverum in south-eastern Norway is situated along a lightly trafficked railway line from Støren Junction via Røros and through the Østerdal (parallel with the River Glomma) to Hamar. Åsta is situated halfway between two stations with passing loops, Rena to the north and Rustad to the south. The Swedish border is not far away to the east. The line involved is known as the Rørosbane; a 384-km (240-mile) single-track non-electrified railway line built in stages as a Cape-gauge line between 1862 and 1877 and converted in stages to standard gauge between 1917 and 1941. Closure of this line and doubling of the electrified Dovre main line was proposed, but after the accident it was kept open with improved train protection and radio.

Changes to the line

The Rørosbane is a typically remote Swedish/Norwegian secondary rail connection. A few trains per day run along an endlessly winding, fairly low-speed line with widely spaced halts, which crosses mountain rivers and cuts through rock formations and large stands of coniferous trees close to the track. As a result, the view forward from the cab is often restricted to no further than the next curve. Due to its light traffic and in an attempt to operate the line more economically by taking out the expensive manned stations with their local signalling operation, in December 1994 the Rørosbane was resignalled with colour-light signals but without ATP. Line control was moved to a Centralized Traffic Control (CTC) centre at Hamar, to be worked by one man from an NSB/ABB-designed, low traffic density (and low-cost) NSB-87 control installation. Further modernisation with electrification and train radio was not foreseen. It is odd that after traffic control modernisation only traditional signal-post telephones at the widely spaced stations were considered as the means of communication between traincrews and the traffic controller. Moreover, whilst the two-track stations (a through track and passing loop) received train-detection track circuits that reported on the CTC screen, the long single-line sections in between had none, except where they served the very few automatic level crossings. But these did not show on the screen and as a result the modernised single line had in effect become 'dark territory' as far as operation of trains was concerned, a strange retrograde step to take in the year 2000. The stations along the line (an average distance of 40 km or 25 miles apart) were de-staffed and changed to request stops. The conductor was relieved of the actual train operation except signalling by radio to the driver that a station stop had been requested or that the station work had been finished and the doors were closed, but the driver became solely responsible for signal observation and final dispatch of the train.

In official corners there was dissatisfaction with these changes due to rather obvious safety implications, but NSB (Norwegian State Railways) took little notice and forged ahead as planned. ATP was slated to be fitted to the entire network after 1995 following the *Tretten* head-on collision on 22 February 1975, when a train failed to stop at a crossing point on the single-track electrified Dovre line and slammed into an oncoming service. The accident involved two loco-hauled 12-car ski trains carrying some 800 people, 27 of whom died. Nevertheless, ATP was not fitted to the Røros line at the time of the accident, due to the earlier talk of possible closure, but installation was in fact planned to commence 5 months later in June 2000. Funds for the ATP trackside equipment and installation had been allocated and both trains involved had ATP on-board equipment fitted. After the accident GSM-R train radio was fitted instead of the soon to be phased out Scanet equipment.

The trains involved

Involved were southbound train 2302, consisting of a locomotive with three carriages, and northbound train 2369, consisting of a 2-car diesel multiple-unit railcar train (DMU). The locomotive of 2302 was a Swedish (NOHAB, Trollhättan, see ***Getå***) licence-built GM EMD G series Co'Co' machine No Di3.625, delivered in 1965 and well past its replacement date. The Siemens-MaK Di6 replacements, however, had failed to meet the stringent Norwegian weight specifications and moreover were technically flawed, so whilst the useless new machines went back to Germany the Di3 machines were overhauled and put back in service again. The weight of a Di3 was 102 tonnes, its power 1,325 kW (1,776 hp) and its underslung fuel tanks contained 3,500 litres of diesel fuel. The coaches were B2/3 25 674, built in Sweden in 1947 (38.1 tonnes with 58 seats); B3 25 598, built in Norway in 1967 (37.6

tonnes with 58 seats); and BF11 21 522, built in Norway in 1966 (38.5 tonnes with 32 seats, a service room, children's compartment, a luggage van and a conductor's office). At the moment of impact there were 73 passengers on board and two traincrew.

Train 2369 consisted of a 49.45 m (162 ft) aluminium-built 2'Bo'+2'2' diesel electric-multiple-unit train type BM 92.14 from 1985 (Duewag, Germany). It had a 714 kW (957 hp) underfloor diesel engine and could carry 1,800 litres of diesel fuel in underslung tanks. Its 59-tonne power car was leading the 39-tonne driving trailer (98 tonnes, with a total of 114 seats). There were nine passengers and two traincrew on board at the moment of impact.

The accident log

Train 2302, the 07:45 Trondheim to Hamar, to connect there with a southbound Dovre main-line service to Oslo, was driven to Røros by Roger Wiggen from Trondheim traincrew depot, who collected his rolling stock and prepped the locomotive at Marienborg traction depot. At Trondheim station the train had had to wait for passengers from a delayed long-distance bus connection from the north, which delayed the departure by 15 minutes, the departure signal having cleared at 07:30. This delay led to out-of-course stops for crossings with train 413 at Marienborg, train 5781 at Heimdal and train 407 at Ler. When train 2302 arrived at Røros at 10:24 it was handed over to driver Stig Juliussen. Juliussen, aged 42, had been driving these locomotives since 1985. He was known as a quiet and reliable man, who had passed his last rules and traction repeat test with good results in 1998, without any marks against him on his file. His conductor was 43-year-old Jan Anders Arneberg, a senior conductor since 1985 and known as a friendly and personable man, well liked by passengers and also with an unblemished record. They left Røros 21 minutes late but were only 7 minutes late when arriving at Rena, using the request-stop character of the line to good effect. Passengers boarding with luggage and children in buggies at Rena caused a slight loss of time there, but at 13:07 they were on their way again.

Egil Lodgaard, aged 58, had been a driver since 1977. Having booked on at Hamar depot at 11:30 to work train 2369, the 12:30 from Hamar to Rena, he telephoned the Hamar traincrew manager at 12:09 to report his mobile phone number as requested but made a few mistakes whilst doing that, which, however, were corrected. Known as a steady and safety-conscious driver and a bit of a stickler for correctness, he collected and prepared his DMU set well in time. After the accident his watch was found still running and it was only 12 seconds out, so this cannot possibly have influenced his way of working his train. His conductor was 32-year-old Erik Stårvik, who had been a senior conductor since 1990. He had been off sick for a month, but on the day in question joined train 2369 at 12:00. Whilst the southbound train had reduced its delay by two-thirds between Røros and Rena, DMU 2369 had departed Hamar on time and steadily worked its way northwards to Rustad.

So it was that both trains stood at opposite ends of the single-track section on which Åsta was located. At 13:07 southbound train 2302 departed Rena approximately 8 minutes late past a proven green signal (railwayman on local repeater display, railwayman on location and the actions replay facility on the main display at Hamar CTC centre). Northbound train 2369 departed Rustad at 13:07 after only a 24-second stop, not waiting as booked until its departure time of 13:10 to cross with train 2302. Train 2369, therefore, departed Rustad 3 minutes early without receiving a green signal aspect. It also split the turnout where the station passing loop became a single-line section, the turnout having been set and locked against it as the road had been set for train 2302 to enter. Clearly, both Lodgaard and Stårvik failed to notice their 3-minute early departure and, worse, the red signal during the dispatch ritual. Lodgaard, with an ample 40 seconds to check aspect and his turnout whilst moving from his stopping point to the departure signal, also did not pick up on the fact that the turnout was against him.

On investigating after the accident, the train trip recorder indicated that train 2369 braked later and more sharply than normal when entering Rustad station, as if Lodgaard was not going to stop there (it was a request stop) but then finding that he had to pick up a single passenger. He stopped so far back in the platform that the rear of his 2-car train, which was only 50 m (165 ft) long, had not even cleared the track circuit behind him. Which prevented the southern-end turnout and southbound starting signal being reset for train 2302 to pass through until 2369 had moved forward to clear that track circuit or, in this particular case, had left the station. Could it be that, although it was a booked move in the timetable, the crew of 2369 did not expect 2302 to cross here? What had happened to make this otherwise competent traincrew so recklessly hurried? It had all the appearance of them working on the assumption that the crossing with the delayed 2302 had been moved from Rustad, where they had just stopped, to Rena where their train would terminate. But no known instruction to this effect had taken place as far as the telecommunications logs were concerned. Another (remote) possibility is that one of the traincrew or a passenger had become ill or urgently wanted help to be available at the termination point, but no further indication to back this up can be found. Nevertheless, issues concerning traincrew sickness were at the root of fairly recent accidents in such places as *Amersfoort, Barendrecht* and ***Roermond*** in The Netherlands and *Waterfall* in Australia.

The Rørosbanen traffic controller at Hamar CTC was also in charge of the much busier line from Hamar to Eidsvoll and matters on that line were occupying his attention at 13:07 when both the trains involved started their final trip. On completion of his work concerning the Eidsvoll line he turned his attention to the Røros line display at approximately 13:11, as by now the crossing between trains 2302 and 2369 at Rustad was imminent despite the delay to the southbound service. At 13:12:58 he noticed a stationary line of 16 mm-high red lettering at the bottom of his video screen with the line display, which indicated that at 13:09:28 a malfunction had occurred with the northern-end single-line turnout at Rustad. During the post-accident investigation it transpired that this was the moment that train 2369 had split the turnout open on its unauthorised departure. Unfortunately, to save on expense, NSB had not installed audible and visual fault alarms in the equipment, which meant that about 3 extra minutes in which action might have been taken had been lost. For which, the accident report notices, the controller could not be held responsible. The controller had not been made aware by his equipment that train 2369 had departed from Rustad, where it should have been awaiting the crossing with the oncoming train at 13:12. The inevitable conclusion was that train 2369 had left Rustad on a SPAD against a red starting signal and that a head-on collision between both trains was imminent. How to stop them?

Neither ATP nor radio telephone equipment to call the trains had been installed along this line. Finding this telecommunications omission hindering train running, in order to enable contact with traincrews in the 'dark territory' between the widely spaced signal-post telephones NSB had previously entered into a commercial UPT contract with mobile telecommunications provider Telenor to run a service in which control could call private GSM mobile phones on trains by using a special prefix alpha number service. The train number was part of this alpha number and so a designated member of traincrew on an individual train could be directly called. A few months earlier, however, Telenor had cancelled this service and, from that time on, traincrew were required to call control previous to departure on a particular service to give the numbers of their mobile phones. Both traincrews had, in fact, done that – call record tapes exist – but the person who took those calls at control failed to enter the numbers on a list against the train numbers, as required before being relieved by the present operator at the end of the shift. This omission precluded any possible action to stop the trains; it now was a matter of waiting for the report of the accident to come in.

After departure at 13:07, both trains headed towards each other for 5½ minutes before colliding head-on at 13:12:35 at an estimated impact speed of 160 km/h (100 mph). The welded aluminium body of the leading traction vehicle of the DMU came apart at the seams on impact and was completely destroyed, but its trailer escaped largely unscathed and remained on the track, rolling back a few metres in the direction from which it had come. On impact the locomotive fell out of the track, the head-end of the front carriage behind it buckling sideways and rearwards into a J-shape as it followed its overturning locomotive, the second carriage derailing but staying upright, and the third carriage remaining undamaged on the track. Virtually on impact a large fire broke out near the locomotive. Survivors initially attempted to put it out using fire extinguishers from the coaches, which proved futile. Even after the fire brigade reached the site and started work using foam, the fire continued to increase in size and intensity against all attempts to extinguish it, eventually engulfing all vehicles. They were hampered by the fact that the centres of the fire were inaccessible as they were covered by heavy wreckage, and by the limited amount of water available. The River Glomma was too remote, water had to be trucked in.

The train burned with a number of the immobilised casualties still inside, until snow started to fall in sufficient quantities to help with extinguishing the fire hours after the crash. Firefighters and police officers had to abandon boarding the wreckage for further rescue attempts and then had to watch victims burn at windows, reminiscent of the level crossing accident with a road tanker transporting 15,000 litres of petrol at *Langenweddingen* in Germany in 1967. That resulted in a train fire that killed 94 passengers.

Diesel fuel fires

The ***Ladbroke Grove*** accident report dealt with the reason for such diesel-fuelled fires on wrecked trains and heavy road vehicles. Unlike petrol (gasoline), diesel fuel is fairly difficult to ignite when cold and in fully liquid condition as it does not easily evaporate at ambient temperatures and is not mixed with oxygen except at its very surface contact point. However, in the event of a vehicle involved in a violent collision, the fuel tanks are likely to be deformed, compressed and ripped at the seams. As a result, fuel squirts out in a well-aerated mist creating conditions for ignition if sufficient heat or sparks are near. Once the initial fire is ablaze, spilled diesel fuel rapidly heats up and starts to give off fuel vapours that in turn sustain and increase the fire. Such fires are known to be very hot and difficult to extinguish and are a well-understood risk accompanying accidents with diesel traction.

The inquiry

The accident report was less than impressed with the way NSB dealt with the safety side of its operations. None of the thorough changes in signalling, traffic control and operational safety appeared to have been risk-assessed or designed with relevant safety protocols in mind, including fallback procedures in case of failures. The lack of ATP and of decent telecommunications arrangements, as well as the less than perfect signalling control equipment, triggered serious criticism, as did the age of the coaches. Wood used in their interiors had increased the intensity of the fire.

Another report touched on the existence of departmental fiefdoms in the organisation (see ***Buttevant***). This was to an extent always the case with the national rail operators of old, but decent overall management usually ensured that the various departments would not take safety-critical decisions ignoring the competencies and needs of other departments. In Norway the separation of train-running operations from the infrastructure network provider Jernbaneverket caused these fiefdom characteristics to be reinforced, however, as indeed was the case in most other European nations that went through this process. In this case it led to providing convenient modern train control installations but leaving out the bits for train-running safety that were considered peripheral and too costly. That, also, was not an exclusively Norwegian situation. I have witnessed senior signalling people elsewhere in Europe state in all seriousness that train drivers simply had to stop for red signals and that the demand for safety systems stood in the way of efficient modernisation.

Comment

Given that it is very unlikely that the signalling control equipment failed wrong-side and that a proceed aspect was shown at both ends of the single-line section between Rustad and Rena, the finger of blame must point at the crew of train 2369 at Rustad. Southbound train 2302 was proven to have had the road from Rena into Rustad.

The accident report states that the crew of train 2369 was neither informed of the delay to the southbound train nor of the traffic controller's ruminations about moving the crossing to Rena, their termination point for that trip (which indeed had been discussed around the time when train 2302 departed from Røros with a delay of 21 minutes). I am entirely convinced that Messrs Lodgaard and Stårvik were well aware of that delay to train 2302 at Røros and that when they departed from Hamar they were certain that the crossing had (at least tentatively) been moved from Rustad to Rena as a result. They departed Hamar very probably with advice to call Hamar control (because, as it later turned out, control could not have called them!) when nearer Rustad, in case of further changes. It would have been utterly normal behaviour for traincrew to phone, or more likely in their case, to walk over to the CTC traffic controller's room at Hamar to ask about the situation along the single line between their booking on at 11:30 (an hour after the other train had departed Røros) and their departure towards Rena at 12:30. The considerable making-up of lost time by train 2302 had taken everyone by surprise, so either Lodgaard or Stårvik, calling by mobile before reaching Rustad to ask for an update, would then in all likelihood have been told of the radically diminished delay to that train. However, no mention was made of the important fact that the mooted change of crossing point to Rena had not been implemented. The trip-recorded capers at Rustad therefore then become quite a logical consequence.

The investigation commission appears to have been aware of this possible issue, because it recommended fitting security locks on relevant control centre doors to prevent unauthorised persons from entering. This in turn makes it look like the commission also had its doubts about whether everyone's co-operation with the investigation was altogether open and honest. But after a crash of this magnitude the commission should have worked from the premise that not everyone was going to admit in detail to things that potentially could be damaging to their career or even continued employment. Were, for instance, all those private mobile phones checked concerning the calls and SMS messages placed in the run-up to the accident, notably those of the Hamar traffic controller and those of Juliussen, Lodgaard and Stårvik?

The inquiry found that under certain conditions it was possible for the departure signal of train 2369 at Rustad to show no aspect, to black out, for a few seconds. It could have been a reason for their action at that signal, compromised as they appeared to be by their haste, ignoring rules about signals not or incompletely displayed. An issue not mentioned in the report, though, but which I include from my own experience, is that railwaymen know that a green signal aspect in a colour-light signal with tungsten light bulbs is the least conspicuous of them all. The red and especially the yellow aspects easily outshine it. So if a signal is not showing much light, a distracted driver might possibly accept it as a green signal. But there was no bright daylight at that time out there in the wintry Norwegian outback, so any colour-light signal would have been conspicuous, as they are on overcast days or at night. It was, in fact, one of the reasons that ghosting of the signal could safely be ruled out. Yet especially driver Lodgaard had ample time to see that red signal aspect during the stop and whilst rolling towards it. Any experienced and on-the-ball train driver entering a single-line section will always take a good look at his departure signal before passing it, just in case it has come back to red. His life depends on it, certainly when another train is known to be nearby and heading his way.

If everything went as surmised – and I candidly admit that it is no more than surmise –, then following the Sherlock Holmes principle (see ***East Robinson***) it is a baffling example of railway people wrong-footing themselves through accepting tentative delay-management ruminations as alterations to the schedule for their trip. For this accident to make any sense at all, alterations to the booked workings must have been communicated to the northbound crew prior to their departure, with the advice to phone in later to hear about developments. When they did that they were told about the diminished delay, but not that the normal timetable would now operate. As a result driver Lodgaard interpreted the situation as him holding up 2302 and suddenly he was in a hurry.

If this was not the case, and there really were issues with either signalling or instructions after all, then NSB cannot be commended for the thoroughness with which it implemented and recorded those issues. The number of questions surrounding this accident for which there is no unequivocal answer is rather more than should be expected for modern-day operations. What is also unpleasant to notice is the haphazard way in which changes with safety-critical implications for crew management, telecommunications, traffic control and operations were introduced without serious risk-assessment. These resulted in the opening of traps for distracted staff to step in to, which is precisely what happened. Foreseeing and closing such traps is what safe, modern train operation is all about – to catch out and correct crews before accidents happen (see ***Brühl*** for example).

Immediately after this accident, train drivers boycotted this line due to a lack of trust in its signalling, which in turn ensured that ATP and GSM-R train radio were installed forthwith (see also ***Pécrot***). I do wonder, however, what was omitted from the accident report.

***

Brühl, Germany, 6 February 2000 at 00:13

(Derailment due to excessive speed)

The accident

A derailment was caused by excessive speed at a set of turnouts because of mistakes made by an inexperienced and poorly trained driver. A lack of route knowledge also played a part, concerning some characteristics of the track and signalling layout, taking the train from main line Platform 2 into slow loop line Platform 3 at Brühl. It caused extensive damage to five vehicles, two of which suffered body deformation against the platform side and canopy stanchions, whilst the locomotive went down an embankment. The derailment left nine people dead and 149 injured.

The location

The accident happened on Sunday 6 February 2000 at Brühl station on main line 2630 from Köln via Bonn to Koblenz, also known as the 'Linksrheinische Hauptstrecke' (Rhine left bank main line). Coming from Köln, Brühl station limits encompass a freight yard and a passenger station within a distance of 3,167 m (2 miles), operated by a local signal box from a 1969 Siemens-built, NX-operated panel installation. The passenger station consisted of two main lines with platform faces, and two platform loops to allow trains to overtake each other, one off each main line. Signalling was the traditional German 2-aspect Hp/Vr colour-light system, which in 1971 and 1977 had been extended to allow reversible working between Hürth-Kalscheuren and Sechtem, approximately half the distance between Köln and Bonn. Of importance with regard to this accident is the extent to which design restrictions in the reversible signalling contributed:

1. Few intermediate crossovers in both directions had been installed, which in this case required reversible working all the way from Hürth-Kalscheuren to Sechtem, as there was no other set of crossovers available to gain the proper track again after Brühl. (See the remark in point 3 below.)
2. Signalling disconnections due to turnout renewal from the main line to the freight yard allowed no automatic NX signalling control, which meant that no main signal aspect could be used to enter Brühl station from Köln under reversible conditions. That is why turnouts had to be set manually and an Ersatz position

light signal aspect was operated to call the train on towards the passenger station. The Zs 1 Ersatzsignal (a small triangle of three white lights on the signal post under the main aspects) instructs a train driver to pass a main signal at danger without the obligatory written 'pass signal at danger' order and to travel at a maximum speed of 40 km/h (25 mph) so as to be able to stop before any obstruction, until a) a next working main signal is reached (in this case the platform loop starting signal), or b) the last turnouts of the layout have been passed. This last item requires detailed local route knowledge, which the driver involved in this incident did not have. An Ersatzsignal is really meant for one or a few individual services having to pass, for example, a failed main line signal awaiting repair; it is not intended as a replacement for non-available signalling indications to work a series of booked trains throughout the night past an engineering worksite.

3 When working reversibly through Brühl station, trains had to enter the platform loops at no more than 40 km/h (25 mph) and leave at no more than 60 km/h (35 mph), as only these platform loops had starting signals for the reversible direction; the main line platform signals did not. This in/out speed limit difference, incidentally, was based on the fact that turnouts negotiated reverse in the facing direction are subject to higher wear than turnouts used reverse in the trailing direction. The official background to this somewhat peculiar design restriction was that for a crossing manoeuvre during single-line working, the normal-line trains (in this case from Bonn to Köln) were signalled as usual through the station at line speed of 160 km/h (100 mph) whilst the reversibly working trains, running slow and out of course anyway, were looped out of the way. As a direct consequence of this set-up, however, all reversibly working trains had to be sent at severely reduced speed through the platform loop, even when there was no train crossing necessary, as in this case. That was a time-wasting, wear-inducing and, as shown by this accident, a dangerous manoeuvre under certain conditions. On the other hand, because this was an existing operational issue repeated elsewhere on the network, train drivers ought to have been familiar with this feature from their route knowledge.

Unofficially, the reason for this unusual layout design was to reduce signalling equipment costs. This point is also remarked upon in Erich

Preuss's book *Eisenbahnunfälle bei der Deutschen Bahn*.

The investigation

The root cause was established as non-adherence to the 40 km/h (25 mph) speed order as attached to the use of a Zs 1 Ersatzsignal proceed aspect. Contributory causes were: a) serious lack of education for the role of express train driver, b) confusion about permitted speed through an unusually long engineering site, c) inaccurate and incompetent signing of the temporary speed restriction as well as inaccurate information in the relevant notices, d) Insufficient route knowledge on the part of the driver, and e) no provision of PZB overspeed protection for temporary reversible working under control of a Zs 1 Ersatzsignal through an unusually long engineering site.

The train, rolling stock and traincrew involved

The train involved was the international night train D203 'Schweiz Express' from Amsterdam Centraal in The Netherlands to Basel SBB in Switzerland, with through coaches to Brig in Switzerland. The train consisted of DBAG day coaches (coaches one and two: Emmerich–Basel SBB, coaches three to six: Amsterdam CS–Basel SBB) plus NS-owned (but acquired from DBAG) sleeping/couchette vehicles (coaches seven to nine: Amsterdam CS–Basel SBB–Brig). Train weight without locomotive was 435 tonnes, and its length without locomotive was 237.6 m (784 ft). The maximum permitted train speed was 130 km/h (80 mph) instead of 140 km/h (85 mph) due to brake equipment isolation on one coach. This was common practice and played no role in the accident.

The locomotive was Bombardier-built electric Bo'Bo' No 101 092-5 weighing 87 tonnes, 19.1 m long (63 ft), with a power rating of 6,400 kW (8,580 hp), a design speed of 220 km/h (135 mph), fitted with LZB & PZB ATP and with AFB automatic driving and braking controls (including cruise control). Since 1996, 145 units had been introduced.

The 28-year-old driver, Sascha B of Köln depot, was working diagram Köln 4652 starting on Saturday at 18:52. The job entailed working train 2 from Köln Hauptbahnhof to Emmerich, where the locomotive would be detached and run round, before being attached to D203 with two coaches to travel via Köln to Mainz. After a break he would finish by bringing D202 from Mainz back to Köln, ending on Sunday at 05:47. D203 out and D202 back were popular overnight shuttle trains between the Dutch North Sea coast and the Swiss Alps, patronised by an international clientele on their way to and from ski holidays as well as being used as a convenient night-time connection to destinations along the Rhein (Rhine). The train, therefore, was well occupied.

The accident log

Coming in from Amsterdam via Utrecht and Arnhem in The Netherlands, D203 arrived on time at Emmerich in Germany, a border station where the overhead line equipment was switched between the Dutch 1,500 V dc and the German 15,000 V ac at 16.67 Hz. At Emmerich D203 went through the necessary crew and traction change, and DBAG locomotive 101 092-5 plus two DBAG day coaches for Basel SBB were attached by driver Sascha B, who then took the train forward to Köln Hauptbahnhof. The rather leisurely night-train timetabling allowed the imposed 130 km/h (80 mph) speed limit to cause no delay, so there was no particular time pressure.

After right-time departure from Köln onto the Rhine left bank main line from Köln via Bonn to Koblenz, the train received restrictive signals at the crossover turnouts at Hürth-Kalscheuren where D203 was crossed over from the normal right-hand track at 40 km/h (25 mph) to the left-hand track, after which the driver accelerated again to around 100 km/h (60 mph) although he was allowed 130 km/h (80 mph). This was possibly the first sign of uncertainty generated by unfamiliar circumstances. At Brühl the reversible distant signal slowed the train down again and, as expected, the reversible line home signal for Brühl showed a red aspect. At the signal post, however, Ersatzsignal Zs 1 was illuminated.

This signal, being associated with the reversible home signal, permitted the driver to pass the red signal at danger without a written permit and continue on his way at 40 km/h until reaching the next main signal, which was the Brühl reversible Platform 3 loop starter. This particular local situation cancelled out another rule governing the use of an Ersatzsignal, which concerns running at that speed from a Zs 1 until the end of the station or yard limits as defined by the last set of turnouts. As illustrated by this accident, that is a rather unclear rule (as mentioned in the accident report) but it nevertheless played a major role in the way the accident unfolded, as by now D203 was passing the brightly illuminated worksite on its right-hand side where turnout 42 was being renewed. The plain track ahead appeared to be clear through a left-hand curve (to the passenger station), and D203 passed a less restrictive temporary speed restriction sign on its right-hand side showing '12' for 120 km/h (75 mph). For reasons explained earlier, however, D203 still had to pass through the platform loop at 40 km/h entry speed and 60 km/h exit speed.

Things began to go wrong. Shortly after the Ersatzsignal, and against all rules and regulations, the driver of D203 opened up the powerful electric locomotive that quickly accelerated its train to 122 km/h (76 mph) as recorded by the OTDR. As a result, on reaching reverse lying turnout 48 with its speed limit of 40 km/h where the loop line branched off the through line, the train almost completely derailed.

The locomotive and the two following coaches from Emmerich were deflected by the platform edge down the embankment, the locomotive coming to rest against a dwelling at the bottom. The rear of the third coach, being subject to its own momentum as well as forces from the following part of the train, was forced up onto the platform against the canopy stanchion and was folded into a V-shape, whilst the fourth coach turned onto its right side and was similarly folded against the third coach. The fifth and the sixth coaches, deflected by wreckage, went down the embankment to the left of the front of the train, the fifth coach on its left side and the sixth upright but badly crumpled at its front end. Two of the three Dutch sleepers for Brig derailed but stayed in line and upright, the third remaining on the track undamaged.

The alarm was raised immediately and the rescue and clear-up effort soon began. The subsequent investigation was not aided by the fact that the driver, on his first turn on such a long-distance express train, was suffering badly from shock. He was under sedation, could not be interviewed and later communicated through his lawyer only. Which, in view of the attitude of his employer, was understandable.

Comment

This was one of those accidents the cause of which initially seemed rather obvious. DBAG's chairman, Hartmut Mehdorn, clearly stressed by this development so shortly after the massive loss of reputation for DBAG at ***Eschede***, was in no doubt who was to blame. The signalling had worked fine and if only the driver had adhered to that Ersatzsignal-imposed speed restriction, all would have been well. He also pointed out that 10 other drivers travelling in the same direction that night came through Brühl safely without ending up in a mess. It later transpired that he was being somewhat disingenuous, as he would also be at later press conferences, but at the time he appeared to have a point. The main question to be answered was why had this driver not observed that speed restriction?

The accident sequence had, in fact, started on the day before the accident. Driver Richard P, a Dutch DBAG driver who should have worked Köln diagram 4652, reported sick with a bout of flu. Finding short-notice cover for such a winter weekend night job was usually very difficult. The traincrew resource managers were put under stress as volunteers to cover short-notice work at such a time were extremely thin on the ground. However, the driver that had been found was a new man, who had previously worked only as a local freight train driver. By rights he should still have been under intensive conversion training for long-distance express work.

Sascha B was a qualified electrician but had always wanted to be a train driver. He had applied to the old Deutsche Bundesbahn years before, but had failed twice due to examination nerves. He then applied with private rail freight operator Häfen und Güterverkehr Köln (HGK), where he did pass his exams in 1999 and became a main-line freight train driver. At that time his employer was mainly doing work with low- to medium-power diesel traction in the extensive Rhein river port areas around Köln. As a result, he found himself busied with shunting freight trains in yards whilst occasionally doing last-mile pick-up and delivery work or bits of main-line transfer, tripping to the freight yards at Kalk, Gremberg and Nippes. None of that would remotely prepare him for zipping along with a state-of-the-art 8,500 hp electric locomotive hauling intercity and international 'EuroCity' express trains full of important passengers. After five years of slow freight work he once more followed his ambition and applied for main-line passenger driving again, but with DBAG Reise & Touristik (Travel & Tourism) this time, the main German operator of long-distance and high-speed trains. He was taken on partly because DBAG badly needed passenger traincrew anyway, but also because he was now qualified as a driver.

At that time the state railway operator was in a bit of a pickle. It was undergoing a major organisational upheaval in connection with the ongoing unification of the former East and West German state railway organisations, DR and DB, which was compounded by splitting up the new combined operator DBAG into all sorts of operating businesses, as was happening everywhere else in Europe. One of the bigger issues on the road to unification was that the combined rules and regulations of the two former German rail administrations had been collected in a large and at times rather confusing rulebook, with rules, signals and signs that at first glance sounded and looked similar but out in the wilds could have some surprisingly different meanings and interpretations. Many train drivers were now crossing what once had been the Iron Curtain dividing Germany on a daily basis and needed urgent additional training (in connection with Expo 2000) to deal with this issue, so classrooms were filled with traincrew on conversion courses and educational staff were very busy indeed. Consequently, the necessary conversion training needs of the new man (and others) were more or less put on the back burner. Whilst his traction conversion tuition went ahead as planned, the necessary operational and route training courses that would have ensured his competence to drive passenger trains at considerable speeds were simply not available. Even the EBA accident report was uncharacteristically understanding on this issue.

At the same time, DBAG was subject to the unrelenting, ideologically driven push to be listed on the stock exchange under its supremo Hartmut Mehdorn. It was an exercise accompanied by cost-cutting, an efficiency drive, the creation of a positive business image and, most of all, profitability. The money men in Frankfurt, London and New York would be made to love DBAG, the one railway organisation in Europe that would succeed where all others had failed. By now it was becoming obvious that European rail operators, gaining experience with running 'commercially operated' passenger railway operations, had at times gambled with the welfare of their passengers. Rail safety throughout Europe suffered unmistakably due to such matters as business reorganisation and a different appreciation of cost (certainly in relation to safety) in the relentless drive towards getting higher value for as little money as possible. Many of the European accidents from about 1995 until 2010 bear testimony to that development. DBAG's ambitions for such commercial glory had received its first dent during the major accident with an ICE 1 high-speed set at ***Eschede*** in June 1998.

Much to the dismay of DBAG top management, it was not the driver who was to blame for the 101 dead and scores of casualties at ***Eschede*** but a set of operating policy issues that, to put it mildly, showed DBAG in a bad light. For that reason DBAG was already engaged in a major damage limitation exercise when the Brühl accident occurred. What was worse, however, was that critical external and media investigations into the ***Eschede*** accident began to reveal links with other, often equally baffling, incidents. They pointed to a disregard for time-honoured maxims governing safety-orientated (and therefore often rather expensive) railway management. Traincrew training time allowances had been shortened by a third, maintenance regimes had been 'simplified', and rolling stock overhaul periods had been extended, all without much in the way of risk assessment and testing whether these changes would impact negatively on safety. Even when testing was undertaken, cost-cutting exercises allowed the regular use of service trains with people on board rather than closely watched night-time specials to experiment with new materials and maintenance procedures. Monitoring what was happening was not undertaken with modern equipment on a frequent basis. All this was to save time, effort and money.

These dubious cost-cutting practices linked ***Eschede*** and Brühl with other incidents. The last one in this series was the derailment of an ICE 3 set at *Köln Hauptbahnhof* on 9 July 2008, due to a fractured axle resulting from metal fatigue. The engineers basically went too far with reducing the hollow-axle metal thickness in order to save weight, but this lethal threat to the safety of these trains, capable of 300 km/h (185 mph), had been known for some years before when similar troubles hit ICE TD DMUs running on the same axles.

The investigations into the Brühl accident revealed a lack of something that foreigners tend to think Germany has in abundance – 'Gründlichkeit' (thoroughness) – as outlined below:

1 The lack of conversion training revealed itself in Sascha B's confusion at the Ersatzsignal, causing him to speed up. Although the investigation report mentioned that it is not inherently wrong for a driver to start his career shunting and doing local work before progressing to more demanding jobs, it can be argued that a few intermediate steps up the ladder, with sufficient training and monitoring of competence at each step, would benefit safety. But the possibility of such intermediate steps had vanished with the splitting-up of the operation into various separate railway undertakings. The need for rigorously objective-referenced, well-organised and achievement-monitored training had, therefore, become more crucial than ever, but was not provided.

2 Accurate information about track and signalling layout peculiarities on the route, especially during degraded operations, should have been part of the abovementioned instructions. The EBA report pointed this out as well, as that omission was notably at the root of this accident. As a matter of fact, Sascha B's unfamiliarity with station layout peculiarities when travelling reversible under degraded signalling conditions has a strange ring to my British railway-trained ears, where acquisition of thorough route knowledge during driver training comes second only to sound rules and traction knowledge. In Germany, however, that was not the case (as also later in The Netherlands, after a head-on crash in *Arnhem*, and in Belgium, at ***Pécrot***). The Dutch DBAG driver who should have worked D203 later told me that

a similar type of layout somewhere else had dangerously confused him. Of Mehdorn's 10 other drivers who presumably passed the site safely before D203 crashed, four drove locals that stopped as booked in Brühl's Platform 3, and of the six others a number revealed that they had not been impressed by the quality of the published information and signing of the speed restriction. Checking train trip recorders, according to the EBA report, revealed speeding of up to 48 km/h (30 mph) into the loop. I was told of speeding of up to 60 km/h (35 mph), but the sort of brake applications that must have preceded coming through this turnout and should have been recorded on the OTDRs were not discussed in the report.

3 Technical equipment to enforce adherence to the 40 km/h (25 mph) speed restriction, in the shape of a temporary PZB train protection magnet, was readily available at a cost of approximately €2,000 for fitting. In fact, its use was discussed, together with placing reminder boards, in view of the unusually long temporary speed restriction section past the worksite. But despite the 69 trains booked past the engineering site that weekend it was declined on the grounds that the 40 km/h off the Ersatzsignal would cover all speed restriction issues. However, such equipment is put in place for safety's sake in order to catch those drivers who, for whatever reason, do not adhere to the temporary restriction, for the protection of passengers and staff in such a foreseeably dangerous location. At best it reveals a questionable lack of practical insight, probably engendered because the situation was discussed only by track staff; no operational people were present at those meetings. Railway's all-rounders had disappeared after the split-up of DBAG, like they had elsewhere on Europe's railways.

4 The notices issued to infrastructure staff and traincrew detailing the temporary speed restrictions, the works, their date and protection measures (the 'La' and the 'Betra') were inaccurate to a considerable degree. They confused the locations of the freight yard and the passenger station in the descriptions of the worksite, and the extent and character of the speed restrictions, and also used the incomplete acronym Bf instead of Bhf for Bahnhof (station). Bf at Brühl was the code for the local signal box (Brühl Fernstellwerk). The 'La' inaccurately mentioned a temporarily restricted speed of 120 km/h pertaining to trains past the worksite, and temporary speed boards to that effect must have

been present at the site. This, however, was meant for the opposite direction to the line that D203 travelled on after the actual renewal works had been finished, to protect engineering crews tamping and lining the turnout. That temporary speed restriction would, therefore, only come into effect later that night and for that reason the speed board indication had been turned away from the track it was meant for, but now faced the opposite track along which this train approached. The EBA report does not consider this to have affected the accident driver in his decision to speed up, as it was dark at the time, but I would question that, given the blazing lights at the worksite as well as headlights of trains being directed to the lineside to shine on retro-reflective board indications. Between Hürth-Kalscheuren and Brühl, Sascha B did not use his full speed allowance by a considerable margin, but after passing the Zs 1 signal he suddenly accelerated to 122 km/h (76 mph). He had been looking out for confirmation of what speed he could do, and the incorrect information in the notices as well as the wrongly placed temporary speed restriction signs, as far as I can make out, most definitely contributed to the way the accident unfolded. This point, in fact, was also the moment that the great whitewash by the various DBAG bodies began, as according to German and Dutch media reports those incorrect temporary speed limit boards suddenly disappeared later in the night of the accident. Several relevant people were aware of the errors in the publications but failed to have them corrected, which points to an erosion of the safety ethos.

5 Additionally, in order to ensure compliance with the temporary speed restrictions, a request went out that train drivers should be contacted and informed by the signal box via radio telephone. Unfortunately, earlier that day telecommunications equipment had developed a fault and setting up calls had taken a long time in a number of cases. Earlier in the night, however, calls had been exchanged without problems, but none to Sascha B on his 'EuroCity' service.

The aftermath

From early on the German media, distinguishing themselves by a critical stance and diligent investigatory work to unearth further relevant factors responsible for this spate of rail accidents, showed that they no longer took the word of DBAG as gospel (no doubt ***Eschede*** played a role here) and started to question the amount of blame falling on the train driver. Sascha B was fined €7,000 for his failure to adhere to the 40 km/h indication given by the Ersatzsignal. The people involved in setting up the temporary speed restriction were taken to court on the basis that they could have easily prevented the accident had they put that PZB magnet in place. They received fines between €7,000 and €20,000. DBAG was obliged to improve safety when setting up engineering sites. The loop line at Platform 3, suddenly considered surplus to requirements, has now disappeared, the main line through the station has a starting signal for reversible working, and a sound wall to protect the houses below has been erected.

The big question as to why the driver of D203 acted in the way he did after passing that Zs 1 Ersatzsignal was not answered for a long time. Sascha B spent some considerable time in hospital and underwent further medical care for his mental state. From my own knowledge of colleagues who experienced similar situations I know that the process of having to relive your mistakes and endure a severe grilling at an inquiry is one of the most dispiriting experiences a human being can go through. Your self-respect, self-confidence and ability to engage with society are utterly destroyed. With conjecture, resulting from my experiences as a driver as well as having been involved in training drivers, however, I think that the clue lies in Sascha B's history and the way in which he probably saw his work. He had wanted to be a train driver for a long time and, despite his nerves letting him down before, he showed tenacity in his struggle to achieve his goal. Finally working such a red-liveried Class 101 electric locomotive with its nine coaches for the first time on his own marked his coming of age in his chosen profession. In the situation of taking possession of the cab of such a train he would have been keen to be seen as a good driver by those around him, and that usually means being good at keeping time as that is generally what everybody considers. To what extent does timekeeping influence driver behaviour? This keenness to keep time in a beginner is illustrated by the number of times that I, in my later days as a trainer, had to tell trainee drivers in a station to take a look at the signal ahead before even thinking to look at their watches. Very probably the situation was no different with Sascha B.

As long as the situation was normal Sascha B gave an exemplary ride, but the cracks appeared once uncertainty set in along the first stretch of unfamiliar wrong-line working. My guess is that he lost the plot from the moment he saw that triangle of white lights at the red signal. In the glare of the works-illumination the track ahead beyond the worksite appeared clear, the end of the freight yard was passing him on his right-hand side and temporary speed restriction boards then showed a '12' for 120 km/h. With his lack of route knowledge, training and experience and with the delay to all those passengers on his mind, his left hand pushed the power controller stick forward and he experienced the surge of acceleration that this locomotive was capable of. No doubt he felt liberated. In the situation at Brühl, however, the invitingly clear-looking track through the lazy left-hand curve ahead was not the only stretch he was going to travel. There was that obligatory trip through the 40 km/h platform loop line that was not even visible yet due to lineside vegetation. And there was no train protection. A really nasty trap.

Late in 2001, Sascha B issued a statement in which he declared himself morally responsible for the accident and said that he had made the mistake of taking the stretch of line between the freight yard and the passenger station at Brühl as being the open line again (the rule about passing the last turnouts of a yard when travelling on the authority of an Ersatzsignal!). He had seen the location where the turnout was being replaced and had seen the temporary speed restriction signs with the '12' at his right side passing the engineering location, as mentioned in the 'La' for services in the opposite direction along the track he was on. In addition, he – mistakenly, as it turned out – believed that his train would be monitored by PZB train protection.

Sascha B should not have driven that train on his own (see ***Pécrot*** and ***Ladbroke Grove***) and he was wrong-footed by the way the engineering work had been organised. The blame, therefore, lies heavily with his employer and the track authority. Erich Preuss in his book *Eisenbahnunfälle bei der Deutschen Bahn* quite appropriately cites Judge Heinz Kaiser who, when closing the trial, ironically noted that he 'had rather missed DBAG management representatives when interviewing witnesses'. EBA and DBAG learned the hard way, like a number of other rail operators in Europe and Asia did, that spending money on safety is crucial to staying in business as a public transport rail operator. Passengers like to arrive in one piece at their destination, whatever people working in stock exchanges may think of that in terms of the effects on the profitability of a business.

***

Voorst, The Netherlands, 16 June 2000 at approx 13:46
(Level crossing collision)

Introduction
All accidents are tragic, but this was a bad one, seeing that an entire family of two parents and three children was wiped out. It happened at a time when the Netherlands Government was reassessing its policy on road safety, and level crossings were very much part of that process. The accident tally at that time was significantly higher than in neighbouring nations (see below) due to the high frequencies of both road and rail traffic at level crossings.

The accident
A Volkswagen MPV (multi-purpose vehicle) with an attached caravan was in collision with a 3-car DEMU train at an automatic open level crossing on one of the few single non-electrified railway lines in The Netherlands. To give an indication of the train frequency: even on this minor railway line the crossing would close every half-hour for a train to pass by in one direction or the other at a speed of approximately 100 km/h (60 mph).

The location
The municipality of Voorst on the railway line between Apeldoorn and Zutphen was the scene of the tragedy. The countryside was level agricultural land with relatively small stands of forest as well as individual clumps of trees and farms occasionally obscuring the distant view. The level crossing was situated in a country lane 5 m (16.5 ft) wide that was busier than normal as a result of traffic diverted from nearby main roads. This lane had been widened from 3.5 m (11.5 ft) in 1969, but the usable level crossing road deck had been left at only 4.35 m (14.3 ft). Due to the type of deck material used – standard concrete (Stelcon) plates in a steel frame – this could fairly easily have been adjusted during the widening, but that was subject to disputes between the various owners and operators of the level crossing, as explained later.

The level crossing was at a slightly higher level than the surrounding countryside, although during the course of the road widening the approach ramps had been eased to allow low-loaders with farm plant and machinery to pass safely. From the direction of approach of the road vehicle involved in this accident, the lane made a lazy turn towards the level crossing and crossed the track at an angle of 90°. The view of the level crossing warning equipment was blocked by (obsolete) road signs and the view of approaching trains was partly blocked by lineside vegetation.

The train and vehicles involved

The train was a Netherlands Railways 1960s-vintage DE3 3-car diesel-electric multiple-unit set, a type which has now been taken out of service. It suffered only light damage and was not derailed. It collided with the Volkswagen Transporter semi-van, the front half of the vehicle being a double-cab MPV with passenger seats and the rear half a van. It was towing a caravan at the time. Both vehicles were comprehensively destroyed.

The accident log

Although the road across the level crossing, Broekstraat, was being used as a diversionary route due to road works on main traffic arteries in the area, the vehicle involved in the accident was not using Broekstraat for that reason. At 13:44 the MPV with its caravan was 1 km (¾ mile) away from the level crossing and accelerated to the permitted speed of 70 km/h (45 mph). At this speed it would have covered the distance in 50 seconds, but nearer the crossing there was some traffic congestion due to an agricultural tractor with a hay-turning machine moving significantly more slowly and occupying much of the width of the road, which prevented overtaking. Two private cars were behind the tractor as the MPV and caravan combination joined the queue. At a distance of 125 m (410 ft) before the level crossing, in order to let another farm tractor coming in the opposite direction pass, the tractor driver at the head of the queue pulled to the side of the road and stopped, all the following vehicles also moving over to the right and staying behind it. On moving again, all the vehicles in the queue followed the tractor, but the two vehicles behind the tractor then turned left into Oude Broekstraat that ran parallel with the railway line. Whilst this was happening at the front, further back the other tractor that had caused the queue to stop, passed the virtually stationary Volkswagen MPV and caravan slowly because of the width of both vehicles. The MPV then accelerated and began to catch up with the tractor in front. Now the white flashing light of the AOC extinguished, the two wig-wag red warning lights illuminated and the bells started to sound, indicating the arrival of a train in about 20 seconds. The MPV driver was close behind the agricultural vehicle again as they both approached the level crossing.

A number of factors may have prevented the MPV driver behind the tractor from noticing the level crossing warning – e.g. preoccupation with the slow and wide tractor directly in front of him, the curve in the ascending road leading up to the level crossing, and the narrowness of the road ahead – but it is debatable whether he was actually aware that he was approaching a level crossing and thus expecting that a train might pass at any moment. An additional factor may have been that the sun, although high in the sky, was directly in front of him and may well have affected his view of the situation ahead.

When the first tractor driver was 60 m (198 ft) from the level crossing and climbing slightly to track level, he spotted another tractor approaching from the opposite side of the level crossing. Adhering to a road sign governing priority over the crossing, he pulled over to the right again and stopped 20 m (66 ft) before the crossing to wait for the approaching train and give way to this third tractor after the train had passed. In all likelihood, the holidaying MPV driver with his caravan in tow wrongly interpreted this move as a gesture of courtesy to him, which presupposes that 1) he missed the traffic sign regulating road traffic priority across this tight spot as well as 2) missing the level crossing warning. With the tractor then blocking his view of the warning lights, he put his foot on the accelerator and moved past the stationary vehicle onto the level crossing. It was at this point, when he decided to 'go for it' and climbed toward the track, that he spotted the obstacle of the waiting tractor at the opposite side of the level crossing. No doubt his concentration on registering and deciding how to deal with this new situation then rendered him completely oblivious to the warning lights and the sounding horn of the approaching train. When the train driver saw what was unfolding, he made an emergency brake application.

Just as the MPV drove onto the level crossing the train arrived, doing 98 km/h (60 mph), and hit the vehicle just behind its front doors with the train's centrally mounted automatic Scharfenberg-type coupler, jutting forward 0.7 m (2 ft 4 in) above the roadway. The MPV broke into two on impact, which was evident from the divergence of the front and rear tyre skid marks on the level crossing deck. The front part rotated wildly on moving away from the crossing, ejecting both the parents from the front seats. The woman died on impact and her husband when on the way to hospital. The rear part of the vehicle, together with the attached caravan, was dragged 400 m (1,320 ft) by the slowing train. All three children were found lifeless inside it. The MPV and its attached caravan were both completely destroyed.

The investigation report

The accident report commented negatively on this automatic open crossing and its features. To start with, there were restrictive sight lines for road users along the lane, making it difficult for them to spot trains bearing down on them. The visibility of the railway line, the level crossing and approaching trains in the landscape was impaired by vegetation growth, perhaps further impeded by the glare of sunlight (see ***Kerang***). In the Dutch accident two obsolete traffic signs also played a role, as one indicated a severe hump with grounding risk that had been taken out years before, and the other indicated priority rulings that would have been unnecessary if the level crossing deck had been extended to the width of the road. These road signs partly obscured the view of the level crossing equipment, so their removal might have allowed the MPV driver to see the crossing warning lights.

As far as maintenance of the level crossing was concerned, the rail infrastructure provider (Railned at the time, ProRail nowadays) maintained the track, level crossing deck and the warning installation, whilst the local highway authority was responsible for everything leading up to it, including the signage. There was no single authority tasked with upholding of level crossing safety. As a result, the maintenance of the roadway leading up to the crossing was rather haphazard, but more importantly, the failure to widen the level crossing deck had created an obstacle at a point that already carried a definite risk. Incidentally, this widening of the crossing deck was a matter of priorities, the rail infrastructure operator ranking the situation in this rural area very low in relation to providing resources for work that was necessary elsewhere on the network. In fact, the world over work on this sort of very local issues is generally done when a whole branch line is tackled with a maintenance blitz.

A further and less savoury aspect of this situation at Voorst was that the local authority had already asked for a bridge a long time before the accident occurred. Netherlands Railways then – and the rail infrastructure authorities that took over later – tied this request to an agreement to take out all other level crossings in the municipality, without further compensation measures, as they were keen to reduce their level crossing maintenance bills and to improve on delay and safety figures. These discussions had become mired in difficulties about funding, so various level crossings in the area had remained unchanged, some in an unsatisfactory state with regard to safety. In fact, government funds were available to alter automatic open level crossings into automatic half barrier crossings, but the €1.5 million necessary to provide a split-level crossing at that location was a different matter again. Another issue discussed was the risk of derailment to trains following level crossing collisions, with consequent danger to rail passengers and freight. On average, there were 4.5 derailments per year as a result of level crossing accidents in those days. With the high frequency of train services in The Netherlands (an average of 6-7 trains per hour in each direction; 12-14 train service occupations per hour) a train derailing towards the opposite line stood a very good chance of being hit by an oncoming train. It did happen occasionally, although comparatively rarely considering the prevailing circumstances.

The report also compared level crossing safety in The Netherlands with that in Germany and Britain. It was found that manually closed full barrier level crossing installations, especially those interlocked with railway signals, were the safest because there were eyes checking whether the crossing surface was clear for a train to pass safely. Actively protected AHBs were rated second best, and AOCs, although not much good, were better than passively protected fully open crossings.

Actively protected level crossings in
Germany, Britain and The Netherlands

	Manual half, full and double barriers	*AHB*	*AOC*
Germany	45%	40%	15%
Britain	60%	30%	10%
The Netherlands	0%	66%	34%

The Netherlands' high level of automatic installations was obvious, but there had been 130 fatalities per 1,000 level crossings in the previous 10 years compared to 40 such fatalities per 1,000 level crossings in 10 years in Germany, and 10 in the same circumstances in Britain. The much higher average frequency of rail traffic, as well as the dense population of The Netherlands, could only be partly to blame.

The following figures shed some light on the road/rail interface and its accompanying dangers on the three networks. In Britain there is a level crossing for every 2 km (1¼ miles) of railway, in Germany a level crossing for every 1.5 km of railway and in The Netherlands a level crossing for every single kilometre of railway. Exposure to level crossing danger in The Netherlands is therefore high. What does not seem to tally is that in Germany, where not all the level crossing barriers and covering signals are interlocked as they are in Britain, and where the level crossing keeper is indicated as the 'risk factor human being', such crossings are deemed safer than automatic versions. Could that be because German motorists behave more responsibly, as the crossing keeper might report them to the traffic enforcement authorities? However, railway crossing keepers worldwide have indeed caused accidents more than once and are certainly not considered the safest option. The one positive function of the crossing keeper that has not been replicated by automatic machinery – checking that the level crossing surface is clear – is at present being taken over by automatic detection installations such as radar. The feedback to approaching train drivers, and the distance and time needed for them to come to a stop, however, undoes much of the time gain that automatic level crossings give to road traffic.

The outcome in The Netherlands

Resistance of trains against derailment after hitting a normal private car or a light van on a level crossing was substantially improved by fitting obstacle deflectors under all but the oldest types of trains. Since then derailments have occurred only when such smaller road vehicles have slammed into the side of a train. Road freight vehicles on level crossings, however, continue to pose a major risk factor to rail operations. Level crossing safety has been substantially improved by altering all motorised traffic AOCs into AHBs, a process that is now complete in The Netherlands but is still ongoing in other European countries. Level crossing accidents in The Netherlands have dramatically diminished following this action, and the Dutch rail infrastructure provider ProRail has been looking at improving safety on user-worked and public non-motorised open crossings with simple barrier installations. High risk crossings are going to be replaced with either tunnels or viaducts in a rolling programme.

Comment

My inescapable conclusion is that the MPV driver was irritated by the hold-ups caused by the three tractors, despite being on a leisure trip on a minor country road early in the afternoon of a pleasant sunny day. His judgement deserted him in his determined attempt to make sure he would get past the tractor in front of him, with devastating results for him and his family and severe emotional shock for the traincrew, passengers and others closely involved. Strangely enough, the accident report hardly commented on the responsibility of the driver of a road vehicle for his own safety and that of other road and rail users – to adhere to rules and regulations, to anticipate changing traffic conditions, to notice features of the road ahead in time and to safely drive the vehicle accordingly. It is almost as if the report had either given up on motorists or was engaged in a political tussle with the railways, looking for faults in order to allocate blame.

The moment that the MPV driver decided to overtake the stationary tractor he was facing a rather unclear situation ahead that elsewhere on the road would probably have made him think twice. By ignoring the level crossing warning signals this motorist accelerated into the unknown, and by blindly doing so, sealed the fate of his family and himself. Had he let the delay caused by the three tractors overrule his common sense? Aggressive haste by private motorists after hold-ups, causing inexcusably stupid behaviour, is actually fairly common and is often explained as a normal reaction to an obstruction to swift progress, let alone three obstructions in sequence. It is clearly impossible to provide protection to all road users, certainly those who act from a sense of frustration rather than from a considered assessment of, and adequate reaction to, the situation in front of them. Which is the chief reason why many road accidents happen in the first place. It must be borne in mind, however, that this accident took place before action by the Dutch Government to improve significantly safety on level crossings had had much impact on the figures. For that reason, this accident would in all probability not have happened in recent years. The driver would have seen the barriers with their red flashing LED lights descending across the road and would have been given multiple indications that he was approaching a level crossing that was about to be crossed by an approaching train.

It is clear that this was a true accident – a stupid mistake was made. A more galling situation is one in which rail passengers die in a catastrophic derailment as a result of someone committing suicide in their car at a level crossing. (Seven died this way at *Ufton Nervet* near Reading in November 2004, and eleven died at *Glendale* in California in July 2005.) Level crossings have become the main obstacle to safe train travel, and the ultimate solution is not yet forthcoming other than their total removal and replacement with bridges. Unfortunately, not even that would stop those mindless people who seek to cause a train crash for fun.

***

Hannover Langenhagen Airport, Germany, 29 June 2000 at 10:10

(Head-on collision following SPAD)

Introduction

This is one of these irritating accidents caused by a train driver who, although doubtless aware that he had to stop for a red signal aspect, apparently did not really understand what signalling was for and thus what was expected of him from a safety point of view. No doubt he had been told that timekeeping was of the utmost importance.

The accident

A driver committed a SPAD at the starting signal, having had to start ahead of the signal, but was stopped by the PZB train protection system. Instead of calling control, however, the driver waited for automatic train protection reset with the release of the brakes and departed anyway, resulting in a head-on collision with an incoming set.

The location

On 28 May 2000, a month before this accident occurred, Hannover Langenhagen Airport station had been opened. S-Bahn line 5 from Hannover city served it via Langenhagen Markt, the spur from Langenhagen Markt via Langenhagen Mitte to the airport being a single-line section. The airport area was served by this single-track electrified line going underground in a curved concrete trough and tunnel and then serving a two-track island platform under the airport terminal. The trough and the tunnel had been laid out for double track, but the second track had not been installed at the time. The curve in the tunnel and trough reduced the view ahead on the track to a significant extent.

The signalling

One home signal with a distant signal and two starting signals (one for each platform) were provided, with one turnout to split the single line into the two platform lines. This minimal layout was controlled from a Siemens electronic installation by clicking a screen cursor on the few graphically presented items that needed to be set for arrival or departure of trains.

The airport service

During the day there was a half-hourly service to town, which allowed for all incoming sets to depart 6 minutes after the driver had changed ends. Most of these services were worked with single 4-car articulated electric multiple-units indicated as a 'Kurzzug' (short train), although some services ran with two coupled units that were described on the indicators to passengers as a 'Halbzug' (half train). These could all depart from behind the platform starting signals. In the rush hour a 15-minute frequency of departures would be in operation and the trains would be made up to 'Vollzüge' (full-length trains) of three 4-car units. These, however, were too long to fit behind the platform starting signals (see ***Pécrot***) and would therefore start with the cab some 10 m (33 ft) ahead of the signal, but would be protected by a PZB 2,000 Hz magnet to stop the train in the event of a 'starting against signal' SPAD. Another issue with this 15-minute frequency was that the manner of operation at the airport station changed. An incoming train would not now depart after 6 minutes and clear the single-track section to Langenhagen Markt before another service was scheduled onto the single line, but would have to wait for the next incoming train already on its way in the single-line section between Langenhagen Markt and the airport.

Rolling stock involved

The 4-car electric sets involved (15 kV ac 16.67 Hz) were Class 423 EMUs, Bo'2'2'2'Bo', built by Alstom-LHB, with a power output of 2,350 kW (3,150 hp) and aluminium bodyshells, each cariage with three double-leaf entrance doors on either side. They were specifically developed for short distance, short dwell time and high acceleration work with a top speed of 140 km/h (85 mph). Many similar series of articulated units from various manufacturers are in service in Germany for this so-called S-Bahn city-rail type of operation, as well as recently in The Netherlands (second-generation 'Sprinter' EMUs). The latest versions have a top speed of 160 km/h (100 mph). The Class 423 units are 67.4 m (220 ft) in length and are worked under driver-only operation (DOO) conditions, without a conductor.

The accident log

Service 5711, a 3-train 'Vollzug' set (and therefore extending ahead of the departure signal), had arrived at 09:53, 4 minutes late, on Platform 1 track 3711, and after a quick change of ends the train was ready to depart. As a 15-minute frequency schedule was in operation, it was required to await the delayed arrival of incoming train 5712 (expected at 10:04) on Platform 2 track 3721 before departing. Its destination for the coming trip was Hannover Messe/Expo station.

Train 5712 was on its way to the airport along the single-track line from Langenhagen Markt with a 7-minute delay. It had departed from the last station before the airport, Langenhagen Mitte, at 10:06 instead of 09:59. There were five people in the cab: the booked driver and four road-learning drivers. One of the road learners was at the controls, against regulations, but it happens everywhere and is not normally of consequence (as in this case with regard to the accident). The only problem when road learners are driving is that there is usually a slight time-loss penalty due to a lack of familiarity with the road; they are, of course, learning the road in order to familiarise themselves and be able to keep time safely. When the traffic controller accepted the train into Platform 2, this caused home signal 37A to display a restricted aspect with two illuminated digits under it, a '5' and a '3'. This allowed entry into the station to the buffer stops at 50 km/h (30 mph) until a certain marked point (a 30 km/h speed board) and from there at 30 km/h (20 mph) until the stop blocks. Train 5712 then came down the ramp into the tunnel.

At its booked departure time, train 5711 departed against its unseen signal indication, thereby committing a 'starting against signal' SPAD. The 2,000 Hz PZB magnet, however, stopped the train as intended. Then, after a rather quick time-out of the PZB intervention, the driver got his brakes back and against all rules, regulations and expectations, accelerated away again without contacting the signaller. He cut through the turnout onto the single line and had barely reached the curve into the tunnel before he slammed into oncoming train 5712. The collision resulted in 16 people being injured, the derailment of several bogies and severe head-end damage to both sets. The damage was estimated at €3.6 million.

Comment

This was a typical example of a 'starting against signal' SPAD. Had there been a conductor, it would have been known in Britain as a 'ding-ding and away', referring to the two bells or buzzes a conductor gives on the intercom to indicate 'right away' (start moving) to the driver. That can still be heard today when travelling on a fully crewed old Routemaster double-decker bus in London. Had there been a conductor on board this train, able to see the signal and hopefully rather more on the ball than the driver, the chance of it departing against the signal would almost certainly have been reduced, although the presence of a conductor is no guarantee against such SPADs (see ***Schiedam-Nieuwland***).

These SPADs usually have their root in the urge of the traincrew to be away on time but being distracted from observing the starting signal by the public. People will keep boarding until the very last moment, and it does take insight and choosing the right moment to close the train doors to ensure mishaps are avoided. Obviously, if a driver has to carry out all this dispatch work on his own, he may overlook the signal if he has forgotten the cardinal rule that he should not even consider starting the dispatch procedure until he has received a proceed aspect. This assumes, however, that he can actually see the signal, or is alert to the fact that he is starting ahead of it. Forgive me, but I am not impressed by such arrangements which require trains to start ahead of a signal (see the influence of the similar situation at Wavre during the ***Pécrot*** accident). It is bad signalling practice that negates the very reason why starting signals exist at all, although it can be acceptable in circumstances such as in changed individual operating conditions (e.g. building work at platforms or a very long special) when a train, or even a few trains, temporarily have to start ahead of a signal. The existence of a designed situation like this generally stems from the wish to limit spending. An extra platform length of 15 m (50 ft) under Hannover Airport would have eradicated the problem entirely, but tunnelling is rather expensive.

This structural issue was then compounded by the driver's mistake of not being alert to the changeover to the rush-hour 15-minute departure frequency. In these circumstances, instead of the routine of arriving, changing ends and departing again with a short train that enabled the driver to see the starting signal, it was necessary to await the arrival of the incoming train. At the same time the driver needed to be aware that he was now driving a longer train and consequently would not be able to see his starting signal. In all fairness, this amounted to a foreseeable trap for any driver who allowed himself to run on autopilot rather than keeping his wits about him. Was no risk assessment undertaken?

However, the third and most serious error committed by this driver was his failure to call the signaller immediately after being brought to a stand through an intervention from his PZB train protection system, which was a clear breach of the rules. I would have thought that, especially in this particular case of starting ahead of a signal onto a single line, the presence and the function of those PZB 2,000 Hz magnets at the platform exits would have been drummed into every driver's head, together with a clear explanation of how to deal with an intervention, especially given the apparent quick time-out of the PZB intervention and the foreseeable risk of a mistake occurring as in this case. Apparently not.

There was a similar situation in Britain – called the 'reset and go' scare – when TPWS was introduced in 2002/03. Somewhat surreptitiously the reaction time for an AWS warning had simultaneously been 'brought back to what it was meant to be' according to the official view, but an additional change meant that the reaction to the warning could now only take place successfully within the time that the horn sounded. This was to keep drivers from cancelling a warning on the 'automatic' without establishing what it was they were cancelling for. From the point of view of a train driver, however, the reaction time allowed to cancel the AWS warning on trains faster than 160 km/h (100 mph) had been shortened by at least a third, from approximately 3.5 seconds to 2 seconds, and for a number of reasons this led to quite a few unwanted stops. Initially, teething problems with the new TPWS system were blamed (the interventions of either system could not be distinguished from each other, and both were dealt with by pressing the AWS button), but in due course drivers started to cotton on to the fact that the interventions had little to do with TPWS proper but were triggered by the amended AWS part of the system. AWS was looked on as something of a nuisance anyway; hardly anyone took it entirely seriously, as could be seen from the investigations into various accidents (e.g. ***Cowden*** and *Southall*).

As a foreseeable result of the above, drivers started to ignore TPWS/AWS interventions, failed to consider the reasons for them and took off again as soon as their brakes released. If their train got stopped again they waited for the system time-out and then moved on, as the driver at Langenhagen had done. Initially, this caused no concern until the day that a train was stopped by an unrecognised *proper* TPWS intervention instead of the usual AWS part. The driver had gone past a red signal aspect at a single-line junction without noticing, on yet another double-track line that had been reduced to a single track. After operating the AWS button, which released the brakes for both types of intervention, in time-honoured fashion the driver awaited the system time-out to release his brakes, then upped his anchors and moved off. Unfortunately, this time he was heading towards an oncoming service along the single line for which his signal had been kept at red. After some very fast work with the radio both trains were brought to a stop before they met, but following this caper it was made mandatory to call the signaller after any kind of intervention. That caused quite a bit of costly delay due to the number of times trains got stopped as normal by AWS/TPWS, and the ruling had to be modified again, but the current arrangements are just about right given the characteristics of the train protection systems available to work with work in Britain.

In Germany, in the years between the introduction of PZB in 1994 and 2000, there were no fewer than 22 cases of 'reset and go' incidents, which equates to 3.5 potential collisions per year after the train had first been stopped by a PZB 2,000 Hz magnet intervention at a signal at danger. What strikes me as odd, however, is how little information about these incidents, their causes and remedies seems to have reached our shores and vice versa. From the end of 2002 onwards, Britain experienced exactly the same problem. No doubt the language barrier did not help, but perhaps that was mixed with a lack of interest in things that happen outside your home patch. The quick train protection system reset with release of the brakes after an intervention, no doubt inspired by the wish to get the train going again to avoid delay, was a mistake, nevertheless. If such a system was meant to protect then there should have been a substantial time-out, something like 5 minutes at least. Whilst sitting there awaiting the release of his brakes, the driver would have had a few moments to come to his senses and reflect on what might have caused the stop and what he was supposed to do about it. Perhaps he might even have called the signaller or control to report a fault with his brakes if he missed seeing that red light. As long as he called and did not move off with the train; that would be the most important thing.

***

Red Deer, Alberta, Canada, 2 February 2001 at 20:23

(Derailment with release of dangerous chemicals from tank wagons)

Introduction

Reading about this accident took me back to my days of being taught about handling trains in general as a British Rail train driver. Dealing with a bad spill of dangerous chemicals, such as anhydrous ammonia in this case, was considered very specifically when learning about the rulebook, as it required that the normal course of action after a train-running incident be abandoned. The normal course was that the driver at the front and the conductor at the rear would get off the train to meet midway in the opposite cess to assess the damage and whether the derailment endangered the opposite track and to decide on the best course of action for protection. This was strictly forbidden in the event of certain chemical spills. Anhydrous ammonia was specifically mentioned as the commodity that ought to be handled by moving away from the train in opposite directions and attempting to stop approaching trains as well as warning the traffic controller and contacting the emergency services. Of course, conductors or brakesmen generally no longer travel on brake vans or cabooses at the tail end of a train in Europe, Britain, the US or Canada. On continental European locomotives involved in hauling freight traffic all cabs are supplied with emergency breathing equipment and eye protection specifically for this sort of incident. Such a spill occurred in the Red Deer incident, but notice how differently the situation was dealt with by some of the people involved.

The accident log

On the evening of 2 February 2001, during a hard frost, Canadian Pacific Railway CP 966-02 had arrived on track 1 at Red Deer yard and its crew were involved in using the three locomotives to move 61 wagons at 6.4 km/h (4 mph) in order to set out a vehicle to track 5 and then pick up 12 wagons from track 2 to make up the consist for onward travel. Whilst drawing forward from arrival track 1 in the yard onto the main line for the first shunting move to track 5, the driver suddenly noticed that he had lost his brake pipe air pressure and the train came to a grinding halt. He radioed to the conductor, who was standing at track 5 to stop the movement and uncouple the single delivery, saying that he would get off his locomotives and check the situation. Shortly thereafter, having walked out and hurriedly back again, he reported by radio that he had seen a white steam-like cloud at some position further down the train and that he was not going to get anywhere near it. The conductor was able to determine from his train manifest that this could be escaping anhydrous ammonia and told the driver so, following which the driver in his cab checked the 2000 Emergency Response Guidebook and, having realised the danger of the situation, uncoupled his locomotives and left. The conductor, however, still unaware of what had actually happened but intuitively guessing correctly, went to the yard office about 500 m (1,640 ft) from the cloud, warned everyone present of the potential danger and then summoned the emergency services. Seven minutes after the derailment (as that was indeed what had happened) at 20:30, the emergency response unit at Red Deer was activated and the emergency assistance department of shipper Agrium Inc at Red Water was also warned. They immediately dispatched two of their emergency response units and the dangerous goods emergency team. All these people would be on the site by 23:30, but too late to prevent various local emergency response team members from being contaminated by the inhalation of the very dangerous anhydrous ammonia vapours.

The cause of the accident

Despite the track and the ballast allegedly being in a generally good state (having been inspected the previous day), several of the wooden sleepers had deteriorated and the rail spikes used to fasten the rail to the sleepers (which to my knowledge are no longer used in western Europe) had become loose to the extent that, according to the subsequent investigation, some could be wiggled 6.5 cm (2½ inch) from side to side. On one such sleeper with dodgy fastenings in track 1, one of the rails had moved sideways under a slowly passing heavy tank wagon fully laden with anhydrous ammonia, which had caused the opposite wheel to slide inwards off the rail head and fall between the rails. This pushed the other wheel harder into its rail, spreading the track gauge and so derailing a further four such tank wagons. On encountering the turnout that connected track 1 to the main line the heavily laden tank wagon rode up against the wing rail, broke through the actual running rail at a point weakened by cracks and corrosion around fishplate bolt holes and then tipped over sideways, dragging another with it. The two wagons rolled into a trackside ditch and turned over, hitting the frozen ground hard but luckily also splitting the brake pipe and so stopping the train. In this process, however, one of these two, PROX 88421, had the top dome and its equipment contents so badly damaged that the eduction valve nozzle was knocked off and the weld around the dome on the tank suffered a brittle failure. The vehicle, lying virtually upside down in the ditch, started spilling its dangerous contents of anhydrous ammonia, as it would continue to do for more than 24 hours. Three other such vehicles were derailed but stayed upright and were undamaged.

The emergency response

On his arrival the fire department commander, smelling ammonia but unaware of exactly what it was, ordered the evacuation of the immediate surroundings. He then sent in a team to locate the source of the smell, following which he contacted the Canadian Transport Emergency Centre, CANUTEC, for information about the hazards and for instructions on how to handle them. Suddenly, within seconds, his eyes were swollen shut and he had trouble breathing, whilst the two fire officers undertaking reconnaissance had to be taken away also with breathing problems as they had not initially donned their breathing apparatus. Two ambulance attendants also experienced similar problems and later still three Royal Canadian Mounted Police officers (Mounties) were exposed to the fumes. At 20:40 the RCMP established a safety exclusion zone with clearly defined perimeters and blocked the roads giving access to the site. It was only the next day, however, following an assessment by the Agrium experts, and after becoming aware of the effects that anhydrous ammonia can cause, that a full-scale evacuation was ordered at 08:30 and would be in effect for 37½ hours. Fortunately, no further injuries or fatalities occurred.

At 01:40 the next day, however, when a thorough search of the site and assessment of what happened could finally be undertaken by suitably protected railway and chemical experts, an unconscious railway worker was discovered in the middle of the chemical cloud close to the overturned wagons. It was the railwayman who was to have made sure that the train cleared the turnout before moving it over to allow the shunting movement to set back into track 5. He was quickly taken to hospital where he was diagnosed with first-degree chemical burns to his face, second-degree chemical burns to his body and severe damage to the inside of his mouth and respiratory tract as a result of inhalation. He clung precariously to life until May, when he finally succumbed to pneumonia attributed to irreparable chemical damage to the respiratory tract caused by unprotected exposure to anhydrous ammonia.

In the Canadian Transportation Safety Board (TSB) report there was mention of the amount of anhydrous ammonia moved around per year on trains in North America. It was cited as the seventh most commonly transported high-risk chemical commodity according to the AAR Bureau of Explosives. Yearly on average 60,463 railway wagonloads of the dangerous chemical cruised across the continent, of which 23,802 cargoes originated in Canada. There is no reason at all to assume that the situation in Europe is much different, and it is notable that no such exact figures exist for road transport.

Comment

One aspect of this accident takes us to an issue that played a crucial role in the derailment at ***Hatherley***, Gloucestershire, in October 2005. It concerns the necessary attention to detail that can make the difference between a train arriving without mishap and a train ending up in a disaster. Although these two incidents concerned freight trains, a similar less than robust approach to ensuring that everything was fine with the track and the train before or during departure (e.g. brake tests, to mention just one regular requirement) has led to a good number of passengers being killed or injured all over the world. Whereas at ***Hatherley*** it was the fault of someone not releasing a parking brake and someone else failing to ensure that the train departed without sliding wheels etc, in the Red Deer case it was those who were responsible for ensuring that the track was in a sufficiently good state to deal with the massive North American maximum axle load of 35 tonnes. If on the day after such a track 'inspection' a derailment occurred as a result of the spreading of the track it is clear that someone had not done their job assiduously enough, the inspecting railwaymen having overlooked the dodgy sleepers and rail fastenings in the yard. Never mind the rail breaking at the obvious place where bolt holes for the connecting fishplates are drilled, there must have been a lack of attention to the deterioration of the track there as well. At any station or yard it is easy to spot which sleepers, particularly wooden ones, are in poor condition. Watch a train roll by whilst looking at the movement of the rail in relation to the movement of the sleepers and if the sleepers remain in place whilst the rail moves, the fasteners are loose. A look at whether all fasteners are complete also gives a quick indication of the quality of the attention given to the track.

It is this issue that provides a link with other accidents described in this book in which a lack of thoroughness and attention to detail played a contributory role. Virtually all cases relate to areas of the network where there was pressure to get things done hurriedly between trains, or to places where nothing much ever happened and the people involved appeared to be preoccupied with other things. Such circumstances can be a real killer on the railway (see ***Quintinshill***, ***Wijchen*** and ***Holzdorf***). No rulebook can be of any help here. It is only the innate intention of those involved to do the job correctly, or perhaps the intention to avoid getting involved in trouble, that makes all the difference. For that reason, of course, there is a need for regular inspections to check on the current state of matters, as well as ensuring that measures are taken to remedy any deficiencies found. Failure to do so played a role as far back as the accident at ***Staplehurst*** in 1865, as it did in the Red Deer incident and would do so again at ***Chatsworth*** in 2008. At Red Deer yard, a simple kick against a rail spike in a mouldy-looking sleeper, exactly what the Canadian TSB did at the start of its investigation after the accident, would have made all the difference.

***

Pécrot, Belgium, 27 March 2001 at 08:45

(Head-on collision following SPAD)

The accident

In this incident an inexperienced driver committed a hard-to-believe

SPAD at Wavre station and then took his ECS train forward along a wrong track, causing a collision with an oncoming train. The accident highlighted insufficient training and a low preparedness for emergency situations as a result of the French/Dutch language divide in Belgium.

The location

The collision occurred at Pécrot in French-speaking Brabant, on line 139 from Leuven via Louvain-la-Neuve Université and Wavre to Ottignies. The line crosses the divide between the Dutch- and French-speaking parts of Belgium.

Rolling stock involved

EMU unit 709, working trains L2358 and ECS L2378, was a GTO thyristor chopper-controlled A1'1A' + A1'1A' 2-car unit type AM73/74/78/79 from 1972-80, 47.5 m (155 ft) long, EP braked, weighing 108 tonnes, with 28 first-class seats and 150 second-class seats, and a maximum speed of 140 km/h (85 mph). It was one of a characteristic family of Belgian electric multiple-units stretching back to pre-war days, with inter-unit gangway connections through the cabs, very much like many British 1950s and 1960s types of such units.

Train L6458 was made up of 2-car EMU units 932/921, GTO thyristor chopper-controlled Bo'Bo' + 2'2' type AM86/89 from 1988-89, EP braked, with a combined length of 105.6 m (346 ft), a combined weight of 212 tonnes and total accommodation of 80 first-class seats and 268 second-class seats, and a maximum speed of 120 km/h (75 mph). The units were officially known as 'Sprinters' but unofficially, aptly, as 'Goggles' owing to their front-end design with a horizontal single-pane windscreen in a somewhat rubbery-looking frame.

The accident log

Train L2358 terminated at line 4 of Wavre station (the Ottignies-bound main line) on arrival from Leuven. It stopped at the 2-car mark as required, which meant that the departing cab (after the change of ends) was 4 m (13 ft) ahead of the starting signal. The driver changed ends using the train aisles and gangway rather than walking via the platform, so he failed to spot his departure signal showing a red aspect. He then set up the cab for departure as ECS to St Joris Weert and as train L2378 from there, calling at all stations to Leuven. The starting signal was at danger because the road was set for a freight service on the Leuven-bound main line that train L2378 was booked to use. However, at 08:41 the signaller at Wavre noticed L2378 on his panel running reversible ('contre-voie') – quite a normal arrangement in Belgium, although not set up in this case. This was due to the turnout not having been changed after arrival of the train as L2358, in readiness for the passage of the freight train.

The driver could not be stopped by signals because the reversible working of the line was under simplified bi-directional signalling whereby a driver received a junction signal only to cross over into the wrong-line section and to leave the section but got no intermediate signals. In the case of both simplified and full reversible signalling, however, the signaller needed to operate a direction switch to enable reversible signalling and equipment such as automatic level crossings to work properly. As this had not happened, the offending train was unable to activate level crossing equipment.

To the consternation of the signaller, the driver did not stop his EMU to report a SPAD but continued along the (now) right-hand line he had come in on, closely following the freight service that was running parallel on the proper left-hand line. This was of some importance, as the offending driver was consequently unable to notice the additional clue of open AHB level crossings with regard to his error, as these were being closed by the freight train. However, having overtaken the freight (indicating rapid acceleration and considerable speed on the part of L2378), the open AHBs ahead of him still did not raise doubts about what he was doing in the mind of the errant driver. Luckily, no accidents occurred on that score.

At 08:42 the French-speaking Wavre signaller called his Dutch-speaking colleague at Leuven to report the runaway. Under the stress of the situation he did not use appropriate communication protocols, as a result of which his colleague failed to understand what was required, i.e. to stop approaching train L6458 from Leuven to Louvain-la-Neuve Université and get it out of the way. Some sources indicate that equipment to enable the emergency isolation of the line current was out of use, so at 08:43 the Wavre signaller called Brussel control to get the current isolated between St Joris Weert and Wavre, and 2 minutes later Brussel control called Antwerpen electrical control that dealt with this line. At 08:46 Brussel control also attempted unsuccessfully to contact the traincrew but found it impossible to get through to them. There was no cab-to-shore radio telephone and neither the driver nor the senior conductor had a mobile phone. It was then discovered that the senior conductor of train L2378 had swapped duties without informing control, yet the booked conductor for this service also could not be reached in order to find out who was actually working the train. By this time train

L6458 was past the distant (which showed a green aspect) for the signal that, if at red, would have stopped it from entering the section where train L2378 was moving towards it. All avenues for stopping either of the trains having been exhausted, overhead line traction voltage was switched off at 08:47 all the way from Leuven to Wavre, which was confirmed back to operations control at 08:48.

With the line voltage cut, the regulations required trains to slow down immediately to 40 km/h (25 mph) and continue to coast if at all possible to a station or another location where the passengers could be safely detrained. If a train was moving at speed and not powering, however, adhering to this ruling required the driver to notice that on his desk the line voltage indicator had extinguished or showed as being off, as otherwise he would not notice his traction motors cutting out with his controller open. Due to the noise in the cab, the driver would also be likely to miss other indications that he had lost line voltage, such as motor or static dc control voltage generators and compressors cutting out. ECS L2378 continued at an estimated speed of 80 km/h (50 mph) and, given his lack of perception until then, its driver in all likelihood did not notice that he no longer had line voltage either.

At 08:50 the two trains collided head-on at Pécrot, on an embankment high above the centre of the village, but fortunately stayed on top of the embankment in line with the track. Unit 709 had reared up over the two modern units, the front vehicle of which was almost completely destroyed. Eight people (5 passengers and 3 staff, among whom were the two drivers) lost their lives and 12 people were severely injured. Fortunately, the freight service that had been overtaken by the errant train had been stopped at signals that had been put back to danger, so was not involved in the accident.

The zone controllers were informed at 08:52 and railway and emergency services were dispatched. An emergency train with a crane was requisitioned from Schaerbeek near Brussel at 09:30, whilst 6 minutes later the power was restored between Leuven and St Joris Weert in order to run a shuttle service to clear the stations en route and allow other services on that stretch of line to continue. At 09:50 a crisis control centre was installed at Pécrot, half a mile (800 m) from the accident site, and investigation into the accident started.

The inquiry

The investigation came to the following conclusions:

1 The driver of L2378 had committed a Category A SPAD by not checking the starting signal aspect as a result of starting ahead

of it, before taking his train 8 km (5 miles) along a wrong line without authority, exacerbated by further errors.

2 The failure to act decisively on the part of the Flemish (Dutch-speaking) and Walloon (French-speaking) controllers led to a substantial loss of time and was a definite contributory cause of the accident.
3 There was an apparent absence of relevant emergency procedures or equipment to communicate across the language border.
4 The line was not equipped with cab-to-shore radio.
5 Driver training was unacceptably poor and there had been insufficient monitoring of the competence of the offending driver. Antwerpen-based Koen Heylighen, aged 31, who had passed out in August 1999, had recently been off the track for a considerable period in the light of another Category A SPAD.
6 The poor siting of a signal in relation to the relevant stopping point.
7 The line was not equipped with ATP, whilst the existing 'Crocodile' warning system was of no use as a result of the departure ahead of the starting signal.

The principal cause of the accident was determined to be driver error. Within two days of the inquiry's findings being given, the national Belgian railway operator SNCB/NMBS moved the Wavre starting signal to a position that was visible from the cab of a reversing unit, and cab-to-shore radio was installed in the year following the accident. SNCB/NMBS was also fined €99,175.

Comment

Heylighen was an inexperienced driver with a serious blot on his record already. His lack of understanding of the very basics of his job ought to have been picked up during training, examinations and subsequent post-qualifying checks and was a serious indictment of the flawed training provided by Belgian Railways at the time. Full and simplified bi-directional signalling of double and multi-track lines was widely installed in Belgium, just as in the neighbouring Netherlands, so unexpected reversible running protected by dedicated signalling was nothing unusual. Heylighen must have experienced it several times in the previous years during training and since passing out. Therefore, having departed whilst seeing the freight train rattle by at speed to his left, it is imaginable that he assumed it was safe to travel 'contre-voie'. Moreover, it is profoundly disturbing, in view of his acceleration and speed, to consider that at no time did he appear to ask himself why he had not seen the necessary flashing proceed signal aspect with a 'marguérite' (CD/RA indication in Britain) to depart, plus a 'V' (chevron) junction indicator with a speed indication to go reversible. That was a most remarkable lack of perception in someone who had passed out to drive trains.

Bi-directional signalling in Belgium is clear and specific. The crossover distant signal slows you down and then the junction signal displays a chevron junction indicator and a speed indication digit under the double yellow signal colour-light aspects. Your train will be crossed over to the right-hand track and from that moment on your signals are placed on the right-hand side of your track and, more importantly, all the indications pertaining to your train movement flash, even the red danger aspect. Obviously, as driver Heylighen did not see the departure signal he also cannot have seen any of the other indications that would have authorised his trip. In the case of simplified bi-directional signalling as here, however, another signal would not have been seen until the distant signal for the next set of crossovers. With full bi-directional signalling there would have been section signals en route that could have been used to stop him. What surprises me in the light of the problems to stop the errant train is that it took so long to cut overhead line power. One of the nice things about electric traction is precisely that you can do that to stop traffic and that was overlooked to the very last.

The driver training period had been reduced from 18 to 12 months and additional post-qualifying courses had been stopped as a result of staff shortages and cost-cutting exercises. The critical staff situation also led to refusals to grant annual leave and rest days, a total of 600,000 such days being outstanding according to railway union publications following the accident. The absence of OTDRs, ATP and cab-to-shore radio was largely explained by the fact that much of the Belgian rail transport budget at the time was being swallowed up, for years on end, by the political decision to construct the high-speed railway lines from London and Paris to Brussel and between Brussel and the German border, and to implement the difficult upgrade of the line from Brussel via Antwerpen to the Dutch border simultaneously. It did make Brussel an important hub in the emerging European international high-speed rail network, but did so at the expense of the existing network. Negative effects on operations were the order of the day, but passenger usage also increased quite steeply at the same time so capacity and staffing problems were the inevitable consequence. The hurried training and less than thorough monitoring of driver Heylighen are fully attributable to that issue. The question of staff swapping diagrams without authority can also cause unforeseen problems. In similar situations following accidents, when spouses have been contacted to report the death of their other half, the caller has found that the person concerned has actually answered the telephone.

Readers will notice similarities here with what happened at ***Ladbroke Grove*** and ***Brühl*** in which other poorly trained and inexperienced drivers made errors at signals – one committing a SPAD and then accelerating away against the flow of traffic to cause a head-on collision, the other accelerating in error and causing a derailment. Both cases exposed identical gaps in driver and route training, and just as happened in Britain and Germany, Belgian Railways was rudely made aware of the need for action.

***

Three accidents at Moulinearn, Perthshire, Scotland, 8 August 1953, 5 May 2001 and 8 February 2009

(Level crossing collisions following crossing-keeper and road-user errors, and pedestrian on track)

Introduction

Moulinearn level crossing is a so-called user-worked crossing (UWC). A user-worked crossing usually is a gated crossing of the lowest category where road users open and close the gates themselves. Moulinearn crossing serves a private lane that leads towards a small hamlet of seven houses and two holiday cottages plus agricultural land off the A9 dual carriageway. It has always been somewhat different from most other UWCs in Britain as from its inception until 1997, despite serving a private road, it was a keeper-controlled crossing. It became a UWC in 1997, with the novelty that it was equipped with electrically operated boom-barriers with push buttons to be operated by the crossing users. In both periods accidents happened, for which the blame wholly or partly could be laid at the door of the railway network operator. That is a fairly exceptional situation.

The location

Moulinearn is a very small hamlet on the bank of the River Tummel near Pitlochry. It is a well-known salmon fishing venue that in season attracts substantial numbers of visitors using the private lane without hindrance. The road user census counts vary because of the irregular use of the holiday cottages, but the crossing can sometimes be rather busier than even a normal public level crossing, with visiting road users having to work out how to operate the crossing barrier equipment. The A9 is a busy trunk route through Scotland and generates a fair bit of traffic (with accompanying noise) and is closely parallelled in this area by the railway line linking Perth and Inverness. The level crossing on the single-track non-electrified main line is located on the private lane some 65 ft (20 m) after leaving the trunk road.

The crossing in 1953

At that time the keeper-controlled gates were normally closed against road traffic and padlocked, contrary to normal practice in Britain and Ireland. The crossing keeper controlled no signals for trains and there were no approaching train warnings. Therefore, the gates could be opened for road traffic only after the keeper requested permission to open via telephone from the signaller at Ballinluig North signal box. Following passage of the road-user the gates had to be closed and locked across the road, after which the signaller had to be called to report closure and locking of the gates. When opening them, the exit gate furthest away from the road vehicle had to be opened first and the entry gate last.

Traffic frequencies

In 1953, rail traffic ran at 56-58 trains per day and there were approximately 130 daily road user crossings, more during harvest time and less at other times of the year. About this time road traffic substantially intensified when freight vehicles were used to transport gravel, extracted from the riverbed of the Tummel. For that reason the resident (female) crossing keeper was assisted by a second crossing keeper to keep her hours per day within allowances, and she instructed the relief crossing keeper in working the crossing. At times there were about 14 vehicle crossings per hour on average, so the provision of the facilities as described was questionable in terms of robust safety, as it is today, despite it being a UWC on a private lane.

The accident on 08 08 1953

About 1½ hours after the passage of the previous train, a freight train consisting of Class 5 MT 2'C (4-6-0) steam locomotive No 44796 (a 'Black Five') with 35 empty vans and a brake van with a guard approached Moulinearn level crossing at speed. Only the first six vans of the train were connected to the vacuum train brake of the locomotive (a so-called 'fitted head'), the brake van being used to brake the rest of the train based on the guard's route knowledge and on whistle signals from the locomotive in the event of out-of-course brake applications being needed. A few fast freight trains (usually perishables such as fish and fruit) could be 'fully fitted' and had the vacuum train brake operative throughout the train, but the rest of the freight trains were braked in a fascinating manner by co-operation between the driver at the front and the guard in the brake van at the rear. Speeds of such loose-coupled or unbraked trains were, of necessity, slow.

At about 40 yards (35 m) from the crossing the locomotive driver saw a tractor with a trailer enter the level crossing from the left. He immediately shut off power and opened his brake valve but hardly had time to whistle the signal for emergency brakes to the guard in the brake van before the collision occurred. The coupling between the tractor and the trailer was severed, the tractor being thrown violently to the right and ahead into a stream, its driver perishing in the process, and the trailer being thrown to the left. The locomotive was undamaged.

The investigation

In the course of the investigation the relief crossing keeper, a man whose mental capabilities were described in a way that would no longer be tolerated ('far from bright, slow of thinking and writing'), alleged that he had called the signaller twice (admittedly in very short succession) to request permission for a road user to pass, but that the signaller had not picked up the phone. On his own responsibility he had then opened the gates to let this vehicle through and he was just about to close the gates again when he heard and saw the tractor. He claimed that he had tried to stop the tractor driver but that the gent took no notice and entered the level crossing, with dire consequences. He was, however, contradicted by track gangers who had witnessed the accident and stated that there had been no other vehicle, and also by the signaller at Ballinluig North signal box who was insistent that he had received no telephone calls on the Moulinearn level crossing handset. Unfortunately, no records were kept in the book of these calls, something that changed as a result of recommendations made after the accident. Worse for the relief crossing keeper was that the resident lady crossing keeper, who lived next to the crossing, added that she had reason to believe that he had not been operating in accordance with set procedures but had used his own initiative to open the gates for road traffic. Despite her urgent admonition to adhere to the rules during his training and thereafter.

As far as the tractor driver was concerned, an open or opening crossing gate meant a safe crossing, which is something that will be encountered in other such situations as well. He simply did not keep a lookout, and his vehicle was no doubt noisy enough to blot out any sound of an approaching steam train. The train driver, for his part, knowing that Moulinearn crossing was secured with a keeper, had not sounded his whistle.

Comment

In the book *Level Crossings* it has been shown that the use of crossing keepers is not necessarily the safest means of operating level crossings. Indeed, Swiss Railways abolished them completely after a severe accident with a German excursion coach on a level crossing near the *Zürchersee* on 12 September 1982, which was due to the oversight of a crossing keeper. The disastrous crossing accident at *Langenweddingen* in Germany on 6 July 1967, although triggered by equipment failure, was also due to crossing keeper errors to a substantial extent. Crossing keepers were retained in Britain, but it was generally made sure that signals covering the crossing were locked at danger until the crossing was fully closed against road traffic. Even then, accidents were possible, however, as shown at *Moreton-on-Lugg* on 16 January 2010, where a distracted crossing keeper completely overlooked the presence of a train in section and due to particular technical interlocking shortcomings was able to open the gates, again with fatal consequences. But Moulinearn was different, and this accident was a direct result of that difference. The actions of the relief crossing keeper also illustrate that regular supervision of safety-critical staff at work is necessary to maintain safety levels, hence the event recorders or loggers attached to present-day equipment.

Some history concerning this type of level crossing

When the drive to remove remote crossing keepers started in earnest in the 1960s, the miniature stop lights (MSLs), which at the time were called miniature warning lights (probably a more apt description) were introduced on the small rural crossings with a certain level of road traffic. Sometimes the old gates were retained, but other crossings got hydraulically operated user-worked lifting barriers to obviate the need to cross the track five times in order to open the gates, move the vehicle across and then close the gates again. By 1977, 45 public crossings and 43 private crossings had been equipped in that way, and at all of these the road user was required to get out of the vehicle to open the gate or operate the crossing controls and thus be exposed to the sight and sound of approaching trains. Problems soon arose, however. One particular hazard was that gates were regularly left open after crossing, leading to fatal accidents as at *Naas* level crossing near Lydney on 13 March 1979 when a refuse collection vehicle was driven unchecked past opened barriers into the path of an approaching train. As people largely ignored the small warning lights and freely left barriers open, the British Railways Board decided to stop further commissioning of such crossings.

On 31 March 2000 there were 135 level crossings with MSLs. At Moulinearn the residents had opposed the removal of the crossing keeper, who by then was instructed from Pitlochry signal box and at the time was a Pitlochry box signaller's wife, Mrs Herbertson. But once the new installation was commissioned on 23 March 1997, they quickly learned to appreciate the new situation, as previously the wait at the barriers before permission was given to cross the tracks had at times run to something like 20 minutes. The wait had been shortened to 40 seconds at the most, and the new electric gates were very easy to use for the local residents, many of whom were elderly. Things improved even more when a local councillor, Eleanor Howie, talked Railtrack (the national rail infrastructure operator at that time) into installing roadside low-level barrier operation buttons, based on the argument that these would enable disabled road users to use the level crossing without having to leave their vehicles. A locked box initially covered those extra buttons, and keys were issued on personal title, but a count of disabled road users revealed only five such crossings per year (a disabled fisherman three times and a disabled visitor twice) and following pressure from the residents and Councillor Howie the restrictive box covers were removed. After that, all and sundry used the low-level operating buttons, the trouble of having to get out of the car and go to the buttons at the trackside having been dispensed with. However, this went against the set standards for such crossings as adopted by the railways in Britain, and it is surprising that Her Majesty's Railway Inspectorate (HMRI) agreed to the new situation. After all, the bulbs in those red and green MSLs might fail and the telephone might have been vandalised or otherwise put out of order at the same time. Railtrack's trust in the human instinct to deal safely with level crossing dangers rose to breathtaking levels in this case.

Serious near-misses at Moulinearn level crossing were reported on no fewer than three occasions – 11 April 1998, 4 October 1999 and 25 November 1999 – and, based on the fact that not all train drivers are willing to spend time writing reports about stupidities among the general public if they think they can get away with it, it is quite possible that a few more may have occurred. During that period there were also 39 automatic alarms in the Pitlochry signal box as a result of the Moulinearn barriers being left open. These led to train delays as drivers were instructed to pass the level crossing at greatly reduced speed. Such clear and recorded proof of road user level crossing misuse plus three near-misses previous to a level crossing collision in roughly 1½ years since opening was worrying, yet it triggered no action, such as carrying out a risk assessment.

The crossing in 2001

There were a pedestrian wicket gate with a foot crossing and red and white barriers, complete with skirts, for the vehicle crossing in 2001. Next to the barriers at the trackside were sets of steady (non-flashing) red and green MSLs with explanatory signs informing road users to stop at red, how to lift the barriers and then to cross only with a green light. The buttons to lift the barriers were located near these signs, but, as already pointed out, there were additional buttons on posts set well back from the barriers for motorists to activate the barrier lifting installation without leaving their car. The lift barrier button cover had to be opened and then the button pressed and held until the barriers were fully up. To lower the barriers after crossing, however, the non-covered lower barrier button just had to be pressed once. In case of problems, on either side of the crossing there were direct telephone lines to the rail traffic controller. Like all gates on UWCs, the Moulinearn barriers were operated wholly independently of train traffic (something which is not generally realised – or its potential consequences – by the general public). There was no interlock with train signalling, so the barriers could be lifted at any time, whether a train was approaching or not. The red and green MSLs and a Yodalarm bleeper type of klaxon were used to indicate to road users the approach of a train, the red light being activated by the train 40 seconds before it reached the crossing. At the same time the klaxon started to sound and continued until the train had passed.

Traffic frequencies

On weekdays there would be 10 daily booked passenger trains in each direction plus a varying number of additional freight and departmental services. Line speed was 80 mph (130 km/h) but freights would usually travel somewhat more slowly at 60 mph (100 km/h). Train traffic had dwindled since 1953 but was nevertheless still substantial and a lot faster.

A 72-hour census taken between 11:30 on Friday 25 May and 11:30 on Monday 28 May 2001 showed 176 road user crossings (150 private cars, 12 delivery vans, 5 tractors with attached equipment, 9 motorcycles) as well as 14 pedestrians. The human contents of all these vehicles (plus the pedestrians) were made up of 71 residents, 51 fishermen, 53 visitors, 4 holiday cottage occupiers and 11 workers such as farmers. Of the cars, 27 were driven by people (15 fishermen, 4 holidaymakers and 8 visitors to residents living in the lane) who said they were unfamiliar with the crossing. Whilst the count would normally have been lower, it was also substantially higher when the salmon fishing was good.

The accidents on 5 May 2001 and 8 February 2009

Two families in two cars were involved in the incident on 5 May 2001. The leading car was driven by Jane R, a 38-year-old television producer, with her companion Bruce T, a 45-year-old college lecturer beside her in the front seat. In the rear of the car was 3-year-old Sarah C, the daughter of the family in the second car. That was being driven by James C, with his wife Lorna in the back and Sarah's brother Charlie in the front seat. The weather was fine as the two families drove down the A9 to Moulinearn village to visit friends after an afternoon spent watching a football match at Huntly. Initially, Bruce had been driving but somewhere en route he had changed places with Jane. On reaching the slip road on the A9 for the Moulinearn turn-off, both cars turned into the private lane and reached the level crossing. Jane parked next to the low-level operating buttons on the right and started to push the open button whilst James stopped next to her car, thus blocking the lane for other traffic.

The level crossing event recorder revealed everything in minute detail. The red MSLs had already been illuminated and the klaxon had been sounding for 23 seconds to warn of an approaching train, but no one appeared to have noticed it despite the open windows in the car. Jane had also missed the instruction that the open button had to be held whilst the barriers were raised, and she needed four pushes in 10 seconds before they were fully raised. Completely unbeknown to her, there were just 7 seconds left before the arrival of the 17:40 2-car DMU from Edinburgh to Inverness travelling at 80 mph (130 km/h). When she drove onto the crossing the train must have been about 100-150 yards (80-120 m) or 2 seconds away and its headlights must have been clearly visible had she but had the inclination to look. James and his wife saw the other car with their daughter in it move onto the level crossing and suddenly being rammed aside by the passing train. The impact swung the car round, demolishing the opposite crossing barrier installation, after which the near side with Bruce T in the seat smacked against the passing train that tossed the car aside like a piece of litter. James and his wife later instituted successful legal proceedings against Jane as a result of the emotional and mental stress they had suffered.

After the train had stopped, the on-board catering host walked back along the track and met Jane, who was shaken and injured but apparently lucid enough to speak. She told him that she had raised the crossing barriers because Bruce had told her to do so, in which case he had also failed to be on the alert whilst instructing her. Jane had not looked for danger signals, and what was worse, it was questionable whether she had even realised the safety implications of the situation ahead. At the inquest she said that she had looked out for trains when approaching the level crossing, but inexplicably she had completely missed the approaching headlights of the train. However, was she even aware that trains carried headlights? Jane had had previous experience of this crossing, but only whilst Bruce was driving, and appeared to be somewhat lacking in confidence when driving away from home and rather dependent on others in situations such as this. When the available 'authority', Bruce, did not include the instruction to look out for danger signals she seemed unable to check for such issues on her own initiative.

According to witnesses, the bleeper klaxons were heard and the MSLs were seen working as far as 75 yards (68 m) away from the crossing, yet neither Jane nor Bruce appear to have heard the Yodalarm klaxons whilst driving past a few feet away with their side windows open. However, the event recorder of the crossing confirmed that the installation had been working properly. The post-mortem on Bruce confirmed that his heart showed evidence that he had suffered from either severe angina pectoris or possibly a mild heart attack at the time of the accident, which might explain why he had let Jane drive the car. It may also explain why he did not pay more attention to Jane's behaviour at the level crossing.

On 8 February 2009, Moulinearn UWC once more made it into the media in Scotland. While walking across the track, 94-year-old Frank Clifford was hit by the 09:18 Inverness to Glasgow train and died as a result.

Comment

The basic problem at Moulinearn was that a UWC was not really meant to be a public crossing in the way this one clearly was. Only local farmers or industrial workers were supposed to be using it, and they, presumably, were aware of how to use it safely. There are gates, but these are never interlocked with railway signalling and could be opened at any time, they were no different from any other farm or factory gate. Open level crossing barriers set a clear trap that an average member of the public will interpret as meaning that it is safe to cross, certainly at sites that mimic 'real' level crossings with boom-barriers, as *Naas* and Moulinearn clearly did. For that reason, similar UWCs in The Netherlands, for example, are painted in different colours instead of the normal red and white blocking in order to indicate that they are not regular level crossings but are more akin to the normal gates giving access to a field etc.

Level crossing safety at UWCs wholly depends on people using their brains as well as showing a measure of discipline. However, too many users the world over fail miserably on that score and accidents occur as a result, some of which also endanger passengers on the train. Tough measures to mitigate risks are now being taken in some nations. In Australia, for instance, if the UWC gates are reported as having been left open three times, the railway operator has the right to take out that level crossing completely after issuing warnings every time such a breach is reported.

Users of Moulinearn crossing should be protected with a normal level crossing because whatever experience they may have with public level crossings, it simply does not apply at Moulinearn. Despite the fact that the UWC deceptively looks like a normal crossing with barriers that close when a train approaches. Most members of the public have never had to work level crossing equipment themselves, certainly not this uniquely tricky equipment, and although it is true that people have acquired certain push-button habits and expectations from their daily lives, it is doubtful whether these had been analysed and incorporated into the design and operation of the level crossing equipment at Moulinearn.

Some level crossing technicalities

If road users know about level crossings (and there are some doubts about that according to experts in various countries) then the following issues guide their experience:

1. *Open and opening level crossing barriers mean that it is safe to cross. If barriers close or stay closed, a train is coming*. The barriers at Moulinearn have the appearance of normal level crossing barriers that come down when a train arrives and open when it is safe to cross. Therefore, crossing is wrongly perceived

as safe when they go up. This, in fact, is what happened, as Jane later admitted during court proceedings.

2 *Level crossings that indicate danger do so with flashing red lights*. Flashing lights are more arresting to the human eye than steady lights, although the MSLs, particularly after they were mounted in square fittings without black backboards, did not impart the same imperative to stop as a red traffic light encountered by people in their everyday lives. Something quite similar was experienced in The Netherlands with the yellow and red 'clearing' lights at some level crossings. Replacement with normal traffic lights, accompanied by proper road markings, solved the problem there. At Moulinearn, the small, steady coloured lights were inconspicuous and conveyed no degree of urgency through lack of movement or linkage with the bleeper klaxons. They also bore no resemblance to wig-wag warning lights used on other types of level crossings.

3 *For people in cars with closed windows and people wearing integral crash helmets, klaxons or bells giving audible danger warnings are as good as useless*. Car manufacturers deliberately minimise the penetration of external sounds to create an atmosphere of comfort, the isolation from the outside world often being further reinforced by music systems etc, as clearly demonstrated by NTSB tests following accidents in the US. Only pedestrians and cyclists benefit from audible warnings, and only then if they actually know what the sound indicates. This appears doubtful in the case of Jane at her open car window wrestling with the button.

4 *Yodalarm klaxons sound like a car or house alarm*. This is something that most of the population has learned to ignore, which in all likelihood had an impact on what happened at Moulinearn. For that reason level crossing alarm bells, as used in the USA and The Netherlands, is a better option.

5 *Reading written instructions in public places is surprisingly uncommon*. Even a local safety consultant managed to overlook the instruction signs as well as the miniature warning lights at Moulinearn on a daily basis. Moreover, most people will not spend time and effort reading and interpreting instructions, however well drafted and clearly displayed they may be. This was clearly the case at Moulinearn.

6 *The relevant authorities should first and foremost consider how they would like the public to behave before drafting any written*

instructions. The fact that the instructions were not suitably placed for the benefit of those using the low-level barrier operation buttons only made this matter worse. To the public the level crossing barriers were merely a hindrance to progress, to be dealt with as quickly as possible, irritated by looking for controls such as buttons to be pressed. It is likely that this tendency for impatience was overlooked during the design of this installation. Minor design errors such as green barrier-lifting buttons and red barrier-closure buttons instead of the other way round, and the odd location of the MSLs and instructions in relation to the most used operating buttons, were of lesser importance. However, the system would have benefited from the input of behavioural scientists, and remedial action should have been taken to deal with the resulting risks once they had begun to manifest themselves.

7 *At a UWC the road user should be forced to get out of their vehicle to operate the closed gates or barriers in order to gain maximum exposure to visible and audible warnings and awareness of approaching trains*. This is ancient railway wisdom contained in traditional UWC design, something that perhaps had been somewhat carelessly tossed aside when the Moulinearn roadside low-level buttons were accepted by HMRI.

As Stanley Hall in his role as consultant during the legal proceedings pointed out on more than one occasion, the low-level buttons were a major mistake and ultimately led to fatal consequences.

So what should be done with these crossings in general, and the one in a rather public environment at Moulinearn in particular? The following points are relevant:

1 If there is a link to train signalling (as with the MSLs) then a train striking in for those lights must disable the barrier-opening button. If a train has struck in and activated the klaxon and the red warning light, it must no longer be possible for the barriers to be able to be opened, certainly not as at Moulinearn, until the train has struck out.
2 Strike-in should not immediately close the barriers, however, so as not to trap road users on the level crossing. Automatic time-out closure (point 5) takes care of that.
3 The barriers must not resemble normal level crossing barriers, so that their different purpose is readily distinguishable. A different

colouring scheme would be a good start, looking more like a normal field gate (e.g. brown and yellow warning stripes) as practised on a new breed of similar UWC boom-barriers in The Netherlands.

4 Roadside low-level buttons allowing operation from inside a car should not be permitted and, where already fitted, should be removed immediately. Road users must be required to get out of their vehicles in order to be exposed to the sight and sound of approaching trains when opening the barriers, and to the sight of the level crossing deck when closing the barriers in order to avoid trapping previously unseen road users behind on the crossing deck. The buttons should be placed in a box with a lid, on which the instructions are clearly written.

5 The barriers must be self-closing after a short time-out in case they are left open. As it must be expected that self-closure will be the road users' preferred method, the time-out should be fairly tight, which would also combat dangers arising from vandalism. Road users should be informed about the restricted length of the opening time.

6 In the event that a farmer, for example, needs a longer time-out period to bring heavy or slow plant across or a herd of cattle, the local traffic controller must be phoned, who can then remotely disable the automatic time-out and provide protection with covering rail signals. Another option is a system of individually issued and tagged mechanical or electronic keys, that disable the automatic time-out. These must be removed again to close the barriers, for which reason the use of such a key must be registered with times and identity on the crossing event recorder, to ensure that individuals are held responsible for their actions in the event that things still go wrong. This technology already exists.

7 The warning lights must either be normal traffic lights or normal level crossing wig-wag lights to ensure that road users recognise the requirement to stop.

8 For a number of reasons (e.g. copper cable thefts) it would be preferable if the crossing installation was a stand-alone, autonomous unit with regard to energy provision, for example through the use of combined solar and wind technology to feed batteries, as commonly seen on roadsides for certain electronic traffic signs. Remote control functions and voice telephone connections to traffic control centres would be achieved with wireless GSM technology, which is also nothing new.

All the matters outlined above point to the fact that at Moulinearn a run-of-the-mill AHB should have been installed. There is no more fitting way to end this story than with the words of PC Gilroy, the police traffic officer who gave evidence in the subsequent legal proceedings. He stated that the crossing design 'had not fully taken account of the human elements of ignorance, stupidity and laziness'.

***

Burlington, Ontario, Canada, 9 May 2001 at 19:04

(Child struck at accommodation level crossing)

Introduction

The reason for including this sad report is fairly personal. From the 10 years I spent in the cab there are a few moments that stand out because of mishaps involving young children and animals that could not be held responsible for their actions. Young children cannot be expected to fully appreciate verbal warnings about the dangers associated with trespass on the railway, because their curiosity is immense and overrules their incomplete understanding of danger simply due to their immaturity. Animals, especially young ones, are subject to the same problems. There is very little that traincrew can do in terms of mitigating the dangers of trespass, and I have known colleagues who were deeply upset as a result of their train hitting a young child or an animal. Because most drivers are parents themselves, and may also have pets at home, they know the emotional turmoil such an accident brings.

The accident

On 9 May 2001, in Burlington, Ontario, a 3-year-old girl slipped away unnoticed during the preparations for an evening church service. She wandered off to a nearby railway line and was fatally hit by a freight train just at the time the congregation were starting a search for her.

The location

The site of the tragedy was at mile 47.95 on the westbound track of the Canadian National Railway's Halton subdivision at Burlington. The weather was good and the visibility clear, at over 24 km (15 miles), the temperature was 24°C and the wind was blowing from the west at 30 km/h (20 mph). The Halton subdivision consisted of a double-track main line, which at the location of the accident was straight. The line was controlled with Centralized Traffic Control (CTC) from Toronto

Traffic Control Centre. Approximately 24 trains passed the location every 24 hours, which amounted to a seriously busy line in a North American context.

The accident occurred on a stretch of line between two overbridges at Guelph Line Road and Queen Elizabeth Way. The bridge at Guelph Line Road was being widened at the time, for which reason the embankment leading up to the bridge had been extended with landfill and a temporary dirt road had been laid west of the widened embankment. This had necessitated the removal of fencing from a church and an attached school accommodated in a former commercial property about 150 m (490 ft) north of the railway line. The fencing had been partially replaced – up to about 60 m (200 ft) from the level crossing mentioned below – with the familiar temporary orange plastic 'safety' fencing, which was found to be trodden down and easily passable in several locations. The previous month, the dirt road had been extended across the railway line with a 3 m (10 ft) wide temporary level crossing 25 m (80 ft) west of the Guelph Line Road bridge. Gates had been provided at either end of the crossing to prevent access to unauthorised vehicles, and during working hours (08:00-18:00) a gate-man was provided to control access, otherwise the gates were locked shut. However, because no further fencing was provided, it was possible to walk around the gates unhindered. In all probability this was how the child had gained access to the tracks, as no leaves or burrs from nearby bushes were found on her clothing.

An arrangement similar to that introduced in New Zealand after the ***Ohinewai*** accident provided site-staff security. Footplate crews were notified of the location of construction work in bulletins, requiring them to call the foreman in residence by radio to obtain permission to pass the site. When this had been obtained, the driver was obliged to use his horn when approaching the worksite to warn all and sundry of his approach. Outside working hours, however, no such actions were necessary. The location was not particularly known as a high trespass area, although illegal footpaths were found during the investigation (as in any residential area) and people from north of the line did cross the tracks to walk their dogs on open land on the south. However, since work on the bridge widening had started, the construction workers as well as railway crews had noted an increase in dog-walkers as well as children from the nearby church school coming close to the railway line or crossing it, including during normal school hours.

The train involved

CN Q-143-31-09 was an intermodal train on its way from the Brampton Intermodal Terminal to Chicago, Illinois. The train was hauled by two locomotives and consisted of 60 loaded wagons weighing 3,190 tons and about 1,280 m (4,200 ft) in length. In the front cab were a driver and a conductor. Whilst the permitted speed for all trains was 80 km/h (50 mph), this train was running at 82 km/h (51 mph) in power notch 1 with the brakes released.

The accident log

Whilst easily rolling westbound along the north main track and approaching the worksite around the Guelph Line Road bridge the crew suddenly spotted a young child on the eastbound south main track. The driver immediately sounded the horn to warn her and then applied the train brakes in emergency, on which the child initially froze and covered her ears, then started walking towards the track along which the train was approaching. At 19:04 she was hit by the left front corner of the leading locomotive and suffered fatal injuries. The train came to a stop approximately half a mile (800 m) from the point where the brakes were applied.

The child had been brought to the nearby church by her father, who was assisting with the preparations for the evening service that was about to commence at 19:00. At about 19:03 it was realised that the girl was missing, although two ushers at the doors of the church had not noticed her leaving the building. A search was immediately started around the perimeter of the church grounds and then extended into nearby commercial plots and streets, and an emergency call was made at 19:10, but somewhat later someone looking over the parapet of the widened bridge at Guelph Line Road spotted the girl's body near the railway line. Police arrived at 19:22.

The church had, in fact, recognised the dangers arising from the construction work and the consequent removal of part of the site fencing and had warned the children more than once about the dangers of going near the track. Nevertheless, small groups of children from the school were seen to visit the construction area on several occasions. The increased use of the trains' horns in the area doubtless also played its part in arousing further interest amongst the youngsters.

Comment

This accident is typical of those horrendous events that over the years have forced many members of traincrew to give up their job. It is impossible to prepare people for this sort of incident. In the course of my train driving career I twice came close to having accidents involving very young children. On one occasion, as I came through Clapham Junction station along the Up Brighton fast line at 60 mph (100 km/h), a young toddler suddenly emerged from the milling crowd towards the platform edge beside which my train shot past. The second occasion was when a young girl, walking ahead of her grandmother who was encumbered with a pram and other children in a park, suddenly ran to a foot crossing across the single line between Weston-super-Mare station and Uphill Junction as my train was approaching. She merely wanted to wave at the train, but did so whilst standing on the edge of the crossing surface, very close to the passing train and not easily visible from the cab of an HST. The only thing I could do was to use the horn, and because I never heard any more in both cases I can only assume that nothing untoward occurred, but those are the sorts of events that can leave a train driver emotionally raw.

Children are attracted to worksites, particularly when there is a regular traffic of freight vehicles carting dirt in and out and interesting machines such as bulldozers etc, as I think we would all admit when recalling similar events from our own childhood. I remember several such instances in the village of Oosterbeek in The Netherlands, where I spent my early youth. At the time the village had been undergoing repairs as a result of damage sustained during the Battle of Arnhem and was later being considerably extended. Kids will be kids, and I have no reason to think that the situation was any different in Burlington, Canada, in 2001. The increased trespass around the worksite had already been noticed.

Concerning the behaviour of the little girl on the approach of the freight train, I think that she must have visited the site before, possibly with other children, and knew the way from the church grounds. Railway tracks lure children, as they do certain animals. The only course of action open to the driver in the circumstances was to brake as hard as possible and sound the horn. He knew from the moment he saw the child that he would never be able to stop in time and that using the horn might cause the child to freeze with fright and so prevent her from walking into the path of his train. Most unfortunately in this case, it did not work and it is highly likely that the little girl was trying to get back to her father when the situation started to frighten her. I remember a case involving a young dog that had escaped from its owner and had got near the track at Kemble in Gloucestershire. By using the loud high-pitched horn as well as giving the infamously noisy Paxman Valenta HST diesel engines the whip, the pup did freeze and stayed where he was, flat on the ground. Hopefully, he returned to his owner a tad wiser.

Until 1995, when it was repealed, section 217 of Canada's Railway Act had required suitable fencing of the railway except where considered unnecessary on a number of grounds. Initially due to land use of the area, and later due to the repeal of section 217, there was no legal requirement for fencing along this stretch of line at the time of the accident, but the situation with respect to land use and traffic was changing. More people lived in the area near the tracks, along which an increasing number of trains were running. At the time that the accident report was being written new legislation was being drafted, based on a) the expected volume, source and pattern of pedestrian and vehicle traffic, b) current use of land adjoining and near the railway premises and the anticipated use patterns for the coming five years, and c) causes of any access problems or anticipated problems. However, these new regulations had not then come into effect and it was unknown when that would happen. The city of Burlington had no by-laws that covered the situation, but Transport Canada had issued a pamphlet on 'Procedures for Prevention of Trespassing' that focused on education, enforcement, fencing, access control, signage, the development of train-running operations and urban planning. Sadly, none of that was much use to a 3-year-old child.

The situation regarding the safety of rail traffic close to my home near Weston-super-Mare is something that could do with a similar holistic approach to equipment and policy matters as well as enforcement. A single-line loop runs off the main line through housing estates from Worle Junction into Weston station and then onwards to Uphill Junction where it rejoins the main line again. On any day, people of all ages can be seen walking on and near the track in an area where trains may run at up to 60 mph (100 km/h), and it would only take a young child following a bad role model at the wrong time for a similar accident to occur. In Canada, the abovementioned pamphlet and its inclusion in 'Direction 2006' did much to halve the number of accidents caused as a result of trespassing.

As already mentioned, before this accident happened, the particular location at Burlington was considered a low trespass area, but that was before the dirt road and level crossing to the bridge construction worksite had been put in. Circumstances had changed and so had the risks. Between January 1990 and June 2001, there had been 23 trespass accidents involving children of 10 years of age or younger across Canada. Ten of these children were no older than 5 years of age and most of these were accompanied by older children or adults. (If anything, these numbers point to the low population density in that vast nation. I do not have the figures for European nations, but my gut instinct is that they are substantially higher for a number of reasons.)

At the location of this accident there was evidence of trespass by older children, as graffiti had recently been sprayed on the bridge abutments. However, between 1990 and the time of the accident in 2001, only two adult trespassers had been involved in incidents at Burlington, one of whom was seriously injured and the other died (a suspected suicide). Both, incidentally, had required considerably more effort to reach the railway line than was the case after the temporary road had been built.

***

Hamina Port, Finland, 3 March 2003 at 19.15

(Derailment of three empty tank wagons on ice)

The accident

Three empty bogie tank wagons derailed on ice-covered, grooved street rail at an unloading point for chemical liquids. This frivolous incident is included merely to show how another railwayman's working day, just after New Year, was thrown into disarray, and how the extreme cold in northern Europe can interfere with operations. It also illustrates how a particular incident can indicate the need for modifications to rolling stock.

The accident log

A set of three bogie tank wagons had been emptied during the night. In order to make the contents of their tanks sufficiently liquid for discharge, the wagons are fitted with a hosepipe coupling each, through which pressurised steam can be pumped into heating elements in the tanks. However, the steam condenses into water whilst cooling down, certainly on a Finnish January night, and in order to drain the system (and also in this case to protect the heating system from freezing up) drain outlets are fitted under the wagons to discharge the still warm condensed water onto the track below. As the vehicles were parked on a hardstand (road freight vehicles were unloaded there as well) the track was made up of rail with grooved heads through which the wheel flanges ran. This allowed the track to be embedded level in the paving, and is the sort of rail that is often used for street-running trams. As a result, the water did not flow away from the track but formed puddles that filled the rail head grooves and then overflowed onto the freezing tarmac. This quickly cooled down and then froze into substantial sheets of solid black ice.

When a Class Dr14 locomotive with a short pick-up train arrived the next morning to collect the empties, it shunted on to the wagons and coupled up as normal, although the shunter no doubt had a problem staying on his feet whilst coupling up, connecting the brake pipes and ensuring that the parking brakes were off. After carrying out the requisite brake test he signalled the driver to move the vehicles forward, but instead of rolling away nicely with the rest of the train the three empty tank wagons slowly and smoothly climbed off the rails onto the rock-hard ice, rather reminiscent of certain scenes in the animation movie *Polar Express*. The shunter immediately stopped the movement, as to all intents and purposes the wagons were derailed.

Comment

This was one of those moments that make life on the tracks so frustrating, or surprising and interesting, depending on your state of fatigue and your mood. What to do? If it had been just one 2-axle vehicle, I would probably have tried to push it back very cautiously into the holes where it had sat before and then asked for a steam lance to try and melt the ice out of the rail head groove. If successful, play innocent and disappear as intended; who needs the hassle? But with three bogie vehicles the risk of further mishap was far too high, and embarrassing evidence of your endeavours would be left behind on the ice.

After this incident, the condensation drain outlets were no doubt moved away from above the track, or drains were installed in the hardstand, or perhaps the ice was cleared from the track by steam-cleaning before moving any vehicles.

***

Roermond, The Netherlands, 20 March 2003 at 11:46

(Head-on collision following driver incapacitation)

The accident

The incident involved a head-on collision at a junction near Roermond in Dutch Limburg, not far north of Maastricht, between a diesel-hauled freight train and a local DMU serving one of the relatively few single-track non-electrified lines in the country. The collision was due to the driver of the DMU falling ill, and the accident was remarkable in that a number of available on-board and lineside safety systems were bypassed. The local train was worked by a Netherlands Railways driver and senior conductor from Nijmegen depot, and the freight by a Rotterdam-based Shortlines driver. Neither company works these services any longer, a private company now operating services on the Maas (Meuse) line (with different DMUs) and Shortlines having been taken over by another company.

Rolling stock involved

The local service, train 16337, was an NS-designed 2-car 2'B'+B'2' DMU of the DM'90 type, No 3405, built in 1996 by Talbot at Aachen. It was fitted with two Cummins 640 kW (860 hp) diesels with Voith hydraulic drives for brisk acceleration and a maximum speed of 140 km/h (85 mph) These sets are nicknamed 'buffaloes' in The Netherlands due to their bulky wide-body and head-down look. A capable and comfortable train, it was one of the last types introduced by a national state railway in Western Europe before the off-the-shelf types from various manufacturers started to take over.

The freight service, train 69875, was a Canadian-built 126-tonne General Motors EMD JT42CWR Co'Co' diesel-electric locomotive with running number PB01, fitted with a 2,385 kW (3,200 hp) GM 710 G38 diesel engine, a type of locomotive otherwise known throughout Europe as Class 66 and designed to fit the restricted British loading gauge. It hauled a loaded intermodal service of 11 wagons, well within its capabilities for accelerating and maintaining speed.

The location

The collision occurred at a junction where the single-track and non-electrified Maas line that follows the course of the River Maas from Nijmegen via Cuijk and Venlo joins the double-track and electrified main line from Eindhoven via Roermond to Maastricht. Both lines, fitted with colour-light signalling and ATP, were remotely controlled from the regional traffic control centre at Maastricht. An understanding of the two types of ATP equipment used in The Netherlands, and the way they interact to provide safety, is important in order to appreciate fully how this accident happened.

ATBEG (Automatische Treinbeinvloeding Eerste Generatie)

ATBEG is the Dutch first-generation ATP, installed at great effort and cost after the accident at ***Harmelen*** in January 1962. Its functioning is based on the transmission of a continuous speed code signal from the track to the traction vehicle. If the driver adheres to this speed signal as indicated on his desk or (in later applications) in his speedometer, he retains full control over his train. If, however, he travels faster than the permitted speed or does not obey the brake criterion (see below), an emergency brake application is initiated and the train is brought to a stop, after which a brake override button has to be pressed in order to regain control over the brake. This is recorded on the trip recorder.

If the system indicates a lower speed or a stop ahead (a target speed indication), it requires a brake application to (theoretically) slow down as appropriate for the speed restriction or come to a stop at the red signal ahead. This brake application, called the brake criterion, is monitored with the fall of the brake-pipe pressure or with the position of the electric brake controller in the case of electro-pneumatic brake systems. For a number of reasons pertaining to the technical state of matters at the time, as well as convenience when passing a signal at red with permission, the lowest speed restriction of 40 km/h (25 mph) allows a train to freely pass any signal at danger as long as the train speed does not exceed the 40 km/h mentioned. In the days before the general application of vigilance (alerter) facilities on the driver's safety device or 'dead man's switch', this position started a vigilance sequence whereby the driver had to press a button regularly to cancel a buzzer.

In short, **this type of ATP does not necessarily stop a train at red.** The 40 km/h pass-signal situation has, unfortunately, permitted quite a few accidents to happen, all at 40 km/h or lower, although where two trains approached each other at that speed impact might still take place at a combined speed of 80 km/h (50 mph). In this particular accident the freight train was permitted to travel at 80 km/h, so the potential impact speed could have been up to 120 km/h (75 mph) had it been given clear passage through Roermond station. Nevertheless, no one had died in accidents on ATP-protected lines in The Netherlands until this particular incident. The ATBEG protected signalling is currently being fitted with wire loop beacons at signals, which will stop trains when the signal is showing a red aspect (ATB-vv). The investment, running into tens of millions of euros, precedes massive future investment to convert all ATP in The Netherlands to ETCS.

ATBNG (Automatische Treinbeinvloeding Nieuwe Generatie)

In the 1990s, when all main lines and many secondary lines had been fitted with ATBEG, there were a number of non-fitted secondary non-electrified lines remaining, whilst fitting of ATP to all freight lines was also decided on. The first-generation ATP was considered obsolete and the 40 km/h pass-signal situation was no longer acceptable following a string of accidents. It was therefore decided to select a more modern type of ATP with enforced stop at red for further application, and to design the breaks between both systems in such a way that they both armed and disarmed automatically at these locations.

The Belgian-designed intermittent ACEC TBL2 system was selected (nowadays marketed by Alstom), the same system that is fitted on a number of lines in Belgium and formed the basis of the Great Western main line BR-ATP system which I had experience of working. Speed limit, signal aspect and additional information from trackside cabinets is beamed from the track into the train with balises (beacons, tags) or infill wire loops – with which the system can be made semi-continuous in locations – and is processed by on-board computers in a two-out-of-three configuration. These display their instructions to the driver on the dashboard speedometer – good equipment that introduced continental advances such as a decently illuminated speedometer dial with a 4-position light-dimmer switch in the absurdly austere British HST cabs of the day. In general, ATBNG has the same functionality as BR-ATP and would be familiar to a driver working the Great Western main line. It has the same LED dots to indicate permitted and target speeds at the speed numerals on the dial and the braking curve calculation facilities work similarly, although tighter, at 3 km/h (2 mph) intervals instead of 5 km/h (3 mph). The Dutch version, however, has a few interesting extras for anyone who has experienced how BR-ATP can slow things down due to its cautious settings as far as calculating braking curves is concerned:

1. There is a 'Glad Spoor' (slippery track) button that is used to alter the normally more daring calculation of the braking curves (which takes the far superior braking power of modern rolling stock into account) to the more cautious setting as used in Britain. The driver uses this button if he has experienced wheelslide when braking or has been instructed to use it.
2. The calculated braking curve is indicated with an increasing, red illuminated ring on the outer edge of the speedometer, the driver being required to stay ahead of it with his speedometer needle. In Britain it is necessary to guess what the system wants as far as deceleration is concerned.
3. The ATBNG display has a restricted distance movement authority bar graph that enables blind driving up to a red signal. About 3 km (2 miles) away from a stop point the LED bar graph lights up and, on approach, bars initially extinguish per 100 m, then

10 m, thus indicating the distance to the stopping point. Doing a blind ATBNG or ETCS approach to a red signal using the bar graph is an impressive experience. You stop with one bar illuminated, the blinds are pulled and there in front of you is the red light.

The interaction between both systems

The switchover between both systems was something that impressed Stanley Hall and me when we travelled this particular line whilst researching ATP for publications on the subject following the ***Ladbroke Grove*** accident. We had to be alerted to this switchover by the escorting Dutch signalling engineer otherwise we would have missed it. Along the track there were two boards on one post, indicating that ATBEG ends and ATBNG starts, whilst on the display an LED extinguishes and another one comes on. That is all there is to it. It takes knowledge of both systems to see that the functionality has changed (e.g. ATBNG has more finely graded speed authority indications). Incidentally, between Arnhem in The Netherlands and Emmerich in Germany, ATBEG and PZB similarly change supervision automatically in trains fitted with both systems under one display. At the break points at Venlo and Roermond, the ATBNG and ATBEG systems change role again in a similar unobtrusive manner, which brings us to the Roermond accident.

Although the Maas line is fitted with ATBNG, with train stop at red, the junction signal and the main line are fitted with ATBEG without enforced train stop at red. This means that between the junction signal and the signal preceding it, the systems have automatically changed supervision and the train will no longer be stopped if it has a SPAD at the junction signal. This had been recognised as a risk and changes had been proposed, but no action had yet been taken. It should also be noted that the junction signal for the single line was placed on the left-hand side of the track because the reversible junction signal of the main line, situated to the left of its track, occupied the space that should have been taken by this Maas line signal.

The accident log

On 20 March 2003, train 16337 had come in from Nijmegen to Venlo, before departing Venlo station at 11:25. There was nothing remarkable about the trip, other than the fact that the senior conductor had noticed that the driver was not as talkative as usual. The conductor's job was to sell and check tickets and to use the public address (PA) system when approaching the main stations to announce where it was possible to change trains for destinations on other lines. The driver's job involved using his access to the PA to announce the minor stops, and he had faithfully done so throughout the trip thus far.

The Shortlines freight driver had brought his fairly short intermodal train of 11 vehicles loaded with maritime containers off Born intermodal yard to Sittard, where he had to run his locomotive round the train to reverse direction and gain the main line in the direction of Roermond and ultimately Rotterdam. After coupling up and brake continuity testing he departed northwards on clear aspects and was booked to run non-stop through Roermond. However, any experienced freight driver in Europe knows that passenger services usually have priority and therefore no such thing was ever certain, even when booked. At 11:30 the driver was approaching Roermond at a fair speed, running approximately 5 minutes early.

Roermond station consisted of a long single-face platform along the main track in the southerly (Maastricht) direction and a long island platform between the northbound through line and a platform loop. The platform loop, however, itself had a bypass loop on the outside of the platform loop track. In true Dutch style, there was a scissors crossover between the bypass and the platform loop tracks halfway along the length of the platform to enable incoming or outgoing trains from their end of the platform to travel past berthed stock along the other end of the platform. The freight was booked to use this bypass loop track, whilst the arriving local was booked to arrive at the northern half of the platform loop. Their conflict point started from close to the junction, between the turnout where the bypass track came back into the main (platform) loop track and ended at the turnout at the junction where the single-track main line came into the double-track main line. The freight had to negotiate a crossover from the loop track into the northbound main line again, but these were all fairly high-speed (80 km/h or 50 mph) turnouts.

The traffic controller at Maastricht control centre who was in charge of the Roermond site had had a fairly quiet morning. His automatic route setting (ARS) equipment had been well able to deal with the traffic, most excitement having been generated by work going on with the overhead line equipment and dealing with a p-way crew working on a faulty turnout. At 11:42 he saw freight 69875 approaching Roermond, whilst also noticing that the local train had departed Swalmen station, the last stop before the junction. He then had to decide which of the two was going to get priority through the abovementioned conflict point. He began by disabling the ARS equipment for the freight train and manually giving it the road from signal 12 to the end of the bypass loop 304B at signal 88, which was, in fact, the planned route for the freight. If the freight driver understood this move correctly he would slow down to creep through the bypass track and would probably be given the road completely after arrival of the local train in platform loop track 303B. The signaller had to consider that the freight would be running late if it had to come to a stop and wait for the local train, which at that moment was still about 4 minutes away from its junction signal 104.

The freight driver had passed the last automatic signal before the station (P819), which was showing a yellow aspect with a white-illuminated '8' under it, instructing him to reduce speed to 80 km/h at the next signal. The next controlled signal, which was signal 12, showed him a single yellow aspect, which instructed him to reduce speed to 40 km/h (25 mph) and to expect a red aspect at the next signal. British readers, no doubt, will recognise similar functioning in a double yellow, single yellow to red aspect sequence, whilst Belgian readers should similarly recognise their single yellow, double diagonal yellow and red aspect sequence. Shortly before passing signal 12, doing approximately 25 km/h (15 mph), he could read through to his next signal, signal 88, which was showing a red aspect, but the signaller had by now come to a decision and had manually entered the next section of the road for the freight through the junction (precisely as at ***Harmelen*** in 1962). Consequently, signal 88 changed to a green aspect, the continuous operating ATBEG equipment updated to 80 km/h (50 mph) and the driver, seeing the signals change and understanding that another train must be near but that he had been given priority, expeditiously went through the 8 notches. The locomotive issued its (externally) civil growl and picked up speed rapidly with its light load.

When local train 16337 departed from Swalmen it was carrying about 45 passengers. As it approached signal 104, which was showing a red aspect to cover the passage of the freight service through the junction ahead, the senior conductor entered the driver's cab to use the PA and announce the approach of their final destination at Roermond. Much to his consternation, he noticed the approaching freight and the fact that they were going to split the junction turnout if they kept moving, realising that they must have had a SPAD. He shouted at the driver but got no reaction. The driver was sitting upright with both hands on the dashboard. The conductor initially hurried back into his train but then returned to the cab to warn the driver. He noticed a slight reaction from him, but the train was now only about 50 m (165 ft) away from the junction turnout where the freight train was approaching. Running back into his train again, the conductor warned a couple sitting to his right and left the driver alone in his cab. The impact occurred about 5 seconds later. The traffic controller at Maastricht noticed that the track circuit occupied indication at signal 104 had disappeared from his screen and also that turnout 97 had developed a fault, initially suspecting a turnout failure as had occurred elsewhere on this patch earlier that morning.

As the driver of the freight train accelerated past his Roermond starter signal 88 at green he saw the local train near the track that he was going to use, but was unable to make out whether or not it was moving. However, he was not perturbed because he was familiar with the situation of trains waiting at junction signal 104 showing up at approximately that place. It was when his ATBEG display changed from '80' to '40' as a result of the local train now occupying the same track circuit that the freight train driver realised that the local was moving on a collision course and that impact was imminent. He therefore shut off power and placed his brake controller in emergency, whilst confirming that the turnouts ahead past the approaching local were set properly for him. (That was a very recognisable characteristic – ensuring that the fault did not lie with him.) About 100 m (330 ft) before impact he saw the flashing danger signal (all red tail and white headlights illuminated) at the front of the local train turn to steady. When impact followed at 11:46, the trip recorders showed that the freight was doing 32 km/h and the local about 36 km/h (approximately 20-25 mph). The freight locomotive demolished the front DMU vehicle over a length of about 7 m (23 ft), the mass of the freight train pushing the local back some 20 m (66 ft). Externally the freight locomotive was barely damaged, but internally it suffered extensive damage.

The freight driver, physically unhurt, attempted to use his train radio telephone to make an emergency call to the traffic controller but the equipment was out of order. He then used his private GSM mobile phone to make the call before vacating his cab to see whether he could render first aid. Meanwhile nearby residents had contacted the police and emergency services. The traffic controller had attempted to use the train radio telephone to contact both drivers and had also got no response but then received the GSM call from the freight driver and instigated the emergency procedures as per instructions.

The driver of the local was found in the four-foot under the train, having died as the result of a myocardial infarction (heart attack), but despite the heavy damage to his cab he had suffered only a broken foot and abrasions, which might as easily have been caused when he fell onto the track as the floor beneath him split under the violent impact. The senior conductor, running away from the cab, was hurled back into the sharply crumpling front of his train, breaking his pelvis badly and having to remain in hospital for a considerable time after the accident. Sixteen passengers on the local train were also seriously hurt, eight critically.

Comment

As this accident happened whilst I was celebrating my birthday in nearby Oosterbeek, and in view of the reasons why I had had to relinquish my driving duties the year before, this collision made a more than usually deep impression on me. Two days after the accident I was allowed to ride a service from Nijmegen via Venlo to Roermond and back on a similar DM'90 DMU with a Dutch instructor driver as companion. It struck me how the atmosphere at Nijmegen traincrew depot, where the crew of the local train were based, closely resembled that at Bristol Temple Meads after the ***Ladbroke Grove*** collision in 1999.

My own experience of a first light heart attack was mainly that of not having much of a clue as to what was actually happening. The sensation was one of characteristically radiating pain from the chest to the shoulders, heavy perspiration in the face and upper body, feeling rather ill and experiencing very noticeable physical weakness. However, the seriousness of the situation was not readily apparent, especially when a moment later everything had cleared and you felt fine, wondering what on earth had happened. Therefore, I can easily understand how the driver, despite feeling unwell earlier on, probably thought that a headache tablet at Roermond would solve his problems. On many occasions I took medication for a headache before departure for similar reasons of feeling unwell, without suffering anything as serious as a heart attack, especially when returning from Paddington to Bristol dog-tired after a very early morning trip from, say, Swansea, Cheltenham or especially Hereford to London. It comes with the job and there is not much use in being squeamish about it.

What is noteworthy in this case is that none of the available on-board safety systems, such as ATBNG and ATBEG, vigilance, the driver safety device or radio telephone, were instrumental in avoiding this collision. The changeover from ATBNG to ATBEG before reaching junction signal 104 took away the enforced stop at red, and what surprises me is that ATBNG was not extended to this signal and the changeover to ATBEG into Roermond station installed beyond it. What surprises me even more, however, is that the otherwise often irritating need to cancel a warning from the vigilance switch about every minute or so (to prove that you are awake and functioning) failed to stop this train. The driver, dazed by his illness as he must have been, must still have reacted as required on the blue light or the buzzer requesting the vigilance cancelling action by the releasing and depressing of the 'dead man's pedal'. It was, however, probably indicative of his determination to reach the station in order to get assistance or perhaps of how ingrained the cancelling of this repetitive warning had become, as with some types of vigilance, and certainly with AWS in Britain. It does raise the question whether a system as fitted on some train types in Britain, whereby the vigilance does not ask for cancelling when the driver proves he is still alert by operating controls such as the windscreen wiper, horn, brake controller and power controller within the inter-warning period, is not more effective. In this instance a vigilance warning becomes more of a surprise and might, therefore, in the circumstances of this accident have been missed and so stopped the train. That is something for the safety people to mull over.

The other question is whether traincrew who suffer anything that feels like a heart attack or something equally disabling should be told that the only acceptable course of action in these circumstances must be to stop the train (as you would a road vehicle) and use the emergency radio facility to get assistance, or seek help from other traincrew or passengers. A stopped train is safe, covered by signals and ATP, whereas a badly controlled train on the move endangers many. This was graphically proved at *Waterfall* and later still at *Barendrecht*, when it was found that the train had already been stopped twice en route as a result of the driver not cancelling his vigilance in time. It was only a very sharp passenger train driver who took timely action on seeing unusual features ahead that prevented his train from storming into wreckage from the head-on collision between two freight trains that had happened just seconds before.

Back on the Maas line, as we approached the junction where the accident had happened, it struck me that the position of junction signal 104, with the two main-line signals in line next to it, must have presented a problem. A Dutch driver, especially a badly stressed one, will instinctively glance to the right of his track in the event of more than one signal being located next to each other, in the way that I would always look to the signal on the left first, as that is where they belong in Britain. I found that a problem on multi-track lines of other networks with right-hand travel as standard. That may have happened with the desperately ill driver in this incident, although the reversible main-line junction signal he would have looked at showed a red aspect for the same reason that signal 104 showed a red – to cover the freight train. It was never a proceed aspect on the right-hand signal that wrong-footed him, but it possibly did distract him from observing his proper signal on the left. The accident report found that it played no decisive role, but from experience I am not entirely convinced of that.

Concerning the collision performance of the rolling stock involved, the report pointed out that the DM'90 cab collapsed under impact as designed. That may be so, but annihilation of 7 m (23 ft) of carriage, encompassing virtually the entire first-class saloon, is a rather excessive amount of crumpling as it included passenger survival space. The front bogie had been moved back by about 2.5 m (8 ft). I understand, however, that the design philosophy of collision crumple zones on passenger trains has entered a new area of thinking in the meantime, and it is to be hoped that this situation has now been addressed satisfactorily.

The Class 66 locomotive and its predecessor, the Class 59, the cabs of which were designed to give the driver maximum survival chance under impact, proved again what excellent machines they are in such circumstances. At *Great Heck* in February 2001, one of the occupants of the cab of a Class 66 'Freightliner' coal train was removed alive after a collision with the derailed but still fast-moving Newcastle to King's Cross train at an impact speed rather greater than 200 km/h (125 mph). The Class 66 drivers of three head-on collisions in The Netherlands – at Roermond, *Arnhem* and *Barendrecht* – all came away from their machines alive, the freight drivers in two of those incidents being instrumental in raising the alarm with the traffic controller and thus the emergency services. At *Barendrecht* the ERS Class 66 came through the heavy head-on crash far better than the two MaK/Vossloh machines of the opposing DB Schenker train, one of which had literally crumpled off its frame. Moreover, with a centre-cab design the driver has nowhere to go in order to escape the impact.

The aftermath

After the Roermond accident the ATP situation at signal 104 was changed, and now a train is stopped in the event of a SPAD. The monitoring of physical health of Netherlands Railways drivers was tightened up considerably as far as checking the condition of the heart was concerned, although unless every driver is subjected to X-ray angioscans or MRI scanning, the true condition of the heart cannot be accurately assessed and the risk of heart attack evidently remains. The accident at Roermond had followed another serious accident (when buffer stops in a bay platform were overridden) at *Amersfoort* in The Netherlands on 5 December 1996 due to a driver being partly incapacitated by sickness. Among other severe accidents caused by operator incapacitation as a result of acute illness was the *Waterfall* incident near Sydney in January 2003, and a number of people died in a metre-gauge tramway vehicle in *Wien*, Austria, in December 1979 after the driver collapsed and the tram became a runaway, derailing and overturning on a curve. There was also a French derailment in August 1974 at *Dol-de-Bretagne* for exactly the same reason. In most cases there was no vigilance equipment coupled to the driver safety device (which is now standard equipment) and in all these cases this equipment did not stop the vehicle as it was designed to do.

After three angioscans, an MRI scan and a triple-bypass the extent of my own heart problems is well-known, but many of my colleagues work trains whilst the condition of their heart can only be guessed at. The human being is increasingly becoming the weakest link in safety-critical situations like flying planes and driving trains, buses and coaches. This has been tragically proved by the few cases in which a loaded passenger aircraft has been used as a means to commit suicide by deranged pilots, who feared for their coveted jobs after failing to meet medical standards.

***

Lahti, Finland, 28 May 2003 at 23:42

(Derailment caused by seized hot axlebox)

The incident and location

The incident involved the derailment of the last vehicle in a freight train due to a seized and disintegrated hot axlebox. One kilometre (¾ mile) of track and three turnouts were damaged behind the train and a catenary post and its base were pulled over, tearing down overhead electrification wires. Direct damage was in the order of €220,000. The incident happened on a stretch of electrified double track leading into the yard near Lahti.

The train involved

The locomotive was a Russian-built Sr1 type electric Bo'Bo' with Finnish electronics. It was hauling 36 freight wagons.

The accident log

As the Joutseno to Tampere via Riihimäki freight train passed by in the darkness, a local signaller spotted sparks flying from below the solebar of the last vehicle and used the radio to warn the driver and stop the train. On checking, the driver found the front bogie of the last vehicle derailed with one axle. On closer inspection he found that the complete axlebox assembly including much of the primary suspension had disappeared on one side, and of the axle tap that protruded from the axle into the axlebox bearing only a short stump remained. When the axlebox cover with the severed axle stump in it was eventually found back along the line, the assembly had been so hot that it had scorched the wood against which it had come to rest.

The brass outer wheel-bearing ring inside the axlebox had failed, which had caused the inner bearing ring to fail through overload, and that assembly in turn began to seize up in the axlebox. In such circumstances the increasing friction heats up the axlebox assembly progressively until the axlebox is red hot, which causes the metal of the axle end inside to glow and expand, usually resulting in the axle seizing up and not rolling. This causes both wheels to slide but at the same time exerts a very high rotating force on the seized axle tap. This may also cause the hot and weakened axle tap of a heavily loaded wagon to fail and be wrenched off (similar incidents also happening under passenger vehicles). As a result, the wheel is no longer carrying the bogie frame and the bogie usually derails on a curve through unloading weight from the damaged wheel, making it climb the rail head, particularly on sharper curves as when negotiating turnouts coming into a yard.

Comment

Hot axleboxes with seized bearings have been the primary cause of derailments throughout railway history, but since modern roller bearings took over from the old-fashioned plain bearings the incidence of hotbox problems has dropped markedly. Nowadays railwaymen may spend their entire working life without ever having to deal with one, yet they do occasionally still occur and are then as dangerous as ever. Following the demise of the people with the skill and knowledge of how to spot them and deal with them (hotboxes usually smoke and stink because of the boiling or burning bearing lubricant, sometimes they literally glow red. On other occasions they emit squealing or grinding noises that are guaranteed to upset the neighbourhood through which they are passing), the answer is the hot axlebox detector. That equipment measures the average temperature of the passing axleboxes and issues a detailed warning in the traffic control centre if a hotter than normal axlebox is noticed, enabling the controller to stop the train at a red signal for inspection.

In reality, dragging brakes on wagons or faults in the system, and occasionally the hot firebox of a steam locomotive, causes the majority of alarms. If a driver is told that he has a hotbox on his train, he must be at the indicated place within 10 minutes after coming to a stop and check the relevant box and those around it. That is often achieved by first cautiously putting the back of the hand close to it and, if the box is obviously hotter than normal, spitting on it and watching what happens. Nowadays there are wax sticks with a pre-set melting temperature that must be used on the inner surfaces of the axlebox. If the wax melts then the axlebox is too hot and may be on its way to seizing up. Before moving off, the axle must first be checked to see whether it still rotates and the train must then be worked off the main line into a station or siding at a much reduced speed and then taken out of service. The wheels of the offending axle will then be put on wheelskates and the train will be taken slowly to a place of repair or the bogie will be exchanged locally for a fresh one.

***

Schrozberg, Germany, 11 June 2003 at 12:03

(Head-on collision caused by signalling errors)

The accident

This was a head-on collision on a single line with a very minor signal aspect failure as its root cause but with the subsequent questionable handling of the situation by one of the signallers as the main contributor to the accident. There were six fatalities (both train drivers, three children and one accompanying adult passenger) and 24 people sustained injuries.

The line

The accident occurred along the non-electrified single-track main line at km 30.360 between Niederstetten and Schrozberg, in the area between the distant and the home signal of Schrozberg station, on line 4953 from Crailsheim to Lauda via Bad Mergentheim. The location is inside a triangle with the cities of Würzburg, Nürnberg and Heilbronn at its three corners. The accident occurred on a sharp curve, the view ahead being inhibited by forest and dense vegetation close to the track on both sides of the line, although the line itself at the location in question was on an embankment 8 m (26 ft) high. Line speed was 80-90 km/h (50-55 mph).

Both Schrozberg and Niederstetten stations were laid out as a single through (main) line with a passing loop, both lines serving platform faces, with speeds into and out of the loops varying from 60 km/h (35 mph) to 40 km/h (25 mph). Niederstetten was fitted with colour-light signals of the Hp/Vr 2-aspect type and was worked with a Siemens NX electro-magnetic relay-operated panel of 1975 vintage. All signals had Zs 1 Ersatzsignale fitted (as at ***Brühl***). Close to Niederstetten station was a level crossing at km 38.058. The signaller manually controlled this level crossing and, as in Britain, it was incorporated in the signal interlocking. Schrozberg, however, takes us back to the mechanical era, with semaphore signals and a muscle-power-operated locking frame (1955 vintage) with turnout and signal levers. As in Britain, the home signals normally defined the station limits – the signalling control span for the signaller – but the starting signals, in contrast to British practice, had their own distant signal near the home signals. All stop signals were also fitted with Zs 1 Ersatzsignale.

The single-track line was worked as a fully signalled 'Streckenblock' (block section) between the two signal boxes involved. Whereas in Britain communication between the signal boxes was handled with various bell codes and block status indicators, in Germany offering trains was done by means of prescribed messages by special telephone, in both cases the times and events of train traffic management having to be entered in the train service book, as in Britain. A window in the block instruments changing to red from white indicates reservation of the block section for a particular service after acceptance, and another window indicates the occupation of the section ('Train on Line'). Both changed back to white if the train was off the relevant track circuits and had cleared the main line, after which this system automatically gave 'Line Clear' ('Rückblocken'). In the event of failure of this automatic unblocking of the single-line block section (as in Britain or anywhere else in the world) it was important that the signaller ensured that the tail-light of the train was visible in the passing loop before unblocking the road, to ensure that no vehicles had inadvertently been left behind in section. However, in this particular accident this unblocking, which should have occurred automatically when the line track circuits were cleared and a treadle (track-switch) was operated by the train, failed due to a faulty signal aspect and had to be bypassed to enable trains to depart. Unfortunately, the signaller assessed the situation incorrectly and thereafter made further errors that finally led to the accident.

As far as signalling at Niederstetten station was concerned, it is of importance to explain the following. Because the main line through the station was laid in a curve that impeded sight of the starting signal in the direction of Schrozberg, a so-called 'Vorsignal-Wiederholungssignal' (literally a distant signal repeater) had been used on the approach to the starting signal. In fact, such a signal was a distant signal aspect repeating the aspect of the actual distant signal but to which a small white light had been added to indicate that it was closer to the stop signal than the braking distance from line speed. (In Britain, and on many other networks, a banner repeater type of signal would perform this function.) The root cause of the accident, therefore, had to be the intermittent failure of either the yellow distant signal lights or the small white light of its repeating signal, possibly through contact problems. This in turn caused the track circuit to continue to show as occupied and therefore not releasing the line when the train actuated a treadle to enable renewed acceptance of trains. This was, in fact, the means designed to indicate signal aspect failures to the signaller. Which therefore meant that he should first of all have recognised this repeater signal failure from the symptom of route release failure and then have taken appropriate action to remedy the situation, working his way around it until repair had taken place.

One way to deal with this problem was through use of the Zs 1 Ersatzsignal, the small triangle of white lights under the signal head mounted on the post. Working such an Ersatzsignal, potentially as dangerous as the British failed signalling release key, was subject to strict regulations for use by the signaller and the driver, and its use was registered on a recorder that had to be backed up by a description of the problem in a special book. It is also important to bear in mind that a signaller-controlled level crossing was incorporated on this stretch of line, a failure of which would have caused quite similar non-release effects as far as the signaller was concerned but would have been indicated differently on the panel, near the buttons that controlled the level crossing closure. That was a matter of knowledge, but it is what wrong-footed the signaller when attempting to assess why the route had not unblocked.

German absolute block signalling

As mentioned above, trains in Germany are regulated not with bell codes as in Britain but with telephone messages – 'Zugmeldungen' (train reports) – through a dedicated telephone line using prescribed text. The initial messages have to be repeated with 'Ich wiederhole' (I repeat) and the messages end with the expression 'Richtig' (correct). These train reports are repeated on loudspeakers in, for example, crossing keeper's huts to keep others aware of what is happening (see ***Genthin***). All these train reports must be entered as time-stamped messages in the signal box train report book, as happens all over the world.

Such train reports concern 'Anbieten' (to offer a train), 'Annehmen' (to accept a train) and 'Abmelden' (to report 'Train on Line' from 5 minutes before departure time). The receiving signaller has to report clearance of the line on arrival of that train in his station, by so-called 'Rückblocken' (giving 'Line Clear'). In cases of more modern block section signalled lines this is automatically done by train detection equipment on leaving the block section (track circuits, treadles, axle counters) and is registered in the departure signal box as well. In the event that signalling equipment failure hinders automatic clearance, reminder collars or magnets indicating the need to adhere to the manual line clearance procedure must be fitted to the levers or over the panel buttons.

Normally reporting 'Line Clear' through this manual procedure is indicated as the 'Räumungsprüfung' (train clearance check) to indicate that the complete train has cleared the section. If clearance systems have failed, however, this clearance must be reported back to the originating signaller by telephone, using the tail-lamp or tailboard of the arriving train as proof that the entire train has cleared the section. With such systems the tail indications of a train are still very important. The locked signalling (as here) can then be unlocked using the 'Fahrstrassenhilfstaste', one of the release key types of instruments the use of which is strictly regulated and logged. As in Britain, the route is set and the signals are cleared through working levers in a locking frame. However, after turnouts and shunt signals have been set appropriately a route-proving lever is moved, and if that action as well as visual ascertaining by the signaller does not find occupied track circuits in the route to be set up then it locks the turnout levers. The necessary main signals can then be cleared, which in turn locks the route-proving lever. The passage of the train through section normally clears the locks. During offering and accepting a train the route is secured after agreement through 'Erlaubnisabhängigkeit' (acceptance/permission) which is indicated on an NX panel with illuminated direction arrows and locks all signals except those needed to enable the train to depart. After offering and acceptance of a train between two signal boxes, the signaller claims the route of a departing train along a single track by 'Vorblocken' (pre-blocking). This normally automatically locks all signals at either end as soon as the starter signal comes back to danger after the departing train.

At Niederstetten there was another fault on this system that made manual pre-blocking impossible in the event of certain other failures.

When a train arrives at the next station from its trip through a single-track section it normally releases the route through section automatically, at the moment of 'Rückblocken'. If, however, this automatic line clearance fails, the signaller has to ensure that his section is cleared and obtain a report from the next signaller that the train has arrived complete at his station ('Rückmelden'). This again is similar to obtaining a 'Line Clear' indication in Britain. If there is any doubt about this section clearance then a train clearance check has to be done before reporting 'Line Clear', or a train has to be sent through section in the same direction as the previous train, travelling at sighting speed of 40 km/h (25 mph) and able to stop for any obstruction, in order to ascertain that the line is indeed clear. A similar local test to ensure that a required path through the station layout is clear is called an 'Abschnittsprüfung', usually undertaken by looking out of the window or by getting out and taking a look at the actual location. Had the Niederstetten signaller gone out and looked at his signals, he might have noticed the problem with the small white light of the distant signal repeater, although its repeater on his panel should also have revealed the problem.

These three issues are indicated on block instruments with three windows per instrument that are white when clear and red when blocked. Obviously, when blocked this instrument also locks the signalling. One of the available manual overrides that can be used to keep trains moving during failures is again the Zs 1 Ersatzsignal, which is actually meant to get a train past a signal displaying a red aspect without the signaller having to come out of the box to hand the driver a written pass-signal order. Any out-of-the-ordinary actions, such as use of the Ersatzsignal or manual unblocking, are registered on counters and the numbers of these counters must be entered in the repair book, accompanied by a written explanation as to why the procedure was used.

Rolling stock involved

Regional service RE19534 was made up of a DHMU consisting of powered vehicle No 628 285 (trailing), and driving trailer (remote control car) No 928 285 (leading), pretty much in the style of the 2'B'+2'2' DEMU involved in the ***Åsta*** accident in Norway. In fact, both types of diesel unit shared much of their technical background and were built in Germany, introduction of these units having started in 1987. Both DHMU vehicles, which were heavily damaged in the Schrozberg accident, had a Daimler-Benz/OM 550 hp (410 kW) diesel prime mover that drove wheels through a Voith gearbox and Gmeinder final drive, a transmission set-up that found its way all over Europe and the rest of the world for similar units. The German units had a permitted top speed of 120 km/h (75 mph).

Regional service RE19533 comprised 80-tonne 2,500-hp (1,865-kW) B'B' diesel-hydraulic locomotive No 218 285 with four passenger carriages, the one at the far end from the locomotive having a driver's cab (remote control vehicle). (Note that this accident constituted a very rare case where the locomotive and the DHMU vehicles of the colliding trains all had the same order number, 285, within their respective class numbers.) The locomotive was a member of the formerly large (900-plus) group of single-engine second-generation West German B'B' locomotives, Class 215-218, designed and constructed during the mid-1960s and 1970s as a follow-on from the well-known 1950s-designed V200 twin-engine type that in Britain spawned the BR Western Region 'Warship' and 'Western' classes of diesel-hydraulic locomotives and in the US the German-built Southern Pacific/Rio Grande diesel-hydraulic C'C' units of the mid-1960s. The locomotive overturned on impact and rolled down the embankment following the accident.

The reason for the disappearance from the German railway scene of the Class 215-218 machines (apart from their age, all having exceeded their designed 30-year lifespan by a considerable margin) is because their normal work in passenger service has been taken over by various types of more economical diesel multiple-units, and on freight duties by newer and more economical types of diesel-powered as well as electric locomotives. Nevertheless, quite a few of the more powerful Class 218 units with electric train heating are still working on freight, express and regional train services.

The accident log

A reconstruction of events from the signal box train reporting books, the signalling fault and failure books and the train OTDRs has brought the following to light on the day preceding the accident. After train 19543 departed from Niederstetten in the direction of Schrozberg, the departure route from Niederstetten station (route P2) did not clear automatically as it should have done. When the train arrived at Schrozberg the signaller immediately offered train 19540 to Niederstetten, which the signaller there accepted despite his problems with the still locked departure route for the previous train. As the signalling equipment did not allow him to give the required acceptance of train 19540 to the Schrozberg signaller, the manual line clearance procedure was instigated to allow trains to be moved by other means. The arrival of train 19543 at Schrozberg had been dealt with in this way at 18:10 according to the books. The release key at Niederstetten had as yet not then been used to clear the still locked route there. Due to the failure of the system, acceptance could not be given and the dispatching signaller at Schrozberg could not clear his departure signal. He therefore gave train 19540 permission to pass the departure signal at danger with the Zs 1 Ersatzsignal and the train departed at 18:11 from Schrozberg. Whilst this train was still on its way, at 18:16 the signaller at Niederstetten used his release key to unlock his signalling and was thus able to receive train 19540 under normal signal indications at Niederstetten. These actions were recorded in the train register and the signalling faults and failure books, but there was no apparent understanding of the cause of the fault at Niederstetten. The next train from Schrozberg to Niederstetten was worked in the same manner under referral to the previous problems and the way they were dealt with. It is, however, unclear how the level crossing en route was secured and whether the signalling at Niederstetten still suffered from the fault.

The following day, 11 June 2003, was characterised by a heat wave, the temperature peaking at 35°C. At 05:30 the fault with the distant signal repeater aspect had apparently been noticed, although the failed signal was indicated in the fault book as the starter signal, not the repeater. The subsequent accident report referred to this as 'in all likelihood a writing error' but it does indicate that the signaller's familiarity with the system he worked had unacceptable limitations. The signallers then started the manual protocol to work trains through section. Three trains – freight train 52245 and locals 19533 and 19534 – had to be worked this way. After the freight train (a Russian-built Co'Co' diesel electric locomotive No 232 677, nicknamed a 'Ludmila', with one wagon) had departed from Niederstetten at 11:34 with a 13-minute delay (which was nothing unusual), the departure route did not release because of the faulty Schrozberg distant signal repeater. The signaller should have noticed this from the route locking indication and from the miniature signal repeater on the panel, but as a result of this omission he failed to place reminder markers on the panel, carry out a local clearance test by looking out of his window to see whether the track had been cleared, and then work the release key for further departures with the use of signals as normal. The freight train travelled through section and arrived at Schrozberg at about 11:44. It is rather strange that it stood for 2 minutes at the Schrozberg home signal according to its OTDR but was still reported as having travelled the section in 10 minutes and arriving right time, whereas in fact 13-14 minutes were necessary. The accident report gave no explanation for this, other than that reporting of 2-minute delays did not have to be entered.

Niederstetten then offered the slightly late-running local 19533 to Schrozberg at 11:49 and it was accepted also at 11:49, both signal boxes entering 11:54 as the departure time. Niederstetten cleared the route for this train, arriving from Laudenbach, into Platform 2 (through road), although it was booked to arrive into Platform 1 (loop). This was done either to keep the arrival and departure speeds higher, as the maximum speed through the turnouts to Platform 1 would have been 40 km/h (25 mph), or because the system did not want to set the route into Platform 1 as the departure route from there was still locked. This alteration was acceptable according to the rulebook, however.

The signaller then found that he was unable to set the route for train 19533 to Schrozberg as the system was still locked following the departure of the freight train. This indicated that he still had no real idea what was causing the problem. Thinking that the level crossing was faulty, he then tried to close it in order to deal with the fault. However, the level crossing would not co-operate until the signaller had used yet another system to close it, the 'Dauereinschalttaste' (long-period closure switch), used, for example, when a shunting convoy needed to be on the crossing for an extended period. This switch was not interlocked with the signalling, and the level crossing closed. The signaller should then have handed a written order to the driver of train 19533 to stop at the level crossing and ensure that it was closed, but he failed to do this. Train 19533 arrived at Platform 2 at 11:54.

As the signalling had failed, the signallers then began the line clearance procedure because the automatic release by arriving trains was no longer working. The start of the procedure and the arrival of freight train 52245 at Schrozberg was reported at 11:54 and entered in the books. The application of reminder devices such as the signal lever collars at Schrozberg and magnetic collars on the NX panel at Niederstetten to prevent unsafe actions was not carried out in either box.

In order to allow train 19533 to depart, the signaller at Niederstetten pulled off Ersatzsignal VS1 at the starting signal of Platform 2. At 11:55, the train passed starting signal P2 at red with the authority of the Ersatzsignal, about 6 minutes late. The line was still not released after this train, but more to the point, the train's presence was not reported on the single line 'Vorblock' due to the existing situation with the signalling. It cannot be proven that the signaller at Niederstetten tried to do this manually, but it is known that in the Niederstetten panel the manual pre-blocking system was not working properly. At 11:56 he tried to establish the change of claim on the path between Niederstetten and Schrozberg but this could not be done from the panel as the path was still locked. The signaller finally recognised the still existing locking at 11:57 and worked the release key, which released the departure path along starting signal P2. More importantly, it also released the 'Erlaubnis' instrument, the claim of the pathway changing from Niederstetten to Schrozberg signal box, where the relevant indicator changed from red to white. The section block system was then ready to accept a train from Schrozberg to Niederstetten, despite a train from Niederstetten to Schrozberg still being on its way in section.

Train 19534 arrived at Schrozberg's Platform 2 one minute later. The signaller at Schrozberg offered this train to Niederstetten at 11:59, despite not having reported train 19533 as having arrived. Niederstetten therefore asked him where it was, which was followed by the 'Line Clear' report for train 19533. There was, however, no entry in the book that he reported the line as clear whilst the train had not yet arrived. In the accident report it was assumed that either the signaller at Schrozberg thought that train 19534 in his platform was 19533 from Niederstetten (but the train manager of 19534 also reported during the inquiry that he had been told that he would be departing on an Ersatzsignal aspect, which more or less rules out this error with train identities) or that train 19533 had departed onwards to Blaufelden already. As far as an on-time departure was concerned that certainly should have been the case, but train 19533 was running 6 minutes late at that time and was still on its way to Schrozberg. The other possible explanation for this terrible chain of errors is that the Schrozberg signaller thought that train 19533 was still at Niederstetten and that the change of his 'Erlaubnis' indicator was due to the Niederstetten signaller clearing the road for him. In any event, the signaller at Niederstetten failed to notice that train 19533 was being reported off the single-line section within 5 minutes whereas the timetabled run was 9 minutes.

The signaller at Schrozberg cleared Ersatzsignal N2 at 12:01. This again was a remarkable way of handling the situation, as he could have worked his main signal following clearance of his block indication. As a result, train 19534 departed immediately. Sadly, the signaller had not had the gumption to wait for the crossing with oncoming train 19533, and the traincrews did not have to be informed of altered crossings along the single line. Had that been obligatory, perhaps the signaller would at last have come to his senses and seen what a trap he was inadvertently setting. Yet at 12:01 he called Niederstetten to cross out train 19533 in the book as that train had not run. It was 2 minutes later, on a sharp curve surrounded by dense woodland, that the head-on collision took place.

Comment

Anyone who has worked in safety-critical positions on the railway and has ever been in a situation where training and experience have fallen well short of what was required to keep the trains rolling safely will recognise what the Niederstetten signaller, and later the Schrozberg signaller, must have gone through during this chain of events. The equipment had played up and neither of them had a clue as to what was going on or what safe options were available to them in order to correct the situation. Their insight and knowledge evaporated by degrees even as their confusion increased.

The truth is that until the 1980s, on virtually all networks, fault and failure indications of signalling and traction problems were never indicated directly, such as with a red fault light with a clearly understandable name under it, but indirectly by means of a single general fault light coming on in response to a sudden failure. The BR Southern EPB units were full of those sorts of puzzling faults and failures (but they kept running), as most equipment was until the 1990s, when increasing introduction of electronics in signalling centres and on trains made it possible to have screens with proper indications and suggestions to deal with the problem. Then things went to the opposite extreme and drivers were harassed with indications of empty toilet water tanks or failed water boilers in the buffet car etc. However, even today, the celebrated HST high-speed diesel electric sets on Britain's railways have just one nondescript traction fault light in the cab, and the driver has to figure out what the fault is by looking at his gauges in order to identify the problem and what can be done about it, which mostly comes with experience. In the case of these two signallers that experience was lacking, and the only way to try and instil some experience before such disasters as the one described here take place is the simulator. There is no other way, since many faults and failures occur very rarely indeed and cannot be taught before the real thing threatens signalling integrity or train safety as it did in this incident. The accident report did, in fact, point to this. The signaller at Niederstetten had been placed in an impossible situation, whereby his lack of training and experience to deal with extraordinary situations had made it difficult for him to analyse what was wrong and deal with it appropriately, albeit he did not help himself by his failure to use what was available to him, such as the reporting reminder collars.

On the other hand, it is questionable whether a single, second-ranking signal light bulb failure should have interfered with traffic control processing in the way it did, and whether a clearer fault indication in the signal box should have been provided, which would doubtless have been possible with the electronics available even then. The second fault, with the 'Vorblocken' manual route claim system, had existed ever since the panel was installed and was well known. The fact that no one had taken the trouble to eradicate it systematically from all these installations makes the German railways of the period just as guilty as the unfortunate signaller at Schrozberg. Incidentally, he ticked all the boxes as the sort of person born to carry blame. He was not particularly bright and had only just managed to scrape through his examinations. Moreover, he was also rather poorly trained as far as the time allocated for him to learn the job and gain experience was concerned, having spent only 40 days in Blaufelden box and 22 days at Schrozberg working trains on his own before the hand of fate took him by the throat. Unfortunately, the railway worldwide tends to have somewhat of a problem with such people, who just about make the grade and then need constant monitoring to ensure safe working.

***

Hammaslahti/Tikkala, Finland, 16 July 2003 at 15:48

(Derailment following buckled track)

The incident

A track buckle due to hot weather caused the derailment of 14 freight wagons between Hammaslahti and Tikkala in eastern Finland. The line was a single-track, electrified secondary line running through marshland.

The line

Give or take a few millimetres, Finland has the same broad-gauge track that Russian Railways have been laid with, so exchange of vehicles is unimpeded. It is only in certain port areas in Finland that standard-gauge vehicles can be received off railway ferries, the rail vehicles then being lifted and re-bogied with broad-gauge bogies that are of the same manufacture as the standard-gauge ones. Normally the variation of internationally accepted bogies is limited to the French Y25 or German standard bogies, unless bilateral agreements exist for other types between operating networks. The latter allows Russian 3-piece bogies into Finland, for instance, as US and other 3-piece type of bogies are not normally accepted in western Europe due to their high unsprung weight with the potential to cause excessive track wear. Only Britain accepts them without restrictions and, as a result, is one of the very few western European nations where US-designed freight wagon bogies can be seen in daily service. An additional issue is the fact that Russian rolling stock is fitted with the Willison-type SA3 automatic coupler whereas normal Finnish rolling stock is fitted with the standard European buffers and screw-link coupler, but all sorts of ways have been found to couple both together. In fact, many Finnish freight locomotives have been fitted with the SA3 auto coupler.

The cause of the accident

Steel rail expands when heating up in strong summer sunlight and contracts when cooling down, and even more so when exposed to freezing temperatures in winter. The summer sun can cause the hot and over-expanding track to buckle sideways or even upwards. Track expansion buckles in the summer (and frost contraction breakages in the winter) have caused many serious accidents all over the world, especially when fast-running passenger trains have been involved.

The accident log

Freight train T 7526, made up of three Dv12 light diesel-hydraulic all-purpose locomotives of VR Finnish Railways and 45 wagons of both Russian and Finnish ownership, was on its way from Joensuu and Uimaharju to Niirala, rolling along at 73 km/h (45 mph). Suddenly the train swayed wildly and the 14 Russian vehicles in mid-train positions left the track, tearing up 10 overhead line equipment posts, 400 m (1,310 ft) of track and the wooden deck of a user-worked crossing and becoming seriously damaged themselves in the process.

Although the track buckle was the direct cause of the derailment, the track was not in a very good state, part of the problem being that the subsoil under the track was marshy and soft. This meant that it was less able to take the stresses of expanding track and hold it in place as a result of the pounding it had taken from heavily loaded freight trains. Other drivers had reported the same area of rough riding that later led to buckling of track to the signallers, who had passed on the reports to the track maintenance people, but no action had been taken. Direct damage was in the order of €400,000.

Comment

Whilst no further mishap occurred in this case, apart from damage to vehicles and track, a derailment can potentially cause complete loss of control over a train and may therefore lead to very serious accidents on double- or multiple-track railway lines due to intrusion into the loading gauge of adjacent lines, especially when high speeds are involved. Unfortunately, such derailments also have a habit of occurring in places where the train is crossing a viaduct or is inside a tunnel.

Russian vehicles invariably run on the 3-piece bogies that are so well known in the Americas and in Britain for example. As a result of the track buckle in this case, the relative ease of vertical movement of the axleboxes at the outer ends of their sprung bogie side frames versus stiff bogie rotation may have been unable to follow the quick changes in horizontal direction of the buckle but did enable the wheel to climb over the rail head instead and so derailed the vehicles. What caused the stiff bogie rotation? An issue that Finnish Railways had with them is that the swivel action of the Russian bogie frames was too often inhibited by lack of greasing of the pivot between the underframe of the vehicle and the bogie frames. In fact, pictures were included in the reports of the incident that showed rusty and scaling equipment filled with water where there should have been plenty of grease. The fact that the track was also substandard only made it that much easier for the rails to give way under the assault of a heavily loaded vehicle with non-turning bogies, and severe rail head wear and derailment were certainties from that moment onwards.

Other similar incidents in Finland

An interesting incident occurred in *Jyväskylä* yard on 14 May 2003, when a Dv12 DH locomotive and 19 Russian wagons were involved in a derailment due to a wagon spreading the track and thus causing its own derailment plus that of the vehicles behind it. Apart from the greaseless bogie joints, the wheel treads of the offending vehicle were so badly worn that they had hollowed out false flanges at the opposite side of the in turn badly worn real flanges. That this vehicle had got as far as it had was a small miracle, derailment of vehicles in that condition on turnouts etc almost being guaranteed.

At *Harjavalta* on 7 July 2004, two Russian container carrier wagons transporting sulphur dioxide tanks derailed on a sharp curve while being propelled during a shunt move. The track on the curve had started to stretch and kink, forcing the wheels to '50 pence' around the curve, a move which the stiffly turning bogies could not perform. Some 50 m (165 ft) of track in the yard were damaged and wheelsets had to be replaced.

At Eskola on 27 April 2005, the last vehicle of a train of 29 wagons carrying Taconite pellets hauled by three Dv12 locomotives derailed due to greaseless bogies. Obviously, the track was poor as well. This incident led the Finnish rail accident investigation authorities to call for the lifting of incoming Russian wagons at the border in order to check the grease levels of the bogie pivots. There was also trouble the next morning at *Heinävesi*, when five out of 19 Russian wagons arriving behind a Dv12 derailed in the yard.

The last incident happened in *Heikkilä* yard near Turku on 8 February 2008 at 09:53, when a derailment took place involving a Dv12 DH locomotive with 13 Russian wagons. Three Russian tank vehicles and two vans were also involved, tearing up 70 m (230 ft) of track. Poor track (rails fastened with rail spikes in wooden sleepers) was the other part of the problem, in which the ruinous effect of stiffly rotating bogies spreading the track on a curve was exacerbated by some of the vehicles being overladen.

***

Holzdorf, Germany, 28 September 2003 at approx 13:00

(Head-on collision following procedural errors)

Introduction

This fatal accident was in many ways a carbon copy of the accident at ***Winsum*** in The Netherlands. It involved exactly the same kind of traffic, the same breed of rolling stock and the same kind of 'dark territory' radio-based dispatcher traffic control system without signals. Even the red and green lines to indicate occupation and clearance of the single-line sections in the traffic controller's logbook were in use. And the reason why the accident happened is exactly the same – a driver departed without awaiting permission from the traffic controller – although in this case the services were worked without a conductor and the reason why the driver departed in the way he did appears to become somewhat clearer. The thought-provoking issue, however, is that the accident in The Netherlands happened 23 years before this one and clearly exposed the dangers associated with this means of controlling train traffic, whilst in an Australian accident report, their version of this type of traffic control with so-called radio-issued warrants was also noted as being not very secure. In The Netherlands this situation led to a thorough revision of the single-line signalling system, but it is surprising how little rail networks appear to learn from problems similar to their own on networks elsewhere.

The location, railway line and signalling

The line involved was the 25-km (16-mile) single-track non-electrified branch from Weimar Hauptbahnhof via Weimar Berkaer Bahnhof, Holzdorf and Bad Berka passing loops to Kranichfeld, line 6681. The accident occurred between Holzdorf and Weimar Berkaer station, close to Holzdorf, on a curve under a motorway viaduct. The drivers were unable to see each other's trains virtually until the moment they collided. The line is comparable to the one between Groningen and Roodeschool in the ***Winsum*** accident. Apart from the signalled curve between the two Weimar stations with two platforms and where the traffic controller was located, there was the single-line section to Holzdorf with its passing loop, the single-line section to Bad Berka with a passing loop, and then the single line to Kranichfeld single-platform terminus. There were also a few platform halts along the single-line section between Weimar Berkaer Bahnhof station and Holzdorf.

The turnouts at Holzdorf and Bad Berka were of the sprung type that had to be opened by the departing train but would send the arriving train to its proper track in the passing loop (in this case the right-hand one). Of course, these turnouts had indicators to show their position, and there were fouling markers to indicate that an arriving train had cleared the single line completely as soon as its tail-lights had passed them into the loop. An additional German feature was signal So 5, the 'Trapeztafel' (trapezoidal panel) sign, standing where the home signal for the branch line stations with a loop would be. It indicated that the train must come to a stop for operational reasons, standing clear in the loop (i.e. to call the traffic controller and request permission to continue the trip). It replaced the home signal and the starting signal with their distant signals.

The line was operated under no-signal radio control with an open channel radio system (precisely as at ***Winsum***), the German name for this method of operation being 'Zugleitbetrieb'. The single-line sections between stations were called 'Zugleitstrecke' and a loop a 'Zuglaufstelle', which could be compared to the US expression 'control point'. Messages used for the exchange of arrival and permission to depart information were fully prescribed and use of 'telephone banter' was not permitted. Records of the telephone calls were available but these were not time-stamped.

In the cab of any German or Austrian traction vehicle there ought to be the 'Buchfahrplan' and the 'Fernsprechbuch'. The 'Buchfahrplan' (timetable book) is a mix of timetables for the service worked, a distance chart, route instructions, driving instructions and local information such as train meeting points and instructions with regard to signalling. It can be quite prescriptive. For example, drivers may find instructions concerning where to power up to what speed, and where to switch off and coast, assuming that the train is running on time. The 'Fernsprechbuch' (telephone book) contains the driver's record of exchanged radio messages and instructions, and has to be time-stamped to ensure that the time that instructions are received can be compared with a similar record in the traffic controller's office, the 'Zugmeldebuch' (train reporting book). It is, in fact, the sort of record you would expect to find in any well-run local signal box. It is imperative for safe operation that the driver and the traffic controller enter the radio exchanges only when the exchange has actually taken place. For that reason, the telephone book has to be in the cab with the driver.

Trains had been fitted with the OTDR at the time of the accident, which was quite different to the earlier accident in The Netherlands. Together with the call-recording facility, this allowed comparison between what was written in the telephone book and the train reporting book and what the OTDR showed. As far as protocol was concerned, on this line the first arriving service in the loop reported its own arrival, then reported the arrival and clearance of the fouling markers of the opposite train and then requested permission to continue. The second train could hear this exchange, and its driver reported and then also requested permission to depart. Due to the nature of the line and the frequency of the trains, the timetable was the same virtually throughout the day. A train arriving at Holzdorf would generally have to await the arrival of the oncoming train to clear the single line ahead unless a train service had been cancelled for some reason, although late services would keep running without crossing at Holzdorf as the timetable was thinned out at that time of the day. The existence of the unchanging daytime routine, however, helps to explain why the collision occurred.

Rolling stock involved

The accident involved a Class 641 single B'2' DMU vehicle, a representative of the Siemens-Duewag 'Desiro' family, various types of which are in service throughout Europe, electrically as well as diesel powered. Within Germany there are 2-car articulated B'2'B' Class 642 units of the same family in service, but the line discussed in this report was operated with the single-car version only. The prime mover of the diesel version is an MTU 370 hp (275 kW) engine that drives the axles through a 5-speed automatic hydrodynamic gearbox, Cardan drive shaft and final drive. This modern drive system caters for automatically blended hydrodynamic braking in the gearbox and disc friction braking with electro-pneumatic (EP) brakes. The bodyshell is constructed from aluminium. Train 26506 from Kranichfeld to Weimar Berkaer was worked by unit 641 024-5, and train 26507 from Weimar Berkaer to Kranichfeld by unit 641 030-2. Outbound services from Weimar had odd train service numbers (headcodes), the incoming units to Weimar carrying even numbers.

The accident log

During the investigation it was found that the OTDRs of both units showed a time difference of 100 seconds. This was normal and can be found on virtually all trains in the world, requiring adjustment of the noted times to the common time as indicated in traffic control centres and on stations, which is not normally a very difficult process. For example, when a download is made, the download time has to be recorded as accurately as possible to enable proper timing of the recordings; many lineside features such as timed track circuits and hot axlebox detectors time-stamp their record of the passing trains and on the OTDRs the locations of places where the driver cancelled warnings etc are logged. It is therefore easy to make up an accurate log of the train trip. In this case, due to the absence of any such recording features, or features that imprint on the recorder except for a few speed boards with PZB vigilance function, there were very few issues that made identification of features along the line possible, for which reason the OTDR times of both units were kept as they were recorded.

Unit 641 024-5 had arrived at Kranichfeld as train 26505. From there the driver reported as follows:

> Dvr: 'Weimar B, come in.'
> *TrC: 'Yes, traffic controller Weimar B.'*
> Dvr: '26505 at Kranichfeld.'

Note that the driver did not use the obligatory prefix 'Zuglaufstelle' (control point) to start his message, to identify the nature of the call. The train number changed to 26506 for the return trip, but something rather odd occurred now. The traffic controller repeated the arrival message but then gave permission to start, as if the driver had asked for it, by repeating the driver's virtual request, but the driver had not requested clearance for departure. The driver similarly reacted as if the traffic controller had cleared him to run to Holzdorf.

TrC: 'I repeat, 26505 at Kranichfeld. Can 26506 run to Holzdorf?'
Dvr: 'I repeat, 26506 may run to Holzdorf.'
TrC: 'That's correct.'
Dvr: 'Out.'
TrC: 'Out.'

To save radio time, only fractions of the prescribed radio message exchange are actually being used, but they do adhere to protocol language for those fractions that are used. It should be looked at as a rebus text with lines omitted, to be filled in by the reader. Nevertheless, both knew what they were talking about, and it was not the reason for the accident. It just points to 'streamlining' of the procedures.

Train 26506 departed from Kranichfeld and was recorded by its OTDR as having stopped in the loop at Holzdorf between 12:57:32 and 12:59:59. But, as at ***Winsum*** two decades earlier, the train then departed without waiting for the clearly indicated meet with oncoming train 26507 and without asking permission (equivalent to a SPAD). It accelerated to 52 km/h (32 mph), following which the speed dropped to 40 km/h (25 mph), after which the OTDR registered a cancelling for the speed increase warning board to 50 km/h (30 mph) at km 9.100. At 12:58:52, around km 8.610 and doing 51 km/h (32 mph), there is a sudden and very substantial drop in the compressed air brake-pipe pressure, indicating that the driver had applied the emergency brake. Some 40 m (130 ft) after that there was a sudden standstill, the speed at the time of the collision being 44 km/h (27 mph).

Train 26507 departed with recorded permission from Weimar Berkaer station at 12:52, accelerated to 53 km/h (33 mph), passed through Obergrunstedt request halt at 26 km/h (16 mph), accelerated again to 53 km/h and called at Nohra request halt from 12:56:49 to 12:57:21 before accelerating once more to 53 km/h on its way towards Holzdorf to cross train 26506 there. At 13:00:27, at km 8.520 and doing 50 km/h (30 mph), there was the same sudden venting of the brake-pipe pressure as in the other unit, indicating that the driver had also applied the emergency brake. Some 40 m later the collision occurred at a speed of 39 km/h (24 mph).

One person died, 14 were seriously injured and 13 others sustained less serious injuries. Both drivers survived and were able to make radio telephone calls to the traffic controller from their damaged units after the accident. The call from the outbound driver of train 26507 went as follows:

Dvr: [Indistinct]
TrC: 'Yes?'
Dvr: 'We have collided, near the motorway viaduct!'
TrC: 'Near the motorway viaduct?'
Dvr: 'Yes, we have collided.'
TrC: 'That is the 6 and the 7?' [26506 and 26507]
Dvr: 'Yes.'
TrC: 'How is that possible?'
Dvr: 'I don't know. We have collided near the motorway viaduct and we need medical assistance. There are a lot of injured people, can't say yet how many.'
TrC: 'It's okay, it's okay.'
Dvr: 'I'm telling you!'
TrC: 'It's okay.'

The call from the inbound driver of train 26506 went as follows, with the sound of the traffic controller making a phone call:

Dvr: 'Weimar B, come in, [first name TrC].'
TrC: 'Yes, [first name Dvr].'
Dvr: 'We are right near the motorway, we need a number of ambulances quickly.'
TrC: 'Yes, I asked for those just now.'
Dvr: 'And the units, both badly damaged.'
TrC: 'Both badly damaged, it's okay.'
Dvr: 'Get them here as fast as possible, there are many injured here. Right under the motorway!'
TrC: 'Yes, it's okay.'
Dvr: 'Shit, man!'
TrC: 'Yes.'
Dvr: 'Okay, hurry up!'

A few calls later this driver made a very unpleasant discovery:

Dvr: 'Weimar B, come in.'
TrC: 'Traffic controller Weimar B.'
Dvr: 'Tell me [first name TrC], I haven't called you from Holzdorf…'
TrC: 'No'
Dvr: [Indistinct]
TrC: [Indistinct]

Dvr: 'I haven't reported! How could I have departed, man?'
TrC: 'Well, you could come as far as Holzdorf.'
Dvr: 'I did come as far as Holzdorf, but this was under the motorway bridge!'
TrC: 'Okay, don't blow your top.'

The investigators found this driver's telephone log in the rear cab of the offending unit, well out of reach from the front cab, but with the never given permission to continue from Holzdorf, complete with the departure time already filled in. However, that permission was noted 5 minutes ahead of the same entry by the traffic controller in his book. Clearly some further 'streamlining' of this tedious job had been taking place in the form of entering the messages ahead of actually making the trip. And the driver wasn't the only one; the traffic controller was also already a few trains ahead of what had actually travelled so far that day, complete with the red occupation lines under them. He only had to enter the green wiggly line to indicate that the section was clear again, but his oversight of what was actually going on, who was where, in the event of an emergency for instance, would have been found in his head only.

During a number of post-accident random checks into the use of this system among others on this same line two months later, on 20 November, the imprecise use of the radio system was still noticeable. A national check revealed similar problems. The main culprit was said to be the checks on driver competence, which had not been thorough enough. But who in his right mind expected no deterioration of standards on such utterly boring local jobs like these, up and down a 25-km single-line branch quite a number of times per day? It was concluded that other technical means had to be found and implemented in order to secure safety. And that was similar to what Netherlands Railways had done two decades before.

Comment

From my own experience of working certain types of duty as a train driver, the first thing that struck me about this case was the type of job involved on this short stretch of line every day. I realise that there are millions of jobs in the world that hover between boring and lethally boring, and it is obvious that no employer has an obligation to provide work that is interesting. But the nature of a job like this, in conjunction with the type of control used and the fact that people's lives were at stake, make it no surprise to me that it eventually went wrong somewhere. The traffic control arrangements on this line took no account of human fallibility, and it was almost inevitable that such an accident would occur, given the monotonous nature of the job with its requirement for constant and disciplined attention to procedural detail in order to ensure that safety standards were maintained.

There is also the matter of splitting up available work by line between various operating companies (a possibly unforeseen outcome of the present-day privatisation drive in Europe), which leads to people having to work a single type of diagram along a single route with a single type of rolling stock, day in and day out, until the next round of tendering a few years later. Absolutely unsurprisingly, albeit indefensibly, it also leads to people behaving as they did at Holzdorf – making the job as easy and undemanding as possible by removing all those grating edges requiring serious attention. Hence the books that were filled in during a quiet moment, well before the actual trips took place, and the stilted versions of the radio procedures. Note also the evidence from random tests a month later, which found that people were still doing the same thing, as was the case all over Germany. Given that these unsafe practices apparently had never been reported – instead, drivers and traffic controllers all sailed through their exams with good marks – the competence of those responsible for the management of these people must seriously be called into question.

The eventual fatal oversight was therefore almost inevitable. There was nothing in the cab to remind the driver of his having to call the traffic controller, no action like getting the telephone book out to register the clearance/permission call as part of the station loop routine. All this driver had was his own personal routine. That is why it was remarkably easy for him to get tripped up when circumstances changed and his attention was momentarily drawn elsewhere. (Was it his last trip of the day, perhaps?) In any event, he overlooked a crucial part of his duties. Whilst I have to blame the driver of train 26506 for the trap he set himself (the same kind of trap that driver Bob Sanchez would prepare for himself 5 years later at ***Chatsworth*** in California), the kind of duties this driver worked under the level of safety provided by the traffic control system and the failure of management to keep an eye on his working methods brings a term such as institutional incompetence to mind.

***

Berajondo, Queensland, Australia, 15 November 2004 at 23:55

(Derailment caused by speeding on curve)

Introduction

This accident, fortunately without fatalities or serious injuries, is included to show the importance of route knowledge as well as illustrating another side of driving trains – the risks associated with working through the dead of night in a dark and fairly featureless landscape. The behaviour of the modern coaches in this crash should also be noted in comparison with the derailment at ***Aitrang*** in 1971. The train was on the 1,655-km (1,030-mile) round trip from Brisbane Roma Street station to Cairns and back, for which a scheduled journey time of 24 hr and 55 min was booked in each direction.

The accident and location

Around midnight on 15 November 2004 one of the two diesel-driven tilting high-speed passenger trainsets of Queensland Rail (QR), named *Spirit of Townsville* and used for thrice-weekly services between Brisbane and Cairns, derailed at km 419.493 on a severely speed-restricted curve just north of Berajondo on the Bundaberg-to-Gladstone section of the North Coast line of QR. The leading power car and all seven passenger vehicles derailed and overturned, the trailing power car derailing its leading bogie in the direction of travel but remaining upright. At the time of the incident there was no moon or any other background light; the train's headlight, the AWS track magnets to warn of approach to stations and the speed restriction boards were all that was available to navigate the train by. Although there were no fatalities or severe injuries, a few people did required lengthy hospital treatment.

The line

The North Coast line is part of the QR's Cape-gauge 1,067 mm (3 ft 6 in) network and therefore not part of Australia's standard gauge Defined Interstate Rail Network (DIRN). The line is a single-track, long-distance corridor with passing loops and is partly electrified with 25 kV 60 Hz ac. The signalling is of the route-indicating multiple-aspect colour-light type, pretty much developed from British practice and operating standards. Traffic control is provided from a remote North Coast control centre and electrical control for the section involved is at Rockhampton. Following the contours of the coast, the line has a steady exchange between high-speed straight sections and curved and undulating stretches. As a result, sections with permissible speeds of 150-160 km/h (90-100 mph) are interspersed with sometimes quite severe permanent speed restrictions of 50-60 km/h (30-35 mph) through curves. The accident happened at the first of such a string of curves leading up to the bridge at Cabbage Tree Creek. Despite the provision of local speedboards, a high level of awareness was demanded from a driver at this location, as braking had to commence well ahead of the speed limited section. No warnings for speed restrictions were provided at the time, although that was one of the subsequent recommendations made by the Australian Transport Safety Bureau (ATSB).

Despite its narrower track gauge, the specification of the line was impressive. It was laid with 47 kg/m and 50 kg/m rail on pre-stressed concrete sleepers spaced 670 mm (2 ft 2 in) apart on ballast 250 mm (10 in) deep and fastened with Pandrol clips. The permitted axle load was 28 tonnes and the maximum permitted speed for the tilting trains was 160 km/h (100 mph). No issues with the track were deemed to have been material to the cause of this accident.

Rolling stock involved

Having had positive experience with two Class 300 6-car electric trainsets with Hitachi technology between Brisbane and Rockhampton since 2000, QR ordered two 169.8 m (560 ft) high-speed diesel-hydraulic tilting trainsets from EDI Rail at Maryborough, Queensland, which were brought into service in June 2003. The two 160 km/h (100 mph) trains, with the proven Hitachi body-tilt technology, were made up of two single-ended non-tilt power cars with turbocharged and intercooled 1,350 kW (1,810 hp) 12-cylinder diesel prime movers hauling seven intermediate tilting trailers. Tilt is fitted for passenger comfort only, not for increased safety on curves. As a result, the sets were permitted to run at 25% increased speed in comparison with non-tilting passenger stock. The names of these two sets were *Spirit of Cairns* and *Spirit of Townsville*, the latter being the train involved in this accident. The set consisted of leading power car 5403, luggage van coach A 7408, seating coaches B, C and D, 7409, 7410 and 7411 (accommodating 39 passengers each), club car 7412, seating coaches 7413 and 7414 (accommodating 28 passengers each) and trailing power car 5404. The train identity was VCQ5.

The set was not fitted with either a global positioning system, an automatic tilt authorisation and speed supervision system (TASS) or an ATP system that would monitor and automatically enforce adherence to the permanent and temporary speed limits. The driver drove the set based on posted speed limits, route knowledge and an AWS-type warning for certain locations. After this derailment, during which the set behaved very well and protected its occupants, it was deemed repairable and is back in service again.

The driver controlled traction and braking with a Combined Power/Brake Controller (CPBC). This 'joystick' device, also quite common on British passenger rolling stock since the 1990s, has a centred 'off' position and controls power in one direction and braking in the other. In itself it was nothing new, as many electric trams with solenoid electro-magnetically or electro-pneumatically controlled brakes were driven with that type of controller from well before World War 2. The chief advantage was that the brake could not be applied during powering, and power was always off during braking. In the case of electro-pneumatically braked passenger stock it was a useful manner of traction control, although a quick-acting parking brake or hill-start brake button had to be fitted to avoid rolling back whilst traction configured if the train was dispatched along an uphill graded railway line. When a driver was used to it, there was no reason to assume that this kind of control influenced the occurrence of accidents in any way.

The footplate crew

The driver and co-driver both had experience of the line and the traction. Whilst the co-driver was not medically screened for drugs or alcohol after the accident, the driver tested negative on both counts. The driver also considered himself well rested, and QR operated its rostering through special FAID fatigue assessment software that monitored work-related circumstances under which excessive driver fatigue might set in. For that reason, although micro-sleep issues may have influenced this accident by causing loss of location awareness, officially driver fatigue is most likely not the root cause.

Only experienced footplate crew were taken on to work these trains. The driver was competent on electric tilting trains along this route since November 2000 and on diesel tilting trains since February 2003, and in the month before the accident had done this particular journey three times. During his 5 years' experience of the Bundaberg route he had been involved in three SPADs – on 28 May 2000, on 31 May 2000 and on 1 December 2003. No indication is given as to the circumstances in which the SPADs took place or what the causes were, but he was still subject to post-incident monitoring at the time of the accident. The co-driver had been an electrical apprentice with QR and passed out as a driver in 2000. His files were remarkably free of incidents, so it appears that he was a careful man.

The accident log

Train VCQ5 left Brisbane Roma Street station on the dot at 18:25 and, following a wholly unremarkable journey, arrived at Bundaberg at 22:58 where the crew who would be involved in the accident relieved the Brisbane crew. At 23:11 *Spirit of Townsville* departed from Bundaberg with 150 passengers, five crew members and two footplatemen. The Bundaberg-to-Gladstone section was a single-track electrified line and had two passing loops with stations at Berajondo (km 413.5) and Baffle (km 424.2). An AWS-type warning that had to be acknowledged announced both stations. The line was built in a wide and shallow cutting with soft earth embankments and livestock-proof fencing, young eucalyptus trees growing close to the fence. The night was moonless, quite hot at 24.5°C and very dark, so the beam of the headlight was all the crew had to navigate by. The headlights had been the subject of complaints about their lack of power, but a new lens to improve the light beam had been fitted to the headlight of power car 5403. As the driver had complained about the quality of the hot drinks at Bundaberg, shortly before the accident occurred his co-driver left the cab to brew up in the galley behind the cab and was thus not available to keep an eye on the driving in the run-up to a severe speed restriction of 60 km/h (35 mph). This was to materially affect what would happen.

At 23:51:41, and doing 72 km/h in a 75 km/h (46 mph) section, the OTDR logged the driver alert as he cancelled the Up direction warning at km 415.470 for Berajondo loop and station. He then maintained this speed, although the line speed increased to 100 km/h (60 mph) but then fell again to 80 km/h (50 mph), which was an example of rather sensible driving. Opening up to full power and accelerating to 105 km/h (65 mph), at 23:54:52 he throttled back to 60% power and from there the train maintained a speed of 100 km/h (60 mph). Again, that is how it should be done; minimal shifting of controls for just the right effect. Spurting away or braking at every move of the controls takes a horrendous toll in fuel use and brake wear, so good train driving practice involves planning well ahead based on route knowledge. The driver quickly acknowledged the mid-section AWS warning and was vigilant whilst maintaining a speed of 111-113 km/h (about 70 mph). At 23:55:24, 13 seconds after cancelling the warning, the OTDR showed that the power/brake controller was quickly moved to 'off' and then through to emergency braking. The accident happened at approximately 5 minutes to midnight, some 40 minutes after departure from Bundaberg.

Power car 5403 rolled to the right out of the curve and dragged its entire train off the track, apart from the rear power car. There was rapid deceleration as the leading power car slid for 108 m (355 ft) on its right-hand side, demolishing three OLE catenary posts and bringing the overhead wires down. Luggage van A rolled over and overturned coach B, which slewed on its right-hand side at a 40° angle to the track. The couplers failed from there, with all further cars becoming uncoupled in the process of overturning and sliding to a stop. Seating coaches C and D jack-knifed on their right-hand sides, virtually at a 90° angle to the track, coach E lying on its side parallel to the track, coach F lying on its side at an angle of 90° to the track, whilst coach G derailed to the left of the track. Rear power car 5404 remained upright in the track but with its leading bogie derailed.

At 23:57 the electrical control operator at Rockhampton noticed that the 25 kV ac traction feed circuit breaker tripped approximately 419 km (260 miles) north of Brisbane. He reinstated power as per instructions but there was another trip and he contacted North Coast control, who immediately linked the occurrence to VCQ5 and attempted, unsuccessfully, to set up radio telephone calls to the train. At about the same time a passenger used his mobile phone to contact the emergency services and report the accident, following which the emergency services control room contacted North Coast control and a full alert for a major accident was declared at 00:02. The situation was helped by the fact that a major road ran parallel to this otherwise remote railway line. The police were at the scene by 00:44, others following from then on, severely injured casualties having been evacuated by 02:28 and all others by 05:55.

After the accident the PA system in the train was out of use due to cables having been severed, and the emergency evacuation lighting was inoperative in most coaches. Many people experienced difficulty in opening emergency and regular exit doors, although four people had been thrown from the train but fortunately did not end up under the sliding coaches. All things considered, these modern coaches had protected their occupants remarkably well.

The investigation

Based on evidence collected at the scene of the accident and on the download of the OTDR it was quickly established that the train rounded the curve far too fast and rolled out of the track. That it did so immediately, rather than climbing the outside rail and then derailing (as at ***Granville*** many years before) and overturning when rolling into the soft cess verges of the track, was determined from the damage to the track and the train. A train climbing the track and derailing normally leaves characteristic scoring on the gauge corner and top of the rail head, the sleepers and on the wheel-flanges. None of this was evident.

Moreover, the driver was awake and attentive, as was proven by his driving, his cancelling of AWS warnings and the fact that the power/brake controller was quickly moved through 'off' to emergency braking just before the accident. In view of the fact that the train was 3 minutes ahead of schedule it could be thought that the driver was taking risks, but looking at the way he was driving, as revealed by the OTDR download, he was by no means chasing the edge of the timings before getting his braking wrong. As at ***Aitrang*** some 30 years earlier, the train was being driven utterly sensibly and economically, which also means that any possibility of an intention to deliberately derail the train can be ruled out. It is far more likely that a mistake was made, but what sort of mistake and why?

The train was found to be in a well-maintained condition and nothing directly pointed to bad maintenance such as worn wheel profiles or equipment failure having been involved. Based on the detailed time/distance/event graph from the OTDR it was possible to re-create the events leading up to the derailment with 'Vampire' software that calculates the powers necessary to tip any particular train out of this particular curve. It found that for a power car that would occur at 97 km/h (about 60 mph), a speed that was far exceeded (as the OTDR had revealed that the train had taken the curve at 111-113 km/h or about 70 mph). During the investigation it was found that in one of the front power car bogies the radius arm that pivots from the bogie frame to guide the axlebox had snapped off. This fault was considered as a possible cause of the accident, but based on the evidence it was decided that the break was the result of the equipment being subjected to force overloads that it was not designed to handle during the derailment.

Technically it is clear what happened, but why it happened is another matter. There were no distracting radio or telephone messages, the driver having left his mobile phone at home. Although co-operating willingly, the driver could not account for the accident, having apparently suffered memory loss as an after-effect of being unconscious for some time. When questioned about train speeds by a police officer he said that he saw the square 110 (HST) over a round 90 km/h (55 mph) (all other trains) speed restriction boards, something that made no sense at

Berajondo but could be seen at another location further down the line at Baffle. The driver in all probability had mistaken his location whilst staring into the beam cast by his headlight in the total darkness, and perhaps subject to a moment of micro-sleep, had confused the warning horn of the AWS for Berajondo as pertaining to the one at Baffle. His co-driver, brewing up in the galley, was not in the cab to put him right by drawing attention to his speeding, as was his job in fact. Something else that might have contributed to the driver's distraction was his bag with his sandwiches in it. He had placed that under the co-driver's seat instead of his own and was unable to reach it easily, having to partly get out of his seat to do so. It makes sense to assume that the sandwiches would be combined with the impending brew. It was only when he looked forward again that he saw the rapidly approaching sharp curve ahead of him.

Comment

QR was in the throes of fitting ATP and was not very impressed with it, as the system used turned out to be expensive and unreliable. Adversely affecting train performance, it was never going to endear it to any operator. But if a better ATP or a TASS system had been fitted, this accident would not have happened. Even a fairly simple British TPWS type of speed trap would have prevented it. Perhaps speed restriction warning boards would have helped; the ATSB certainly thought so. Those boards repeat the speed message and give a warning at braking distance from the actual speed restriction. All the networks that I have had experience on use them.

The following safety measures had been put in place by QR: two drivers, a vigilance system, AWS station approach protection, and speed restriction warning boards and section boards. Let's consider each in turn:

- ***Two drivers***

  I have quite a bit of experience with this in different countries, whether when being piloted or piloting someone else in the event of AWS failure, or when having been invited for a ride, or when sitting behind the cab in a passenger saloon listening to the proceedings behind the door. The vast majority of drivers take cab discipline seriously, especially when they are second-manning as a substitute for a failed safety system. However, the amount of talk and laughter I have heard through the bulkhead, or whilst trying to concentrate on the road ahead, led me to conclude a long time ago that second-manning as a structural safety measure is subject to being compromised by human nature, and this accident proves it. Whilst doing his mate a favour by brewing up, the co-driver was missing just at the time when he was desperately needed, so technically speaking 50% of the safety system in the cab had been switched out whilst the other 50% (the driver himself) suffered from a serious fault. Any real safety system would in that case have reported the fault and emergency measures would have been taken, such as slowing down to a prescribed safe speed or even stopping. In this case it went unnoticed until it was too late.

• ***A vigilance (alerter) system***
Such a system is very useful in ensuring that a train is stopped in the event of sudden incapacitation on the part of the driver, but in this case it was of no use because the driver was alert but mistaken in his actions.

• ***AWS station approach protection***
This is valuable for navigation on long lines with just a few stations in daylight when you can see where you are. In this case, however, it would have been of little use, as the warning for one station is similar to another and the train was travelling through pitch darkness. That could be overcome by placing more than one magnet in a certain order that communicated a code for a particular station for which the driver had to cancel. This driver would then have known that he was approaching Berajondo with its lower speed restriction rather than Baffle with its restriction of 100 km/h (60 mph). AWS types of equipment with a simple horn warning for signals, stations, speed restrictions etc have often failed to prevent accidents and incidents for that particular reason.

• ***Speed restriction warning boards and section boards***
Concerning the route-learning book for this particular line, there is a questionable habit (it occurs in Britain as well) of placing far too many speed boards. The basic idea should be to give the driver an overall safe maximum line speed and then put in a few speed restrictions in difficult places. Not to have a variety of short-distance speed sections tacked on to each other, many of which cannot actually be achieved as the speed variations outreach train performance. This driver had come through a 60-100-80 section; he did 60 km/h, slowly accelerating to 75-80 km/h. The famous stretch

of line beside the sea wall between Dawlish Warren and Teignmouth in Devon is just such a line. Just do 60 mph (100 km/h) and ignore the speed boards until after zipping through Teignmouth on the way to Newton Abbot where you may do 90 mph (145 km/h).

Moreover, having gone through the route-learning book, the line actually looked like a tricky bit of road to me. There were sharp speed restrictions in between very fast stretches, and many speed restrictions for curves that were capable of derailing a train at overspeed. In such a case Murphy's Law inevitably kicks in: If it can happen, it will happen. (See the mess created by non-ATP-protected reversible running in The Netherlands at ***Schiedam-Nieuwland*** and similar crashes at *Halle-Buizingen* and ***Pécrot*** in Belgium.)

Conclusion

Something a bit more potent than two faces in the cab was required for credible safety with this sort of operation. Night jobs that start between 22:00 and 04:00 are the very worst ones as far as fatigue is concerned. They begin when the people concerned should actually be in their deepest sleep cycle, on top of which they had to sleep in the daytime, which is often well-nigh impossible. As a result there are inevitably fatigue-related concentration problems to contend with, whether or not a computer software system has been introduced to track and warn for sleep-deprivation and their fatigue-inducing tendencies in the roster section. From 10 years' worth of experience in the cab there is no doubt in my mind that driver fatigue must have played a role as the train travelled through a featureless ink-black night. Something quite similar with dense fog was referred to in the section on the earlier ***Harmelen*** collision in The Netherlands. That this Australian driver perhaps had a tendency to let his mind wander and give in to distractions might also be true, given his eyebrow-raising SPAD record.

The sad truth is that one crucial mistake can eliminate all the value of someone's experience, skill and knowledge in one moment. A human being is always prone to making mistakes, at any time, but perhaps especially so on a long train-driving job at midnight. And as, for example, German experience proves, that problem is not addressed solely by implementing formal protocols for staff to adhere to; the person concerned has to understand the risks involved and must be willing to act positively in order to maintain safety. On a railway such as this, the above conclusion means installing external speed supervision with an ATP system, or something similarly powerful, to monitor and enforce adherence to signals, signage and speed restrictions. ATP now having been fitted on Queensland Rail, the co-driver is no longer freely allowed to vacate the cab but has to scan the road ahead and call out all signs and signals, and has the means to stop the train if he deems it necessary to do so.

***

Hamilton, New Zealand, 16 Februaru 2005 at approx 15:00
(Level crossing incident)

Introduction

Under different circumstances this could easily have been a collision similar to the level crossing accident at ***Boxtel*** in The Netherlands. Under the circumstances as they were, however, it almost became a farce in which, fortunately, no one was hurt and no damage was sustained other than a few scratches on equipment. The actions of the Te Rapa signaller and the drivers of the train and school bus involved were considered and appropriate to the situation, but this incident would doubtless have triggered a major safety of the line investigation almost anywhere in the world these days, as indeed it did in New Zealand.

The incident

The school bus, operated by Pavlovich Coachlines, became stuck under a barrier at Norton Road level crossing at 15:00 during the afternoon school run. The incident resulted from the layout of the level crossing and the fact that the bus driver could neither proceed on the crossing nor back off it due to traffic around him. When the crossing warning equipment started to operate he did not dare to shoot across the seven tracks as he was unable to establish from which direction trains other than the stationary one visible to him were approaching. He moved his young passengers to the back of the bus rather than allowing them to leave through the front entry/exit doors onto the track, which points to a lack of an escape route at the rear of the bus and would not have prevented injuries or fatalities in the event of a hit at speed. In this case, however, this was not of major significance as the local speeds were low and the track on which the bus had stopped was irregularly used. (In fact, the bus driver said he thought it was out of use as the track was rusty. Uncommonly observant!) The train involved was stopped at a signal at the time of the incident.

The location

The level crossing at Norton Road was on an important bypass for road traffic from Hamilton city centre to State Road 1, avoiding the business district. It was, therefore, a well-used minor crossing, a count several days after the incident recording 263 motor vehicles of all categories between 10:20 and 20:30. Moreover, rail traffic amounted to a combination of North Island Main Trunk as well as East Coast Main Trunk traffic, to which considerable local shunt traffic, resulting from traction changes, had to be added as electric traction from Wellington gave way to diesel traction for the onward runs. There were also shunt moves to and from the container terminal at Hamilton. The controlling signal box was Te Rapa box (see also ***Ohinewai***).

Norton Road from the east (city side) approached the level crossing almost parallel with the railway tracks and then made a sudden and sharp 90° turn to the left to cross the tracks, something that made good observation of the crossing quite difficult. The crossing itself was 37 m (120 ft) long and crossed seven tracks, one of which, at the far western end, had been taken out of use and disconnected. At that end of the crossing there was only 65 m (210 ft) of road before a signal-controlled junction with Lincoln Road, State Road 1, the main trunk road running the length of the North Island. This busy road junction caused regular tailbacks on to the level crossing, for which reason a separate escape lane had been added to allow traffic caught on the crossing when a warning activation started to access the road junction and so clear the crossing. Misuse of this escape lane by left-turning traffic, potentially clogging up this facility, had been noted in the past. The crossing was equipped with lights, bells and half barriers and it could be used as an automatic if a train had been signalled through non-stop, or a semi-automatic if a train stopped at the signals prior to reaching the crossing, the distance to the crossing from the front of the train being about 20 m (66 ft). The road signage of the crossing was deemed to have adhered to the prescribed standards, although there were no crossbucks (St Andrew's crosses) at its western entrance.

As line speed for non-stop rail traffic was 50 km/h (30 mph) on the main lines and 25 km/h (15 mph) in the loops and sidings, with an excellent view ahead for train drivers, the risk of fatal collisions was deemed low. However, at the time of the incident this substandard level crossing was slated to be replaced by an underpass. There was already an underpass 700 m (2,300 ft) to the south as well as an overbridge 400 m (1,310 ft) to the north. Clearly, the Hamilton municipal traffic commissioners were taking steps to end this unsatisfactory arrangement.

The rail service

Express train 312, a northbound freight from Wellington to Mount Maunganui yard, was about to have its traction exchanged from electric to diesel power. The train consisted of 28 wagons, weighed 846 tonnes and was 465 m (1,525 ft) long, and was standing at Te Rapa controlled signal 149. The locomotive was a member of the only current type of electric locomotive on New Zealand's railways, the Brush UK-built EF class, a 106.5 tonne, 3,000 kW (4,020 hp) Bo'Bo'Bo'. The class had been introduced from 1988 for the electrified section of the North Island Main Trunk line from Wellington to Hamilton, to climb the steeply graded interior of the island. It was therefore fitted with multiple control and rheostatic dynamic braking. The locomotives, unlike the US pattern diesels fitted with two cabs, sported 25 kV 50 Hz ac phase angle cut thyristor control with six nose-suspended dc traction motors, which at the time was reasonably modern and quite reliable. In 2004, 17 out of the 22 constructed remained in service, two of them assigned to the long-distance passenger operator Tranz Scenic and the remainder operated by Toll Rail on freight.

The accident log

On arrival at Hamilton from Wellington, train 312 came to a stop at signal 149 as booked. The driver immobilised the train, climbed off and then blew out and split the brake pipe, disengaged the automatic Buckeye couplers between the locomotive and the train, and applied parking brakes on a number of wagons. After that he was to bring the light locomotive to Te Rapa motive power depot from where diesels would come back to take over the train on its continuing journey to Mount Maunganui yard.

After climbing back into the cab and reactivating his controls, the driver called the Te Rapa signaller by radio to report that he was ready for the move to the depot, in response to which the signaller set the road and the crossing equipment activated. However, despite the level crossing making all the right noises, signal 149 took its time to come off to a proceed aspect. Noticing that he was waiting longer than usual, the driver on his locomotive looked around for a possible reason and saw that the school bus was stuck under the barrier at the eastern end of the crossing, unable to reverse off the railway line due to the cars behind it. It was protruding onto the infrequently used siding to the container terminal, but it was clear that his locomotive would miss the bus by a considerable distance when he moved forward to the depot. He also saw that the bus driver was shepherding the 16 occupants to the rear of the bus.

Seeing that there was no immediate danger, the train driver called the Te Rapa signaller again and asked if he could see the signal coming off. The signaller confirmed that he could not, after which the train driver explained the situation and added that he was about 20 m (66 ft) away from the bus. On receiving that information the signaller gave him permission to pass the signal at danger and continue cautiously to the depot. The driver released his locomotive brakes, opened up his power controller, rolled across the level crossing and disappeared into the depot, the automatic half barriers opening again once the locomotive had cleared the level crossing closure track circuit. And that was all.

Comment

How had the school bus come to get stuck in the first place? The later ***Boxtel*** accident springs to mind, with the problems associated with manoeuvring a large road vehicle on to a level crossing due to sharp curves in the road immediately before entering. As the bus driver approached the sharp bend before the crossing, he first had to slow down and wait for traffic in the opposite direction to come off the crossing and clear enough space for him to be able to nose out and turn sharply left to ensure that his rear axle would clear the level crossing equipment. This took time. The bus driver was concentrating on spotting a sufficiently large space in the oncoming traffic flow, and by the time he was able to move ahead on the sharp turn he noticed to his consternation that the level crossing warning had activated and that the barrier was on his roof.

Now it gets interesting. First he saw that the track he was blocking was rusty and that no train was approaching along it. Then he considered that evacuating his vehicle through the door at the front onto this track could be more dangerous than shepherding everyone to the rear. Then the single electric locomotive in front of him moved past, the level crossing barriers opened and, after a moment of regaining his composure, he moved his bus across. Had the school bus been on the level crossing past the barrier, which then could have descended completely, the signal would have come off. In the case of this particular level crossing that would still not have been too dangerous, according to New Zealand's Transport Accident Investigation Commission (TAIC), as the train was stopped, speeds along the sidings were very low (25 km/h or 15 mph) and the speeds along the through lines were low (50 km/h or 30 mph), whilst most trains stopped anyway and the view of drivers ahead was very good.

On YouTube there is a video of a level crossing accident near the entrance to a port area in Turkey, where speeds presumably are low as well. A railway shunter is seen waving a heavy road freight vehicle across the track ahead of his approaching propelled intermodal train movement, which is consequently pushed into the road vehicle at a fair clip (I would estimate approximately 40 km/h or 25 mph), causing massive damage to the train, the vehicle and, unfortunately, the shunter. Had Turkish Railways not set speed limits and handling procedures for propelled freight train movements along public roads? This quite unbelievable occurrence would in Britain have very probably resulted in the dismissal of the shunter and the train driver. That same – still rather moderate – speed seen in the Turkish accident, however, would very likely have caused injuries and possibly death had that road freight vehicle been a school bus. School buses, at least the North American version of them, have proved rather vulnerable during collisions with trains at level crossings, even at fairly low speeds of 35 mph (60 km/h) as, for example, at *Fox River Grove* near Chicago in October 1995 when seven students died. And why are there no emergency escape exits at the rear of school buses in New Zealand?

The measured way in which this incident was handled by all directly concerned illustrates their acute awareness that there was no risk to life, limb and property. It was no more than a nuisance, not least to the school bus driver who had been caught out in a difficult situation due to a very unsatisfactory level crossing layout on an important and busy road connection across the busiest piece of main-line railway in the whole of New Zealand. The somewhat casual approach taken by the TAIC to the shortcomings of this level crossing in relation to the amount of traffic it handled was perhaps governed by the knowledge that it was slated to be replaced by an underpass. What more could you want for safety?

***

Amagasaki, Japan, 25 April 2005 at 09:19

(Derailment due to speeding on curve)

The run-up to the incident

Although I have used newspaper cuttings about transport matters and accidents very rarely as a main source of information for this book, one particular newspaper article is worth mentioning because it is the proverbial exception that proves the rule. It dealt with the state of matters after the break-up and privatisation of JNR, the Japanese National

Railways Corporation, and appeared in the Dutch daily *NRC Handelsblad* on Friday 15 March 2002 under the title: 'Als een speer, de succesvolle privatisering van de Japanse spoorwegen' (Like a javelin, the successful privatisation of the railways in Japan). The newspaper's correspondent, Hans van der Lugt, did an excellent job in describing the nation's social characteristics in relation to the management of its railways before and after privatisation. At that time, however, of necessity Mr Van der Lugt was oblivious to the fact that three years later that influence on rail safety would be exposed following the accident at Amagasaki. The article was the main reason for the selection of the accident for inclusion in this book, troublesome though research turned out to be.

The location

The derailment occurred on the Fukuchiyama/Takarazuka line served by JR West, part of the 1.5 kV dc electrified Cape-gauge 1,067 mm (3 ft 6 in) double-track city railway network. It happened on the sharp curves between straight sections of track on the approach to the flying junction at Amagasaki station in the city of Osaka. The train was a Tozai–Gakkentoshi line rapid (semi-fast) service from Takarazuka to Dōshishamae. The speed limit on the straight sections in this area was 100 km/h (60 mph) and into the curve where the accident happened 70 km/h (45 mph).

An older type of automatic train stop system (ATS) was fitted, which reacted to approaching and passing signals at danger but did not supervise or enforce maximum line speed and proper deceleration for the speed restriction. In fact, the Japanese Transport Ministry had issued JR West with a severe warning a month before the accident, on 28 March, as a result of repeated cases of SPADs and trains overshooting station stops due to excessive approach speed. JR West had been ordered to draw up and present plans for expeditious remedial action based on ATP or ATO.

Rolling stock and staff involved

JR West operated the 207 series Hitachi/Kinki-Sharyo/Kawasaki and JR West Goto-built modern single-deck high-density 7-car 1.5 kV dc GTO-thyristor chopper-controlled EMU for commuter traffic. With a maximum speed of 120 km/h (75 mph), these sets of 1991 vintage were comparatively light stainless-steel, high-density suburban trains with fast acceleration.

The driver was 23-year-old Ryujiro Takami, who had passed out as a driver 11 months earlier after having worked as a guard. He had been off the track for one month for a 'nikkin kyōiku' re-education session (see below) following a 100-m (330-ft) overshoot of a station. As a guard he had been on such correctional sessions twice, due to causing delay.

Rail operators after Japanese privatisation

JR West is one of the seven privatised parts of the former state-run Japanese National Railways Corporation (JNR), which had been broken up and the constituent parts fully sold off to private interests in 1987. The privatisation process in Japan differed fundamentally from that in Europe, where the governments broke up the former state railway company into many rather limited size train running operations that could be bought in the case of freight operations or tendered for by train operating companies (TOCs) in the case of passenger services. These companies were separated from operating the rail infrastructure (horizontal integration). By contrast, JNR was privatised into six regional passenger companies and one national freight company. They all acquire their own rolling stock and operate their own infrastructure (vertical integration), yet they are still presented as one entity to the travelling public – the JR Group – especially where the famous high-speed operations are concerned.

Contrary to current European practice, therefore, these JR Group operators do not have to re-tender for operations every five years or so, the situation resembles that in the US far more than anything in Europe. The European rail operating privatisations are regarded by many in Japan as undesirable due to the high externally imposed fixed costs and the myriads of external or part-controlled responsibilities forced upon the operator, that distract from efficiently and reliably running a sufficient number of trains. Furthermore, in Japanese eyes the frequent re-tendering periods demand too much in the way of funds and management attention, whilst the usual short-term focus (4-7 years) is not conducive to private investment in the operation that *de facto* should be the primary reason for privatisation. Another reason for the different approach taken in Japan was that JNR had ¥37,000 billion in debts. Since privatisation, however, the three biggest JR Group operators have become major corporation tax payers. The other, less successful, regional operators receive operations support for socially important local services, but they have never again been the sort of drain on the government purse that they once were. On the other hand, it should be noted that the former state-owned JNR's lack of profitability was caused to a substantial degree by local politicians who insisted on hopelessly unprofitable lines within their constituencies being maintained. That was something, incidentally, that was also demonstrated in Britain during and after the so-called Beeching years. On reading the above, however, perhaps one should bear in mind that it is fairly unlikely that under the European system the described accident would have happened. See e.g. the UK government interventions in the dissatisfactory operations of Connex and the West Coast Railway Company.

In addition to the abovementioned seven ex-state train operators there are in Japan additionally some 190 private train operators with their own local networks. But unlike Switzerland, for example, which has a similar multi-operator situation yet is strictly run as a national network, Japanese companies, especially in urban areas, are in competition with each other. After this accident it led to the wisdom of that system being questioned, as it clearly stimulates the commercial imperative, at the expense of the safety-first imperative. In the aftermath of this accident, similar incidents of drivers speeding in an attempt to recover lost time on other ex-JNR and private networks came to light.

Lastly, an important part of the reason for privatisation in Japan was to break the stranglehold of the once all-powerful train driver union Kokuro, in achieving which objective the Government and the railway companies were completely successful. When the new JR Group operators were established, all staff had to go through a full severance procedure and re-apply with their new employer, the opportunity being taken to remove those with an excessively negative strike record. Many lost their nerve and left the Kokuro union about a year before the privatisation took effect. Despite 30 years having passed since privatisation, many of the present ex-JNR managers appear not to have forgotten the humiliations inflicted by traincrew unions. To western eyes Japanese rail management appears downright hostile towards their staff, a typical indication of this attitude was reported in the aftermath of the accident. Two train drivers on their way to work were travelling passenger on the accident train and survived unharmed. When they phoned their supervisor to inform him of the reason for their delayed arrival, they were ordered to make their way to their depot as quickly as possible and to begin their duties. Later, one of the drivers publicly apologised for abandoning the accident site and not helping out with the rescue work there, which by then further damaged his employer's reputation in the public eye. Apart from the questionable merit of requiring staff, who had just experienced the fourth worst rail accident in Japanese transport history, to immediately work trains, this management reaction undeniably pointed to outrageous ill-will. Even in Japan there was fair agreement on that score.

The accident

The train's trip recorder showed it as running at 108 km/h (67 mph) through a 70 km/h (45 mph) curve, whilst survivors – mostly regular commuters familiar with the running of this service – also pointed to extraordinarily fast running after the previous stop at Itami. Of the roughly 700 people on board in the late morning rush-hour conditions (about 100 passengers per coach), 106 passengers plus the driver were killed, mainly in the first and second heavily damaged carriages, one person dying in the third carriage, and 555 suffered injuries of varying severity.

The first carriage derailed and fell at speed on its left side, shooting in a tangent out of the curve into the first-floor parking area of a 9-storey apartment block, beside a corner of which the curving line passed within a metre. On its way into the building, the carriage scooped up a parked car and flattened it against the rear wall, the carriage body itself crumpling from 20 m (66 ft) to 7 m (23 ft). It would take several days to remove this wreck. The second carriage followed the first into the parking area but, swinging through 180°, fell onto its right side and under the pressure of the following train folded itself up against the same corner of the building, partly flattening itself in the process. The third carriage derailed and parked itself, not too badly damaged, next to the second, with the fourth derailing its front bogie and coming to rest straddling both tracks next to the third. The fifth, sixth and seventh carriages came to rest on the track.

According to an interesting remark in one article (in view of the ***Eschede*** and ***Granville*** accidents) the building had not been specially reinforced or otherwise protected against this sort of impact, despite being only a metre away from the kinetic envelope of a sharp curve on a busy railway line. It was explained that the railways were implicitly trusted in their safety integrity to such an extent that this sort of measure was seen as superfluous. The fact that this apartment block came through the accident as unscathed as it did may perhaps be ascribed to the way in which Japanese buildings are constructed to deal with earthquakes.

Following the first stages of the subsequent investigation, an expert working on the case publicly stated that this curve should have led to derailment only at a speed of 140 km/h (85 mph) or more, a speed that this type of train could not even attain. There would therefore be an investigation into whether vandals had placed stones on the track, rock dust having been noticed on the rail heads. The flattened car between the front of the train and the wall in the parking area had led to suggestions that it was a level crossing accident from further back on the line, but no such incident had been reported. The reason for the derailment at the speed given was due not so much to the inherent instability of the train on the narrower Cape-gauge track, but rather to diminished stability caused by the infamous Japanese peak-hour crush conditions – the dynamic human overload (18 people approximately equalling a tonne) that swung to the outside under the influence of excessive centrifugal forces when the speeding train entered the curve. It was similar to what is considered by many to be the reason why the English warship *Mary Rose* sank in Tudor times. The ship's cargo of battle-ready soldiers in their armour had collected on her port side to watch the approaching French fleet when a strong gust of wind from starboard made her heel further over to port. That caused more crew and heavy weaponry to be tipped to the port side, nearly capsizing the ship and causing her to take in water through the opened lower port-side gun ports. As a result, she settled and quickly sank on an even keel. This load-dynamics threat to stability is also the reason why on European double-deck trains no standees are allowed on the top deck, the justification of that being borne out in the book *British Tramway Accidents*. That describes runaways with derailment and overturning of double-deck tram vehicles. Standees on the top deck were swung en masse to the outside under centrifugal curving forces acting on them, the main cause of overturning of the unit.

Returning to the Japanese train, an additional reason was that traction and air-conditioning plant was located on the roofs. Under these conditions the claim that only a speed of 140 km/h or more would have caused this derailment cannot be substantiated. Incidentally, the stone residue on the track had come from the apartment building when the derailing train hit it.

The cause of the accident

The root cause of the accident was overspeed through a curve. The trip recorders on the train showed a speed of 108 km/h when the train derailed on the 70 km/h curved section on the approach to Amagasaki station — even higher than the 100-km/h speed limit along the straight section the train just passed. Attention soon centred on determining the reason for this excessive speeding. Investigation of signalling as well as on-board recording equipment, plus interviews with staff and surviving passengers, revealed a SPAD about 25 minutes before the accident, during which the ATS system had intervened. That was the start of the disaster sequence, as no doubt the SPAD shook Takami who probably foresaw another spot of 'nikkin kyōiku'. This was followed by a 40-m (130-ft) platform overshoot at Itami station, indicative of the onset of panicky hurry. The overshoot was reported by the guard as only 8 m (26 ft) – a clear attempt to reduce the company wrath on his stressed colleague. By then Takami must have been fighting his nerves and was probably no longer fit to drive the train. The 90-second delay caused by the SPAD and having to propel the train back into Itami platform upset the important 09:20 cross-platform connection with a local train at Amagasaki. Speeding reveals his attempt to make up lost time and maintain the connection. Indeed, 30 seconds would have been recouped had the train made it into the station. Unfortunately, the absurd final piece of bad luck occurred just before entering the sharp right-hand curve towards the left-hand curved flyover into Amagasaki station, when Takami received a radio-phone call from control that later was established as non-essential. Here Takami, probably fearing questions about his delays, let the call fatally distract him in this precarious situation. But, noticing his overspeed on entering the curve, he still did not use the emergency brake to slow down as that would have been an additional reason for 'nikkin kyōiku'. That, then, is how and why this train derailed.

Timetable implications

JR West had successfully recast its entire local timetable to provide departures every 5 minutes, with tightly timed connections between locals and semi-fasts at selected intermediate hub stations, among which was Amagasaki. The scheduled headways between heavily loaded trains in peak hours, in places where a semi-fast would connect with a local, had been cut to 28 seconds (at line speed) to enable expeditious cross-platform interchange during a 1½-minute stop. This enabled passengers to depart from stations on local trains, connect into semi-fast trains at these hubs for a fast trunk trip and, if necessary, then connect at similar interchange hubs into an all-stations local again for the final leg of the trip nearer their destination.

This smart timetable provided travel times and passenger throughputs that congested road traffic could never hope to match and that, more importantly, the competing operators found a hard act to follow. However, operating this timetable relied on trains being exactly on time all the time and platforms being cleared of passengers before the next interchange between two trains would take place again. If passengers from a delayed train had to wait on the platform for their next connection, their addition to the flood of people coming off the next stopping train 5 minutes later would upset the booked dwell times due to overloading of both platform and train capacity. Nor would the next calling services necessarily have the same destinations, so the congestion would involve other lines as well from that moment on.

Comment

Down-to-the-second timekeeping is never discussed as anything other than an essential ingredient of success with train operations in Japan. It is taken for granted by the travelling public, and the Japanese railway operators are proud of that. Consequently, the network is geared up to this kind of delivery as far as train performance and network capacity are concerned. Infrastructure, signalling and train reliability allow maximum throughput in many ways. The passengers, for their part, are aware of the need to co-operate, which indeed goes a long way towards maintaining efficient running. Another advantage is the comparative absence in Japan of the infantile vandalism and outright criminal activity that blights service delivery efforts in Europe and North America.

The secret of success is a mix of on-train, signalling and infrastructure-based systems and station layouts that are sophisticated but less complex in comparison with Europe, and are technically more advanced and therefore more reliable. Japanese trainsets have a high threshold of equipment redundancy that keeps them operating despite faults. Robust fault reporting procedures and sophisticated diagnostics help timely clearing of faults in the off-periods on depots, whilst train construction very much takes into account the need for short maintenance down-periods, and the electronics architecture with its generous redundancy allows rapid in-service exchange of failed control modules with replacement ones. In many cases main-line tracks leading into a station have their own platform rather than going through turnouts in a station throat to distribute trains into the various platforms. In fact, no turnout is built in that is not strictly necessary, and those needed for service delivery go through strict preventive maintenance routines with dedicated crews who are held fully to account if problems do occur. Finally, there are no trains, signalling and track systems built to a price in Japan; instead they are built to be successful in doing a specific job. The national rail hub in The Netherlands, Utrecht Centraal station, is being rebuilt to this same sort of standard.

In these circumstances it is rather worrisome that human beings are needed to move the trains. Everything works only as long as all drivers match all requirements for strict on-time performance every single second of their working day. In the eyes of someone with experience in western Europe, traincrews become the weakest links in the chain this way, and it is here that we see what went wrong at Amagasaki.

The 'nikkin kyōiku' programme

All sources mentioned the JR West regime of 'nikkin kyōiku' courses for 'underperforming' drivers as the probable reason for the accident. The company needed drivers to be continuously compliant with exact timekeeping demands, so the focus of the programme was to retrain drivers apparently unable to deliver, spurred on by the need to maintain the tight timings and by the competition with another local Osaka urban railway company. This correction system had been devised to incentivise drivers and guards to perform almost robot-like, basically by giving them a nasty time if they made mistakes. They were subjected to financial penalties, endless copying out of rules and writing reports about the need for timekeeping. There were instances of staff being publicly subjected to verbal abuse by their superiors, being ordered to perform menial tasks in full uniform such as weeding gardens, cleaning offices and public toilets, being made to sit at a clearly designated punishment desk in public offices and having to stand on platforms in their spotless uniform (with the white gloves) to salute colleagues departing on time.

This system was obviously designed to be punitive rather than as an objective-referenced correctional measure, the 'success rate' of which was researched on more than occasion by the very few surviving independent rail staff trade unions. They reported during the inquest that no less than a third of JR West's drivers (1,101 out of 3,025) had at some time or other been put through these 'nikkin kyōiku' sessions. As an aside, the company's need to inflict this amount of aggressive corrective action on its traincrews seems a hapless response to the government's demand that JR-West took more effective action with regard to SPADs and dealing with platform overshoots. What is striking, however, is that this 're-education' scheme appears never to have been risk-assessed. If it had been, it would have highlighted the excessive stress imposed on safety-critical staff, with the consequent risk of further failure of operational safety, precisely as happened at Amagasaki. Whilst most interviewed Japanese drivers agreed with some form of corrective training following incidents, to a man they claimed that the 'nikkin kyōiku' methods were used for no other reason than to instil fear. An infamous precursor was the suicide in 2001 of 44-year-old Masaki Hattori, a driver of 20 years' experience, after having been involved in a delay lasting a minute. During a station stop en route he checked a fault indication in his rear cab and was put through the 'nikkin kyōiku' grinder for that reason, only to hang himself at the end of that week.

Driver Takami had in his time as a guard at the rear of a train been involved in a delay that had merited similar 'nikkin kyōiku' treatment.

During his eleven months as a train driver (i.e. he was an absolute beginner in terms of experience) he had been put through the wringer again after a more severe platform overshoot of 100 m (330 ft), and it seems fair to conclude that after a Category A SPAD, a platform overshoot and with a 90-second delay on this disaster trip he could not face having to go through it for a third time. Therefore, the assumption that he was desperate to make up this 1½-minute delay does not seem far-fetched. Additionally, during the investigation into the accident a colleague claimed that Takami had told him that it was only after quickly recouping a delay that you could consider yourself a true driver, and that he had been shown how to do it. All things being equal (faster acceleration is impossible on automatically switching trains like EMUs, later and harder braking giving only very limited scope to gain lost time) this remark can only point to his committing speeding offences and being aware of where that could be done based on route knowledge, whilst avoiding the use of the emergency brake controller position as that was another reason for 'nikkin kyōiku'. For a driver of such inexperience that is absurdly dangerous talk. An indicator that things were not quite what Japan believed about the cast-iron safety of its railways.

One cannot escape the idea that JR West (and the other companies, no doubt) put commercial imperatives well ahead of safety, which is a major mistake for any public transport operator. This appeared to be confirmed when Masataka Ide, the man who designed the new high-density timetable and was responsible for the accompanying corrective regime, resigned from his post in June 2005. He was followed by others, most notably Takeshi Kakiuchi, the president of JR West, who resigned on 26 December 2005, the day after another fatal accident had occurred on sister company JR East. Officials denied any connection and all members of management who had resigned or been dismissed soon found employment within other JR Group companies. Vice-president Masao Yamazaki assumed the presidency of JR West.

The aftermath

Following the Amagasaki accident JR West managed to make itself look thoroughly unsympathetic when, as related earlier, it was revealed that they had ordered safety-critical footplate staff travelling on the derailed train to report for work immediately. They then issued almost offensively prescriptive instructions to all staff regarding what they were allowed to say about the accident in public and to friends and family members, in an attempt to limit damage to the company's reputation.

Understandably, any company involved in a major public mishap would want to avoid further problems resulting from staff being tricked by the media into making damaging statements. But a simple instruction to staff to tell reporters to contact the public relations department, as my employer did after the *Southall*, ***Ladbroke Grove*** and *Ufton Nervet* accidents, would have sufficed. In Britain staff adhered to it; why would it have been different in Japan? The news then broke that JR West office parties, planned for the day of the accident and the following days, had gone ahead as booked. This was followed by news that the company had tried to bribe members of the government transport accident investigation authorities to obtain the contents of their reports before publication, with the clear intention of influencing the conclusions. It led to soul-searching in the Japanese press about national characteristics and their relative merits in the 21st century. Even the need for tight timings on trains and concern about 6-second delays were questioned, given the frequencies of departures. They pointed to Europe, where trains could be 5 or even 10 minutes late on busy networks and no one so much as batted an eyelid. Elsewhere in the media, however, it was remarked upon that the Japanese were indeed changing. Instead of stoically enduring the rigours of 'nikkin kyōiku', train drivers began to develop mental problems and become suicidal.

On 8 August 2009, president Masao Yamazaki, the only remaining supreme representative of JR West from the time of the accident, was charged with allowing corporate negligence to develop. On 11 January 2012, articles appeared in the western media that Mr. Yamazaki, the by now former president who had resigned from the JR West company to attend to trials following allegations that under his tenure public safety on his company's trains had been sacrificed to punctuality, had been cleared by Judge Makoto Okada at Kobe District Court. He was found not guilty of presiding over a company with public safety responsibilities that had been negligent and had consequently caused the fatal train crash at Amagasaki. The accident was considered to have been not sufficiently predictable to merit finding the operator JR West liable. However, the company was criticised for incompetent risk assessment when determining the speed limits through this section (which had, in fact, been designed and worked fine under JNR). As a result, the speed limit on the straight sections was lowered from 100 or 120 km/h (60 or 75 mph) to 95 km/h (59 mph) and on the curved sections from 70 km/h (45 mph) to 60 km/h (35 mph). Additionally, ATP to supervise adherence to speed limits was installed, thereby making it impossible for a driver to act in the way Takami had done in the run-up to this accident.

In all honesty, proving corporate failings with regard to safety – blaming inadequate corporate 'safety cultures' rather than those individuals directly involved for negligence leading to fatal accidents – will always be difficult in a court of law. Many attempts to sue companies whose trains, planes and ships were involved in accidents have faltered as it could not be proved that anyone representing the company at a sufficiently high level was specifically responsible for having adversely influenced safety to the extent that the accident could be considered to have happened as a direct consequence. Given the usual chain of events, that is generally true. In this case, however, the introduction of 'nikkin kyōiku' as a remedy for driver timekeeping failure, clearly indicated as a main contributory cause during this and other accidents, had not been properly risk-assessed, whilst the person responsible for its introduction, Masataka Ide, could be named. How the judiciary failed to consider that fact, how Masataka Ide escaped some form of sanction, and how, on top of all that, he was re-employed in management functions by the railways is a matter of some wonder to me.

Conclusion

It is interesting to consider what can be done to avoid excessive timekeeping stress on platform staff and traincrew whilst maintaining operations with reliable, useful and realistically timed connections at hub stations. Closer to home, what do the Dutch and Swiss, for example, do to encourage people to travel by train in the numbers they do on those networks?

1 *What does the average customer actually want?* First, create national public transport networks by providing seamless, e.g. smart-card, through-ticketing on taxi, rail, ferry, tram, bus and metro services, as in The Netherlands. This mimics door-to-door travel as a credible alternative to using the car, as hardly anyone lives and works within a minute's walking distance of a station. Then create the circumstances under which reliable service delivery can take place. This requires investment in inter-company financial organisation and ticketing facilities, signalling upgrades, infrastructure extensions, rolling stock acquisition and increased service capacity.

2 *Integrate public transport capacity.* Provide easy interchange arrangements and sensibly co-ordinate arrival, departure and connection times of all public transport whenever possible. Do not overlook the bicycle and provide for it, both on stations (hire

and safe stabling facilities) and trains (if the stations are too small for safe stabling and hire) if the intention is to avoid dwell time delays.

3 *Never compromise safety in the pursuit of profitable public transport operations*. Doing so will inevitably lead to an accident that damages the operator's credibility and endangers the continued existence of the company. For example, ATP or PTC are not merely options, but essential tools. Required to maintain customers' confidence in travel safety and keep them sufficiently satisfied to come back for more.

4 *Provide achievable, realistic timetables*. Frequent departures and good connections should be provided throughout a large part of the day, but operation of the timetable should be possible without strain. If the available equipment doesn't match the good intentions, cut back the timetable to preserve reliability. The JR West timetable did address travel issues for customers in a way they clearly appreciated, but at Amagasaki it fell apart for the reasons discussed. As an aside, an unused resource in making the timetable more robust is the on-board trip recorder, as it stores detailed, accurate and time-stamped information about all train service operations and the external influences working on them all year round.

5 *Maintain reliability and capacity*. Provide trains, station terminals and a rail network with a sufficient measure of reliability as well as elasticity in their capacity. Provide sufficient track and platforms and sufficient spare train capacity. Strategic diversionary routes must be identified and equipped with adequate throughput capacity. Platforms used by specific booked services must have sufficient length to take the entire trainsets that serve them, as double stops or people sitting in the wrong coach to leave the train at a short platform cause delays.

6 *Manage speed and track capacity effectively*. Speed enhances track throughput capacity by quicker release of network sections and junctions for other services, as well as by enhanced availability of rolling stock for subsequent duties. It is, however, more a matter of identifying where trains have to run slow (curves, station layouts, junctions, etc) rather than increasing speed in sections where trains are fast already. Even out the average track speed profile and make track occupation times predictable and less dependent on driver skills to get there on time.

7 *Incentivise customers to co-operate*. Japan scores here and the Swiss are also good. Just one tardy rush-hour passenger, muscling up against the stream of boarding customers when trying to alight, may easily cause 15-30 minutes impact delay on other train services, certainly when luggage is involved. So tell them to please stay awake and alert. If you are an easy sleeper, let other passengers or the traincrew know your destination to ensure readiness to alight on arrival.
8 *Invest steadily in public transport*. Specify achievable requirements for future public transport growth within a framework of larger objectives in the longer term, put realistic delivery dates on them and ensure that those dates are met. Germany, Austria, Switzerland, Spain, France, The Netherlands and Belgium are all prime examples of this approach. It is expensive, no doubt, but is cheaper and more effective in the long run when compared to emergency upgrades needed on an overworked network infrastructure that is falling apart. Hasty initiatives forced by circumstances rarely lead to properly maximised, well-planned and economically satisfactory results. See the upgrade of the West Coast main line in Britain as a prime example of how not to do it.
9 *Management and work force should play their part with honesty and social responsibility*. Powerful key-staff unions ruthlessly exerting a stranglehold on every new development by absurd strike action is as bad as management treating their traincrew and platform staff as little more than menial robots and bullying them in the way and for the reasons that JR West did. No railway worker wants delays; it is not in their own interest as it means they are late for their breaks or the end of their shifts. Moreover, delays create hassle and the customers will soon let them know of their displeasure in those circumstances. Apart from that; Unstinting 100% accuracy and safety scores are impossible when people are involved. People will make mistakes sometime or other, certainly in an environment like public transport.

Incidentally, none of the above points is mentioned here for the first time. What happened to JR West and its customers at Amagasaki was that the demands for service delivery on staff exceeded what could be reasonably provided by the system as it was operated. At Amagasaki this situation ran its course at an extreme cost in lives, injury and damage.

***

Horsham, Victoria, Australia, 11 August 2005 at approx 12:13
(Level crossing collision)

The accident
This was a very recognisable accident of the type that is repeated time and again the world over on a daily basis. It shows graphically how distraction, absentmindedness and the consequent lack of attention endanger anyone who uses a level crossing or, indeed, drives a car. It also confirms what is widely known worldwide, that in the majority of level crossing accidents the road user involved lives nearby or uses the crossing on a daily basis. Furthermore, it reveals how a 127-tonne locomotive may be endangered by a small car of much less than one tonne weight.

The location
The incident took place at Edith Street automatic open level crossing in Horsham, Victoria. Edith Street approaches this crossing at an angle of about 40° but then turns slightly right to cross the railway at right angles. In this right-hand curve leading up to the level crossing Edith Street also crosses Palm Avenue, which runs parallel with the railway line. Beyond the level crossing Edith Street connects at a T-junction with a major local traffic artery called Dooen Road. The traffic situation for a road user could be considered as somewhat complicated as a result, but due to the open nature of the neighbourhood and the clear priority indications it is not particularly difficult to negotiate. The character of the area is clearly suburban, consisting of many stand-alone bungalows on their own plot of land.

The level crossing
The level crossing is on the single-track non-electrified Melbourne to Adelaide main line, part of the national standard-gauge Defined Interstate Rail Network (DIRN) and as such quite regularly used by trains. The Australian Transport Safety Bureau (ATSB) report mentions an average of 14.4 trains per day, mainly between 06:00 and 10:00 and from 14:00 to 18:00. The accident, however, happened with an unscheduled working just before 12:13.

The deep-ballasted track was laid with 60 kg/m continuously welded rail, fastened with Pandrol clips and baseplates on wooden sleepers spaced at intervals of 685 mm (2 ft 3 in), allowing a permitted train speed of 115 km/h (70 mph). Signalling was automatic 3-aspect colour-light with track circuits that were also used to work the automatic crossings. During the subsequent investigation the level crossing was found to be in good working order, as confirmed by witnesses to the accident. The crossing was 85th in the Victoria ranking of urgency in the level crossing upgrade programme and was not identified as being high risk. However, the Australian Rail Track Corporation (ARTC) identifies only vandalism, power failure and equipment malfunctioning as issues that define risk at level crossings.

In fact, there was a precursor to this accident. On 24 February 1951, a very serious accident occurred at one of the *Horsham* level crossings, when a train hit a bus. Eleven people died in that collision.

The train and its crew

Train 0783 was an unscheduled trip with a light locomotive, G class G535, for route-refreshing purposes. The crew – a driver, a co-driver and an instructor – had booked on at Portland, Victoria, at 07:30 following two days of route refreshing between Maroona and Portland on the Ararat-to-Portland branch and were on the way back to Portland. Incidentally, this type of activity is very common on systems with the British inspired route-indicating signalling system along trackage that is used only rarely or only for diversions, its purpose being familiarisation with signalling, route characteristics and local yard issues. The trip from Portland to Murtoa loop had been entirely uneventful.

The locomotive was a typical Australian Clyde Engineering machine, a 127-tonne Co'Co' with a 2,238 kW (3,000 hp) General Motors EMD diesel engine and electric traction equipment, with a permitted top speed of 115 km/h. The machine was in perfect working order and the three crew members in the cab, rested and concentrating on their work, were very experienced on the type of traction and the route. The machine was fitted with a version of the Hasler speed recorder encountered a number of times in this book, and its record was of material use in establishing events and backing up other findings during the investigation. After the accident the crew members were checked for alcohol misuse and illegal substances as standard procedure and, as expected, all gave negative results.

The car and its driver

The vehicle involved was a Hyundai Getz 3-door 1.4 XL with a 4-speed automatic gearbox. On investigation of the wreck it was found that prior to the accident the car had been in a fully roadworthy condition. The 65-year-old female driver had moved to Horsham from Warnambool about 18 months before the accident and lived in a bungalow about 90 m (295 ft) from the level crossing. The road sign indicating the approach to the crossing stood near her drive in such a position that she could not actually see it when backing out of her drive and turning towards the level crossing as she did on the day of the accident, but she had often used the crossing more than once a day and there is no doubt that she was well aware that it was frequently used by rail traffic. After the accident no traces of illegal substances or alcohol were found in her blood, and she had been regarded as being in a state of sufficient physical health to be allowed to drive a car. There was no record of traffic violations or involvement in road accidents, but evidence emerged of possible recent upsets in her life that might have caused her to be preoccupied at the time of the collision.

The accident log

At approximately 11:38 the locomotive in question passed Murtoa loop at 115 km/h and continued onwards to Horsham under clear signals. The weather was good, visibility was excellent and the temperature was around 12°C. All the locomotive headlights were illuminated and on approaching Horsham the driver sounded the horn several times, first for Rasmussen Road level crossing and then for the one at Edith Street. Traction was shut off to allow the speed to drop to 100 km/h (60 mph) for the permanent speed restriction through Horsham station a bit further down the line. At 12:11:30, 638 m (2,090 ft) from Edith Street level crossing, the driver activated the locomotive bell and flashing wig-wag headlights, introduced following US practice after a spate of level crossing accidents, to draw extra attention to his approaching train. It was at about this time that the driver in her little car backed out of her drive and positioned it to continue forward to the level crossing. She accelerated to approximately 40 km/h (25 mph) and soon followed the right-hand curve past the junction with Palm Avenue to enter the level crossing. She was driving steadily and at no time was she seen to speed up or to use the brakes.

The locomotive was coasting towards the level crossing where a line of trees at the side of the track ended and allowed a view of Palm Avenue and Edith Street ahead to the right. The crew in the cab noticed the car approaching, not slowing down, and realised that if it did enter the crossing they were going to hit it hard. For that reason the driver moved his train brake valve to emergency, added the straight-air (independent) locomotive brake and sounded the horn again. At 12:12:50 the car entered the level crossing and was struck by the locomotive. It hit the car at its left-rear quarter and bent its front in a 90° angle against the left-hand side. The wrecked car wedged itself under the locomotive's front end and was violently dragged along, kicking up ballast and hitting the fencing of the pedestrian level crossing. Whilst still moving, 220 m (720 ft) past the impact site the wrecked car connected with the casing of the machine for the turnout into Horsham station passing loop. It broke the locking of the turnout blades, which moved under the locomotive, sending its front bogie along the main line but its rear bogie into the loop. This fully derailed the locomotive and brought it to a grinding stop, still upright, from 83 km/h (52 mph) some 330 m (1,080 ft) past the point of impact. The driver immediately contacted ARTC traffic control to report the accident and request assistance from the emergency services and within 8 minutes the first emergency vehicles had arrived at the scene. On smashing into the turnout machine casing, the wrecked car had dislodged itself from under the derailing locomotive and rolled down a slight embankment, coming to rest against the track boundary fence. A nurse among the witnessing bystanders immediately walked over to render first aid, but the occupant of the car was beyond help.

Despite the derailment, the locomotive suffered only a few scratches along its left-hand front end and a cracked fuel-tank sight glass. It was re-railed with two cranes and moved to Horsham sidings where its two bogies were replaced (a normal precautionary measure after a derailment) and it was then moved under its own power back to Portland to continue its duties from there. After track repairs the line was reopened at 03:36 the following day.

Comment

The car driver had entered the level crossing without reacting to any of the warning signals concerning the approach of the train, having apparently failed to notice the warnings given by the crossing equipment or by the approaching train with its flashing headlights, clanking bell and blaring horns. It is possible, however, that she entered the crossing on purpose, the accident report mentioning that she may have recently experienced some sort of upset. She was involved with more than one local social volunteer organisation, but it is not known whether she had just received an upsetting phone call, for instance.

She knew the level crossing well and was aware that it was frequently used. In fact, when indoors at home she must have regularly heard the engine noise and ground rumble of passing trains. It is true that at the time the accident happened there would normally have been a lull in train traffic, but this particular route-refreshing trip had been booked in that lull so as not to hinder regular traffic, which caused this train to run outside the hours in which locals would expect trains. Personally, however, I do not think it played much of a role. Many people are unaware of such details if they have no interest in trains or never use them, and I doubt that a 65-year-old living nearby for only 18 months would have had that sort of detailed knowledge. Many people simply drive their cars in a state of absentmindedness and lack of alertness to the job they are doing behind the steering wheel.

Distraction as a result of preoccupation and lack of attention has been identified earlier in this book as a common cause of accidents, and expectations about certain traffic features ahead that wrong-foot people crop up regularly in accident reports. If someone expects that a train might appear at any time from experience, then that person will look for such signs when approaching the crossing. If, however, a level crossing hardly ever activates and trains are very few and far between, then it is easy for a person to develop the habit of ignoring the presence of the crossing in the road ahead and for absentmindedness to reinforce this situation, as so graphically illustrated in this case.

There are many examples of such incidents in some of the more remote places in the world. It has been found that people often tended to underestimate the amount of train traffic passing the level crossing by a factor of two or three (e.g. thinking there were two trains a day when in fact there were four or six), an error of perception that was reinforced every time they used the level crossing and there was no train to stop them. It illustrates the potential dangers of seasonally reactivating museum or freight branch lines that lie dormant for most of the time. Finally, the road configuration at the accident site near the junction with Palm Avenue and the T-intersection with Dooen Road may have been to blame for distracting the driver's attention from the level crossing. In other words, was she looking out for cars rather than for trains?

***

Hatherley, Gloucestershire, England, 18 October 2005 at 05:20
(Derailment due to poor train preparation)

The accident
One wheelset of an empty 2-axle open wagon seized and was damaged due to the parking brake not having been released at the departure yard. After 68 miles (110 km) of travel the wagon derailed on a rail slide joint with both axles, towards the opposite track. As a consequence it fouled the loading gauge of that track in several locations and damaged a 4-mile (6-km) stretch of line, including turnouts and signalling equipment.

The location
The derailment occurred just south of Cheltenham Spa station at Hatherley. The line was the Down Birmingham-to-Bristol line of the former Midland Railway, for which reason the Up and Down line directions are opposite to all the others in the vicinity, being former Great Western Railway territory. Why that particular detail was perpetuated after all the track simplifications had taken place is beyond me, but it is one of those links with history that can make the railways in Britain so irritating, baffling or interesting, depending on the person concerned.

The double-track non-electrified line, fitted with track circuit operated colour-light signalling, is a very well engineered line with long, straight stretches and lazy curves that allow fast running, the maximum line speed generally being 100 mph (160 km/h). Only the well-indicated 40 mph (65 km/h) permanent speed restriction through the awkward station layout at Cheltenham Spa is something that would have required route knowledge for safety reasons. At the location where the train came to a stop at approximately 05:20 the line speed was 95 mph (150 km/h). The controlling signal box, at Gloucester, was an electro-mechanically NX operated panel box. The temperature was 11°C and there were light winds with occasional showers of rain.

The train and rolling stock
The train headcode identity was 6V19, the consist being locomotive No 66221, one of the uncomfortable but redoubtable 126-tonne EMD JT42CWR Co'Co' Class 66 diesel-electrics that taught European rail freight operators what reliability is really about. It was hauling dead 129-tonne Co'Co Brush diesel-electric No 60018, five empty BYA-bogie open wagons of 26 tonnes each and 14 empty SSA 2-axle open wagons

of 15.5 tonnes each. One of the latter, SSA 470028 in the 14th position, was the vehicle that derailed. English Welsh & Scottish Railway (EWS), nowadays known via DB Schenker UK as DB Cargo UK, operated the locomotives, wagons and staff involved (other than the signallers).

The SSA is an open van for scrap metal, the steel underframe of which originally came from self-discharging vehicles and had been fitted with a steel roofless box superstructure (gondola) later. Their suspension is quite unlike any other type of mainframe axle suspension with horn guides and axleboxes, spring hangers and leaf springs. This vehicle has a so-called pedestal suspension, which has a two-part steel fabrication that keeps the axleboxes located and dampens suspension movement of four coil springs per unit through friction. This unit, seen only in Britain, covers most of the view of the wheels.

The accident log

Train 6V19 arrived at Bescot from Immingham, 77 minutes late and 20 minutes before its scheduled departure time, to drop off and pick up wagons as required. As things turned out, the entire consist from Immingham was to stay at Bescot and a new consist of empty open wagons for Margam in South Wales was waiting for onward transport. So the two locomotives were uncoupled from one set and ran off and then back on to the new set on another track to be coupled up. The train departed on time to its next pick-up point, Washwood Heath in Birmingham, but as nothing was waiting there it passed through and continued on its way to Gloucester and Severn Tunnel Junction at its maximum permitted speed of 60 mph (100 km/h).

The train did not stop after its departure from Bescot (as confirmed by the download from the OTDR) until the Gloucester signaller stopped it, so possible vandalism can be ruled out as far as what happened next is concerned. Whilst the driver quietly concentrated on his job, on what seemed to be a totally uneventful trip, the Gloucester signaller suddenly noticed that a set of turnouts for Cheltenham Lansdown loop gave a fault indication, after which one track circuit went down and failed to return to unoccupied when 6V19 left it. Then the Churchdown hot axlebox indicator activated and, adhering to set procedure, Gloucester 50 signal (the Down junction signal at Barnwood Junction) was kept at danger whilst the signals protecting the opposite track were put back to danger to allow the driver out on the ballast either side of his train without the danger of being hit by passing trains on the adjacent track. On slowing down to come to a stop, the driver opened his side window and heard loud and rapid banging somewhere behind him, as if a vehicle in his train had massive multiple wheelflats; he could also see sparks in the darkness. Having been told by the signaller that axle 59 of his train had activated the hot axlebox detector, the driver walked back and found the 14th wagon standing on the ballast, nicely in line with the train, on seeing which he contacted the signaller again and reported the situation. When investigators and other railway officials turned up, he put a track circuit clip down on the Up line to ensure protection by signals for people on both tracks.

The line would be out of normal running for the next eight days to repair the damage caused. Four miles (6 km) of track had been damaged, two sets of turnouts had been pushed apart (hence no detection and the fault indication in Gloucester signal box), a slide joint for continuously welded track had been damaged (which was, in fact, where the wagon derailed), TPWS grids at signals had been destroyed, track circuit cabling had been cut (hence the track circuit that did not clear in Gloucester signal box), the hot axlebox detector equipment at Churchdown had been destroyed (hence the alarm in Gloucester signal box) and 10,000 concrete sleepers required renewal. The wheels of the leading axle of the derailed wagon had 12 in (30 cm) flats with notches that tightly fitted the rail head ground into the wheel treads. In fact, this notch had 'sub-flats', the axle having tried to turn but unable to do so. Its wheel flanges at the bottom had worn away through having rammed the sleepers for a considerable distance, the common British expression 'absolutely knackered' accurately describing the poor condition of this particular wheelset. The following vehicle had superficial damage from flying ballast but luckily enough no overbuffering had taken place.

Despite its derailment, the wagon had kept its rightful place between the other two vehicles. It was even more fortunate that at the time of the accident there had been no traffic passing along the adjacent track. Had there been, the derailed wagon would very likely have scraped along the passing train in the usually tight British loading gauge. Only between Tuffley Junction and Standish Junction would there have been sufficient space, the former ten-foot between the old Midland and Great Western tracks now being in use as the six-foot between the remaining two tracks. However, had the wagon derailed to the cess-side (outside) then in all likelihood it would have hit an overbridge somewhere, resulting in substantially more damage than there would have been had the vehicle not drawn attention to itself and been routed through the reverse-lying turnouts at Barnwood Junction, where it very probably would have overturned and derailed other vehicles in the train.

The investigation

The Rail Accident Investigation Branch (RAIB) soon identified the culprit that had caused the damage. The parking brake of the wagon had not been released, as shown by the indicator, and unfortunately there was no interlock between the parking brake and the train brake. With more sophisticated (and more expensive) equipment the train brake cannot be released when the parking brake is applied as, usually, the applied parking brake opens a valve in the brake pipe. Moreover, had this vehicle had the traditional British long parking brake handle along its side, the shunter would probably have noticed the situation more readily, even in the darkness and rain at Bescot. The parking brake mechanism was found to be very stiff, and the train preparer, if he did check it, most likely thought that the brake was off and did not look at the indicator.

When the train was leaving Bescot yard someone should have watched it roll by and looked for anything untoward such as sliding wheels etc. In the weather conditions prevailing at the time, however, it is unlikely that, if anyone did watch the train depart, they did so out in the open and close to the passing train, although at such low speed and in the darkness even then the signs of distress of these sliding wheels would not have been obvious because of the vehicle's pedestal suspension. A possible later opportunity to check the train at Washwood Heath did not present itself, since the train was not required to stop there.

Comment

Although it was a messy situation, the people involved were incredibly lucky that the incident did not get out of hand, due to the vehicle staying upright and not spreading the track at the turnouts sufficiently for the vehicles behind it to derail and also because of the dearth of other traffic, especially passenger services, at that time of the day.

In my opinion, the train was not fully prepared in accordance with the prescribed method. I have experienced quite a number of non-released parking brakes (the RAIB incident report mentioned this as well), which is what leads me to believe that the train preparer at Bescot did not carefully walk along the train and ensure the release of every parking brake, even the stiffly operating ones. Given the 20 minutes between arrival and departure there was simply not enough time for that when train 6V19 finally landed in the yard. It is also rare to see anyone giving a look at wagons that are standing around, so they were unlikely to have received any attention there either. Incidentally, Murphy's Law is again seen at work, in that it had to be the vehicle with the difficult-to-operate parking brake that needed to have it released.

A digression concerning a sliding axle

I experienced a seized axle under the leading motor bogie of a classic Gatwick Express service to London Victoria at about 18:00 on a dark evening in December 1993, passing East Croydon station. The consist was a Class 73/2 electro-diesel Bo'Bo' with five Mk II passenger coaches and a Bo'2' remote control motor vehicle and luggage van called a GLV.

Coming through the left-hand curve just past East Croydon in the direction of Clapham Junction along the Up fast line, I had got into the habit of opening the cab side-window to look back for evidence of dragging brakes or worse. That evening I saw sparks flying from under my front left bogie side, but the third rail with the rather sparky traction current collector shoe was on the right side, so I stopped the train and warned the signaller by radio. I then took out my surprisingly ineffectual British Rail traincrew hand lamp (the Bardic lamp) and checked in the darkness. The axleboxes were warm as normal, the brake blocks were off and their rigging was slack except where I had applied the parking brake in the first passenger vehicle to hold the train. The problem was definitely not with the brake rigging.

It took some time before I noticed in the torchlight that the rail head was shiny before the first axle and dull grey behind it. Based on that clue and on closer inspection, I found a 10-cm (4-in) wheelflat. The inevitable conclusion was that the axle had dragged at speeds of up to 75 mph (120 km/h) between Gatwick Airport and East Croydon. I called the signaller to obtain permission to take the train cautiously back into East Croydon station with the locomotive and that was the end of the trip for that train. The passengers were detrained and at 03:00 the following day the axle stood on skates and the GLV was moved off to Stewart's Lane depot for repair. The Brighton line commuters had experienced another rush hour with misery on their way home that evening!

In fact, the problem had not been seized brakes or a seized axlebox – hence my difficulty spotting what the matter was – but the failure of one of the support bearings with which the nose-suspended (axle-hung) traction motor rested on the axle. That in turn explained the traction faults I had experienced at Gatwick Airport when I had had to isolate the traction control cut-out switch due to persistent overloading. It was a definite and worthwhile lesson learned, and would never again happen to me. Trains, however, can be infinitely tricky and you never stop learning whilst working them.

***

Three accidents near Wolfurt, Austria, 29 August 1988, 30 August 1989 and 29 December 2006

(Two head-on collisions, and people struck following train hitting unidentified object)

The location

These three accidents occurred at the western end of the Arlberg railway line, a typical single-track and electrified transalpine main line for most of its length, running from Innsbruck in the Austrian Tirol to the German lakeside port of Lindau in Bayern on the Bodensee (Lake Constance). From Innsbruck West station – where the scenic Mittenwald line branches off northwards to Garmisch-Partenkirchen – the Arlberg line climbs westwards via Ötztal and Landeck to the 10.3 km (6½ mile) Arlberg Tunnel and then descends via Bludenz – where the local Montafonerbahn branches off southwards to Schruns, via Feldkirch and Bregenz to the shores of the Bodensee. The last 7 km (4 miles) of the line into Lindau, almost all of which is in Germany, run along the lakeshore and via an artificial dam to the island on which the station is located. At Feldkirch a line branches off and leads westwards to Buchs in Switzerland via Vaduz in the tiny principality of Liechtenstein. Between Riedenburg and Lauterach an important line branches westwards into Switzerland to St Margrethen via a triangular junction.

The line is electrified at 15 kV 16.67 Hz ac, the old ac system that is shared between Sweden, Norway, Germany, Austria and Switzerland and allows convenient through-running of traction in central Europe, albeit a narrower type of pantograph having to be used in Switzerland where train protection is provided by a different system called Signum ZUB. From Lindau, although located in Germany, there is no further electrification into Germany as yet; electric traction used is Austrian or Swiss only, and German diesel traction has to be used onwards to München. Electrification for that route is foreseen to be commissioned in 2020. The Austrian section at the western end is electronically controlled from Wolfurt traffic control centre with modern colour-light signals, whilst the last 5.9 km (3½ miles) into Germany are mechanically signalled from Lindau signal box with semaphore signals. Train protection throughout is provided with the Austrian/German PZB system, a development of the earlier German Indusi system by using ac magnets with variable pulse frequency settings (500, 1,000 and 2,000 Hz) capable of being interpreted by an on-board computer as speed instructions. PZB also monitors adherence to the received speed limit information.

The accident at Wolfurt station on 29 August 1988
The incident, involving a head-on collision following a SPAD, illustrates a signalling trap that has caused SPAD incidents and consequent accidents all over the world.

The trains and traction involved
The collision occurred between train Ex160, the 'Pfänder' express service, travelling from Linz and Salzburg to Lindau in Germany, and train E641, a semi-fast service from Lindau to Linz. Both trains were hauled by Bo'Bo' locomotives of the 1044 series – Ex160 by 1044.051, and E641 by 1044.096. Built from 1974 onwards as the first series of Austrian electronically controlled (ac phase angle cut thyristor control) heavy traction, they weighed 83 tonnes and had a power output of 5,400 kW (7,240 hp), with a maximum permitted speed of 160 km/h (100 mph). The inspiration for these machines came from the purchase of 10 Swedish-built ASEA Rc2 machines that were later sold back to Scandinavian operators.

The accident log
The 'Pfänder' express had been diverted onto the reversibly signalled left-hand track of the double-track section between Dornbirn and Wolfurt due to a collapsed retaining wall and would have crossed back to its normal right-hand track at Wolfurt. On approach to the crossover point the distant and the stop signals therefore showed the aspect 'clear road with a speed restriction of 60 km/h (35 mph)'. This was modified by an illuminated '7' added to the signal aspect, which indicated a maximum speed of 70 km/h (45 mph) through the crossover turnouts. Train E641 was approaching from the opposite direction at the time and was coming through a temporary speed restriction of 80 km/h (50 mph) due to track deficiencies at Lauterach station. As the express had the higher status it had been given preference, and the semi-fast would be required to stop at signal T to let it cross over first.

The driver of E641 had worked his train all the way according to this restricted speed and cancelled the PZB warning for the distant to the stop signal T that was showing a red aspect. Almost simultaneously as he cancelled this warning, however, he passed the termination board for the section of the temporary speed restriction and, instead of braking for the stop signal at danger ahead as he should have done, he opened up the power controller. The power of modern electric traction was once again exemplified by the fact that 20 seconds later the train was already doing 92 km/h (57 mph). Now the PZB computer, monitoring the speed as a result of the cancelling of the warning at the distant, activated and intervened with an emergency brake application at 95 km/h (59 mph). However, this was not sufficient to bring the train to a stop before reaching the red signal. The driver then saw the red stop signal ahead and applied emergency braking himself. As he passed signal T at 84 km/h (52 mph) the PZB then again triggered an emergency brake intervention. When train E641 passed the insulated block joint between the two track circuits at the crossover still doing 74 km/h (46 mph) the driver of the oncoming express, doing 66 km/h (40 mph), saw his 'clear road with restricted speed' signal aspect go back to danger, but because he had already noticed the SPAD of train E641 he had applied the emergency brake. When the trains hit at an impact speed of 90 km/h (55 mph) the first coach of E641 was pushed heavily onto its locomotive, which telescoped into it. Very severe damage resulted, as well as five fatalities and 53 people being injured.

During the investigation into the accident it was noticed that overhead line equipment masts would have hindered the clear view of signal T by the driver of train E641. The driver also claimed that he passed the distant signal at green, although this was proven not to be the case as the OTDR trip recorder showed him cancelling the PZB warning for the distant at caution.

Comment

Whilst travelling in cabs throughout Europe I have experienced this particular chain of events more than once. After passing a distant signal at caution, a driver on his way to a red signal aspect then inexplicably and wholly erroneously reacted to, for example, the termination of a temporary speed restriction by opening up the power controller and accelerating away toward a red signal aspect. Situations such as departing from a station stop between two signals at caution and danger, or going through a junction and completely forgetting about the passed signal aspects, have also led to such incidents on more than one occasion. In virtually all the cases I experienced this was dealt with by ATP, in the way that PZB ought to have done at Wolfurt had an extra magnet been installed 250 m (820 ft) before the red signal, but in four other cases I had to warn the driver – once in Germany and once in The Netherlands. On the two other occasions it was not too much of a problem as I was the instructor with a trainee driver.

This situation is by no means rare, a serious accident with the same background occurring on 17 October 1991 at *Melun* in France when the driver of an incoming freight train reacted wholly competently to a permanent speed restriction warning sign to reduce his speed from 90 km/h (55 mph) to 60 km/h (35 mph) but on reaching the start of the restricted speed section released his brakes again, completely overlooking that he had also passed a distant signal at caution. As a result, he committed a SPAD, which brought him into a violent head-on collision with a night train from Paris. The first vehicles of the freight train disappeared under their locomotive and lifted it off the track, whilst the badly damaged first couchette coach of the night train jumped onto the roof of its locomotive, the rest of the train escaping serious damage. Nevertheless, it resulted in a death toll of 16 and many others being injured. The fitting of KVB ATP on the railways in France was extended as a result of the accident.

Network Rail, the national rail infrastructure operator in Britain, had already identified this particular trap and required that the termination of a speed restriction be extended to the next signal if that would frequently show a red aspect because it covered, for example, a danger point such as a junction, to ensure that drivers reacted properly to the distant at caution. In the case of automatic plain line signals, however, this would not always be possible, so under these circumstances this type of trap would of necessity remain. Such a situation demands alertness from drivers if they are to prevent a station stop or termination of a temporary speed restriction from influencing their actions rather than the aspects of lineside signalling.

The accident at Riedenburg Junction on 30 August 1989

Erich Preuss has described this accident, which occurred a year later, but I have been unable to find any further information about it. It concerned another head-on collision, a low-speed event between two freight trains that nevertheless caused massive damage. It happened as a consequence of cost-cutting during extension works on the freight yard at Wolfurt. The junction at Riedenburg for the line to St Margrethen in Switzerland had, against specification, not been changed to a flying junction. As a result, it temporarily required trains proceeding towards the Arlberg Tunnel to travel on the left-hand track between Bregenz and Lauterach whilst tying in the second track to the junction. Sloppy handling of train regulation between traffic controllers during degraded signal working was the cause, not unlike what happened between ***Warngau*** and ***Schaftlach*** in 1975.

Any money saved by not installing the flying junction had been lost by the 50 million Austrian schillings of damage that the accident caused.

The accident at Lochau-Hörbranz on 29 December 2006

This accident was caused by a failure in communication (partly due to issues concerning international radio network channels) and a lack of clarity about competence of regional and local control functions with regard to emergency measures.

Train and traction involved

The accident involved train EC196, a 'EuroCity' service from München to Zürich. A DBAG Class 218 diesel-hydraulic B'B' locomotive with provision for head-end electric power to run the air conditioning of the seven EC coaches (the set weighing 334 tonnes and being 185 m /610 ft in length) was used to bring the train from München Hauptbahnhof to Lindau. From there a Swiss SBB Re 4/4-III Bo'Bo' electric locomotive took over for the short trip from Lindau through Germany and Austria into Switzerland, and then onwards to Zürich Hauptbahnhof. The train was booked to call at Bregenz 10 minutes after departure from Lindau and then to disappear at St Margrethen into Switzerland about 15 minutes later. An Austrian driver was booked to work it from Lindau to St Margrethen, and on the day of the accident the Swiss driver who was to relieve him at St Margrethen was also in the cab from Lindau for route familiarisation.

SBB Swiss Federal Railways had allocated its Re 4/4-III type of electric locomotives to SBB Cargo, who converted a number of those for anticipated work into and within Austria and Germany. This freight work had not materialised to the extent foreseen, however, and when the München via Lindau to Zürich service once more had to revert to locomotive-hauled coaches in 2004 following technical problems with German ICE TD diesel units, it was decided to use these Swiss electric freight locomotives between Lindau and Zürich. That is how Swiss-owned 421 397-1 came to be involved in this accident in Austria. The Re 4/4-III was a Bo'Bo' of 80 tonnes weight and 15.4 m (50 ft) length, built from 1964 onwards, the gear ratio between traction motors and axles being changed in 1971 to better enable freight work through heavier pull with a reduced top speed. Its power output was 4,700 kW (6,300 hp) and it had a top speed of 125 km/h (75 mph). For their work into Germany and Austria they received PZB train protection, extended train radio and one pantograph with a wider German/Austrian contact strip. Available electric train heat facilities to provide head-end power were also recommissioned for this work.

Signalling and control issues

Three signallers or local traffic controllers at Wolfurt control centre worked the electronic control equipment but made no train operation management decisions. That was down to the Regional Traffic Control Centre West, based at Innsbrück, where the regional traffic controller had two assistant controllers at his disposal to carry out all communication with frontline staff. Drivers contacted this regional traffic control centre, rather than the local controlling signallers, directly via train radio.

An incident and emergency controller was also based at Innsbrück whose function was to act as the single point of contact between the emergency services, law enforcement officials and the railway in the event of a serious incident. He had no authority, however, to deal with train operational issues, as was normal at that control level.

The accident log

At 05:49 on 29 December 2006, the driver of train 5602 reported to the assistant regional traffic controller at Innsbrück that he appeared to have hit something. There had been a muffled thump somewhere under his train just before Lochau-Hörbranz, at km 8 or thereabouts, but he did not know what he had hit. The assistant traffic controller then asked the driver of following train 5604 to keep a lookout in that area. The report from the driver of train 5604 was negative, unsurprisingly, given the darkness at that time of the morning in December, his headlights not having picked up anything. Wolfurt station staff had also been informed, who contacted the local police. They searched the railway line around the reported area but, unable to find anything, called off the search at 07:12.

At 08:50, a dog walker reported spotting a mutilated corpse close to the railway line and partly hanging in a fence. The police informed their regional control centre which in turn informed the emergency services about a body at 'Langen Stein'. It was only some 15 minutes later that the emergency controller at Innsbrück control centre was notified of the situation and what actions the police were taking. When he contacted the regional police control centre at 09:14 he was told that police were at the site but not actually on the line and that there was no obstruction to rail traffic. They also gave him the mobile phone number of the man in charge at the site. At 09:26 the Innsbrück emergency controller called this man and agreement was reached that the trainset and the driver of train 5602 would be made available for investigation into the accident. Again, the police officer stated that they were not actually on the track and there was no hindrance to train traffic. The emergency controller did not take any further action, such as contacting the local box or, even more importantly, to dispatch a local railway worker to the site.

The police officer in charge at the site contacted the emergency controller again at 09:44, requesting that train traffic proceed slowly past the site between 'Strandbad' and 'Langen Stein' as they needed to go onto the track. When the controller asked just how slow, the police officer said that he wanted trains to be able to stop in time, and after some deliberation was given a sighting speed emergency restriction of 30 km/h (20 mph) by the emergency controller, who also wanted to know the accurate site reference as he was not familiar with the local names referred to. He asked the police officer for a kilometre post number, which after some delay was given as 7-10. This, as it turned out later, was taken from an overhead line catenary post, meaning OLE post 10 in kilometre 7, but the emergency controller mistook it as the kilometre/hectometre post indication. He then asked the regional traffic controller to implement the emergency speed restriction at km 7.10, but omitted to inform him which services were subject to this restriction and during what time period it should be in force. This man, looking at his train describer monitor, saw train 5760 waiting ready for departure at Bregenz with the signal off, so he immediately called the driver and informed him of the emergency speed restriction and its location. Six minutes later, the driver called back to report a wrong kilometre location – it was km 7.8 to km 7.6, at the home signal for Lochau-Hörbranz. Almost immediately, at 09:55, when the local traffic controller at Wolfurt signalling centre called in to report a 3-minute delay in section to train 5760, he was given a digital delay code to indicate responsibility (for allocation of the delay to an owning manager in his report) but was not informed that until further notice all services would be delayed by a similar amount of time. Neither was he informed of the reason for that decision or the fact that there were people on the track. There was also no request made to inform all other drivers about the emergency speed restriction.

In the meantime 'EuroCity' service EC196, having sat at Lindau for some time while the locomotives were exchanged, departed at 10:02 with a 7-minute delay. The assistant regional controller attempted to call the driver several times via the train radio but was unable to make contact, as in Germany the radio is linked to DBAG channel A74 which the Austrian railways were not permitted to use. Only at km 5.9, where the train entered Austria, would the driver switch channels to Austrian channel A65, at which point the train would be only 1,760 m (just over a mile) away from the site. As a result, the driver of train EC196 was not informed about the temporary speed restriction.

Train EC196 approached the site at 86 km/h (about 53 mph), coasting downhill in preparation for the stop at Bregenz whilst coming through a right-hand curve along the Bodensee shoreline. On the left-hand side was a fence and behind that the main L190 road, busy with morning traffic. On his right-hand side was a fence behind which dense vegetation and trees impaired his view through the curve ahead. Suddenly the driver was confronted with a small cluster of people on the track only 90 m (300 ft) ahead of his train, some of them wearing high-visibility vests and most with their backs towards him. He had just about time (4 seconds) to assess the situation and apply the emergency brake before ploughing into the group. The train came to a stop at 10:08, about 200 m (660 ft) from the place where the brakes had been applied. Two police officers and an employee from the undertakers called in to remove the body of the first victim were dead, but one other police officer and another employee of the undertakers had been able to jump clear of the track in the nick of time.

Comment

From reading the accident report it is clear that the people working in the Regional Traffic Control Centre West at Innsbrück made the serious mistake of handling this incident themselves instead of handing it over to the local traffic controllers at Wolfurt. Whilst the local traffic controllers (signallers) based nearby at Wolfurt were likely to know all the local names on their patch of line, the regional controllers that handled this matter were based at faraway Innsbrück, across the Alps, and were unfamiliar with names such as 'Strandbad' and 'Langen Stein'. This led to considerable time being lost in establishing exactly where the incident site was located. Nor, oddly enough, was assistance from local signallers requested, which was another missed opportunity. Also overlooked was how many people were on or near the line, with what sort of authority or responsibility, whether they were in endangered positions and whether there were lookouts – things that any local signaller worth his salt would have asked immediately. If it had been within the authority of the local traffic controllers at Wolfurt to act, they would have called Lindau signal box (their fringe box in that direction, with which they would have been in regular contact) to stop the

'EuroCity' service from departing and speak to the driver to discuss non-acceptance of his train from Lindau signal box or use their own signals as a means of keeping the train away from the site. As things were, neither they nor the driver were aware of this very urgent matter that directly involved them. Does this perhaps point to centralisation having gone a tad too far, or competencies not having been worked out properly? In any event, to railway workers who are used to the way things are handled in Britain and the nearby European continent this has a distinct 'I don't believe it' ring to it.

Nowadays there is a ruling in Austria that, as virtually everywhere else in Europe, traffic is stopped as soon as the emergency services go anywhere near the track, let alone onto the track. Why was that rule not implemented here? Was it because the police officer in charge did not ask? A non-railway person should never be asked to decide about such matters, or to state exactly where they are along the track; they should be told to stay away from it. Traffic should then be stopped, a qualified railway representative sent in (another serious failure in this case) and, as soon as that person has reported back, consideration can be given to possible alternatives to stopping all traffic. Which is precisely what this ruling is all about. Incidentally, this would not have taken hours in view of the very short distances involved at that location. The bartering about the emergency speed restriction and the failure to realise that it was not restricted to just one train were serious blots on the record of the railway operator, even if the root no doubt lay in the wish to avoid delay where possible.

Incompetence was also evident in the way in which an alternative means of communicating the urgent warning to the driver of train EC196 other than via the train radio was overlooked. It ought to have been known that the radio would be of little use due to national radio channel restrictions on the very short distances involved. In Germany, channel A74 was used for only 6 km (3¾ miles) along this line, which prevented the Austrian regional controllers in Innsbrück from contacting the driver of train EC196 in time. But why did the regional traffic controller or the emergency controller not think of simply calling nearby Lindau signal box by telephone? Train EC196 had been stopped there for a considerable time, so platform staff could easily have informed the driver of the emergency speed restriction or asked him to call the Austrian regional traffic controller, which, with a simple change of channel on his radio, he could probably have done from his cab. Was it perhaps a matter of the old European short-sightedness with regard to national borders?

This accident understandably blew up in the face of the Austrian Federal Railway and led to attacks in the media concerning its perceived incompetence. Official explanations had to be issued to defend the way in which it dealt with accidents, and working relationships between emergency services and the railway were also soured for some time, despite the fact that the local police had not handled the situation that very well either. Unfortunately, it takes this sort of occurrence to focus everyone's attention on the risks involved and on the need to sign up to protocols in order to safeguard lives.

An Austrian reader of this text remarked on the fact that the driver of train EC196 should perhaps have used his horns to warn people along this twisting track with its impaired view ahead. In this respect it should be borne in mind that on most European rail networks the use of train horns is increasingly subject to a number of restricting conditions, such as its prohibition (especially at night) in residential areas, except at places indicated by a whistle board and otherwise in the event of a potential emergency only. The shores of the Bodensee are densely populated, and I have no doubt similar restrictions on the use of train horns were in force in that area.

With regard to train 5602 having hit some unidentified object, this occurs on innumerable occasions and is the cause of many delays. It happened to me late one dark evening when driving an Up Gatwick Express close to the Stoats Nest crossovers at Coulsdon, when I heard a hard bang against the flat, tinny front of the ancient GLV remote control vehicle and saw something white flashing past the second man's windscreen. In these circumstances you do wonder for a moment whether a call is worth the trouble, but then make it anyway, having stopped your train (in Britain you are no longer allowed to make a call whilst on the move), later trains in both directions also being stopped and requested to check the line. But the way it was handled in Austria reflected the way it would have been handled in Britain at that time and no doubt everywhere else in Europe.

***

Auckland, New Zealand, 6 October 2005 and 31 March 2006

(Two incidents with brake systems of Auckland SA/SD suburban sets)

The incidents

On at least two occasions the newly conceived SA/SD diesel-electric locomotive-hauled push-pull sets were involved in serious brake failures due to faulty equipment. In both cases these were discovered and dealt with appropriately by traincrew working empty stock on positioning trips, which prevented any mishaps from occurring.

The location

The incidents took place in the city and the area around Auckland, the terminus of New Zealand's original North Island Main Trunk railway line. Local passenger rail transport in and around the city is focused on the excellent Britomart Transport Centre (a former Victorian main post office building that after thorough reconstruction operates very much in the way a city centre station in Europe does) located on the south shore of Auckland harbour. Apart from bus, coach and ferry services, three regional passenger rail services were worked from there at the time of these incidents: the Western line to Helensville; the Southern line via Newmarket to Otahuhu and Pukekohe; and the Eastern line via Glen Innes to Pukekohe and Otahuhu, the latter being the North Island Main Trunk line encountered elsewhere in this book.

This system since has undergone major changes that were brought into use in 2015, including electrification with 25 kV 50 Hz ac, new CAF built electric units and doubling of lines to boost frequency and speed. Further extensions are proposed, there are well-developed plans for a line from Avondale on the Northern line via Auckland Airport to Puhinui, which would become a major junction interchange if plans to open the Manukau city centre branch from there also become a reality. Between Onehunga on the new Avondale–Auckland Airport line and Penrose on the North Island Main Trunk line the former single-track Onehunga branch line is being considered for reopening with a full upgrade to allow a direct and very logical connection from Britomart to Auckland Airport. There is also a proposal to construct a light rail line from Mount Eden on the Northern line through a tunnel under the harbour to the north shore, at present a barren area as far as rail services are concerned.

The Auckland area passenger rail organisation at the time

Since 2002, Auckland metropolitan and regional rail transport had been organised under the auspices of Auckland Regional Council, much as in similar major conurbations in Europe. The main authority was Auckland Regional Transport Authority (ARTA), which specified and purchased public rail transport services, owned passenger rolling stock, specified upgrades of the rail network and managed stakeholder relations.

The providers of train services for ARTA at the time of these incidents, hired in or otherwise, were as follows:

- Veolia Transport Auckland Ltd was the franchised passenger train operator that hired the ARTA rolling stock, provided the on-train and station staff, collected the fares and undertook the marketing of the regional passenger services.
- Ontrack NZ was the state rail infrastructure organisation that managed and regulated the technical and capacity side of access to the network, provided train traffic control and track maintenance, and managed track network development.
- Auckland Regional Transport Network Ltd was the station management authority, regulating the access to the Britomart transport hub and other stations in the Auckland regional area, carrying out station development and managing station facilities.
- Toll NZ Consolidated Ltd, an Australian-based transport company, was at the time the operator of all privatised New Zealand rail facilities under the name Toll Rail NZ and was involved in traction hire and lease, access to maintenance facilities, stabling of traction, provision and training of footplate crews, and rolling stock refurbishment and development.

In July 2008, the New Zealand Government bought out Toll Rail and after October that year the state again operated the railways, under the brand name KiwiRail. As a result of this development, the relations as outlined above changed slightly, but because the name Toll Rail appeared in the incident reports it is important to explain its place within the operational relationships.

The history of the Auckland SA/SD push-pull sets

In the early 2000s, ARTA was instructed by Auckland Regional Council that rail had to be developed further as an alternative to road traffic. A contract was signed in 2003 to convert part of a fleet of about 100 imported British Rail Mk II Intercity vehicles from the late 1960s and early 1970s, declared redundant in Britain, into fourteen 4-car push-pull sets as a quick solution to the problem of expeditiously acquiring an inter-urban passenger fleet good enough to provide 25 years of service. These sets consisted of an SD-type driving trailer/generator van with a driver's cab to remotely control the locomotive at the opposite end of the train and to provide hotel power to run the heating and the air conditioning on the set, with three SA-type trailers of typical suburban set-up that sported two wide sets of entrance lobbies at the one-third and two-thirds positions in the body and were fitted with power doors, instead of the end-entrances with slam doors as originally in Britain. At the opposite end there was a refurbished DC-type of A1A'A1A' diesel-electric locomotive, hired with a driver from Toll Rail to work this 4-car set with a maximum speed of 100 km/h (60 mph). The locomotives, built in Canada in 1955, had been repowered and refurbished in Australia in 1975. They were originally DAs, delivered during the 1950s from Generals Motors in London, Ontario, as a modified G12 model. In the late 1970s these locomotives were sent to Clyde Engineering in Australia for engine conversion and upgrading to DCs, and during the refurbishment in Australia the brake equipment pipes should have been renewed. It is highly likely, however, that a considerable amount of the existing piping was retained. In the late 1990s withdrawal began and the locomotives were stored. When the SD/SA push-pull formations were formed a number of the stored DCs were overhauled at the Hutt works of Alstom Transport New Zealand Ltd and made ready for this type of traffic. DC 4375 was one of the original four to be taken to Hutt works near Wellington in 2004 for such an SD/SA overhaul, which amounted to a 10-year life extension. Toll Rail had advised that air brake pipes had been inspected and tested for air leakage, and the locomotive had successfully passed five different types of brake functioning tests since its recommissioning.

The incident log from 14:40 on 6 October 2005

ECS for train 4356 departed in push mode from the Otahuhu depot backshunt line via the original double-track North Island Main Trunk route to pick up its booked passenger services from Britomart. The train consisted of SD driving trailer 5811, three SA coaches (5719, 5873 and 5835) and DC class diesel-electric locomotive 4375. Shortly after departure the driver slowed the train to 40 km/h (25 mph) to negotiate the turnouts for the route via Glen Innes at Westfield Junction as normal, and after coming through the junction he accelerated the train to the permitted line speed of 100 km/h (60 mph). At approximately 14:50 the train passed an advance warning board for a possession along its route, which carried the obligation to come to a halt at the conditional stop board.

A conditional stop board protects a worksite. In this case contractors were installing a communication cable between Meadowbank and Orakei. At the conditional stop board the driver was required to contact the site protector by radio to obtain authority to pass the stop board and receive additional instructions with regard to speed and other issues. In a way this procedure is similar to obtaining permission to pass a signal at danger from the traffic controller. In the case of this particular worksite, however, Ontrack had also provided an emergency protection person (EPP) who had been stationed at the conditional stop board by the person in charge of the possession (PIC). This was an additional safety measure to ensure that in the event of a train not stopping at the conditional stop board the worksite was warned and could be evacuated in time.

The driver made an initial brake application to adhere to the instructions of the outer warning board and then attempted to move the brake valve handle to the service application position. He noticed, however, that he could no longer move the brake handle as it was jammed but he was able to release the brake. The train manager, who just happened to enter the cab at that moment, saw the driver 'bang' the brake valve handle a few times. The driver, knowing that he was not going to be able to stop at the conditional stop board, operated the emergency brake plunger, which was an independent method of applying the train brake in the event of an emergency. The train came to a stop 170 m (560 ft) past the conditional stop board, some 330 m (1,080 ft) short of reaching the worksite, the EPP observing the train passing his site at the conditional stop board fairly slowly before coming to a stop. Instead of using his radio he phoned the PIC at the worksite, who in turn contacted the driver by radio and was told of the train's brake failure problems. Shortly afterwards a train operations manager arrived and, together with the driver, was able to free the handle slightly, but it took considerable effort on the part of the two men to move the handle beyond the initial application position. As the train was now delaying several others at the start of the afternoon rush hour it was driven slowly to Britomart station, but on arrival in the platform the doors released and opened. This caused people to start boarding, which prevented the train from being taken out of service. However, the locomotive brake valve of the now-hauling DC 4375 was working all right, so the train was taken to its destination at Papakura and after an uneventful journey was taken out of service there and brought back to Westfield depot. The OTDR confirmed the driver's claims with regard to the failure and his consequent actions.

The investigation

The driver was an experienced man, having worked 24 years on the railway. He had been a grade 1 train driver for 17 years and held current certifications. At the time of this incident he worked the DC/SA/SD sets very regularly and knew them well, which included the use of the type 26-C brake valve which incorporated a graduated brake application as well as release function. One strange detail is that this driver had the habit of carrying lubrication oil to apply to the brake valve detent quadrant (notch ring) in the event of stiff operation, something his tutor drivers had passed on to him early in his career. I have never heard of that before, and it does indicate that the type 26-C brake valve as fitted had its apparent quirks.

Driving trailer SD 5811 had been in service for 5½ weeks since commissioning and had no history of brake problems. When the SD carriages were being converted from Mk IIs with some urgency, not enough driver brake valves were available so a batch of reconditioned used equipment was sourced from the American Railroad Equipment Corporation. The valves that arrived were from two manufacturers – Westinghouse Air Brake Company (WABCO) and the New York Air Brake Corporation (NYAB) – both brake valve types being interchangeable. The problem was that a number arrived without their handles, because Toll Rail did not know that these had to be ordered separately, although three arrived complete with plastic brake handles (which undermines the claim of the spare parts provider to a certain extent). As time was pressing, Toll Rail ordered wooden brake valve handles from a firm in Dunedin, New Zealand, which was supplied with drawings that had the plastic brake valve handles as the template. The new handles were made of wood, and as the drivers did not like the feel of the US-delivered plastic handles these were all subsequently replaced with wooden handles. Once these had been fitted a static function test was done, after which the vehicles were signed off ready for traffic.

Inside these brake valve handles there was a so-called detent plunger, a sprung 'finger' reaching out from inside the handle and slotting into notches cut out in certain places along a notch ring on the top of the valve assembly. This set-up enabled drivers to move the brake valve handle to certain fixed application and release positions by going through the notches and so connect or disconnect the various brake air ducts – a standard set-up for brake valves the world over. When investigating the incident it was found that the locally sourced handles had received detent plungers that had not been manufactured in the shape of the original ones; a small bevel at the bottom front edge was missing and the characteristic rounded front edge had not been copied. This gave the brake valve handle a tendency to get stuck in the first quadrant ring notch, the initial application position. And that was all there was to it. Later, a similar incident occurred on 3 November 2006, which was not reported to New Zealand's Transport Accident Investigation Commission (TAIC). This time it was the brake valve handle on locomotive DC 4254 working train 2151 that got jammed.

The accident log from 05:10 on 31 March 2006

ECS for train 4306 from Britomart made its run from Westfield depot to Otahuhu to reverse and then travel along the North Island Main Trunk line via Glen Innes to Britomart, the same trip that the service mentioned above was making. When the train departed from Westfield depot at 05:10 it was in push mode with the remotely controlled locomotive propelling. During the short and slow (25 km/h or 15 mph) journey to Otahuhu the driver noticed that the train brakes were extremely poor when he slowed down to stop in the Otahuhu backshunt to reverse direction. He shut down the cab of SD 3199 and changed ends to set up the cab of locomotive DC 4375. Walking along the train, keen to find out about his brakes, he noticed that the brake blocks of the locomotive were not applied, as they should have been after shutting down and blowing out the brake pipe. He climbed into the cab of his locomotive, set up its controls and, after the brake pipe had been charged and the brakes released, he made a static test application of the train brakes and noticed that the locomotive brakes registered zero on the brake gauge. He increased the brake pressure to full service application and saw that the brake cylinder reading increased only to 200 kPa (29 psi). He kept releasing and reapplying the brakes in various positions but consistently noticed that 200 kPa was all he could get on the locomotive and below it there was no reading of an application at all. He correctly deduced that the locomotive brakes were malfunctioning and failed the train. Using the locomotive straight (independent) air brake he brought the train back to Westfield depot, where the locomotive was taken off the train for examination. Following a preliminary examination TAIC impounded the locomotive and took over the investigation.

The investigation

The OTDR revealed that the train had been driven as required and that the straight air brake on the locomotive had functioned as required. It also revealed that the locomotive had not consistently reacted as required to train brake applications. On 4 April 2006, a full air brake test was performed but no fault was found. The examination then moved on to the brake control valves outside the cab under what is termed the short hood under the cab at the front of the locomotive. There was evidence from a number of brake equipment-related incidents that pointed to moisture in the brake piping. Moisture collects when air is compressed and then cools down again, for which reason locomotives and multiple-unit trains have several types of equipment to get rid of this moisture. Among these are air-dryers and, in Britain for example, the moisture discharge valves of standing locomotives are commonly known as 'spit' valves due to the sounds these make. One of the negative effects of this moisture is that it makes the piping corrode and flake inside, out of sight, and fine-graded dust that has passed through the air filters and been ingested into the piping is mixed with the water into a paste that may cause valves to stick. The same, incidentally, goes for oil from some types of compressors that enters the air pipes.

When TAIC went through the history of various locomotives it became clear that some machines had far fewer problems than others, but no record had been kept of who had done what sort of work on the brake system of the machines, and when, that would give a clue to possible problems with methods of work in the various plants. Problems had also been imported with the used equipment and spares arriving from the US. With regard to this incident, it was found that the problem had resulted from the 26-F quick-release valve on the locomotive having been prevented from closing because its spindle had become stuck in the bore due to contamination. Because the train had travelled slowly as an ECS movement toward a headshunt no mishap had occurred, but had the train been in service from Britomart as booked, with the locomotive pushing and running at maximum permitted line speed, the risk of a serious accident due to brake equipment failure would have been extremely high indeed. TAIC extended the investigation in order to avoid similar risks arising in the future.

Comment

These problems with the braking systems were caused by the fact that reconditioned rolling stock with reconditioned braking equipment was used and that the condition of the braking system equipment apparently had not received the sort of scrutiny it deserved, considering its important safety role.

The report about dirt inhibiting the functioning of the locomotive or multiple-unit brake by making an internal valve in the brake-frame stick points to a familiar problem with ageing rolling stock – moisture contamination and consequent corrosion of air pipes and valves. This should have been addressed during overhaul, both during the upgrade in Australia and when the locomotives were retrieved from storage for the Auckland push-pull work. When reconditioning existing rolling stock (midlife overhauls etc) it is standard procedure to take care of the condition of the braking systems and problems with compressor oil in the pipes, ingested dirt due to insufficient filtering of dust from the air intakes or moisture corrosion problems attached to heating and cooling down of the compressed air. It appears that the brake system conversion work (the addition of driver brake valves in the driving trailers) had not been taken sufficiently seriously by the parties concerned to ensure that new or reconditioned, and properly tested, equipment was available on time. But who in his right mind, as an equipment retailer, dispatches incomplete train brake valves to a customer? It was obvious that Toll Rail would need those handles with the valves they ordered. And who in his right mind, as a buyer, accepts incomplete brake valves in this situation? Then, instead of reordering the missing bits, he hurriedly gets someone to manufacture the handles from drawings made up from one of the complete brake valve handles – a manufacturer who does not copy the brake valve handles exactly but delivers an interpretation of the detent plunger contact patch shape and so causes confusion and potential accidents. There is little more frightening in a cab than the realisation that the train is not going to stop in time.

A further personal observation

It was an inspired solution by the transport authorities in Auckland to acquire tailored rolling stock by reconstructing suitable imported second-hand vehicles, the reuse of reconditioned and upgraded older but not inherently obsolete carriages being a pragmatic approach. Although rolling stock is normally considered to have a useful economic life of 30 years, in daily practice there are many examples of vehicle types that have been kept in service successfully for 50 or 60 years.

The Class 73 ED locomotives and the HST sets in Britain might reach that sort of age, and the DC class diesel electric locomotives in New Zealand, originally built in the 1950s, were still going strong in 2010. The principle of life-extension is dependent on the continued provision of spare parts, timely overhauls and intelligent upgrades. The British HST sets, to mention but one example, are rather better trains now then when they were built. In the case of the SD/SA ex-British Mk II sets in New Zealand, the re-engineering surgery inflicted was thorough, bearing in mind the problems encountered when altering modern stressed-skin integral types of coaching stock. Their bodies were slightly shortened by the removal of the entrance lobbies at the carriage ends and new headstocks were fitted. Cape-gauge bogies of Japanese origin were fitted and the bodyshell layout was altered by fitting new mid-body exit lobbies with power doors. The result was a practical, attractive and modern-looking suburban train, fit for purpose and with the great advantage that when the Auckland rail electrification scheme is put into service the old DC diesel-electric locomotives can finally be retired and there is the option to replace the diesels relatively easily with 25 kV ac electric power cars. As things are all existing stock was retired and replaced with nice Spanish CAF-built EMUs and a few DMUs for the Pukekohe line. The increasingly unreliable DCs finally bowed out.

Examples of that creative approach to stock-shortages can be found all over the world. This situation can be seen in Italy, where various types of semi-suburban coaching stock were overhauled, refurbished and through-wired for push-pull multiple-unit operation, some of the older types receiving modern bogies to improve riding comfort. They were then made up into fixed-formation 6-coach sets with a driving trailer at one end and worked by a dedicated new fleet of modern single-cab Bo'Bo' electric locomotives at the other end. Later, a new generation of double-deck suburban carriages was conceived to fit in with this push-pull concept and could be slotted into these sets or formed into complete double-deck sets as required, whilst later still, sets of double-deck EMUs were constructed that could be coupled up and worked in multiple with this locomotive-hauled stock. Needless to say, everything was coupled with Scharfenberg-type automatic couplers. It was almost as universal, and effective, as the BR Southern Region operation in Britain until the 1970s. Belgian Railways also did something quite similar with their M4 single-deck and their two types of double-deck locomotive-hauled rolling stock, whilst Netherlands Railways have been carrying out a programme of conversion of their 1989 crop of DD-AR double-deck local units into regional express units since mid-2011. Very nice trains they make out of them, too.

In the light of the fact that a substantial number of technically speaking relatively modern British Rail Mk II and Mk III locomotive-hauled passenger coaches are still sitting around in storage yards whilst passengers (myself included) are crammed into an undersized fleet of overworked and ageing diesel multiple-units, many would no doubt appreciate something similar happening here. There might even be sufficient space for bicycles, to mention but one serious problem with train travel in Britain. It requires an open, pragmatic and, most of all, a long-term vision of passenger train operations, none of which we have been good at in Britain of late. Ultimately the approach taken in New Zealand, The Netherlands, Italy and Belgium is rather more effective than what is happening under the pressure of steeply increasing patronage in Britain. Passenger operators are forced to hire in extra capacity with expensive sets of four rather tired and barely refurbished Mk II coaches with their ridiculously obsolete and narrow slam doors, topped and tailed by two 2,500-3,000 hp (1,865-2,238 kW) DE locomotives to enable reversal without having to uncouple and run the locomotive round, but where one could comfortably do the job. Fixed formation sets of five or six reconfigured and refurbished Mk II and Mk III coaches, with a remote control cab at one end and one diesel-electric locomotive in push-pull configuration at the other, could have served this country very well from the 1980s and 1990s onwards had we but foreseen our potential future needs a little better. Unfortunately, at that time we were engaged in vandalising the network capacity following the Serpell Report. Yet, since major investment in new rolling stock has seemingly ground to a halt again, perhaps we could make some creative decisions with the rolling stock that is still available.

***

Carcassonne, France, 27 February 2007 at 12:40

(Derailment following SPAD)

The incident

This derailment was another permanent way incident, involving a SPAD at a shunting signal following which a 2-axle tamping machine derailed on a derailer and consequently fouled a through line just before a passenger train along the fouled track was due to pass through at speed. It is an example of p-way hastiness and lack of detailed route knowledge.

The location and layout

The station in the beautiful medieval city of Carcassonne is a junction of the single-track non-electrified line to Limoux and Quillan with the double-track electrified main line between Bordeaux, Toulouse, Narbonne, Béziers and Perpignan. The single line from Quillan connects with the two main lines through crossovers but its main track then runs straight ahead into the yard to become road 4, which then spreads into a number of other roads. Home signal 117 along the single line from Quillan protects the yard and the crossovers to the main line into the station; SPAD signal 120 is a single exit signal for all lines from the yard via track 4 onto the single line.

The accident log

A 2-axle ballast-tamping machine, recently arrived from maintenance work on the single line to Quillan, was put away to track 4 in the Carcassonne yard. Due to the failure of another tamper at a worksite on the line to Neussargues via Béziers, however, it was decided at short notice to send this machine from Carcassonne to continue that work. To that end the crew were called into the signal box located about 600 m (1,970 ft) away in the passenger station, where the signaller explained the moves to be made but omitted to draw their attention to signal 120 and an associated derailer protecting the exit of line 4 on to the single line. That derailer was located in an awkward location under a dark and tunnel-like bridge. He did tell the tamper crew that they needed to be on the ball, however, because he had only a short time to get them out on the single line, reverse and then cross over from the Quillan single-line home signal via main line 2 onto main line 1 in the direction of Narbonne. A non-stop fast train was due along main line 2, after which they had to be ready to move from the single line via main line 2 onto main line 1. After walking back to their machine and starting it up, they moved it towards the Quillan single line to get behind home signal 117. From there, as explained, they would reverse direction and move through the crossovers to main line 1 following the passage of train 51839, which was booked at 12:39 but travelling with a delay of about 12 minutes. In their hurry to get behind signal 117, however, ready to spurt away onto the main line, both men completely overlooked signal 120 showing a 'carré violet' aspect (shunt movement stop) and so committed a SPAD. It was only when they hit the derailer, hidden in the darkness under the old and narrow overbridge called Pont Routier d'Artigues, that the driver applied the emergency brake, but his derailing machine moved sideways to the right and fouled through line 2. The derailer had lifted up the left-hand wheel and via the axle had pushed the right-hand wheel over and off its rail head, depositing it near through track 2.

Following the derailment, and knowing that a train could pass at speed along that track at any moment, the driver frantically called the signaller and reported the situation, in response to which the signaller switched all cleared signals for the expected through service back to danger. Nothing further untoward happened after that, although the delays to trains were no doubt substantial for the rest of that day. The tamping machine driver and his mate were taken off the track for a spot of report writing and re-education, and after the relatively simple job of re-railing the tamping machine it did not go out to the worksite, as by then it required a few repairs and the resident machine at the worksite was back at work anyway.

Comment

This narrative makes clear how a simple oversight can give rise to a risky situation, quite apart from causing hassle for the signaller, tamper crew and train travellers at Carcassonne and beyond. An accident had been narrowly averted only because everyone involved was aware of the proximity of that delayed passenger train. The report of the French national transport accident investigation organisation (BEA-TT) berated the signaller for not pointing out signal 120 and the derailer behind it more clearly to the driver when explaining the move. As a British-trained former train driver, my hang-up about decent route knowledge again comes to the fore, but it should not be overlooked that these were track workers who operated all over the network, and that it would be unreasonable to expect them to know about peculiar local details such as derailers hidden under dark bridges. Signal 120, showing a 'carré violet' stop shunting aspect, however, was a full-size signal placed against a dark brick retaining wall at the entrance to the bridge. They ought to have spotted that first thing before even starting their engine. Had they done so and obeyed it, the derailer would not have been an issue. A bit of an ***Åsta*** situation here, I think, with ill-advised haste overtaking the required caution. On the one hand, mentioning the approaching train was perhaps not a good move on the part of the signaller as it created the stress under which the tamper crew did not want to hold up proceedings. They appeared to throw all caution to the wind, despite being in a yard they were not too familiar with. On the other hand, it made them report the derailment in double-quick time.

The derailer had been badly chosen for the job it had to do. According to the report, the type used (a hinged block that could be turned on and off the rail head) was a non-standard type that was meant to protect the single line to Quillan from slow-moving (4 km/h or 2½ mph) runaway vehicles out of the yard rather than for powered vehicles accelerating onto it. A 'half turnout' derailer away from the main lines would have been better in the circumstances, which was one of the two recommendations made by the BEA-TT in its report. The other recommendation was that traffic controllers should inform non-local drivers more concisely about essential details of the yard. Had the expected passenger train materialised, doing the permitted 110 km/h (65 mph) along track 2, then people would in all likelihood been hurt, if not killed.

***

Harmelen (2), The Netherlands, 29 March 2007 at 16:59

(High-speed SPAD due to distraction)

Introduction

This was another incident in this, from a railway point of view, still emotionally charged area of The Netherlands. Physically nothing much happened during the second incident; there was no collision, no one got hurt and there was no damage, but the reason why I have decided to include it is because it illustrates the following:

1. The influence of a fault on a train on the way a driver works it.
2. The influence of both distraction and faulty expectation.
3. How familiarity in a train driver can breed contempt.
4. How ATBEG ATP was bypassed, allowing a high-speed SPAD.
5. How another driver read the situation correctly and brought his 5,000-plus tonne freight train to a stop, thereby avoiding a collision.

The incident

'Intercity' semi-fast service 8861 from Leiden to Utrecht was involved in a 104 km/h (65 mph) Category A SPAD at a junction signal and came to a stop close to the trailing turnout where two parallel lines in the same direction became one line. Heavy freight train 48741 was approaching along the fouled line under clear signals but was brought to a stop in a timely manner at the covering junction signal. Incidentally, the use of the term 'Intercity' in The Netherlands does not indicate the standard of service or speed that the word implies in Sweden, Germany or Britain for example. Everywhere else in Europe the service would at best be called a regional express train.

Rolling stock involved

Train 8861 was worked by a VIRM (extended inter-regional rolling stock) bi-level or double-deck EMU from the large series of present-day standard longer-distance rolling stock in use with Dutch passenger operator NS-R. These were 1.5 kV dc high-acceleration and fast 4- or 6-car Holec tri-phase ac-driven electric multiple-unit trains, originally conceived to take up the 'Interregio' services in a three-tier train service timetable consisting of all-stations 'Sprinter' services, semi-fast 'Interregio' services and longer-distance limited stop 'Intercity' services. Whilst the three-tier domestic timetable idea was eventually dropped and limited-stop fast services were largely taken out in the central and western areas of the country (except the high-spec true inter-city international services such as ICE International and Thalys), the name 'Intercity' was retained for semi-fast services that did not stop at all stations. As a rule, virtually all of these services are reasonably limited stop along the core stretch of the route but stop at more stations closer to the terminal stations of their journey.

Train 48741 was one of the 5,400-tonne coal trains operated by Railion Freight (now DB Cargo) and consisting of the originally German-designed boxy side-discharge bogie hopper wagons. It was on its way from the port of Rotterdam to Germany via Gouda and Utrecht and was hauled by three multiple-united 80-tonne ex-NS Cargo Class 6400 centre-cab Bo'Bo' 1,180 kW (1,580 hp) ac driven diesel-electric locomotives (120 built by Siemens-MaK, nowadays Vossloh, as model DE 1002 from 1988). The locomotives were fitted to work in The Netherlands as well as in Germany to enable through running without a change of locomotive at the border. At present most of these trains use the dedicated, newly constructed electrified 'Betuwe' freight route and are now hauled by two multi-voltage/multi-ATP electric locomotives in multiple operation.

Incidentally, these same MaK/Vossloh diesel-electric locomotives can be spotted at Cheriton depot near Folkestone and along HS1 (High Speed 1) all the way from the Channel Tunnel to St Pancras International in London. Eurotunnel uses a number of them as works and rescue locomotives, the machines having been fitted with exhaust gas scrubbers for work inside the tunnel. During the winter of 2009/10, when snow ingress problems crippled five Eurostar sets inside the Channel Tunnel, a couple of these diesels brought a dud Eurostar set all the way to St Pancras International, proving that HS1 is capable of taking full-size continental rolling stock right into the British capital. A further two redundant ex-NS Cargo diesels recently arrived in Britain to extend the traction stable at Cheriton. The majority of these formerly Netherlands Railways Class 6400 DE locomotives, after the sell-off of NS-Cargo to DB-Cargo and therefore being owned by Railion, have been transferred to Germany to replace obsolete types of medium diesel traction there. Modern electric locomotives were commissioned to take over the heavy coal and ore trains between the coastal ports and the industrial Ruhr area that the diesels used to take in three- or four-locomotive sets before.

The incident log clearly illustrates the problems such a train can create for traffic controllers on the densely used rail network in The Netherlands or on similar networks elsewhere in the world. In order to stay out of the way of the glut of passenger services, these trains have to accelerate, run and brake expeditiously, which goes completely against the grain of working such trains. Hence the amount of power provided for 'only' 5,400 tonnes along what is a relatively flat-graded line with just a few humps when crossing bridges.

Changes resulting from the rail upgrade programme

The core rail network in The Netherlands (mainly the Randstad area bordered by the cities of Utrecht, Amsterdam, Haarlem, Leiden, Den Haag, Delft, Rotterdam and Gouda) is being quadrupled and extended with grade-separated junctions where necessary, to enable all-stations local and semi-fast 'Intercity' services to coexist in the anticipated high-frequency timetable. From 2002, all IRM 3-car units were extended with a fourth vehicle and all 4-car units were extended with two additional vehicles. The consist of new build was put together as per the new specification to make maximum use of train length in the case of 12-car rush hour sets by avoiding space being used for intermediate cabs wherever possible. For that reason the indication IRM was changed to VIRM. A few technical changes were also introduced, such as the preliminaries to enable future working off 25 kV 50 Hz ac overhead power and fleet deployment of a train management system (TMS), a screen-displayed status and diagnostics feature in the cabs that shows faults and failures and suggests possible remedies. In some older units a rather large video monitor was used to the right of the driver, but in newer units and in later series a more reasonably sized plasma screen came on stream. Whilst in itself TMS is a good labour- and delay-saving device, in daily reality it can be rather a distraction when reporting all sorts of minor faults and failures that have little or no influence on the journey but irritatingly requires the driver's attention. It appears that the same issue might have influenced the course of events here, as despite the TMS reported traction faults the unit made up 4 minutes of delay from a late start on a nearly all-stations timetable from Leiden to Woerden, which suggests that at least sufficient traction to maintain timekeeping was available. In fact, TMS was being introduced on these units at the time and a number of warning indications were perhaps somewhat superfluous.

Such VIRM units in 12-car configuration make an impressive sight and carry more than 1,500 seated passengers in appreciable space and comfort, certainly when compared to rolling stock found on similar services in Britain for example. When British Government officials visited The Netherlands they were impressed by this rolling stock and asked why it could not be employed in Britain to solve severe overcrowding on trains. It had to be explained that the traditionally cramped British loading gauge precludes its use there.

On a more personal note; a 4-car VIRM set was the last train I ever drove. That slightly sad event did not happen in my railway-wise 'native' Britain. In 2005, accompanied by an NS-R instructor driver, I took a 10-car IRM/VIRM set from Arnhem to Den Helder, came back with it to Alkmaar (where I rediscovered the joys of Dutch fast-food during the break), took another 10-car from Alkmaar to Amsterdam, dropped 6 cars and finally made my way back with a 4-car from Amsterdam via Utrecht to Arnhem. Interestingly, one of the things the instructor mentioned en route was his concern about the apparent lack of respect that new recruits and trainees tended to show for red signals. That was to prove a remarkable prophecy in the light of the string of collisions that occurred in the following years.

The location

In 2005, work had been completed to change the layout between the junctions west of Woerden via Woerden station to Harmelen Junction considerably from the situation pertaining in January 1962 when the first notorious ***Harmelen*** crash had occurred. The line between the two junctions via the station had been quadrupled, whilst the junctions between the lines to Gouda and to Leiden as well as at Harmelen between the lines to Breukelen and Utrecht were provided with flyovers to eliminate at-grade crossing moves. Although the accident as it had occurred in 1962 was now no longer possible, between Harmelen Junction and Utrecht the line still reverted from four tracks to two, for which reason 140 km/h (85 mph) turnouts had been provided. It was at this junction towards Utrecht, east of Harmelen, that this second incident occurred.

In the meantime the line between Harmelen and the Amsterdam–Rijn Canal bridge at Utrecht has been completely quadrupled and relocated to the top of an embankment, which additionally eliminated a number of AHB level crossings. As a result, it is no longer possible for this second incident to be repeated, because the entire line from Gouda to just outside Utrecht is four-tracked with grade-separated junctions. In the coming years a second bridge will be installed across the major waterway between Amsterdam and the Rhine, which completes the quadrupling effort. This area is controlled from the brand new Utrecht traffic control centre, a modern and fully electronic signalling centre that of course includes systems such as automatic route setting.

A reminder about ATP

The first generation of ATP (ATBEG) in The Netherlands allowed any train to pass a signal at danger at 40 km/h (25 mph) or lower. Any train running under this system only had to prove that the brake was applied after a new and lower target speed came up on the display – the so-called brake criterion – but there was no speed curve monitoring as with more modern systems. In the event of the brake criterion not being satisfied, this ATP system intervened by sounding a warning to the driver and then applying the emergency brake. However, once the brake was proven to be applied the system did not intervene, even if the speed was too high to stop at the red signal aspect. This played an important part in this incident. In the meantime this ATP system has been extended (ATB-vv) with speed traps at signals covering danger points and will no longer permit a train to approach a red aspect too fast without intervention. The present two Dutch ATP systems will eventually be replaced with ETCS at levels 1 and 2, which also has speed-curve monitoring in common with modern ATP systems such as ATBNG.

I found that on modern NS-R trains like the DM'90 and IRM/VIRM the application of the brake to step 2 (out of 7) would slow down the train sufficiently to ensure a normal service stop at a red signal, without losing undue speed or causing unnecessary delay if the red signal aspect came off to a less restricted aspect. This was something that NS-R drivers appeared to be quite concerned about. Whilst their rules state that a driver may not put off a brake application until the ATP warns to start braking, many drivers do actually wait until they are at the signal showing the restricted aspect before applying the brake, which is where the ATP speed-code will change. This runs contrary to how drivers are instructed in Britain, where they are expected to take measures such as shutting off power or even lightly applying the brake on spotting a restricted aspect ahead, in order to mentally prepare for having to stop. This diminishes the chance that the red signal ahead is forgotten, resulting in a SPAD or worse, or that a speed restriction is overlooked.

The incident log

After arriving at Leiden Centraal from Utrecht Centraal with his 6-car VIRM set, the driver, a Den Haag depot man with 24 years' experience and a clean record, changed ends and prepared his set to work back as IC 8861 to Utrecht again. Earlier that day he had already made one return trip from Leiden to Utrecht along this route, so this was his second outward trip to Utrecht. On setting up his cab he noticed that the TMS screen indicated a traction fault. He called the relevant authorities on his mobile phone and ended up doing the usual thing in such circumstances – switching out and rebooting his cab. As a result, his departure from Leiden was 4 minutes late. He worked the train along the largely single-track electrified line via Alphen aan den Rijn to Woerden, calling at all stations. During this trip the traction fault was repeatedly indicated and kept him occupied.

Freight train 48741 was rumbling along the line from Rotterdam via Gouda to Woerden, on time and doing approximately 100 km/h (60 mph). It was being followed by – and was holding up – two slightly late running 'Intercity' services, 2861 (4 minutes late) and 2061 (2 minutes late), that should already have been ahead of it past Gouda. After considering the delays to all the trains involved, the Utrecht traffic controller had decided to stop the freight at Woerden to enable both passenger services for the north of the country to get past. Such a stop would delay a freight of this make-up quite dramatically, so in order to minimise the negative impact for the rest of the route to the border with Germany he allowed the freight to leave ahead of train 8861, which rolled into Woerden on time, just after 48741 had departed. At that moment, however, as booked the freight should have been well past Harmelen Junction on the way to Utrecht and out of the way of train 8861.

After the stop at Woerden the driver of train 8861 departed at 16:56, still busy phoning about his traction faults. He also consulted his diagram card to get an idea about his delay, even if by then he was actually running on time. Whilst doing all this as well as driving his train, he overtook the still-accelerating freight before reaching the Harmelen Junction flyover. In the right-hand curve past the flyover, in between the catenary posts ahead, he vaguely spotted a green signal. Expecting this signal to be his, going by the schedule and the normal course of running this service and expecting that normally he had priority over the freight train, he paid no further attention to signal aspects and got on with his phoning. That was a serious mistake on the approach to a location where four lines become two.

In fact, the green signal he had seen was signal 1014 for freight train 48741 on the line to his left. It was located on a gantry with three other signals, as in good Dutch order all four lines are fully reversible. He had missed that his proper signal, 1016 to the right of signal 1014 on the same gantry, was showing a red aspect. It was only when he got closer and looked up that he realised he was about to commit a SPAD at no less than 138 km/h (85 mph). He immediately put his brake controller in the emergency position, went past signal 1016 at 104 km/h (65 mph) and put out an emergency call, finally coming to a stop 226 m (740 ft) further down the line, close to where both lines come together at turnout 1015. Because he feared that he was fouling the line along which the freight train was approaching, he decided to set back a few metres until he was certain that he had cleared the loading gauge of the line to his left.

It is remarkable that although he missed the yellow distant aspect at the previous signal in his self-inflicted state of distraction, he nevertheless routinely reacted to an ATP warning for a permanent speed restriction ahead by shutting off power and putting his brake controller in the initial brake position, without spotting the restrictive signal and the ATP speed-code change for the red aspect ahead. Unfortunately, that habitual action for the permanent speed restriction resulted in the ATP brake application criterion being satisfied, so the ATBEG did not act to warn of the inappropriate speed for the approach to the red signal.

The freight train driver had the power controller wide open in order to let his three locomotives accelerate the substantially delayed coal train to the permitted train speed whilst running under green signals. When, much to his surprise, he was overtaken by a fast-running passenger train that should not have been moving at that speed if it needed to stop at the turnout where the two lines converged after the next signal, he decided not to take any chances. He shut off power and started to brake again even before seeing his signal 1014 come back from a green to a red aspect when the offending train ahead hit the track circuit at the set of turnouts. In fact, he was able to stop his train at signal 1014 and thus eliminated any chance of a collision. That was an admirable example of a focused railwayman using his insight and route knowledge to excellent effect.

Comment

Neither signal 1014 nor signal 1016 were known as multi-SPAD signals. They adhered to relevant specifications with respect to visibility and sighting and from my own experience I can confirm that they were absolutely not hard to spot. Nor was the fact that four signals were fitted on the gantry a problem; in case of doubt, count them out. But, as between Paddington and Ladbroke Grove, or at similar layouts around the world; as a matter of route knowledge you do have to know what line you are travelling on. And it does require a driver to look conscientiously at such signals to avoid cross- or through-reading of other signals, as clearly happened here. These two signals covering trailing turnout 1015 were high-speed signals, so they had 200 m (660 ft) overshoot space allocated. Note, however, that this is not an overlap in the British sense where the track circuit separation, the insulated rail joint or block joint, is at the end of the overlap well past the associated signal. In The Netherlands, this insulated rail joint is at the signal but space behind the signal is allowed as overshoot space into a potentially occupied block section.

However, due to the length of the line between the signals and the turnout, contrary to normal practice, extra track circuits had been installed from the signal to the trailing turnout. These would put the signal back to red after passage and separate the ATP speed codes to the trains on either line, in order to prevent a SPAD from erroneously picking up the clear line speed code of the other track and so potentially release the brakes again. But for that reason the offending train only put the signal for the freight on the other line back to red when it entered the actual danger zone, the extra track circuit of turnout 1015. This took 13 seconds after passing signal 1016 at danger and that was 13 seconds lost in which to warn a driver who perhaps was less on the ball than the one at the controls of freight train 48741.

There is a touch of ***Ladbroke Grove*** in this situation. The circumstances went beyond what had been considered as potential risk at the design stage and therefore the margins allowed for safety failed to secure it. That was a dangerous situation, but the relevant people no doubt took proper note. In fact, this set-up had been designed in 2005, before all such situations had to be thoroughly risk-assessed as is the case now. Clearly, ATB-vv and the four-track layout now avoid this particular conflict point altogether.

Another issue mentioned in the incident report was the lack of flank protection at these junctions, probably with derailers. That had not been part of the specification for this layout in view of the long overshoot spaces available, but would have eliminated the collision impact risk between the two trains. However, I have my deep reservations about the safety of putting double-deck trains at speed into the dirt, particularly when the line is on top of an embankment or next to a watercourse as is so often the case in The Netherlands.

In 2006, there were 27 SPADs on approaches to danger points with speeds of 130 km/h (80 mph) or higher in The Netherlands, and in the 8 years between 1999 and 2006 inclusive there were 2,618 SPADs. The speed was known in 1,307 of those SPADs: on 19 occasions the SPAD occurred at a speed between 40 and 50 km/h, five were between 50 and 60 km/h, seven were between 60 and 70 km/h, one was between 70 and 80 km/h, three were between 80 and 90 km/h, one between 90 and

100 km/h, and two were higher than 100 km/h. Three SPADs occurred at a danger point with a permitted speed of 100 km/h or more. Long live ATP!

***

Kerang, Victoria, Australia, 5 June 2007 at 13:34
(Level crossing accident)

The accident
This significant level crossing accident is included because it illustrates the negative influence on someone dealing with potentially dangerous situations based on expectations stemming from long-standing routine. It also raises a number of technical traffic matters for consideration, whilst the negative influence of prattling politicians on proceedings during the investigation is of interest as well. Eleven people died as a result of the accident and 13 were seriously injured. One of those injured was the truck driver (or 'truckie' to use the Australian term) from whose point of view the story is largely described.

The location
The accident occurred at automatic open level crossing Y2943 (Fairley) on an unsignalled single-track non-electrified railway line 6 km (3¾ miles) north-west of the town of Kerang, 294.31 km (about 180 miles) from the line's zero point at Melbourne. Permitted line speed was 90 km/h (55 mph). The surrounding land was low-lying and flat with the railway running on a low embankment, which meant that the road climbed slightly on the approach to the crossing. Some eucalyptus trees blocked the sight of drivers of approaching trains from Swan Hill to road vehicles coming from Kerang. The four V/Line passenger trains per day on the line (two outbound and two inbound) were run without lineside signalling, being controlled by radio train warrants (authorisations to enter a section) remotely issued via telecommunications provisions from the Central Train Traffic Controllers (Centrol) in Melbourne. This type of control system bears more than a passing resemblance to the systems used during the accidents at ***Holzdorf*** and ***Winsum***, but did not contribute to the accident in this case. The line between Kerang and Swan Hill crosses the Murray Valley Highway at this level crossing, a two-lane main road for heavy traffic with a maximum permitted road speed of 100 km/h (60 mph).

The automatic open level crossing
The automatic open level crossing was secured with the appropriate road signs and also sets of alternating red flashing (wig-wag) lights, but had no physical boom-barriers that come down to a horizontal position to block the passage of a road vehicle.

A train approaching from the direction of Swan Hill would strike in (set off the level crossing approach track circuit) 685 m (2,250 ft) from the level crossing. From there a lazy right-hand curve began to 106 m (350 ft) before the crossing. The warning time from strike-in until the arrival of the train on the crossing was 25.4 seconds. In the direction the road freight vehicle was travelling, from Kerang to Swan Hill, the Murray Valley Highway approached the railway line at an angle of 338° true, curved toward the crossing at 300 m (990 ft) before reaching it and then straightened out again at 354° true to cross the railway line at a 40° angle. This had a bearing on the accident. A traffic check in 2004 registered 2,316 vehicles using the crossing per day (25% were commercial vehicles with an axle load heavier than 4.5 tonnes) at a speed of 100 km/h (60 mph), and there is no reason to assume that the situation at the time of the accident was much different. With the line being used by only four trains a day, the passage of a train was a rare event, which is an important contributory factor to the level crossing safety problem in Australia. Road users generally do not expect to encounter a train at a level crossing.

The train, rolling stock and traincrew
Train 8042, the last train of the day to Melbourne, had departed Swan Hill at 13:00 and was composed of a GM design-based N-type Co'Co' diesel-electric locomotive, No N460, built between 1985 and 1987 by Clyde Engineering. The locomotive, a very common type on Australia's railways, was 20 m (66 ft) long, weighed 124 tonnes, delivered 1,846 kW (2,475 hp) and had a maximum permitted speed of 115 km/h (70 mph). Due to its setting in relation to wheel-tyre wear, the speedometer of N460 was found to indicate a lower than actual speed, and consequently the train was travelling about 6 km/h (4 mph) faster than permitted, although that had no bearing on this accident. The locomotive was fitted with an OTDR that was downloaded and extensively consulted during the investigations.

The passenger accommodation was provided by 3-coach set N7, which was 88.4 m (290 ft) long, weighed 130 tonnes and consisted of the following vehicles as coupled in order from the locomotive:

- Coach one, ACN21, contained 52 first-class passenger seats with a luggage van and a conductor's office. This vehicle, which was 22.8 m (75 ft) long and weighed 43 tonnes, was not damaged in the collision.

- Coach two, BRN20, contained 66 economy-class passenger seats (21 of which were occupied) and a buffet area. It was also 22.8 m long but weighed 44 tonnes. The road freight vehicle impacted with the right-hand side of the coach (in the direction of travel), 14 of the occupied seats being on this side of the coach. Ten passengers died and another 10 were seriously injured, one of the fatalities occurring on the non-impact side when their seat was hit by the front of the road freight trailer.
- Coach three, BN19, contained 88 economy-class seats (seven of which were occupied) and was the same length and weight as ACN21. One fatality occurred in this coach and three others were seriously injured (two by the detached compressor of the road trailer refrigerator unit that detached itself inside this coach). All were seated on the impact side near the front end of the coach.

All the coaches were from a series built between 1981 and 1983 in the Newport workshops at Melbourne. They were fairly heavy compared to European coaches that are 4 m (13 ft) longer, possibly due to the fact that they rode on rather heavy Commonwealth-type equaliser bar bogies. On the other hand, they had the modern (originally German designed) rubber tube gangway connectors fitted to their headstocks and were coupled with automatic couplers. The accident report commented on the age and the design philosophy of these coaches, although that had made no difference as more modern coaches would certainly not have provided better protection for their occupants in the circumstances of this accident. Both passenger rail vehicles as well as buses are highly vulnerable to this sort of broadside attack, as they are not constructed to be capable of withstanding such an impact. Neither are road and rail freight vehicles, incidentally.

The traincrew consisted of one driver and two conductors, one of whom was responsible for checking the tickets and the usual duties of a senior conductor while the other manned the buffet. The way in which the train was handled was in accordance with the rules, and both the road and rail drivers involved were properly qualified.

The road freight vehicle

The prime mover (tractor vehicle in British English) was a 3-axle US-designed Kenworth K104 Aerodyne vehicle, manufactured in Australia in 1999 and weighing some 9 tonnes. Although certified to move a B double-articulated trailer set of 62 tonnes, it was hauling only a single articulated trailer on this particular trip. The tractor vehicle had a Detroit Diesel power unit of 373 kW (500 hp) and was fitted with an event recorder. Unfortunately, all stored data were lost when battery power was cut on impact, which was an almost unbelievable flaw in its electronic architecture.

The single articulated trailer, a 3-axle 13.57 m (45 ft) Kurtainer insulated vehicle designed to load 22 units of palleted goods, was fitted with a Carrier Ultra XL refrigeration unit. Its tare weight was 9.5 tonnes and it was loaded with 2 tonnes of general merchandise and 14.5 tonnes of MDF board, giving a weight of 26 tonnes and a combined tractor and trailer weight of 35 tonnes. The combination had the usual pneumatically operated brakes throughout, the tractor brakes applying before the trailer brakes did for direction-keeping purposes. The tractor vehicle had an anti-lock braking system (ABS) although the trailer did not, the skid marks found at the site having been caused by the locking trailer wheels. The vehicles had been maintained satisfactorily. For the benefit of readers unfamiliar with Australia, traffic drives on the left-hand side of the road and vehicles are normally driven from the right-hand front seat.

The accident log

Trucker Christiaan Scholl was a native of The Netherlands but had been resident in Australia for 20 years, having driven heavy goods vehicles for 23 years in both countries. He reported for work at his Wangaratta depot at 09:00 on the morning of the accident, having come back from 4 weeks' leave and was due to depart at 10:00 for a regular weekly trip to Adelaide in South Australia. However, owing to a consignee turning up late with his cargo, his departure was delayed by 30 minutes. The sun was shining, the temperature was about 12°C, there was light, scattered cloud cover and the visibility range was estimated to be 50 km (30 miles). However, as a result of the powerful sunlight, the clear, dry atmosphere and the angle at which the sunlight reached the ground at midday, there was considerable road glare that made it difficult to see distant objects.

At 13:00 V/Line train 8O42 departed from Swan Hill on its long trip to Melbourne. It was lightly loaded and was doing about 100 km/h (60 mph), speeding slightly due to the low-reading speedometer already mentioned. It approached Kerang level crossing as booked at about 13:30. The train driver saw two vehicles slowing down for the crossing on his left-hand side and a white articulated truck approaching on his right-hand side. At a whistle board he sounded the loud train horn as required, but only for about half a second, and when the train was about 140 m (460 ft) from the crossing he gave a 7-second blast.

After approximately 3 hours of driving from Wangaratta, trucker Scholl was still about half an hour behind schedule when he approached the level crossing at Kerang. He would normally have passed the spot unhindered at 13:00, but due to his half-hour delay was now at the crossing at the booked time for one of the few trains on this line. As the truck negotiated the slight right-hand bend towards the railway line at a speed of 100 km/h (60 mph) the angle of the sunlight moved away from his central field of vision and the glare lessened to a degree. It was then that Scholl noticed cars stopped on the opposite side of the crossing and saw the flashing wig-wag lights. Looking to his right and left in order to assess the situation, he saw the quickly approaching train on his left. He instantly realised that he would be unable to cross ahead of the train nor bring his vehicle to a stop in the 50-60 m (165-200 ft) of road left. He stood up out of his seat, rammed his foot on the brake pedal and 49 m (160 ft) before the crossing turned his steering wheel to the left. With the trailer wheels locked and smoking, as described by the driver of a car behind him, the combination moved over to the left and off the metalled road into the dirt.

The driver of train 8O42 saw the truck starting to swerve with its driver standing behind the wheel. Reckoning that the truck was either going to hit his train behind him or would pass behind his train, he did not apply the brake, which was a good assessment of the situation and how it should have been dealt with. Nevertheless, he felt an ominous jerk as the truck hit his train and then the brakes applied automatically, indicating that the brake-pipe connection must have been severed somewhere. He turned his brake valve to emergency and prepared for assistance to casualties after coming to a stop.

The truck hit the second carriage of the train at an angle of 32°, truck and train going in opposite directions. The truck was moving at approximately 60-65 km/h (35-40 mph) towards the rear of the train whilst the train was travelling at 90 km/h (55 mph) so the impact speed was about 150 km/h (95 mph). Whilst the train had violently stopped the vehicle and mauled the tractor, the right-hand edge of the trailer front gashed the carriage side and reached deep into the interior above solebar level of coach two, stripping out the plating plus all the seats and their occupants after the impact point on that side. Everything ended up at the rear of the coach and, although the truck was now stopped or being pushed back, the train was still moving fast. After the rear bogie of coach two derailed, it was coach three that the trailer sliced into and started to open up. However, the collision pillar in the headstock bulkhead of this coach, next to the gangway door, held and dealt a massive backward blow to the trailer, tipping the cab of the still-attached but badly wrecked tractor vehicle forward.

Three further things resulted. First, the coupler drawbar shank of the rail vehicle broke off under the impact of this blow, which detached coach three from the rest of the train. Secondly, coach three fully derailed and thus was able to move away from the slicing trailer front, sparing everyone behind the exit point. Thirdly, the compressor of the road trailer refrigeration unit detached on impact and landed in this coach, injuring two occupants of the seating bay it ended up in. The locomotive and its first two coaches came to a stop well past the crossing, with only the rear bogie of coach two derailed. Coach three stood at a shallow angle to the left of its track but still upright on its wheels.

The driver of the locomotive immediately contacted Centrol in Melbourne to report the accident, whilst witnesses waiting in their cars at the crossing called the emergency services. Then the train driver left his cab with a first aid box and went back to see whether he could render assistance. Despite both conductors on the train being badly shaken up, they were nevertheless able to take up similar duties. Given the distances involved, the first ambulance was at the scene remarkably quickly, within 10 minutes of the accident.

Just a few hours later Transport Minister Lynne Kosky provided the standard Australian transport authorities' mantra, publicly blaming the trucker for trying to beat the train to the level crossing. Assistant police commissioner Noel Ashby went one further. When answering a question at a press conference about whether boom-barriers would have made a difference he stated that it made no difference had the Great Wall of China been in front of the crossing. This truck, trying to beat the train, would have hit anyway. Two days later, Scholl was formally told that he was being held responsible for causing death through culpable driving. As a result of the obvious government bias against him, and probably on the advice of his union, Scholl decided not to co-operate with the investigation of the Office of the Chief Investigator (OCI). Meanwhile people had started throwing dirt and rocks at passing trucks belonging to Scholl's haulage company and others. The 'truckie' lore in Australia had received quite a few dents lately. The blame had apparently already been firmly placed, but it was only then that the investigation of the OCI got into its stride.

The investigation

The crash was re-enacted as a computer model based on the locomotive recorder timings, the level crossing recorder timings and skid mark calculations, the results being used to re-enact the events in reality with a trainset and a truck in an attempt to discover what issues could have caused Scholl not to react to the level crossing warning. The weather co-operated, and some interesting things emerged:

- At a distance of 360 m (1,180 ft) from the level crossing Scholl could have seen the outlines of the warning light assemblies but not their warning due to misalignment as a result of the curve in the road and the prevalent road glare. In fact, he declared in 2009 that he did indeed check the warning lights when about 300 m (990 ft) from the crossing and saw no flashing lights so continued at 100 km/h (60 mph). The train would at that point have been a small object, hardly visible between the branches of a row of eucalyptus trees.
- When 260 m (850 ft) away, whilst rounding the curve towards the crossing, flashing warning lights would have gradually become visible, but the trees would have hidden the train from Scholl's vision. The truck was still doing 100 km/h at this point.
- At a distance of 106 m (350 ft) from the crossing the road straightened out again and the train would then have become visible on the left-hand side, had Scholl been looking for it. Both sets of flashing lights at the crossing would now be visible, as would the stationary vehicles on the opposite side of the level crossing, which was the point when Scholl started to react but with only 4 seconds of travel left before impact.

From here the accident report quoted some interesting facts about human response to unexpected events, bringing some of the facts surrounding the ***Ladbroke Grove*** accident to mind. Considering that Scholl had only 4 seconds left before impact, reaction time to urgent stimuli in the case of unexpected problems was generally looked at as ranging from 2.5 to 7 seconds, depending on one's application to the job in hand and one's state of preparedness to react to emergencies, which in Scholl's case was virtually zero. This, then, was the explanation for the measured 50-m (165-ft) braking distance out of 106 m (350 ft) from coming through the curve that had led both Kosky and Ashby to say that Scholl was trying to beat the train to the crossing. He was not! It was just that he had never experienced any problems of this kind before and had to reorientate and react in a split-second whilst doing 100 km/h with a heavy truck and suddenly finding that the situation was very serious indeed. His missing of the clues was due to his central vision being concentrated on coming through the curve in the road glare and to his peripheral vision not picking up the misaligned beams of the warning lights. These only became readily visible during the rounding of the curve, after which Scholl had only seconds left to react.

As a result of the glare on the road ahead, the visibility of various details in the landscape was seriously impaired, particularly if the windscreen was not scrupulously clean, which it would not have been after a couple of hours of driving at speed through dusty terrain. It could not be established beyond doubt whether dirt-impaired vision or other issues with the windscreen exacerbated these problems, but it is probable that they did. The position of the sun had meant that the light was not directed towards the train, which was therefore in a visual shadow and even less conspicuous.

What about the use of the train's horns? The NTSB in Washington DC had previously carried out several tests on the penetration of the sound of train horns into vehicle interiors because many accidents had occurred in the US despite train drivers proving with their recorder that they had used the horns liberally, in the face of statements to the contrary from surviving car occupants. It was found that many road vehicle interiors are actually sound-deadening to enhance the atmosphere of silent comfort. The effect of noises from the wheels on the road, the engine, air conditioning, music or voices on the radio or from other occupants of the car also needs to be taken into account. Lastly, and often overlooked, is the matter of impaired hearing (and sight) suffered by many motorists, as in the *Ban Ban Springs* accident (see below).

Comment

Australia is a massive continent with a unique problem in view of the number of its public level crossings (around 9,400 of them). Other large nations share some of the characteristics, with long straight roads covering enormous distances under a hot sun through arid plains, a remote and often featureless landscape, and the loneliness of the road, all of which affect the concentration and alertness of vehicle drivers, but Australia has them all, and in great quantities. Foreign visitors who intend to drive a car there would do well to inform themselves through government websites or the Australian Automobile Association publications. The section that deals with the infamous 5-vehicle road trains is particularly instructive.

As far as the interface between railway and road operations is concerned, two characteristically Australian issues play a major role in the comparatively frequent high-damage accidents that occur between road freight vehicles and trains. One aspect is the size of those road freight vehicles, and the second is the absence of trains. The relative chance of road and rail vehicles meeting at a level crossing is actually very low and that is exactly what level crossing safety in the outback is all about. This low frequency of possible occurrences certainly played a role at Kerang, and the accident at *Ban Ban Springs* near Darwin, Northern Territory, where on 12 December 2006 the famous long-distance passenger train 'The Ghan' hit a B double-articulated road train, also illustrates both issues nicely.

The driver of that road train had used the level crossing several times per day in the past weeks and had never seen and only once or twice heard a train in all that time. He had also never adhered to the traffic signs that required him to come to a stop at the crossing and to proceed only after ensuring that no train was coming (so-called passive protection). Why did he not do that? Simple; too much effort with the kind of vehicle he was driving compared to the perceived risk he was taking. The ATSB calculated that coming to a controlled stop from the usual road speed of 100 km/h (60 mph) with 210-odd tonnes of a prime mover plus a fully loaded B+2A four-trailer road train would take serious effort and about a kilometre of roadway to start with, but, more interestingly, it was also reported in the *Ban Ban Springs* accident report that it then takes 71 seconds, well more than a minute, from beginning to move the vehicle combination again until clearing the level crossing. It meant that where sight lines were impaired (e.g. a curved track with trees and shrubs beside the line, fogginess, blinding sunlight, etc) it was quite possible for a railway train doing 120 km/h (75 mph) not to be noticed and to hit the road train without anyone being able to do a thing about it. In fact, there is arguably a positive safety aspect attached to having a road train clear a level crossing as quickly as possible (there is a good reason why such a road vehicle combination is called a road train). The way to secure safe passage of such vehicles in these circumstances should be thoroughly reconsidered for the sake of robust rail and road safety, as urged by Australian coroners time and again following fatalities. And, yes, it is going to cost money.

The phenomenon of expectation in road-users also needs to be considered. During investigations after level crossing incidents in Australia and the US it was found that motorists simply did not expect to encounter a train, so they were not concentrating on picking up potential clues and warnings regarding the approach of a train as most people would do in Europe. It meant that the infrequency of seeing a train at level crossings on many routes in Australia led to the routine of simply bumping one's vehicle across them without properly looking out for a train. When this is taken together with the questionable signage and level of equipment at this important Murray Valley Highway level crossing, the various factors that contributed to this accident become evident.

How did that issue influence Scholl's (and others') patterns of expectations on rail traffic frequency as described above? Despite 192 level crossing collisions occurring between 2002 and 2012 in the state of Victoria alone (8% with fatal results), the various state governments as well as the Australian Federal Government did not want to spend money on improved level crossing safety, as so many lines carried a very few trains per day (if that), and people, after all, do have a responsibility to look after their own safety. This, in fact, explains the pre-emptive shots across the bow fired by Lynne Kosky and Noel Ashby. Both knew these questions would crop up sooner rather than later, as the matter was already a talking point in Australia following a number of other such accidents. They had a point with respect to irresponsible road-user behaviour at level crossings, don't get me wrong, but in this case they talked before they were entitled to do that and they were entirely wrong. Moreover they impeded the Kerang investigation proceedings without a second thought for the proper course of justice. And to the best of my knowledge they saw no need to apologise for subverting the investigations later; not to their electorate and certainly not to Mr. Scholl.

As far as the Kerang crash is concerned, would the familiar outline of descending level crossing barriers indeed not have more effectively and quickly alerted Scholl to the warning at the crossing, in view of his statement at the coroner's court at Bendigo in October 2011 that he checked the crossing warning lights from a distance of 300 m (990 ft) distance and saw nothing? I also wonder if the improvements that were later made at this accident site on the Murray Valley Highway – such as distant road signs with flashing lights for the crossing, rumble strips to make the steering wheel shudder, realigned warning signs with LEDs instead of tungsten light bulbs and upgraded road markings and signing – would not have helped to prevent this accident. Following these changes no further accidents were reported at Kerang, whereas on other open crossings in the neighbourhood accidents continued to occur with depressing regularity. The aftermath of the Kerang accident proved that well-equipped level crossings with boom-barriers are substantially safer, something that was acted upon notably in Britain, France and The

Netherlands, and that there was absolutely no need for Great Walls of China to bring about such safety improvements.

In May 2009, Scholl, a man described as a conscientious and careful trucker (and regarded as a bit of a pansy by some of his colleagues) was cleared of the charge of causing deaths through culpable driving. Families of the victims later sued him for damages, but what might have made such legal actions successful?

1 In Britain there is evidence that train drivers resuming work after a few weeks' time off are more likely to be involved in incidents. Scholl had been off work for four weeks; should he have been more careful on this first trip? But in what way? Should he have slowed down for every level crossing? Would it indeed not have been prudent to simply slow down on the approach to a level crossing when under conditions of impaired vision and with a vehicle the size of Scholl's truck? That required awareness stemming from previous experience, however, which was ten years of no train.
2 Did Scholl, in fact, try to beat the train? Thinking he'd have to wait ages for it to pass, but missing the fact that this was a short passenger train that was travelling faster than the average freight train; a familiar accident factor in the US? That is a nonsense argument. Scholl had never seen a train at that crossing in all the ten years he had passed it at 13:00 and no long and slow-running freights used that line anyway.
3 What was the possible influence of the 30-minute delay to the start of Scholl's trip, which resulted in a change to his routine journey against his ingrained expectations after a decade of these trips? Were there additional issues with similar sorts of absent-mindedness that made his driving less safe? Delays and their consequent distraction from set routines are known to have been behind many accidents throughout the world.

On 21 October 2013, he was finally cleared of all claims and charges. On that day he expressed his regret about the accident and offered his sincere apologies to the families who had lost relatives in the crash.

***

Ohinewai, New Zealand, 19 June 2007 at 11:05

(Engineering train accident due to staff errors)

Introduction

The third accident due to permanent way activities in this book, this clearly shows the familiar dangers that accompany working along railway lines when trains are still running through or past the worksite. These dangers arise particularly from the big three unsatisfactory issues that nearly always lead to these p-way accidents: training, organisation and communication. All three played a crucial role in this accident.

New Zealand railway track manager Ontrack and the train operating companies learned a lot from this accident, and a programme of improvements to training and safety was implemented in 2008 and 2009.

The accident

Passing freight train MP2 struck two crane jibs of work train WT22 that was engaged in recovering worn and removed rail from a rail replacement site on the Down main line of the North Island Main Trunk route. On impact the crane jibs swung round and hit one of the track workers. He was thrown off the wagon on which he was standing and under the passing freight train, fatally injuring him.

The location

Ohinewai is located between Huntly and Te Kauwhata, just north of Hamilton and Te Rapa, along the lower top end of the North Island Main Trunk line, which passes between lakes Waikare and Whangape on its way towards Auckland. The line at the location of the accident is double-track, not electrified – the 25 kV ac wires from Wellington stop at Hamilton/Te Rapa (see the ***Hamilton*** level crossing incident) – and is built with New Zealand Cape-gauge track of heavy rail on concrete sleepers fastened with Pandrol clips. It is fitted with route-indicating multiple-aspect colour-light signalling, as can be seen in Britain. The line is straight and level in the Ohinewai area, with good forward view for traincrews, and freight train speed along the line is 80 km/h (50 mph).

The North Island Main Trunk route

The North Island Main Trunk line between Auckland and Wellington was fully opened by Prime Minister Joseph Ward when he hammered the last spike home at Manganuioteao on 5 November 1908 after 22 years of hard construction work through very difficult mountain terrain. The greatest challenge was the steep climb to Waimarino via the Raurimu spiral, including two tunnels, three horseshoe curves and one complete spiral circle, and massive steel bridges were also needed to cross the gorges at Makatote, Hapuawhenua, Mangaweka and Makohine. This important line opened up the interior of the North Island and made settlement possible, whilst travel around the island no longer required braving the Pacific Ocean along New Zealand's dangerous coastline. Incidentally, the line also spelled the end for the massive native forests on the island, although reforestation is now taking place. At the time of the accident, 38 trains were scheduled to pass through Ohinewai each weekday, of which 10 ran along the Up main and five along the Down main between 07:00 and 16:00, the period during which engineering train WT22 would be active along the Down main line.

The traction and rolling stock involved

Work train WT22 consisted of a locomotive with six EWR-type rail recovery wagons operated by Ontrack, numbered from the locomotive as 387, 237, 346, 102, 143 and 171. These vehicles were made up from a US type of bogie flat wagon, upgraded with a centrally mounted petrol motor driven hydraulic crane with an extendable jib and used to load worn rail after re-railing. The normal manning level for these vehicles was two persons – one crane operator and one man on the track to hook up the rails. This train should, therefore, have had 12 gangers plus a person in charge (PIC) and one or two train drivers, but WT22 was operated with two drivers and only six gangers plus the PIC, which was acceptable to Ontrack at the time. If the cranes on the wagons were turned to the six-foot space between the tracks of a double-track line, their jibs intruded into the kinetic envelope of the adjacent running line, which was the root cause of this accident.

Train WT22 was hauled by locomotive DC 4565, a member of the once very widely used DC class of diesel-electric locomotive. The type was a 1978 Clyde Engineering 1,230 kW (1,650 hp) rebuild of the originally 85 examples of the GM EMD model G12, delivered as the DA class A1A'A1A' type from the 1950s, with a maximum speed of 100 km/h (60 mph). By 2003, only 50 were still in service, and the use of this machine on an engineering train was typical of the sort of jobs to which the class, like any obsolete class of traction, became relegated. Nevertheless, reconditioned class members also worked the SD/SA push-pull urban trains around ***Auckland*** until electrification was completed in 2015, the SD/SA coach sets being rebuilds from a batch of expatriate British Rail Mk II coaches.

The problem with New Zealand's prototypically US-engineered diesel-electric 'hood' type of locomotives is that there is a cab at one end only. As a consequence, like in the US or Canada on such local trips with a single machine, there are instances where the long engine hood is leading. This is tantamount to working a steam locomotive with the boiler obstructing half the forward view and for that reason multiple-operated machines are coupled in such a manner that the outer locomotives have their cab ends leading. Singly operated machines, when working with the long hood leading, must be worked with two drivers positioned either side of the cab, like the driver and fireman on a steam locomotive. This was the case with DC 4565 at the time of the accident, but two drivers were a requirement for p-way trains anyway.

The other train involved was MP2, a booked Tuesday-to-Friday-only express freight service operated by Toll Rail from Mount Maunganui yard to Auckland. At the time of the accident it was running 15 minutes early. It is important for the purposes of this narrative to know that the train controller (and consequently a p-way worker and the on-train driver of WT22) for some time erroneously referred to the identity of this train as MP30, which was a Monday-only express freight from Tauranga to Auckland. At the location involved in this accident, train MP30 used the track slot that was occupied by MP2 from Tuesday to Friday, and this had a clear bearing on the outcome. Train MP2 was operated cab-forward with a single train driver.

Re-railing

In order to understand the type of work WT22 was doing, it is useful to explain re-railing for those who are not acquainted with p-way work on the railway. To start with, the new rail is delivered by a special train. In this case two 67-m (220-ft) lengths of rail at a time are offloaded on to the centre of the sleepers from the rear of the slow-moving rail transport train, one rail for each side of the track, length after length. The next move may be done by groups of gangers or worked by an equally specialised re-railing train, rolling on the old rail at the front, rolling on tracked wheels in the middle and on the new rail in the rear, of which machinery at the front undoes the fasteners that hold the rails to the sleepers, tips the old rail off the sleepers to the outside of the track and at the same time puts the newly delivered rail from the centre of the sleepers into its place on the baseplates and puts in new fasteners. In the majority of cases it is normal practice to place the newly delivered rail within the four-foot and the discarded rail after re-railing outside it.

Engineering train WT22 was detailed to pick up the discarded rail lengths from either side of the track of the Down main line, the cranes reaching towards the Up main line for the rail that was discarded into the six-foot between the Up and Down tracks. Old rail is often used for sidings, or secondary track if the rail head allows another run of profile grinding; otherwise it is cut into short lengths and sold as scrap.

Possible organisational influences on the accident

In 1993, the cash-strapped New Zealand Government sold off the operation of the national railway to a train operating consortium. The venture initially looked promising but by 2004 had started to go into the red and the quality and frequency of operations suffered. At the time of the accident the government was involved in renationalising the railways to safeguard the provision of services. Such moments have often had adverse effects on rail safety; Sweden, Britain and Germany are other examples.

The accident log

At 07:22 on 19 June 2007, diesel-electric locomotive DC 4565 departed from Hamilton/Te Rapa traction depot, travelled north along the Up Main line, its long hood leading, and arrived at Te Kauwhata at approximately 08:10. The locomotive ran into the sidings onto the southern end of the six EWR engineering wagons that were to be its train, coupled up and did a brake test. The cab end was now leading. The regional train controller rang the PIC at the worksite near Ohinewai at 07:36 to ask for the precise location of the site and the foreseen duration of the job. The answer was that they would work from the Down Main line to pick up worn rail, but the PIC was not sure about the exact location and said he would call back in about 15 minutes. The train controller then said he would authorise WT22 to depart to Ohinewai and await instructions there. The PIC called back at 08:08, stated the location of the worksite as being Down Main line km 583.6 towards Huntly and then asked what the maximum available time slot between services was to do the job. The train controller told him there was a 2-hour gap after southbound express freight MP3 had passed by along the Down Main line. In the light of what followed, the train controller could at that moment have obtained further clarification of details such as whether just one track would be involved or both tracks. That would have had a major influence on what happened.

At 08:00 a lengthman (p-way man) in road-rail vehicle 46285 called the train controller by radio from a location north of Horotiu to request a slot along the Up Main for him to make his way to the north of Huntly, expecting to clear the track there at approximately 10:00. The train controller confirmed the passage of train TRN138 there and authorised the road-rail move after it. No mention was made of the next Up Main movement. That would be express freight MP2. The train controller then contacted the two drivers on WT22 in Te Kauwhata sidings and told them to pick up their work gang at the site at Down Main line km 583.6. He asked them to call back for a check at around midday, which the drivers okayed. Express freight MP3, which is of no further consequence in this narrative, then rumbled along Te Kauwhata Down Main at 09:12 and, after its signal came off, WT22 departed to follow it at 09:20. Train WT22 arrived at the worksite at 09:25. At 10:04 the lengthman in road-rail vehicle 46285 called the train controller again by radio from position km 567.5 and asked for an extension to his slot on the Up Main line. The train controller approved this extension but told him to be off the line at 10:35.

At 10:28 the Te Rapa signaller informed the train controller by telephone that express freight MP2 was ready to depart. The train controller instructed the signaller to let it go as far as Ngaruawahia and block the train there. He would advise him when the road-rail vehicle was clear of the line at Huntly so that MP2 could go forward then. When the lengthman in road-railer 46285 called the train controller to inform him that he was off the track at km 571.86 just south of Huntly, the controller acknowledged and asked him to call in when MP30 – *he made an important error here, he meant MP2* – had passed. He then notified the Te Rapa signaller by telephone to remove the block on MP2 and let it go. At 10:56 the lengthman on the road-railer called the train controller by radio to say that 'MP30' (actually MP2) had passed his location, the driver on the locomotive of engineering train WT22 at the worksite overhearing this radio exchange and deciding to keep an eye on 'MP30'. Why the driver of MP2, who at that particular moment must have overheard and realised that his train had been erroneously referred to as 'MP30', failed to query this situation is not clear (see the remarkable developments later).

The lengthman from the road-railer called the train controller about his meeting with the Ohinewai worksite PIC and wanted to know what trains would pass by along the Up Main line. The controller told him about 'MP30' (actually MP2) and that it would pass by at approximately 11:00, but the lengthman did not pass on this information to the PIC and disappeared to another site along the line. This was another missed chance by someone, who was well aware of what was happening regarding the intrusion into the kinetic envelope of the other line, to make everyone else aware of what was going on at the worksite and what the dangers were. Because of the block for the road-railer (that was being lifted according to instructions) the driver of MP2 in the meantime had run into yellow caution aspects at Horotiu and at Ngaruawahia. These, however, had not subsequently stopped him at red signals but had been followed by proceed aspects. Like any driver under such circumstances he believed that he was following something. As he had heard radio calls about 'MP30' he assumed that it was the train ahead of him that had caused his signals to be restricted. As he then received green signal aspects, he thought that this train ahead had accelerated to line speed again so he opened up as well. That is how trains are driven.

With WT22 stopped on the Down Main line at the worksite near Ohinewai, one of the drivers had left the cab and positioned himself with a portable radio halfway along the train between the Up Main line and Highway 1, to direct the other driver on the locomotive by radio about positioning moves on request from the PIC. The driver on the locomotive sat on the right-hand side overlooking the opposite track and was able to watch the cranes on the EWR wagons setting about their task of lifting the used rails onto the wagons. However, he became concerned about the jibs of these six cranes as they regularly fouled the Up Main line that, as far as he knew, was still open to traffic. He therefore called the train controller at 09:45 and asked for information about train movements along the line and was told that a train was due at about 11:00. The driver, missing yet another chance to prevent the accident, failed to raise his concerns about the intrusion of the crane jibs onto the Up Main line, the train controller also missing his chance to mitigate the risk by not mentioning the proper train identity number of the expected train (MP2).

After another repositioning move of WT22 to enable further rail lift, the still-concerned train driver called Te Rapa signal box this time and asked whether any train had come through yet. He was informed that a train was just coming through Ngaruawahia and would probably be near the worksite in about 15-20 minutes' time. Again, neither the train identity number nor the train driver's specific reason for the call was mentioned. Ten minutes later the driver on WT22 heard a p-way man call the train controller from a location near Huntly and say that 'MP30' had passed him and could he follow the train with his road-rail engineering vehicle to km 577.0. When the WT22 engineering train driver heard that the train controller had confirmed and authorised the move he realised that the train mentioned was in fact approaching the worksite along the Up Main line. He called his colleague on the ground but got no answer, although he could see him with his radio in his hand and knew that the radio worked as they had had very regular contact. Probably work or road noise prevented him from hearing the handset.

At approximately 11:00, the time given by the train controller earlier, the locomotive driver saw the headlights of an approaching train (MP2) along the Up Main about a kilometre (¾ mile) away and again attempted to contact the driver on the ground by radio to warn him so that he could also warn the crew on the wagons. This time that driver responded and asked him to contact their colleague on the express freight to get him to slow down. The locomotive driver then used the wrong train identity, MP30 instead of MP2, and as a result there was no response. The driver on MP2 did hear the request to slow down for 'MP30', which he thought was running ahead of him, so he did not respond. Coming round the curve at 80 km/h (50 mph) he saw the engineering train. He was aware of such a train being around from his bulletin (in Britain, the notices), but had expected that train in the area between Huntly and Te Kauwhata and thought, therefore, that this had to be a stopped freight. Then he saw the second driver with his radio on the ground on the cess side of his track and realised that it was the engineering train. He sounded the horn in warning but continued at his authorised speed. The driver on WT22 tried to see whether the freight was slowing down (by looking for brake dust) but saw the ditch lights starting to flash to gain the attention of the people along the engineering train. At a distance of about 500-600 m (1,600-2,000 ft) the driver of WT22 heard the driver on the ground shouting at the gang and saw him making slow-down gestures at the approaching freight. As he looked back he saw crane jibs still fouling the opposite track, the crane operators being unaware of the approaching train. A moment later the four intermediate cranes had pulled back onto their wagons but both end cranes were still extending over the Up Main line as MP2 passed his cab.

The PIC, unaware of the approaching train, had been halfway along the engineering train directing recovery of rails that had become jammed between each other and, together with one of the crane operators, was busy undoing fishplates and lifting rail ends off each other. The crane operators were wearing hard hats and ear defenders against the noise of the petrol engines that drove the cranes, so hand signals were used to direct the crane operators from the ground, but whilst there were prescribed hand signals in the regulations, in this case non-standard gestures were used and these did not include signals to indicate an emergency. The PIC had suddenly spotted the rapidly approaching freight and had started to make frantic gestures to indicate that the crane operators should pull in their jibs. It was then that the four intermediate cranes had pulled back, but the operator of one of the other cranes was with the PIC on the ground and the operator of the last crane, 25-year-old Sean Smith from Hamilton, on only his second job on the track since joining the railway, did not hear him and was standing with his back to the frantically gesturing and shouting PIC. When the locomotive of express freight MP2 struck that last jib it spun round several times, slammed into Smith and threw him off the wagon under the wheels of the passing freight.

The driver of MP2 had spotted the hand signals to slow down at a distance of 450 m (1,475 ft) and had made a brake application before spotting the two cranes still fouling his track, although initially he had thought that they had been withdrawn. He jumped out of his seat and walked round the control panel in the centre of his cab to protect himself from flying glass and put his arm over his eyes. The first jib shattered the windows on the six-foot side of the cab before the locomotive then hit the second crane, after which the driver went back to his control position to bring the train to a stop, unaware of the fatality. The distraught driver in the cab of locomotive DC 4565 of train WT22 radioed the train controller to report the accident and to alert the emergency services. The lengthman, on another site nearby, heard this call and immediately walked back. He later said that he would not have stopped traffic but would have arranged decent work slots and protection of the site. He also said that he would have assisted the PIC had he known it was his first time on a double-track main line.

The investigation

During the investigation the following things were determined to have played a role:

1 The PIC had 37 years' experience with p-way work but it was all on single-track lines. He had taken over this section of double-track only three weeks before, due to a change of boundaries, and had received no instruction or training on how to deal with this fundamentally different working situation in which trains could run past his worksite on an adjacent line. On single-track lines, any work of the nature that the PIC was engaged in with WT22 would automatically stop all train traffic as long as his engineering train was in section, so an agreement with the train controller about the duration of the time his work

train was allowed in section was all that was needed to maintain safety.

2 He had therefore completely overlooked to mark off his worksite with the appropriate boards, to warn approaching train drivers and make them to slow down. Nor had he organised lookouts whose job it was to spot approaching trains and give warning of them, although a lack of crew may have contributed to that omission.

3 His lack of knowledge and experience with double-track p-way work also led him to make another fundamental error. When he received from the signaller the 2-hour authorisation to be on the Down Main line after express freight MP3 had passed by, he wrongly assumed that he had possession of both Up and Down Main lines. In fact, the authorisation referred only to the Down Main line. He had no permission at all to obstruct the Up Main line, and was not remotely aware of that, or of the implications for his safety and that of his crew. Nor had he informed the train controller that he needed to intrude into the kinetic envelope of the Up Main line. For his part, the train controller did not take the trouble to inform himself of any track occupation requirements.

4 The PIC – and a number of others with him – appeared not to understand the need to establish all the necessary facts to secure a safe working environment; to ask what needed to be asked to ensure that everyone involved fully understood the situation. It was a typical case of one party not knowing, not having been told, and the other party not being sharp enough to assess the potential risks.

5 The train controller was remarkably uninformed about the nature of the work and its requirements. He was not even aware of the exact location. This means that either the bulletin was inaccurate or incomplete (not something restricted to New Zealand, see ***Brühl***) or the train controller had not scanned his copy of the bulletin thoroughly enough. Given the reaction of the driver of MP2, incidentally, the bulletin was apparently not very clear.

6 The train controller also appears to have been slack when his remarkable lack of attention to detail in identifying trains is considered. His mix-up of express freights MP30 and MP2 during radio broadcasts was a major contributory cause to the accident. Whilst this was a busy line by New Zealand standards, if these had been the standards with which European traffic

controllers work, there would be good reason to be ill at ease with our rail safety.

7 There was a lack of accuracy with the telecommunications as a result of too many assumptions being made by several of the people involved. Both Ontrack and the PIC should have been more proactive with safety on double-track lines. Had the PIC's and WT22 driver's questions about trains been clear, and their reason for asking, the train controller would have known about the crane jibs fouling the Up line. Had the necessary repeats of instructions between the train controller and the drivers been given as required, the controller might have realised his error in confusing MP2 and MP30 and corrected that immediately, enabling WT22's driver to address the oncoming train with its correct identity in his last-ditch attempt to avoid the accident.

8 Another problem with the telecommunications exchanges was the amount of people giving half-messages to each other via mobile phones, radios and landline telephones. The driver on the locomotive of WT22 talked with the train controller and with the Te Rapa signaller but was vague as to the reason for his calls and consequently failed to receive a decent answer from either. Various people at the worksite called the train controller about train movements, but the only usable bit of information they obtained was that at around 11:00 a train would be passing by. Had these people actually talked with each other, then perhaps they would all have realised that at around 11:00 the jibs would have to be within the limits of the wagons and clear of the Up Main line, as train MP2 would be passing. This points to a lack of worksite authority structure with clearly delineated work and management responsibilities and the necessary protocols. The same thing had blighted safety at ***Clapham Junction*** for example.

9 P-way man Sean Smith was new to the railway. Taken on in 2006, he had undergone an induction course in theory and procedures but had been given no supervised practical experience before going out on jobs, so he had not been properly inducted in the way of the railway and how to stay alive in that environment. That day he was thrown into the deep end in his role as a crane operator. As everyone thought that job was a doddle and did not take it very seriously, why should he? He should have had someone experienced next to him who was in the habit of regularly scanning the tracks, which is one thing that

many railwaymen will say has saved their lives at some time or other.

10 Working the cranes on the EWR wagons was a relatively unusual experience for p-way people anyway. It was rare for someone to do it twice, and therefore it was something that people were taught when the EWR wagons turned up at the worksite. However, the danger of extending over the adjacent track was not mentioned, probably because most of the track in New Zealand is single track, so it is doubtful whether many of those supposedly teaching the job were all that aware of those risks themselves. The lengthman who visited the worksite did check on Smith's competence in handling the crane, but was that done in accordance with certain laid-down standards or did it simply amount to checking that he was not an immediate danger to himself and others when manoeuvring the crane?

11 The failure to give serious consideration to safety matters was also illustrated by the way in which hand signals were used. Because hearing defenders had to be worn, the official hand signal gestures should have been used, or approved equipment with failure-indicating radio links should have been employed, but neither was. As a result, confusion arose the moment panic-induced gestures were used that deviated from those normally employed throughout the working day. Nor did anyone know the correct gestures that indicated danger, approaching train, or turn those jibs away on the double.

Comment

The railway is often blamed for being a stickler for correctness and having a stifling amount of inflexible rules and regulations. Indeed, it is one reason why many freight shippers are reluctant to use rail services, as the train never goes as soon as the loading doors are shut but has to wait for a slot in the timetable. Road freight vehicles can be pushed to go that bit faster in order to arrive that little earlier, whereas whole trains are stopped for the most trifling of reasons (e.g. broken windscreen wipers, popped headlights or braking problems on a few wagons). The official report into the accident gives an indication as to why that situation has developed. Airlines are no different, incidentally, for pretty much similar reasons. The lives of so many people are dependent on decent organisation and accurate execution of the work on hand that strict rules and work methods become inevitable. And it is necessary for people to adhere to all these prescribed methods of work, they would potentially lose their jobs if they didn't. In too many cases, road transport operational safety is sloppy in comparison, however much some people in that industry may deny it.

When things go grievously wrong on the railway, as in this case, unsatisfactory methods of work can often be identified as the root cause. And these are often the consequence of bad or incomplete training. The railways have become even stricter in enforcing adherence to prescribed methods of work, as operators can find themselves in the dock facing serious charges if their employees on worksites, in traffic control centres and in driving cabs get it wrong. The habit of making assumptions, which was rife here at Ohinewai, can be lethal on the railways. Probably the most feared thing on the railway is having to look at the mess left after an accident and know that you could have avoided it if only you had asked the right questions. In British railway jargon it is known as 'coming to an understanding' and it is not called 'understanding' for fun. Ontrack/KiwiRail now dedicates a page on its website to its commitment to safety, and this accident is mentioned as the one issue that brought about this change in approach. It is bad that someone had to die first, although that is often what it takes before people accept the consequences that arise from doing things wrong, and that accordingly it takes time, effort and discipline to do things right.

***

Röykkä, Finland, 13 August 2007 at 15:15

(Level crossing accident involving intoxicated car driver)

Introduction

This is yet another everyday example of a level crossing accident, this time on a sparsely protected open crossing in Finland. The reason for the accident was not, however, the lack of protection but the incompetence of a drink-fuelled car driver and the unreliability of his car due to its age and poor state of maintenance.

The location

Korvi open crossing, near Röykkä in the south of Finland, was protected only by the standard red 'Stop' sign, with no crossbucks (St Andrew's crosses). In addition, it was later found that the sight lines from the position of the car to points along the track were not in accordance with the prescribed distances. This allegedly caused the train driver to see the car late and may have influenced the distance judgement of the car driver. The reason for the impaired sighting was, as is often the case with such secondary railway lines throughout the world, vegetation growth along the track. The line is the single-track non-electrified line from Karjaa to Hyvinkää, the line speed at Korvi crossing being 80 km/h (50 mph). The train had departed from Kirkniemi at 14:22 for a trip to Riihimäki.

The train and the car involved

Freight train 3649 consisted of two Dv12 diesel-hydraulic centre-cab shunting and tripping locomotives of 68 tonnes each, running in multiple and hauling 41 freight wagons at 71 km/h (44 mph). The train, which was 723 m (2,370 ft) long with a weight of 1,567 tonnes, sustained no damage.

The car involved was a 4-door Ford Sierra 2.0i Laser. At the time of the accident it was 18 years old and was in a far from roadworthy condition due to lack of regular maintenance, which was a contributory factor in this incident.

The accident log

The car had been travelling along Korventie Road, parallel to the railway line, before turning into Leppälammentie Road and soon reaching the railway line at the open level crossing. The car was stationary at the 'Stop' sign for a short moment, but the driver stalled the car whilst moving off. It stopped again with its front wheels just past the middle of the track and the driver attempted, unsuccessfully, to restart the engine. By then the driver and his passenger should have seen the headlights of the approaching train, but they were both intoxicated with alcohol and made no effort to vacate the car, for which there was more than enough time. In fact, had they been sober and kept their nerve, they might even have had time to push the car back off the crossing deck.

This sort of incident rarely leads to serious injuries or fatalities as the car occupants normally have time to bale out. The driver of the freight train first noticed the car stalled on the crossing when he was 300 m (990 ft) away and on straight track. He used his whistle, but it was only when the train was 150 m (490 ft) from the crossing that he initiated emergency braking, slowing down to 63 km/h (39 mph) on impact. The left-hand buffer of the leading locomotive hit the Ford slightly behind its right-hand front wheel, throwing the car 18 m (60 ft) forward to the left off the shallow embankment on which the track ran. The car passenger, who took the full brunt of the collision, was dead and the driver seriously injured. A bystander who witnessed the accident warned the emergency services from a building in a nearby yard and was able to give the exact location of the accident as neither the train driver nor the traffic controller were able to do so.

The investigation

The Finnish Accident Investigation Board recommended an AHB installation at Korvi crossing. This, however, would not have prevented this accident from happening, as the car driver had actually obeyed the 'Stop' sign but had then made a mess of accelerating away. More to the point, an AHB would, in the case of a very minor level crossing on a secondary railway line such as this, have been an expensively ineffective way to address the issue of blatantly incompetent road users.

The investigation also recommended providing all level crossings with name and location boards. This is an ongoing programme, which also involves adding an appropriate telephone number to the crossing name and its location.

Comment

It is hard to escape the impression that, probably owing to the impaired sight lines resulting from unchecked vegetation along the line, the driver and his passenger saw the headlights of the approaching train only when they started to cross, and that the stall might possibly have stemmed from the sudden fright experienced by the intoxicated car driver. His decision to try and cross ahead of the train instead of vacating the car, something that under normal circumstances would probably have gone without a hitch, was fatally flawed. The question, of course, is whether a driver, even when drunk, should ever take such risks in a car that he knows may fail at any moment. The price was high, to paraphrase F. Scott Fitzgerald.

***

Tossiat, France, 19 December 2007 at approx 09:15

(Heavy road vehicle involved in level crossing accident)

The accident

Until the 14 November 2015 overspeed derailment at *Eckwersheim* in the Alsace, the passenger safety record of the French TGV high-speed operations was only slightly less perfect than that of the Japanese Shinkansen equivalent. That is because, unlike the Japanese high-speed operations that run on a different track gauge (standard gauge) from the classic network (Cape gauge), the TGV sets also widely operate on the classic lines, miles away from the level-crossing-free LGV high-speed network. Consequently, French high-speed sets have been involved in level crossing crashes more than once, with a surprising number of heavy road freight vehicles to boot, although until the Eckwersheim derailment in no such mishap rail passengers were fatally injured. This accident at Tossiat had that same unreal air of ignorant blundering on the part of a heavy road transport specialist that characterised the *Hixon* and ***Kissimmee*** accidents. The outsized and heavy transport involved had not been given permission to use this road, and not one of its four crew members, stuck on an automatic half barrier level crossing, even remotely considered informing anyone in charge of train operations of what was going on in that dangerous situation.

Despite the heavy impact during the collision, only the driver of the road freight vehicle died when, instead of getting clear, he tried to move his vehicle off the line. No serious injuries were sustained other than by the train driver, so the TGV once again stood up to its impressive passenger safety record. Damage to train, track and level crossing was significant, whilst the road vehicle and its cargo were destroyed.

The line

The accident happened on the 1.5 kV dc electrified former PLM (Paris-Lyon-Méditerranée) double-track main line from Mâcon to Ambérieu-en-Bugey, the level crossing being near the town of Bourg-en-Bresse in an area that among gourmets is known for its Bresse blue cheese. The line is signalled with automatic track circuit block and colour-light signals. From Bourg the southbound and eastbound diverging lines head for the Alps, the southbound going to Grenoble and Gap. This accident happened on the eastbound line that takes a train onwards via the Jura Mountains to Bellegarde and Genève along the roundabout route via Ambérieu and the triangular junction at Culoz. From Culoz the route continues via the Maurienne line to Chambéry and Modane into Italy (which is the line on which the ***St Michel de Maurienne*** accident happened in 1917).

The level crossing concerned was AHB 34 at La Vavrette-Tossiat, the first station after leaving Bourg in the direction of Ambérieu. The line speed at the location of the accident is 160 km/h (100 mph) and from Bourg the line approaches the level crossing after passing under the A40/E21 motorway viaduct and through a right-hand curve that was to hinder the sighting of the stalled road freight vehicle on the level crossing until about 300 m (990 ft) from the impact point. As a passenger, when travelling on an express or a TGV that does not call at Bourg it is rather difficult to spot this location. This route, incidentally, was abandoned for the TGV connections from Paris to Genève via Bellegarde in 2010. The rather curvy and partly disused, yet far shorter and rather scenic, single-line non-electrified Haut Bugey freight mountain route via La Cluse was thoroughly upgraded, electrified and taken into use for the TGV services, which cut the travel time from Bourg to Genève substantially.

The level crossing

The crossing was of the SNCF/RFF SAL 2 standard type, the length of the road between the barriers being 18 m (60 ft) and the width of the crossing deck 7.1 m (23 ft). The road is a local main connection, the D64, by the name of Route de Certines. There is an appreciable hump in the road surface where it intersects with the slightly curved tracks, and the catenary wires are also slightly lower than normal (at 5.7 m or 18 ft 8 in rather than the usual 6 m or 19 ft 8 in) due to the lower viaduct under the motorway. For that reason, two portals with a true clearance of 4.7 m (15 ft 5 in) but with road signs giving the headroom as 4.4 m (14 ft 5 in) have been erected across the road on either side of the tracks to warn of the danger of touching the wires. Traffic density on this AHB has been measured as running to 1,220 vehicles on average per day, whilst 127 trains pass the crossing on average every day. Classed as a category 1 level crossing it was not listed as a problem crossing.

The train involved

A TGV PSE set with the operations number 46 worked train 6561. This first generation of TGV sets was put into service up until 1986 (this set in 1982) along the first LGV, Paris Sud-Est (PSE), from Gare de Lyon in Paris to Lyon Perrache. Due to increasing demand for this capability, this particular set had later been modified to work off both French (1.5 kV dc and 25 kV 50 Hz ac) and Swiss (15 kV 16.67 Hz ac) types of traction current. The total weight of this articulated set was 385 tonnes, its power under 25 kV ac being 6,300 kW (8,445 hp) delivered via 12 body-suspended dc traction motors. It was 200 m (660 ft) in length and consisted of two 68-tonne power cars (TGV 23091 and 23092), two half-powered trailers and six non-powered articulated trailers, axle notation Bo'Bo'-Bo'2'-2'-2'-2'-2'-2'-2'-2'Bo'-Bo'Bo'. The maximum permitted service speed was the standard 300 km/h (185 mph).

An emergency brake application at the speed at which the train was travelling at the time of the accident, 153 km/h (95 mph), required a stopping distance of 700 m (2,310 ft). The emergency brake was fully applied 210 m (690 ft) before the impact point, making the set speed drop to 132 km/h (82 mph) on impact, fully within the designed braking parameters. The set came to a stop virtually within its own length due to the fact that the entire train derailed and ploughed in line through the ballast between the two platforms at La Vavrette-Tossiat station. It fouled the opposite track but, as luck would have it, no train was passing by at that moment, and traffic was quickly stopped after the collision.

The road freight vehicles involved

The mobile asphalt mixer combination involved consisted of a DAF XF95 tractor vehicle with a 530 hp diesel engine, the axles in a 6 x 4 configuration (the two rear axles were mounted in tandem in the chassis, both of them driven), and a semi-trailer that was made up from an oversize Ermont asphalt machine. This vehicle rested at the front in articulated fashion on the so-called fifth wheel of the tractor vehicle and ran on three axles at the rear with double sets of tyres (12 wheels), the rear axle of which steered with the direction of the vehicle. The weight of the combination was nearly 64 tonnes, the total length was 29 m (95 ft), the width was 3.2 m (10 ft 6 in) and its theoretical height was 4.4 m (14 ft 5 in), with a true maximum height of 4.5 m (14 ft 9 in). This particular trailer vehicle had permission for six trips to work at road construction sites via strictly authorised routes between 9 November 2007 and 8 May 2008. Its permitted maximum speed on motorways was 60 km/h (35 mph), on secondary roads 50 km/h (30 mph) and in built-up areas 30 km/h (20 mph).

Of note with regard to the accident is that fitted to the rear of this vehicle was a hydraulically operated conveyor belt to load the asphalt from the machine into vehicles for onward transport to road construction sites. During the journey the conveyor belt assembly was folded away, in which position it did not exceed the 4.4 m height above the road of this vehicle. However, a part around the raw material distribution silos near the mixer equipment end did exceed this height; it went up to 4.5 m. Another oversize load vehicle in this convoy, a DAF XF95 2 x 4 tractor vehicle hauling a 2-axle semi-trailer transporting a large (empty) bitumen silo used in connection with the mobile asphalt mixer, was not involved in the accident.

The shipper of this mobile road paving equipment was SCREG (Société Chimique Routière et d'Entreprise Générale), which operated as a hire company for specialist road construction equipment, but the tractor vehicles and their drivers belonged to a normal transport contract company, Altead Abram, which had taken care of the certification of this transport. In one of the two Altead cars accompanying the convoy there was a convoy manager, hired by Altead from a specialised scouting company called Société Piloting G. Bonnet (PGB), and in the second car was a technical specialist to operate the asphalt machine. There had been a second scout from PGB, but he had had to leave the convoy behind at Bourg to meet his employer somewhere en route.

Their point of origin was Chastreix (from where they had departed on the morning of 17 December) to nominally reach Vénosc as their final destination according to the issued travel and route authorisation. Late in the trip, however, during a phone call at their rest point near Bourg at 13:00 on 18 December, the final destination for this trip was changed to a nearby town called Certines. This location could not be reached along the approved route, so further authorisation for this alteration should have been sought. As is clear, however, this requirement was ignored.

The accident log

After leaving their rest point near Bourg-en-Bresse on the morning of 19 December, the heavy road freight convoy took a right turn off the authorised route to Vénosc and made its way along the D64 to the new destination of Certines. Having no local road knowledge, they had used the previous afternoon to discuss the route, and the technician and the driver of the asphalt mixer combination had attempted to scout a usable route to Certines. After checking route 64*bis*, which, apart from a narrow AHB level crossing, also had a series of cramped hairpin bends, they returned to their motel having decided not to use that route. From the map and after a short second scouting trip they decided to take route D64, as it was clearly the widest, shortest and most direct. Their initial reluctance to take this road may have had something to do with the chance of meeting police officers, who might have stopped them to check their authority for taking such a convoy along that route, which was more likely to occur on the D64 than on a more minor road.

On reaching the level crossing at Tossiat the combination with the asphalt machine negotiated the crossing first. Everything went well until the tractor vehicle had almost cleared the crossing and the rear axles of the semi-trailer ran up onto the hump in the crossing deck. At that moment the top of the conveyor belt at the very rear of the vehicle hit and damaged the warning gantry on the entry side, a matter of being just 2 cm (¾ inch) over height. The driver of the combination stopped, fearing he was going to hit the electric catenary wires, and after discussion with the others he decided to empty the 12 trailer tyres to ensure safe passage, despite realising that he had only 8 cm (3 in) ground clearance where the conveyor belt reached under the trailer frame behind the rear axles. This could potentially have given a serious grounding problem on coming off the track hump in the crossing deck. That was a foreseeable problem that proper scouting would have revealed.

Train TGV 6561 had left Paris Gare de Lyon for Genève Cornavin about 2½ hours earlier that morning, starting off by negotiating the welter of trackwork to Lieusaint Junction 29 km (18 miles) down the road where the classic PLM 1.5 kV dc route and the 25 kV 50 Hz ac LGV PSE high-speed route part company. From there the TGV is a 300 km/h (185 mph) high-speed train all the way to Mâcon, where it leaves the LGV PSE to again follow the old 1.5 kV dc PLM route to Bourg and beyond.

The train's first booked call was Bellegarde on the Swiss border, followed by its final stop at nearby Genève Cornavin station. The service, carrying about 150 passengers, was not booked to stop at Bourg-en-Bresse but merely slowed for the permanent speed restriction through the station, and after zipping through Bourg the driver opened up again to gain the 160 km/h (100 mph) line speed limit and sat quietly looking ahead for signals and everything else train drivers look for. The set, steadily increasing its speed, raced towards the motorway viaduct at 153 km/h (95 mph), when leaning into the right-hand curve and looking through the viaduct, the driver suddenly noticed that his view of the track some 300 m (990 ft) ahead of him was different from what it should be. Realising that the level crossing was not clear and that the obstruction was big and probably heavy, he applied the emergency brake, which was fully effective 3 seconds later when his train was 210 m (690 ft) from impact. At this point he made his way back into the power car, and another couple of seconds later the right-hand cab side was partly ripped open by the impact with the trailer. It made large dents in the bodysides of the power car, much as in the way the road-trailer frame had damaged two coaches during the ***Kerang*** accident in faraway Australia. The DAF tractor vehicle also did its bit, swinging wildly around and slamming into the side of the TGV set, whilst parts of the disintegrating asphalt mixer dented coaches two and three to the extent that seats were dislodged from their mountings. However, none of the train passengers was seriously or fatally injured.

Following impact the train, completely derailed, threw up ballast but, in the usual TGV style, stayed in line and came to a quick stop in the

Culoz-bound platform at La Vavrette-Tossiat station. A railway manager travelling on board immediately called control, signals went back to red and within 5 minutes of impact all approaching rail traffic had been located and stopped, securing the damaged train from further danger. The front power car was badly damaged but remained on both its bogies and was structurally intact. Even the cab would have sufficiently protected the driver had he still been there. Despite the heavy impact damage at several places along the front coaches of the train and the underbody damage due to the derailment, in essence the train was still in one piece.

When the level crossing warning equipment had started to activate and its half-barrier booms descended on top of the asphalt mixer, the three men trying to get the vehicle off the track were taken completely by surprise. It took a few seconds before they realised just what was bearing down on them, after which the technician and the driver of the second outsize transport hurriedly ran off the crossing deck and back to their vehicles. The driver of the asphalt mixer combination, however, ran to his tractor vehicle, its engine still running, and attempted to get it off the crossing. Just 8 or 9 seconds extra would have made that possible, but given the speed of the train that was too much to ask, and on impact he was thrown out of his cab and died virtually on the spot. His DAF tractor vehicle came through the collision in a surprisingly recognisable state, having been turned through 180°, although it was damaged beyond salvage. The mobile asphalt mixer was completely destroyed, smashed to barely recognisable bits of machinery.

The following day the TGV set was taken apart and lifted out onto road vehicles with two massive road cranes. The vehicles were taken to the TGV central workshop at Bischheim near Strasbourg, where it was decided that the ageing front power car was only good for cannibalising usable parts and then scrap, but despite its age the rest of the train was repaired and reinstated. A spare power car and two trailers that had survived an earlier AHB level crossing accident were used to get this train quickly back into service again.

Comment

Based on this accident and a counterpart at *Domène* in October 2006, French national transport accident investigation organisation BEA-TT remarked that such heavy transport businesses appeared to have a habit of solving problems as and when they occurred, rather than anticipating them at the preparation stage. They seemed to count on having the time to sort something out on site, which this accident (and many others throughout the world) proved was a dangerous policy to follow. Particularly if as ignorant of the correct procedures for dealing with railway intersections as internationally quite a number of heavy transport specialists appear to be.

The crew involved at Tossiat were on a road in an unfamiliar area that had not been properly scouted and was not covered by the necessary authorisation – a situation that BEA-TT, to its surprise, found to be quite normal for this kind of work. It also noted that knowledge of rules and regulations for this type of transport among heavy hauliers involved in such accidents, especially with regard to the use of level crossings, was far from satisfactory. Which played a significant role in the way in which the accidents unfolded. It should have been obvious that the asphalt mixer combination would never be able to clear the crossing in the prescribed 7 seconds, but would need at least 25 seconds even if all had gone well. This meant that, according to published and available guidelines, the rail operators had to be involved in order to safeguard human life as well as the cargo. Moreover, the vehicle driver had ignored the headroom warning at the crossing, indicating that he was not conversant, as required, with the real dimensions of this vehicle.

The failure to secure further authorisation for the last-minute change to the route needed to reach the new destination at Certines was a typical example of the lax attitude towards safety that BEA-TT mentioned in its report, as was the failure to realise that negotiating a level crossing with a heavy and slow-moving vehicle is inherently dangerous without first making contact with the rail operator (see also *Hixon* and ***Kissimmee***). In all these cases it was astonishing to notice the extent of the ignorance these 'specialists' displayed with regard to railway level crossings and the dangers emanating from them. And, of course, the accidents mentioned in this book were not the only ones. Unexpected grounding of low-slung heavy-load road vehicles on level crossings remains an ever-present danger to rail safety to this day.

It will come as no surprise to hear that the recommendations put forward by BEA-TT covered journey preparation and licensing of heavy and outsize transports as well as enforcing adherence to authorised routes – all issues that played such a crucial role in this accident and others like it.

***

Boxtel, The Netherlands, 30 June 2008 at 16:46

(Level crossing collision)

Introduction

This was another typical example of a large articulated road freight vehicle not clearing the level crossing deck and being involved in a collision with a train. What is interesting in this case, however, is the reason why it could not clear the level crossing. This was due to the questionable road layout together with the ineptness of a private motorist on the exit side for the road freight vehicle. Fortunately, no one died, but considerable damage was caused. A motor-scooter rider suffered an open leg-fracture and the motorist sustained bruising, which were remarkably light injuries given the violence with which the collision took place.

The location

The accident occurred in the town of Boxtel near Eindhoven in the south of The Netherlands. Boxtel is an important junction of two automatically signalled double-track electrified railway lines: one from Tilburg in the west – the line involved, with a line speed of 80 km/h (50 mph) – and the other from 's-Hertogenbosch in the north.

This situation has resulted in the location of two AHB level crossings only 50 m (165 ft) apart in the 'V' of the two converging double-track railway lines. The road that crosses both lines at these level crossings is called Tongersestraat and between the two sets of railway lines a narrow local road with the name Tongeren makes a T-junction with Tongersestraat. At the opposite side of the level crossing, Tongersestraat changes its name to Van Salmstraat, which makes a very sharp left turn immediately on leaving the level crossing and then runs parallel and next to the railway line. At this left turn Van Salmstraat is joined in a T-Junction of sorts by Kapelweg, which runs parallel to the railway line and Tongeren on the opposite side of the tracks. Traffic on Tongersestraat/Van Salmstraat has priority over that on Tongeren and Kapelweg.

All roads leading up to this AHB were equipped with so-called clearing lights – traffic lights able to show yellow (amber) and red that would illuminate only 8 seconds before the warning of the level crossing started to hold back road traffic. The one on Kapelweg stood back about 8 m (26 ft) from the junction with Van Salmstraat and had no stop line painted on the road surface. The motorist in the private vehicle, coming from that direction, may therefore have missed this indication.

This level crossing is a busy one. On 2 December 2008, a 10-minute count taken between 14:25 and 14:35 recorded 133 users: 78 private cars, 18 delivery vans, 13 road freight vehicles, two agricultural tractors with trailers, 13 cyclists, eight mopeds and a pedestrian. This equated to 798 vehicles per eight hours on a working day. In order to keep a check on crossing misuse video cameras had been installed, which faithfully recorded the collision and the lead-up to it. It also showed that the AHB installation had worked as it should. This pressure of traffic and the frequent closures of the level crossings had previously generated accidents. In the previous 15 years there had been four accidents, three of which involved road freight vehicles and one a private vehicle. In that same 15-year period there had also been 10 cases of road vehicles damaging the level crossing equipment, three of those being road freight vehicles. It was for these reasons that video recording had been installed.

The train involved

Train 89300 was an ECS movement from Tilburg to Eindhoven in order to work a passenger service from there. It consisted of 12 ICK locomotive-hauled passenger vehicles of DBAG's 1970s second batch of the first-generation UIC-X standard type. A substantial number of these vehicles, redundant in Germany, had been taken over in the 1990s and were refurbished by the German firm of PFA Weiden for NS Netherlands Railways' passenger business in order to alleviate an acute shortage of rolling stock in the face of rapidly growing passenger numbers at the time. At present all these coaches are stored pending disposal and batches are being sold on.

The locomotive was Alsthom-built 80-tonne B'B' 1.5 kV dc electric No 1837 with monomotor bogies, one of the then ground-breaking 4,400 kW (5,900 hp) French standard machines with their typical 1970s 'nez cassé' (broken nose) front of designer Paul Arzens that NS bought in substantial numbers from the early 1980s onwards as Class 1600 and Class 1700. At the time of the split-up into rail transport business sectors and the sale of NS Cargo to DB Cargo (first called Railion and at present DB Schenker) the Class 1600 machines received the class number 1800 to identify the change in ownership. The maximum permitted speed of these machines in The Netherlands is 135 km/h (85 mph), and at the time of the collision the train was doing 70 km/h (45 mph) in an 80 km/h (50 mph) permitted line speed area according to the trip recorder.

The road vehicles involved

The road freight vehicle was a 3-axle Mercedes-Benz tractor vehicle (prime mover) with an articulated Nooteboom Trailers flat-deck low-loader on four axles (the rear two axles steering) with eight small sets of double wheels fully under the load deck to enable taking on bulldozers, wheel shovels and full-size road freight vehicles. The combination was operated by BST Trading of Boxtel. The private vehicles involved were a Volkswagen Passat and a light type of motor scooter.

The accident log

The accident sequence began when the driver of the unladen road freight vehicle approached the T-junction for the right turn from the narrow Tongeren into Tongersestraat. The struggle to make the turn was captured by the level crossing video camera. The tractor vehicle is seen pulling out onto the opposite side of Tongersestraat to enable the trailer wheels to clear the corner before slowly turning sharp right onto the level crossing. This manoeuvre took some time, and while the combination was straightening out on the crossing the warning lights came on to warn of the approach of a train. At the same moment the Passat jumped the red traffic clearing light, shot out of Kapelweg on the other side of the level crossing and parked awkwardly next to a motor scooter that had stopped at the level crossing barriers. The left rear corner of the private vehicle was intruding into the opposite carriageway and was blocking the exit for the articulated freight vehicle, due to the sharp turn in the road. Consequently, the driver of the freight vehicle had to stop and back onto the level crossing again in order to be able to steer away to his right, onto the adjoining cycle lane, to create the space necessary for his trailer to make the sharp left-hand turn off the level crossing into Van Salmstraat. He used the horn to indicate his difficulty to the driver of the Passat, but the only thing that driver did was to pull a few inches forward to the barriers instead of reversing out of the way. Predictably, it was not enough to clear the road for the freight vehicle, the tail end of the trailer still being on the level crossing when the locomotive struck. The driver of the train, negotiating a right-hand curve, had no chance to stop.

The impact was fearsome. The trailer was tossed out of the way, its low-profile tail end shooting under the locomotive frame and damaging the brake rigging of the front bogie. Pivoting on its articulation joint (fifth wheel), the trailer swung around and, acting like one of the toggles in a pinball machine, smashed both the Passat and the motor scooter out of its way, both vehicles ending up well away on the other side of Van Salmstraat. The trailer continued to swing round, dragging the tractor vehicle with it and demolishing the level crossing lights and barrier installation on the Van Salmstraat side.

The motor scooter rider sustained an open leg-fracture, but no one else was seriously injured. The tractor vehicle and its flatbed trailer were surprisingly little damaged, but the Passat and the motor scooter were write-offs.

Comment

This was a classical situation in which an awkward and therefore difficult-to-negotiate level crossing layout, resulting from a historical road situation, demanded more than the normal insight from road users of what their own vehicles and, more importantly, other vehicles needed as far as road space was concerned in order to clear such an obvious danger point. As discussed earlier, that is too much to ask from private motorists if robust road safety is to be assured. Had the articulated vehicle been loaded with the sort of heavy cargo it was designed for, had it instead been one of the large fleet of articulated public transport buses in use in The Netherlands, had the train been in service and had it derailed, or had there been other vehicles waiting at the level crossing, then potentially this story would have had a very different ending indeed.

The municipality of Boxtel had already been considering how to deal with this set of busy level crossings, their replacement with a bridge or even a bypass having been suggested as alternatives. All of them would be costly and damaging to this small, isolated corner of a once quiet agricultural North Brabant landscape. It was clear, however, that something needed to be done, and the level crossing clearing lights were replaced by normal traffic lights with stop markers closer to the corners where the turn onto the level crossing has to be made. Whether any of the more comprehensive measures to deal with this situation will be implemented remains to be seen.

***

Chatsworth, California, USA, 12 September 2008 at 16:22

(Head-on collision due to driver distraction causing SPAD)

Introduction

This accident has been included because it is relevant when attempting to show the diversity of factors involved in a crash. Moreover, it clearly points to the risk of employing someone who is known as not taking the job of driving trains seriously enough yet is not taken aside for the problem to be dealt with.

The accident

A head-on collision on a curved single-line section occurred between a Los Angeles Metrolink locomotive-hauled 3-car push-pull passenger train and a local Union Pacific Railroad pick-up freight with two locomotives and 17 wagons. It happened after the driver of the Metrolink service had departed from Chatsworth station and passed a signal at danger there, where the double-track station loop reverted to a single-track main line.

The National Transportation Safety Board (NTSB) established the root cause of the accident as the driver of the Metrolink train being engrossed in using his mobile phone to exchange SMS texts with a young male 'rail fan', probably arranging to meet him further down the line with the intention of inviting the boy into the cab and letting him work the controls under his supervision. It later transpired that similar gross rule breaches and flouting of cab protocol had previously occurred.

Apart from the injuries caused by the forces acting upon bodies during a collision of this type, most of the fatalities and serious injuries occurred in the first carriage of the train after the Metrolink locomotive badly telescoped backwards about halfway into the coach, its fuel tank dislodging and catching fire. (The telescoping aspect of this accident, incidentally, is strongly reminiscent of what occurred between the double-deck EMU and the rear Motorail wagon of the 'Indian Pacific' service at ***Glenbrook***.) The impact speed with which both trains collided was established as being approximately 80 mph (130 km/h), both trains doing about 40 mph (65 km/h). As the train driver's mobile phone was not retrieved from the wreckage of his locomotive, most of the information with regard to the content of the texts came from three teenage boys with whom he regularly exchanged SMS messages. They fully co-operated with the investigation and proved a major source of information regarding the accident timeline and mobile phone records.

Rolling stock involved

Metrolink train 111 from Los Angeles Union Station to Moorpark comprised locomotive No 855, a 3,000 hp (2,238 kW) EMD F59PH Bo'Bo' of 113 tonnes with a GM 12-cylinder 710-type diesel prime mover and a separate diesel generator engine for hotel power to provide heating and air conditioning on the train. This locomotive either hauled or propelled three double-deck Bombardier-built passenger coaches in push-pull formation, the last coach having a driving cab from which the locomotive could be remotely controlled in the reverse direction. These distinct carriages and locomotives were developed in co-operation with, and were first operated by, the Greater Ontario Rail Transportation Authority (GO Transit) in Canada. The Southern California Regional Rail Authority (Metrolink) was an off-the-shelf follow-on customer, which explains the Canadian-style 'comfort cabs' on these locomotives. The locomotive was badly damaged on impact (no survival space was left for any occupants of the cab), the entwined locomotive and first coach derailing and falling to their right.

Union Pacific train LOF65-12, the 'Leesdale Local', was travelling from Oxnard to UP's Gemco depot near Van Nuys station and consisted of two EMD SD70AC locomotives Nos 8485 and 8491, 4,000 hp (2,985 kW) Co'Co's of 188 tonnes each, sporting a GM 16-cylinder 710-type G3C-ES diesel prime mover and Siemens-designed tri-phase ac traction drives. Their maximum permitted speed was 70 mph (115 km/h). European readers might wonder why two such massive mainline diesel-electric locomotives were employed for such a short pick-up and delivery trip. For these trips standard North American single-cab locomotives are coupled back to back with their cabs facing outwards, to avoid having a long hood leading in one direction (see ***Ohinewai***).

The train consisted of 17 freight wagons of various types, of which the seven nearest the locomotives derailed and jack-knifed in the usual North American fashion. The lead locomotive also derailed, trapping its seriously injured crew in the cab. They were unable to reach the other cab exit door above their heads and had to be rescued by the fire brigade by the expeditious removal of one windscreen as the locomotive cab was surrounded by the fuel tank fire of the Metrolink locomotive and both heat and fumes were entering their cab. One of the subsequent recommendations from the NTSB was that roof escape hatches should be fitted on locomotives.

The line and location

The accident occurred in the Chatsworth district of the metropolitan area of Los Angeles. The line is the Ventura County line, owned and operated by Los Angeles Metrolink services, the impact occurring on a strongly curved section of single-track line with three tunnels (Nos 26, 27 and 28), just outside the 500-ft (150-m) long tunnel No 28 somewhat east of Stoney Point between signalling control points Davis and Topanga.

In this area the line negotiates the Santa Susana Pass with tunnels and cuttings that preclude double track, twisting and turning around rocky hills. For this reason, the driver of the 'Leesdale Local' pick-up freight was unable to see the approaching passenger train until 4 seconds before impact, applying his emergency brake only 2 seconds before the collision. As the Metrolink passenger train driver did not apply his brakes at all, the nature of the line is not relevant when judging his chances of avoiding the accident or mitigating its severity.

Some history

The Southern Pacific Railroad built the line as part of the Montalvo cut-off, starting at the more or less newly established railway town of Montalvo, conceived in the 1880s to allow the Pacific Coast line from San Francisco to encircle the north-west of greater Los Angeles along an inland alignment to Burbank Junction rather than continue along its existing coastal-orientated alignment through the rapidly extending urban areas. The new route from Montalvo to Chatsworth through the Santa Susana Mountains was opened on 20 March 1904, largely as a double-track line but with single-track sections through the most difficult terrain, to link up with an existing and upgraded branch line through the San Fernando Valley to Burbank Junction. Interestingly, part of the old coastal line not too far from Hollywood is now the Fillmore & Western Railroad, which has been extensively used in movie sequences. This well-established tourist railroad is now working on establishing a connection with Los Angeles Metrolink.

During the big mergers in the 1980s and 1990s, the Southern Pacific Railroad was incorporated into the Union Pacific Railroad and that company then sold the Santa Susana line to Los Angeles Metrolink in the later 1990s, but retained running powers (trackage rights) to work its local freight trains to destinations along the railway. The US national long-distance passenger operator Amtrak also has running powers along this line. By US West Coast norms, this is therefore quite a lively stretch of suburban railway, the single-track stretch being booked to be occupied by no fewer than 24 passenger trains each day (including two long-distance Amtrak passenger services, the 'Pacific Surfliner' and the 'Coast Starlight') and 12 freight trains (including some heavy long-distance services). Since purchase of the line in the 1990s, Metrolink has technically modernised and upgraded the signalling with new track, electronically controlled turnouts and colour-light signals that are now controlled from the rail traffic control centre at Pomona. No form of train protection has yet been operationally commissioned but a PTC test installation has been fitted. No derailer turnouts (catch points) were included at the single-line junctions, but these are not a normal feature of North American mainline layouts in any case.

Some explanation is needed in order to understand the location names used. Traffic control location identities, the locations of stop signals and associated turnouts, are known as control points (CP), the names of which are used when locations are mentioned during radio telephone calls between drivers and controllers for instance. The SPAD of the Metrolink passenger train occurred about a mile (1.6 km) away from Chatsworth station at CP Topanga (i.e. Topanga single-line junction), whilst the sequence of cautionary signals leading up to the red aspect at CP Topanga commenced at CP Bernson. The green signal for the 'Leesdale Local' freight train was last confirmed at CP Davis. If put in a British context, on my patch we would have talked about CP Locoshed and CP Kemble when identifying similar signal locations, which at the time were also single-line junctions, between Swindon and Standish Junction, but are no longer in existence, double-track having been restored once again.

The signals were of a 3-aspect type, comprising three lenses in a downward-pointing triangular configuration on a circular backboard. The countdown sequence from green to red was flashing yellow as the advance approach (preliminary caution) and a steady yellow as the approach (caution) aspect. The driver was obliged to call out the aspects he passed to the conductor on the train through his shortwave radio, and in the event of restrictive aspects the conductor had to repeat these back. These exchanges were received and recorded in the traffic control centre at Pomona, where it was noticed after the accident that neither the driver nor his conductor called out the restrictive signal aspects between CP Bernson and CP Topanga as they should have done. As already mentioned, it was established that the driver was busy receiving and responding to SMS texts at that time, so the idea that he did not actually see the signal aspects is not that far-fetched. However, had the conductor enquired about them, it could have been material in avoiding the accident.

Incidentally, with regard to signal location along a line like this, it appears strange to European eyes that, given the mile distance between the Chatsworth platform end and the CP Topanga signal location, an extra repeating or main starting signal was not provided at the platform end. Train 111 had to depart from the station stop against a faraway and not readily visible red signal aspect (except in darkness, tested by the NTSB after claims that train 111 departed Chatsworth station on a green aspect) at CP Topanga, to stop there again and await the passage of the 'Leesdale Local' into the double-track section. In the ***Cowden*** and ***Wolfurt*** sections this situation was mentioned as a known precursor to SPADs at red aspects following, for example, an intermediate station stop. Like Positive Train Control (PTC) that the NTSB mentioned as an improvement to avoid a repetition of this accident, either an extra repeater or a stop signal controlling the station stop might have made all the difference as to whether this accident occurred or not. However, in a situation where a train driver is distracted by using his mobile phone keypad, the extra aspect might not have made that much difference after all.

The accident log

The driver of train 111, 46-year-old Robert Sanchez, was working split shifts on the day in question. He first worked a morning shift during which he drove between 06:44 and 08:53, exchanging no fewer than 45 SMS text messages en route. At 11:30 the crew (a driver, a conductor and a brakeman) for LOF65-12 reported for duty at Gemco depot. They departed with the 'Leesdale Local' at 12:30 to Oxnard near Montalvo to exchange wagons in industrial sidings there.

Driver Sanchez returned for the second part of his shift at 14:00, claiming that he had enjoyed a 2-hour nap, and picked up his train at 15:03 to work the 15:35 from Los Angeles Union Station to Moorpark via *Glendale* (another Metrolink accident site), Downtown Burbank, Burbank Bob Hope Airport, Van Nuys, Northridge, Chatsworth and Simi Valley. Mobile phone SMS activity picked up virtually immediately. At around 15:30, 5 minutes before departure from Los Angeles at 15:35, Sanchez used his mobile phone to call a restaurant in Moorpark to order a beef sandwich. Although use of a mobile in the cab was prohibited, this particular call had no serious consequence and was typical of train drivers across the world before they actually depart. Until CP Raymer, train 111 was on double track between Van Nuys and Northridge, the signals being called out and recorded. The Metrolink train waited after the Van Nuys stop at CP Raymer for the single-line section to CP Bernson to be cleared by Amtrak train 784.

At 15:13 the 'Leesdale Local' had finished its work at Oxnard and was permitted to return to Gemco. At 16:11 the incoming 'Leesdale Local' cleared CP Davis and 2 minutes later the route for the Union Pacific freight train was set through the Chatsworth single-line junction, causing the signal for train 111 at CP Topanga to be kept at a red aspect. At about that time train 111 stopped at Chatsworth station for its 16:16

departure to the next stop at Simi Valley. SMS activity was identified when the train left Chatsworth, and the train was recorded on its OTDR as accelerating to 40-plus mph (about 65 km/h), after which it failed to stop at the CP Topanga red signal and cut through the turnout at the single-line junction, causing considerable damage to switch blades and drives. At 16:21:03 Sanchez received a text message and sent a reply at 16:22:01, which was 22 seconds before impact. At 16:22:19 the Union Pacific crew first spotted the approaching local passenger train around a sharp curve, with vegetation shielding the view. Two seconds later the trip recorder registered the application of the emergency brake on the lead freight locomotive and at 16:22:23 the collision took place, Sanchez not having shut off his traction power or applied his brakes.

Comment

Undeniably, Sanchez displayed an inexcusable disregard for his responsibilities as a train driver for the safety of his train and of others. It is questionable from his behaviour whether he had any real understanding of the degree of risk he was running of causing an accident whilst concentrating on his mobile phone at the expense of watching the line and signals ahead. He was not only risking a collision but was also failing to look out for obstacles, whether legal or illegal. Why had he not considered this aspect of his behaviour in the light of his continued employment as a train driver, given that he had a history of being caught doing the same thing, even by his conductors?

This situation points to insufficient monitoring and corrective action on the part of his employers, who appear to have been aware of his behaviour yet did nothing, despite their responsibility for maintaining safety and the quality of the staff they put in charge of their trains. This was a typical situation in which a competent traincrew manager should have taken forceful action knowing that he would receive the full support of his management. On the other hand, it often takes an accident of this nature before traincrew will accept being disciplined and having their freedom to act curtailed. That is reality, but admittedly that is of no comfort for train passengers or for more responsible colleagues.

Furthermore, for reasons important to himself, Sanchez grossly abused his position in view of the attraction that the job of train driver holds for quite a few people. Many would do anything for a chance to sit in a cab of a moving train, let alone be at the controls. In that respect, had I been the father of one of the boys involved and discovered what was going on, I would have had deep reservations about the intentions of this man. In all likelihood, however, it was not quite as bad as that.

What probably happened was that Robert M. Sanchez, a lonely, obese, not very bright, diabetic and HIV-positive 46-year-old, who had long held the rather interesting and often coveted job of train driver, had misused that position to get in contact with and please at least three 15-year-old male 'train fans', with whom he communicated through a glut of rather infantile SMS texts and to whom he gave illegal cab rides and even more illegal stints behind the controls of passenger trains. In doing so he contravened rules and regulations that had been put in place for very good reasons that stemmed from earlier mishaps in railway history.

As a result of his gross negligence, Sanchez caused the death of 25 passengers and injuries to 135 passengers. Some sources put the damage at $7.1 million, but others reckoned it to be more than $12 million. On a personal note, I am sure that I am not the only (ex-) train driver with a certain belief in the value of – and pride in – their job who views this particular sort of accident with a certain degree of exasperation. It is an affront to all train drivers who make a serious effort to do their energy- and health-sapping 24/7 type of shift work with constant regard to maintaining maximum safety for all those people (or freight) on the train behind them. Incidentally, on more than one occasion I have heard aircrew, subject to very similar health- and energy-draining work schedules, use similar expressions to describe their emotions after a galling incident caused by malpractice on the flight deck, such as near *Novosibirsk* In Russia in March 1994 (see Chapter 5).

The use of mobile telecommunications whilst driving a vehicle

Quite soon after the general introduction of this kind of personal mobile telecommunications equipment in the late 1980s and early 1990s, road accident statistics highlighted its negative impact on safety and the various authorities began to issue warnings about the dangers of its use when driving. Nevertheless, it took until the early 2000s before legislation prohibiting or restricting use of mobile phones whilst in control of a motor vehicle was introduced, yet even now it is common to see motorists at the wheel with a phone at their ear or bent forward working the keypad to put a text message together. Not to mention the additional loss of concentration that lighting cigarettes or working satellite navigation and in-car stereo equipment causes. Lack of attention as a serious source of danger has been clearly demonstrated on a number of occasions elsewhere in this book.

On the railway, the use of such telecommunications equipment in the cab was not subject to many restrictions until the dangers of its use on the road came to people's attention. I remember a well-known British transport journalist writing in 2004 that since aircrews on the flight deck used radio all the time, what was the danger with regard to train drivers doing their job? The difference is that in the majority of aviation situations there are two people behind the windscreen, one handling the flight controls and the other monitoring conditions, navigating and handling the hands-free radio traffic. Apart from that, aircrew are not normally bothered by such external things as signals in all sorts of places or other aircraft coming towards them on the same track, animals or people, and damaging misbehaviour high up in the air.

A brief but relevant digression

In my early days on British Rail I witnessed a few strange things that in some ways are comparable to Sanchez's idiosyncratic behaviour. Train drivers could be regularly spotted reading newspapers or books whilst at the controls of fast-moving trains, relying on the audible warning indications of their AWS equipment to redirect their attention to the line and signals ahead. I also witnessed one of my tutor drivers, a man from quite a different period and approaching his retirement, switching out the little bit of safety equipment available in Britain, the AWS, prior to taking out our train from London Victoria to Barnham for Bognor Regis, in order to 'stop that racket'. The same driver went for a few pints in the pub across the road from Barnham station during his break before working a train coming in from Bognor Regis back to London Victoria.

Some drivers also had the habit of switching the direction (master) switch to neutral whilst belting along at speed in order to give their arm a rest, as the driver safety device had then been neutralised (Southern EPB stock!). Admittedly, pressing down the dead man's handle on the power controllers for extended periods did cause repetitive strain discomfort in the right elbow. For the same reason, others performed the trick of hanging the heavy, long-handled BR driver's bag from the controller arm whilst powering uphill to East Croydon, not only to give their arm a rest but also so as to have the contents of the bag nearby if they needed to find something, a habit that caused a runaway on the London Underground on at least one occasion. The train would not move off after door closure, so the driver, leaving his power controller open with the bag hanging from the handle, went onto the platform to check. When he found the culprit he kicked the door shut, only to see his train full of passengers move off without him. It was stopped by the train-stop at a red signal. Long live the enforced stop at red!

Incidentally, 'locking up' through selecting neutral on the direction switch whilst moving once happened in my presence before reaching

East Croydon, where the train was not booked to stop. From there the line descends quite steeply all the way to the terminus at London Victoria except for the hump of the Grosvenor railway bridge across the Thames. It would be interesting to know just where the train would have ended up if that driver had been on his own and developed a serious medical problem. And I did hear similar stories from drivers on other networks in Europe as well. In The Netherlands I was told about a driver who performed the trick on older electric multiple-units so as to be able to visit the toilet whilst his train continued on its merry way at line speed. Train drivers and their need to visit toilets is one issue, incidentally, that after shunting mishaps and people jumping in front of trains litters the lore of the train-driving fraternity the world over. Nevertheless, our present keen safety awareness has a bit of that background to it. Yet it has to be said that accidents were rare, nevertheless.

Much of what is mentioned above is technically no longer possible anyway, as on-board control systems have been altered to make this sort of practice impossible. On top of that, switching out of safety systems is now recorded and train performance, as far as reacting properly to signal indications is concerned, is kept in check by train protection systems such as ATP. Things have come a long way, but part of the cure was an intense stream of clear information about the dangers of certain types of behaviour to traincrew. The fact that a number of accidents also took place in which the role of the train driver was questionable, to put it mildly, did its bit to encourage a greater awareness and willingness to listen or change behaviour. No doubt Metrolink is now busy working on something quite similar in Los Angeles.

I used a particular argument to promote the need for responsible attitudes when tutoring colleagues on these matters: ‘It is bad enough for your partner and certainly for your children to have lost their spouse and daddy, but socially it is a thousand times worse for them if it becomes public knowledge that you caused death and pain to many other people simply by being stupid. Do you want them to have to go and face that in addition to your death? Never consider yourself lucky if you caused an accident but survived. Hell on earth only begins to describe what you will have to go through for years to come before the events start to fade away to some degree.’ In that respect, Sanchez could arguably be described as one of the luckier people on his train. He apparently did not even notice the ‘Leesdale Local’ coming towards him until the impact and escaped the retributions.

Conclusion

In the US this accident had the kind of impact on transport policy that ***Lagny-Pomponne***, ***Harmelen***, ***Eschede***, ***Brühl***, ***Clapham Junction***, ***Pécrot*** and ***Ladbroke Grove*** had on their networks. Apart from demanding PTC, as such train protection systems are called in North America, and the NTSB arguing for better rail safety in exactly the same way that similar institutions in Europe had done for a century or more, some of the recommendations to increase safety and avoid further such accidents went very far indeed. In reality, these recommendations will in all likelihood not be welcomed by the train drivers of this world because, in addition to the OTDR trip recorder already fitted, the NTSB proposes cameras and voice recorders in the cab to keep an eye on what the crew is doing, such as making phone calls or texting – things that have been forbidden for some time on most networks but are still very occasionally found at the root of accidents and are difficult to monitor and suppress. The memory carriers with their information should not only be downloaded after incidents but also to monitor competence in the way in Britain the OTDR trip recorder is downloaded occasionally to see how a particular driver works his train. Many operators worldwide have been unable to reach agreement with their train driver unions for the use of OTDR downloads to check competence due to protests about the 'spy in the cab', despite the fact that many kinds of lineside equipment, from timed signalling track circuits to hot axlebox detectors and wheel impact load detectors, pick up and record train speeds as well. My experience is that the chance of an occasional download, together with overt and covert inspection rides on trains, has improved driving (and diminished fuel use) in a most remarkable way. On the other hand, looking at the cost of implementing such a system as well as the impact it can have on train throughput, the question as to whether it should be implemented across the entire US network as opposed to only where passenger trains share the tracks is valid.

It was the death of those passengers at Chatsworth that made the US Government finally act on the matter, as has also happened all over Europe, although accidents and their costs had never seemed to be that important when passenger traffic in the US was almost non-existent. I have to admit that the thought of cameras in the cab videoing people in the course of their day's work makes me shudder, but I acknowledge the harsh yet justified logic behind it, for which conclusion even the NTSB apologised in a postscript to the accident report. On the other hand, I just wish we could impose the same sort of control on road vehicle drivers, to remove their anonymity and personal isolation behind the wheel when committing offences. It would record what had happened in detail and so establish their responsibility in the event of a mishap, which would no doubt help to bring incompetent driving, road rage and obnoxious misbehaviour to an end, assuming, of course, that the justice system finally played its part and started meting out serious punishments where warranted.

***

Viareggio, Italy, 29 June 2009 at 23:48

(Derailment with consequent fire)

Introduction

This was a derailment and fire that re-opened the fierce debate throughout Europe about the safety of rail transport for dangerous cargoes, as well as questioning the state of maintenance of rail freight vehicles under the present set-up of separate rail equipment leasing companies and freight train operators. This followed a number of disruptive and high-damage brake failure and derailment incidents that had occurred with freight trains on various European networks.

The train and location

At 23:40 on a pleasantly warm and clear Mediterranean night, light freight train 50325 from Trecate to Gricignano near Pisa rumbled more or less unnoticed as one of the many nightly freight trains that passed through Viareggio station in Toscana on the important coastal line from Genova via La Spezia to Pisa. The train was hauled by FS Cargo electric locomotive E655 175, a typical Italian articulated Bo'Bo'Bo', originally Class E656 but re-geared for higher tractive effort once handed over to FS Cargo and reclassified E655. They were 120-tonne 3 kV dc electrics rated at 4,200 kW (5,630 hp). Its train consisted of 14 tank wagons loaded with LPG, of which ExxonMobil and ERG Oil Refineries (where the cargo originated) were the shippers. The tank wagons of the short freight train were leased vehicles, the property of US-owned GATX Rail Austria, operating as KVG Kesselwagen. One vehicle was registered with Polish Railways' PKP Cargo and the others with German operator DBAG Cargo. At the time this latter company still operated as Railion Deutschland, but nowadays it is known as DB Cargo again after a period of DB Schenker Logistics. The company operates rail, air, maritime and road freight services throughout Europe, including Britain. Registry of European rail freight vehicles is interesting these days, major lessors of railway freight vehicles sometimes have their roots in minute but enterprising former local private rail operators, to which owner interests from the rest of the world have been added.

The accident

After entering the station at about 40 km/h (25 mph) and making its way through the platform, the first wagon behind the locomotive suddenly derailed on a piece of plain track and dragged the next four out of the track. Two more wagons derailed but stayed upright, the other seven remaining on the track. The derailed wagons crashed into nearby dwellings alongside the railway line, the tanks of two wagons rupturing on impact and leaking LPG. The pressurised liquid rapidly converted to gas and cooled as soon as the pressure vented. This expanding cool gas is heavier than the surrounding air, particularly when the ambient temperature is high, so spread along the ground (see ***Ufa***). Sparks generated by the steel vehicles scraping along the station platform and track were enough to ignite the gas into a violent explosion with a fireball (a so-called BLEVE event, Boiling Liquid Expanding Vapour Explosion), immediately affecting nearby houses and quickly spreading along the adjacent roads.

The explosion was powerful. Two buildings along the tracks collapsed straight away and the resultant fire, following the escaping streams of gaseous LPG along the ground, initially burned ten mostly residential buildings nearby that were on fire the entire night. In these early disastrous events seven people were killed when buildings collapsed, an eighth person dying whilst riding a scooter on a nearby road. People started to flee the area in panic, during which two parents loaded their youngest child in their car, went back to retrieve the other two and were then caught in the cloud of gas catching fire. The youngest child was later found completely carbonised in the burnt remains of the car. By the end of the night, 1,000 nearby residents had been hurriedly evacuated out of danger. Nevertheless, there were 31 fatalities and 17 people were injured, four seriously. Surprisingly, the two traincrew members came away with nothing more than minor injuries despite the eruption of fire right behind their locomotive.

Both the explosion and the subsequent fires had damaged large areas around the station of this seaside town. An entire street paralleling the track close to the station had been obliterated and an estimated 100 people had been left homeless. This was recorded as the worst rail accident since April 1978, when two trains collided in the neighbourhood of Bologna at *Murazze di Vado*, killing 48 people. When

Prime Minister Silvio Berlusconi (who was not particularly popular with the labouring classes) made his way to Viareggio after the situation had been made safe again, in order to – as his press release declared – take control of the emergency, he was asked to go home.

The investigation

The railway accident investigation directorate of national railway operator FS Ferrovie dello Stato Italiane soon started to look into what possibly had gone wrong. Police initially issued a statement that there may have been a fault with the brakes on the train or with the track, both of which were somewhat unlikely in view of the fact that the train involved had not been recorded as having brake problems or going too fast and that previous trains had gone through without any noticeable or reported problems. Furthermore, there was no evidence of track damage other than that clearly caused by the derailing wagons. Italian railway union CGIL was the first to start blaming the supposedly decrepit state of some of the rail freight vehicles used, but no problems had been reported during the trip south and GATX was quick to claim that the vehicle in question had been recently overhauled in Germany and been fully certified as roadworthy.

Shortly thereafter Transport Minister Altero Matteoli was the first to indicate the failure of an axle on the first derailed wagon, but he did not specify how or why the axle had failed. Given that he was close to the investigation, however, an axle failure was indeed likely to have been the cause (as was later confirmed). The reason for the failure of the axle was as yet unknown. Then, on 29 July 2009, an extraordinary meeting of European national safety authorities was convened to discuss fatigue problems with Type A standard freight wagon axles (a link with ***Meudon*** there!) and to provide information on this matter to owners, operators and lessors of rail freight rolling stock.

It would be useful here to provide an explanation about rail vehicle wheel and axle loading. Axles and wheels on the railways form one unit, which revolves in the normally outboard-located axle bearings in the wagon frame or bogies. Under static conditions the weight of the wagon rests on the axle taps *outside* the wheels, to explain this matter in a manner of speech, this would make both wheels taper inwards from top to bottom and the *inside* axle section between the wheels come up like an arch. When a train departs, a dynamic element is applied through the rolling of the axles, the axle steel between the wheels then experiencing continuously alternating tension, bending and compression moments with every revolution. This causes failure if an axle is insufficiently strong, particularly when the track is in a less than perfect state and adds its dynamic stress peaks to the job of the axle.

At the end of 2010, a statement was issued by the Swiss railway authorities that, from 1 January 2011, Swiss Railways would ban any vehicle from its lines that did not show a valid Entity in Charge of Maintenance (ECM) code as agreed by EU rules. This ECM code gives clear information about the owner of the vehicle (the lessor) and the operator (the lessee), and who schedules and pays for its maintenance and performs the maintenance. It also indicates when the vehicle was last overhauled and to what sort of standard. This, however, is not entirely new. It is something that in those simpler days of national railway undertakings was comprised in the vehicle revision data visible in the REV box painted on the frame, but in those days decent maintenance was more or less taken on trust, apart from the fact that from other data on the vehicle it was immediately clear who owned, operated and maintained the wagon in the event of incidents. Without much further clarification, the Swiss statement mentioned the accident at Viareggio and its findings as one of the leading reasons for this measure, which suggests that a number of issues regarding vehicle integrity were not quite as had been claimed immediately after the accident. Therefore, the Swiss, seeing countless freight trains crossing their nation without adding much to their economy but potentially exposing their population to some serious dangers, had taken the lead among European operators to minimise risk arising from incompetently maintained rail freight vehicles.

The German federal railway authority EBA found that wheelsets were responsible for around 20% of the reported wagon defects there. Clearly, the Viareggio accident was the tip of the iceberg. An insufficiently robust Euro-standard freight wagon axle design, however, was something that could not be known about until incidents of a public nature started to occur. Unfortunately, this is still not something that everybody with dubious safety experiences immediately brings to international attention. Although this is due largely to potential financial implications, it also results from a continuing lack of interest in what happens on other networks. This typically causes a network that has experienced problems with this type of axle to consider it as a one-off and no more than a fluke. Therefore, it takes an accident on the scale of Viareggio, and the consequent trawling through international archives, for fundamental Europe-wide action to be taken. In that respect the Viareggio derailment and its resulting fire were positive influences on the safety of European rail freight operations.

Inevitably, the impression remains of people being aware of that unsatisfactory axle design but failing to remedy the situation – whether through lack of interest, influence or funds – until the problem could be evaded no longer. That was usually after many people had died and politicians had to be seen to be doing something about it. But one wonders what else such people are aware of but fail to act on. Then again, it took the *Titanic* (see Chapter 5) to founder with appalling yet wholly avoidable loss of life for the relevant authorities to accept that vessels are never unsinkable and thus should have sufficient lifeboats for all on board. After the sinking of the liner in 1912, that was taken care of soon enough.

Conclusion

Development, mostly with a view to stronger yet not heavier vehicles and to more specialist cargo vehicles with higher payloads, continues apace. In Britain the newly developed large biomass hopper vehicles are a typical example. Hamburg-based German wagon leasing company VTG announced in November 2012 that, with the aim of improving safety margins, it would equip its entire 20,000-odd tank wagon fleet with stronger axles certified for 25-tonne axle loads, which is as yet by no means permitted throughout the whole of Europe. On 5 May 2015, the same company unveiled a new 90-tonne (22.5-tonne axle weight on the 25-tonne certified axles), 25-m (82.5-ft) long tank vehicle for larger liquefied natural gas (LNG) consignments. The vehicle is built in its own Waggonbau-Graaff workshops in co-operation with a US-based cryogenic transport specialist. In this long vehicle, with its double-walled tank, high-density LNG is transported at high pressure in a strong vacuum at an incredible -162°C. LNG ships have transported massive amounts of natural gas this way for some time now, so it is assumed to be proven technology.

***

Stavoren, The Netherlands, 25 July 2010 at approx 23:30

(End-of-line crash with on-track rail grinding machine due to poor preparation and distraction of hired-in train pilot)

The accident

The last incident in this chapter was not an occasion in which people were killed or badly hurt. Two people were slightly injured but were able to walk away from the massive wreckage (estimated at €20 million) when a rail grinding train was driven at 80 km/h (50 mph) through the stop blocks at the far end of the non-electrified line from Leeuwarden into the single-line and single-platform station of the Friesian harbour town of Stavoren. This accident represents the ultimate fear of any train driver who is not completely certain about the location of the end of a railway line ahead. The majority of responsible train drivers would have taken it very easy with speed in such circumstances, but in this instance it all went wrong.

The accident log

The train had been grinding elsewhere and had been brought to one of the three operational permanent-way rail grinding servicing bases in The Netherlands. This one was at a yard called Rotterdam Noord Goederen, where fuel and fire extinguishing water were taken on and further servicing of the mobile grinding installation took place. (Grinders have a tendency to set fire to rubbish and dry vegetation at the surrounding trackside. The last rail grinding action, therefore, is spraying water to extinguish any such potential conflagration.)

The Italian grinder crew had been transferred to a hotel in Zwolle in order to stay within their legal working hours and the hired Dutch pilot driver would board the train there as well. At about this time a change of plans was agreed between back-office staff working on the grinding run that night. The train would go all the way to Stavoren instead of terminating and returning for grinding at Sneek, as a time gain of 30 minutes could be obtained that way. None of the head offices involved were informed about this change of plans. One important change was that the possession of the line would not now take place from Leeuwarden onwards (which was the chief reason for the time gain, as there would be no waiting for the last evening trains to clear the line) but the train would run all the way to Stavoren on track in normal service and start the grinding run from there. This change increased the permitted speed from 40 km/h (25 mph) in a track possession to the permitted maximum line speed of 100 km/h (60 mph), so the grinder could be run at its permitted maximum speed of 95 km/h (59 mph).

Running dead behind a locomotive, the grinding train had been dispatched from Rotterdam to the freight yard at Zwolle Goederen, where it arrived with the grinder crew and pilot waiting to take over. The crew prepared the machine for its onward trip under its own power to Leeuwarden, for which destination they departed at 19:00. Arriving at Leeuwarden, the pilot was informed about the change of plans with regard to continuing to Stavoren. As the Italian grinding crew thought it would be a considerably longer night ahead than planned they were understandably anxious to know just what was going on, but as yet had not been given a definitive answer.

At 22:16 the train, headlights dimmed and cab lights switched off, departed from Leeuwarden along the single track to the junction of the Harlingen and Stavoren lines. In this section the on-board ATBEG-v ATP equipment switched out due to leaving the ATBEG-protected area. On normal service trains the ATBNG equipment would then have switched in as explained before (see ***Roermond***), but this train had no such NG equipment on board and was consequently now running at maximum speed without any train protection supervising the journey. At the Harlingen/Stavoren junction the train took the left-hand track towards Stavoren and stopped at Mantgum station for 26 minutes to let two late-evening trains cross, after which the journey continued. Stops of half a minute each were again necessary at Sneek, Workum and Hindeloopen for signalling purposes. After the Hindeloopen stop at 23:18 there was telephone contact in which the hired pilot announced that he was about 2 km (1¼ miles) from Stavoren.

At this point the (illegally present) Italian grinder operator in the front cab started to ask pertinent questions again about the revised time schedules for the job ahead, for which the pilot had to consult notes and revised time diagrams. The conversation was conducted in halting German, which more than once caused the pilot to turn towards the operator seated behind him to see whether he was understood. At one such moment the train passed the three 'keperbaak' fixed distant boards, situated 70 m (230 ft) apart at a distance of 1,200 m (3,940 ft) from the buffer stops and normally protected by ATBNG ATP, travelling at 95 km/h (59 mph). These went completely unnoticed by the pilot, so no reduction of train speed was made. About 200 m (660 ft) past the fixed distant boards the train passed the well-illuminated 'Koeweg/Kooijweg' AHB level crossing at maximum speed, where several p-way men were waiting to travel back with them, working on the grinder for protection of level crossings against road surface damage. Noticing the excessive speed of the train, they attempted to gain the attention of the grinder crew with arm gestures, but were unsuccessful. The operator seated in the rear cab did see them, however, and used the intercom to the front cab to ask whether they were approaching their destination yet. The pilot thought he recognised the level crossing and expected to see at any moment the speed boards he had selected on his diagram to indicate the end of the line. As he was looking for a kilometre/hectometre post for orientation, all of a sudden he spotted the station platform coming up in the headlights. He shouted, 'Bremsen, bremsen, bremsen' (brake, brake, brake) to the train driver, who immediately put his train brake in the emergency position, but to no avail. The three men had time to leave the cab and position themselves against the rear of the cab bulkhead to await impact just before the train crashed through the stop blocks at the end of the platform at about 80 km/h (50 mph).

After running through the stop block and off the track, the derailed train continued its journey more or less in line across the street, picking up an empty marine diesel-fuel road tanker and smashing it clean through a large watersports and ship chandler's warehouse and shop, demolishing about a third of the premises in the process. About 70 m (230 ft) further on, the train finally came to rest in a zigzag amidst a mass of debris, the rear vehicles still on the track. Fortunately, because it was about midnight no one was in the shops and on the streets near the station when the accident happened. Under different circumstances it could have been far worse, as indeed it could have been at ***Carcassonne***. Unlike Carcassonne, however, the resulting damage was enormous and took days to remove.

The location, route and the trains that use it

Stavoren, in the province of Friesland, during the early Middle Ages was an important Hanseatic trading port and is one of the many ancient European towns steeped in history and full of character where you might easily imagine encountering a medieval merchant. Still a busy port, it is now also an important watersports centre with substantial commercial shipping to the large inland lakes via the Johan Friso Canal. The landscape in this part of Friesland is rather more picturesque and somewhat less remote than the more northerly and easterly lands of the province of Groningen. During the summer watersports season the Friesian canals and lakes, the Waddenzee islands and shoals, and the nearby IJsselmeer attract large numbers of the boating fraternity from many parts of Europe. The area resembles the Norfolk Broads in eastern England but is substantially bigger in size and population.

The railway line between Stavoren and Leeuwarden was built by the state between 1883 and 1885 (rather late in Dutch rail history) and was constructed as a double-track main line to be operated by the Netherlands State Railway Company. A curious but very recognisable train operating arrangement reveals itself here. Even in those days, the Dutch Government built many less profitable but socially necessary railway lines, which were then run by private operators on payment of access fees. In fact, the so-called State Railway Company had little to do with the state; it was just one of those operators. The line, built to provide good access to the remote south-western corner of the province of Friesland, connected with scheduled rail and steamboat services from Amsterdam via Enkhuizen.

As the State Railway (SS), together with the Netherlands Central Railway (NCS), also controlled the overland rail connection from Amsterdam via Utrecht and Zwolle to Leeuwarden and had clearly set out to obstruct development of the Enkhuizen–Stavoren rail and ferry route in order to protect its greater interests in the land route, the government invited the operator of the ferry connection, the Holland Railway Company (HSM), to assume control of the entire Amsterdam, Enkhuizen, Stavoren and Leeuwarden connection after a number of years of wrangling, after which proper operation took off with a vengeance. Thus, in 1890, Stavoren became the eastern ferry terminus for a passenger service employing three ships, and later, in 1899, a freight train on a ferry boat operation, which from 1909 used three somewhat peculiar-looking ships to cross the Zuider Zee (IJsselmeer) from Enkhuizen, north of Amsterdam. However, this lasted only until the unification of railway operations in The Netherlands under the banner of Netherlands Railways (NS) in 1917 during World War 1 (see ***Weesp***), whilst the opening of the road across the nearby Afsluitdijk (enclosure dam) across the mouth of the Zuider Zee (built 1927-35) ended the commercial viability of ferrying passengers through all weathers once buses had started to run. The two remaining capacious railway passenger ferries operated for one more year until 1 April 1936 under railway control, after which private tourist interests took over in the summer season only. In fact, most years it is still a once-a-day, summer-only tourist connection.

In the days of ferry operations the Stavoren quayside station was an important railway venue with a large station building and its own steam locomotive depot, where boat trains in connection with the ferry departed, and a freight yard to marshal wagons off the train ferry into trains. After closure of the regular freight ferry, the line from Stavoren to Leeuwarden was soon reduced to a single-track branch that became part of the group of non-electrified branch lines in the north of the country known as the Northern Diesel Lines (see ***Winsum*** for another accident in this area). At present Stavoren station consists of only one quayside platform with a shelter alongside the single line with a stop block. However, the character of the ferry port with which it was once linked, Enkhuizen, is still reminiscent of former times.

Rail head checking (gauge corner cracking)

At 12:10 on 17 October 2000, the railways in Britain experienced a fatal derailment near *Hatfield* when about 280 ft (85 m) of rail disintegrated under a London King's Cross to Edinburgh express running at 115 mph (185 km/h), approximately 15 minutes into its journey. Regrettably, four deaths occurred in the overturned buffet car, the roof of which was sliced open by an overhead line catenary post whilst the vehicle was still sliding on its side. Catenary posts, often made up of steel profile beams and presenting sharp edges towards passing trains, pose something of a risk in the event of derailments (as at ***Granville*** in Australia and *Hoofddorp* in The Netherlands). The Hatfield accident, however, made rail head checking, gauge corner cracking and rolling contact fatigue household terms. It also brought the potential dangers of ignoring that type of rail wear firmly to everyone's attention, as well as heralding the beginning of the end of that politically instigated peculiar British experiment of having a commercial, stock exchange listed company (Railtrack plc) seeking profit for its shareowners at the expense of serious investment in the clapped-out national rail network where such investment was badly needed. Interestingly, after news of the Hatfield derailment broke, it became clear that railways elsewhere were experiencing similar problems. The North American continent suffered, Germany and The Netherlands were plagued by it, and the French admitted that they too had the same problem. What, technically speaking, was the matter?

Until the 1960s, the vast majority of train wheels rolled comparatively roughly on the track and caused considerable wear on the rail head as well as on the wheel treads. Apart from demanding a substantial maintenance effort, this exacerbated the problems of track top and line deterioration. As a result of the quest for higher speed, concurrent with the increased application of concrete for the track base, all sorts of developments in rail and wheel wear detection, as well as in bogie and wheel suspension technology, took place from the 1960s onwards. The main benefit, other than a much quieter ride, was that the rail head and wheel tread no longer wore down so quickly. However, from the moment that these issues ceased to be a problem, the phenomenon of head checking (to use the American and German term) or gauge corner cracking (the British term) started to come to the fore.

Rail movement under passing trains had been checked with heavier track, better fastening and improved track foundation technology, in which concrete increasingly played an important role. The track was made harder, as it were, and the rail no longer flexed up and down to the same extent as on wooden sleepers in the ballast. This gave a markedly quieter and more stable ride, as well as cutting the traction effort to keep the train moving – which is very important at higher speeds – but in turn this demanded rather more from the bogie suspension. That was addressed by better axle guidance with radius arms, primary coil springs with motion dampers and advanced steel coil or air secondary suspension that was fitted with arrays of motion dampers. As a secondary result of that increased track inflexibility, however, the hardened steel of which the rolling wheel threads and rail head itself were manufactured now compressed minutely at the rail/wheel contact points and then expanded again, similar to a road tyre. This started to cause metal fatigue in the tread rim and on the rail head, which in turn caused very thin slivers to break off the contact surface of the rail head as well as off the wheel tread. Such head checking or gauge corner cracking notably affected the two inner rail head top corners, the gauge corners. The main problem was that under certain circumstances – mostly due to steel rail manufacturing characteristics or at rail welding flaws – the cracking, instead of remaining parallel with the rail/wheel contact surface, might turn down into the head and through the web of the rail to set up a rail break or even a series of rail breaks.

It is what led to the disaster at Hatfield, by ignoring visible and known wear symptoms in long stretches of rail. As a result, a stretch of the most stressed rail, that part on the outside of a curve on high-speed track, simply shattered into fragments under the speeding express. The accident report together with Christian Wolmar's account in *Broken Rails* of how that saga unfolded, media and magazine articles (notably Roger Ford's in *Modern Railways*) and some interesting issues raised in internet forum discussions are all particularly instructive. This problem had no doubt existed before, but the more roughly rolling wheels of the passing trains had continuously ground off the head checking problems through wear. With the harder track and the more stable ride of modern bogies with their regularly re-profiled wheels, head checking or gauge corner cracking – now known by the generic and internationally accepted name of rolling contact fatigue (RCF) – became more of a problem.

The solution was found in increasing the re-profiling sequences of the rail head with so-called rail grinding trains. In one pass such a train grinds off a few millimetres of steel from the rail head to restore the proper rail head profile whilst taking away the potentially fatigue-plagued top. It generally prevented the rail break problem, but did add expensive maintenance work to the bill of running a railway. A few contractors were able to develop the necessary technology to grind the track, the oldest and most well known being the Swiss firm of Speno. They work all over the world and build their own track grinding trainsets. Five such sets are permanently based in The Netherlands, two large ones and three smaller ones, and are virtually continuously employed. This gives an indication of the amount of work to be done on a small but heavily used network. It was one of the two large sets that was detailed to grind the recently overhauled track on the remote single line in Friesland.

The effect of privatisation on the line

When privatisation and the concomitant break-up of the European national railway undertakings hit The Netherlands, passenger train operations on the Northern Diesel Lines network were contracted out to private transport operator NoordNed. This company was later taken over by the British Arriva group, which itself is now part of German state railway operator DBAG Regio. A positive result of that development is that modern Stadler-built DMU railcars from Switzerland now ply the tracks and provide three services per hour instead of the ageing and somewhat dreary Netherlands Railways first-generation diesel-hydraulic rolling stock on an hourly and later half-hourly schedule. The bus and train connections, mostly operated by the same company, are superbly co-ordinated, with tightly scheduled interchanges at stations and midway hubs along the tracks and roads. A typical ‘if only’ experience for the traveller from countries such as Britain.

Route knowledge

To prepare for this job the pilot had selected the track diagram issued by Dutch rail infrastructure provider ProRail, which indicated distances of the features but no curves, and additionally wrongly showed two signalling matters – a speed warning and a speed reduction section commencement board at the end of the line at Stavoren – that had no longer existed for several years already. Unfortunately, because the pilot had failed to inform himself about these details of the local signalling (see below) he selected those non-existent speed boards to find his braking point for the end of the line. During his route-learning trip with a local driver this issue clearly had not been explained, nor had the pilot noticed this important omission.

Train driver route knowledge is a somewhat contentious issue outside Britain, as it is difficult to define accurately what the expression means in terms of what it covers and what its desired effects should be.

All train drivers have their own way of getting to know a route, whilst using route knowledge when driving a train is strongly influenced by the type of train being handled and its braking system. There is a tendency on continental networks to look at route knowledge as being of secondary importance, the argument being that if a driver correctly follows up the speed indications at signals and signs along the route he should be fine. They have a point, but a number of recent accidents in Europe can be officially attributed to insufficiently detailed local route knowledge leading to a misreading or overlooking of signals (e.g. ***Pécrot***, ***Brühl*** and *Arnhem*), remarks to that effect being contained in the official reports.

During discussions I had with IRSE past president Wim Coenraad among others, the conclusion was reached that route knowledge on the European continent fills in where certain, usually historical, anomalies in the local track layout or signalling leave a train driver without sufficient information to run his train safely, which was clearly the case during this incident. Whilst in Britain, with its route-indicating junction signalling, the situation is different, on networks with speed-indicating signalling route knowledge is still of importance. One of the essential issues that perhaps should receive some serious future attention in Europe is the standardisation of signs and signals that cover such clear danger points along the tracks. The Italian driver might then have understood that he was at braking distance from the end of the line and would not have needed the pilot to tell him. Moreover, going by the telephone call from Hindeloopen station, the pilot at that moment appeared to have a pretty good idea of where he was, yet failed to assist in stopping the train on time.

The line from Leeuwarden to Stavoren is, in fact, not difficult to get to know. There are a few station loops on the non-electrified single line, with driver-operated block signalling (key in slot or remotely operated with an infrared light gun from the cab) that is protected by ATBNG ATP at the passing loops, and the line runs through a rather flat and featureless landscape. In conjunction with the track diagram, detailed 1:50,000 or 1:100,000 maps reveal the locations of the stations and the way the line fits into the landscape. There are only three clear left-hand curves and three right-hand curves on the entire line, so even without any previous route knowledge, merely counting station loops and curves would have been one way of keeping track of the train's location. The stations on the line are spaced in easily recognisable patterns. Mantgum, with its passing loop, breaks up the long straight from the junction at Leeuwarden to Sneek, after which there are three stations in a close group, Sneek Noord, Sneek (also with a passing loop) and IJlst. After IJlst there is a right-hand and a left-hand curve, after which come Workum and Hindeloopen stations, the line then skirting the IJsselmeer coast. A further left-hand curve brings you to Koudum-Molkwerum station where you need to start paying attention as there is then less than 5 km (3 miles) to go. The end of the line is imminent when an impossible-to-overlook almost 90° right-hand curve leads into the platform at Stavoren. It is clear even from the maps that after this turn there are just two more level crossings, which would be difficult to miss even on the darkest night, with their bells sounding, red lights flashing and crossing illumination, before finally coming into Stavoren station with a dead straight into the platform.

As has already been said, however, the pilot knew where he was at Hindeloopen, so at that moment route knowledge was not an issue. It was only when distraction set in just before the end of the line that he lost his bearings. But knowing that he was near the end of the line, why had he not immediately requested the driver to slow down?

The signalling at Stavoren

There were no colour-light signals to indicate the end of the line at Stavoren, only three retro-reflective signs that would light up in the headlights of passing trains. However, this in itself was no different from the situation in places elsewhere in Europe. The normal procedure to indicate the end of a running line on many European networks would be to show a distant signal at caution, either as a yellow colour-light signal or as a retro-reflective board (e.g. showing the distant arm at caution – a fixed distant in British railway parlance) and then have the red signal or end-of-line sign mounted on top of, or just before, the stop blocks. Curiously, the situation was different at Stavoren, which warrants a little further explanation.

The signalling feature used at Stavoren was a set of three so-called ‘keperbaak’ boards, which were white vertical oblong boards with respectively three, two and one black chevrons as a countdown sequence, the last ‘keperbaak’ also having a white triangular board on top with a sort of Zorro Z-slash. The function of this set of boards was that of a distant signal, instructing the driver to brake to 40 km/h (25 mph) or less as necessary in order to be able to stop for the following red signal or for a stop board. The wording in the driver signalling manual alone points at some form of route knowledge being necessary to handle a train safely in the situation thus officially created.

The history of these ‘baak’ boards is interesting. In the days of steam, when the footplate crew on a locomotive were widely exposed to external noise because of the open cab, many nations (including Germany and The Netherlands) employed a distinct set of warning posts on the approach to a distant signal that covered an important stop signal. In The Netherlands each post consisted of two or three uprights situated as per the signals on the right-hand side in the direction of travel, parallel to the tracks, on which two or three yellow painted planks with black diagonal lines were fitted that rose diagonally from a lower level to a higher level in the direction of travel and were tilted towards the train. In The Netherlands three of these assemblies in a row, each 70 m (230 ft) apart, would have been provided, the point being that they attracted attention in the daytime but in the event of poor visibility would reflect the rolling noise of the train back into the cab on three occasions with a distinct hissing noise of rising pitch, making them as good as unmissable to a crew intent on locating distant signals.

After the final demise of the steam locomotive in The Netherlands in January 1958, when train drivers were no longer exposed to noise from outside as they were on the noble machines of old, these contraptions were increasingly changed to a set of three yellow-coloured normal boards with diagonal black lines but which nevertheless kept the old name of 'baak'. Furthermore, a white version of the boards, with chevrons replacing the diagonal lines, was introduced that instead of drawing attention to the approach of a distant signal actually replaced that signal and so saved considerably on expense. A drawback, however, was that the boards did not resemble anything to do with signalling. This may point to the reason why things went wrong here, these being the type installed at Stavoren, protected with an ATBNG balise, which, as explained earlier, was not compatible with the ATBEG-v (simplified) equipment of this rail grinding machine.

Stavoren is the only location along a line operated at higher speeds (100 km/h or 60 mph) with passenger traffic in the entire country where this signalling solution had been applied, the other four locations being found at the end of low-speed freight branch lines. The Dutch pilot driver, unfamiliar with the line and thus with the function of these three signs despite their presence in signalling section 4.3 of his handbook, overlooked them on his diagram and opted to go by the speed warning boards he was familiar with. These, however, had been removed from the lineside a few years ago, as they duplicated the speed instructions of the 'keperbaak' fixed distant signal boards that had been installed, but when the old diagrams were scanned into a computer database for easy amendment and publishing they were erroneously retained on the maps. No one thought it necessary to report this anomaly and have it changed, which is typical in a case of this kind of infrequently used literature.

Comment

A rather neat trap had been set up as a result of various unsafe situations running parallel with each other. But notice how many safety issues, each of which was capable of preventing the accident from happening or mitigating its impact, had to be bypassed before the situation developed as it did:

1. The Italian train driver, in the prevailing European signalling and signing circumstances, could not be expected to have relevant route and signalling knowledge in The Netherlands, especially such an unusual distant indication for the end of the line. The Dutch pilot, however, who had been contracted with the express purpose of assisting in running the train safely, had no useful local route knowledge either and had done very little to gain it to a sufficient extent. He thus managed to overlook the unusual distant indication.
2. His employers, aware of that assignment, took no action to ensure that the man was sufficiently well informed to do the job safely. It was neither the first nor the last time that an employer had shown a disregard for the use of route knowledge. However, steps to clarify route knowledge responsibilities were taken in the wake of this accident.
3. As with any meeting between different European nationals, it first had to be established which language would be used to communicate (not necessarily fluently). It turned out to be German, but it should be noted that English is designated the international lingua franca on the tracks. Things being as they are, however, the time when that is the case throughout Europe is probably still a long way off.
4. Despite the route-learning trip, the pilot driver was not familiar with the unique 'keperbaak' fixed distant signal boards at the end of this particular line. Ironically, provision of a recognisable type of distant signal was something the Italian driver would in all likelihood have understood from experience and reacted to under his own initiative.
5. The official route diagram showed the 'keperbaak' boards but erroneously also showed a speed reduction warning board and a speed reduction section board that had been removed a number

of years earlier. The document was also marked 'Not to be used for safety-critical purposes'.

6. The pilot overlooked the indeed rather innocuous-looking keperbaak fixed distant boards whilst concentrating on locating the non-existent speed boards. His knowledge of signalling apparently did not stretch to telling him that any stop signal in The Netherlands (also the red board or light on a set of buffer stops) is preceded by some form of distant signal rather than by a speed warning. This omission in his knowledge prevented him from querying what he saw on the track diagram, or from looking it up in his copy of the handbook.
7. The local driver with whom he had scouted the route did not point out this unique signalling feature, nor the absence of the speed boards. The local drivers apparently had not considered this anomaly important enough to report to the national network provider to have it corrected on the track diagram. Familiarity breeding contempt.
8. ATP, having been installed at great cost on the train in one version (EG-v) and along the track in another version (NG), was fully switched out per design on leaving the ATBEG-protected area just past the junction with the Harlingen line, instead of going to partial supervision and so at least keeping the permitted speed to 40 km/h (25 mph). From then on neither system was of any use in preventing this high-speed accident.
9. At a crucial moment, when perhaps the situation could still have been saved or the violence with which it occurred could have been mitigated, the pilot allowed himself to get distracted and look back several times for prolonged periods. Thus he failed to notice either the 'keperbaak' distant signal boards or the last level crossing, and also missed the p-way people standing there attempting to gain his attention.

The 60-year-old pilot driver was initially charged with endangering rail traffic and potentially faced a year's prison sentence if found guilty. His employer, Spoorflex, went bankrupt two weeks after the accident happened. In late 2014, however, it was decided not to prosecute the pilot driver as too many issues had been incorrectly dealt with when the companies involved had set up the job, and there had been too many faults with the maintenance of the indications along the track and the way they were presented to rail staff. According to the prosecuting authorities, the overall attention to safety had been seriously flawed.

5: Non-railway accidents mentioned

The Comet crashes in the 1950s

The UK launched the commercial jet age on Wednesday 27 July 1949 with the maiden flight of the first production airliner, Comet 1. However, due to previously unidentified issues with the airliner's speed and the higher altitude at which it operated, which led to increased stress of both the environment on the aircraft as well as of the various parts of the aircraft on each other, a series of incidents resulted in fatal crashes. First, in March 1953, there were problems due to V1 over-rotation on the runway during take-off causing a stall, which were solved by making alterations to the wing leading edge. Secondly, a crash occurred in January 1954 with a plane that lost both its wings in flight during bad weather, which led to identification of excessive stress caused by high speed with too responsive a manipulation of the power-assisted control surfaces. This was cured by stricter speed and flight management rules and the introduction of an artificial 'power feel' in the stick that warned of things going beyond acceptable limits, comparable to the present-day stall shudder. This was another first for these aircraft. Then, thirdly, in April 1954, a plane inexplicably broke up in flight, followed by similar crashes with other Comets. An epic investigation into the cause led to the conclusion that the large square picture windows and the punch-riveted navigation window caused metal fatigue in the fuselage, which was already highly stressed by fuselage air pressure at the altitude reached and the high speed of the aircraft. The solution was found in alterations to the size and shape of the windows and to their fixing in the fuselage. However, by the time that these problems had been solved and the plane in its uprated and extended Comet 4 configuration had proved itself completely, the products of Boeing, Douglas and Convair from the US (all of which thoroughly benefited from the conclusions of the Comet crash investigations, but were not subject to the fear of catastrophic failure that surrounded the British product) had cornered the market.

***

The Grand Canyon collision

On Saturday 30 June 1956, TWA flight 2, operated by a Lockheed L-1049 Super Constellation on a delayed flight from Los Angeles to Kansas City, was hit in mid-air by United Airlines flight 718, an equally delayed Douglas DC-7, from Los Angeles to Chicago. Having departed just 3 minutes apart, both aircraft ran into a thunderstorm in the Grand Canyon area. The DC-7 flew at 21,000 ft with permission, whilst the Constellation flew on a slightly converging track at 19,000 ft and requested to climb to 21,000 ft in order to avoid heavy altocumulus clouds ahead. This request was denied, but the captain received permission to fly at cloud top plus 1,000 and was advised that another flight was close by. The Constellation eventually flew at 21,000 ft as well, to avoid the weather problems ahead, and it was only a matter of time before both planes flew into each other, coming from either side of an altocumulus cloud tower. A wing of the DC-7 ripped off the Constellation's tail and its starboard outboard propeller gashed into the fuselage of the Constellation, causing explosive decompression. The Constellation went into a steep and uncontrolled descent, smashing upside down into the rocks of the Grand Canyon about where the Little Colorado and the Colorado rivers meet, the DC-7 spiralled into the opposite side of the canyon. A total of 128 people died in the crash, this accident being the first in which more than 100 passengers had lost their lives in a single event. The main finding was that air traffic control and its procedures were still operating as in the 1930s, with large areas of uncontrolled air space (comparable to 'dark territory' in railway terms) between a few centres of controlled air space. Technology – most of all radar – and procedures were drastically upgraded as a result, and controlled air tracks between hubs, indicated with radio beacons and with radar-enforced aircraft separation, came into being. At present the collision site is a National Historic Landmark and on sunny days some of the remains of the aircraft can still be seen due to the extremely difficult terrain that made it impossible to retrieve them.

***

The Boeing 727 crashes in the 1960s

In another example of the need to 'be there first', the Boeing 727, a first-generation commercial jet based on the successful Boeing 707, was quickly developed to take the jet age into the realm of short flights from smaller airports. Rolled out in 1962, it was a different plane from the 707 in that its three jet engines were mounted at the rear of the fuselage instead of in pods under the wing. As a result, its stall behaviour was quite different from that of jets with engines in or under the wings; the heavy rear tending to fall and increase the stall. Furthermore, in order to aid approach to short runways with often a steep glide path for noise abatement reasons, the plane had flap settings to 40° (there were no lift spoilers on the wings in those days) for a quick controlled descent. Both these characteristics, however, required that the jet engines should quickly spool up from idling to provide sufficient thrust in the event of a stall setting in. Additionally, jet engines fitted aft on the fuselage require fuel ducts through the fuselage from fuel tanks in the wings, which potentially aggravates the result of a crash-landing by bringing fuel fire straight into the passenger space, as at Salt Lake City for example (see below).

Things suddenly went wrong with the Boeing 727 four times within six months: on 16 August 1965, into Lake Michigan near Detroit; on 8 November 1965, at Cincinnati; on 11 November 1965, at Salt Lake City (with fire); and finally on 4 February 1966, into Tokyo Bay near Haneda Airport. Slow spool-up of the three jet engines was identified as the possible main cause by the only surviving pilot of these four crashes. Surprisingly, this was ruled out by the investigating US Civil Aeronautics Board (CAB), yet similar problems occurred in 17 more incidents. The CAB, possibly preserving Boeing's interests as much as justifying its own certification of the plane, came to the conclusion that all the crashes had been due to pilot error. However, pilot error in 21 rather similar accidents seems far-fetched, and a plane inducing pilot error at that rate may well be classed as dangerous. After the slow spool-up was dealt with, the Boeing 727 became a successful aircraft until the advent of the even more successful 737 family in the same role, with 40° flaps, engines under the wings and lift spoilers on top of them.

***

The collision at Tenerife Los Rodeos Airport, Canary Islands

During dense fog, and following overloading of the airport's traffic control systems due to diverted traffic from nearby Las Palmas Airport after a terrorist bombing on Sunday 27 March 1977, an apparent misunderstanding of instructions from the Los Rodeos Airport traffic controller led the hurried (working time constraints due to the terrorist action and consequent diversion) captain of a KLM Boeing 747 – despite a warning from his co-pilot – to commence take-off whilst a fog-shrouded Pan Am Boeing 747 was still taxying across its path on the runway. In the ensuing collision and fire 560 people lost their lives. This remains the highest death toll of any single air accident.

***

The Boeing 747 crash at Mount Osutaka, Japan

On Monday 12 August 1985, a Boeing 747 on domestic flight JAL 123 from Tokyo's Haneda Airport to Osaka's Itami Airport suffered serious structural malfunction 12 minutes into the flight, causing it to go out of control and crash into Mount Osutaka, killing 520 people including all 15 crew. Only four passengers survived. The malfunctions had been caused by incompetent damage repairs by Boeing representatives, which had been needed as a result of the tail of the aircraft hitting the runway during a previous rough landing.

***

The Boeing 747 crash at Lockerbie, Scotland

Not an accident but the bombing of Pan Am flight 103 from Frankfurt am Main to Detroit via London Heathrow and New York on Wednesday 21 December 1988. The flight was operated from London by a Boeing 747-121, registration N739PA. The seventeenth 747 built, it was 19 years old at the time of the bombing. The crash killed all 243 on board plus 11 people on the ground at Lockerbie. In 2003, Libya admitted responsibility for organising the bombing and subsequently paid compensation to the families of the victims. The name Lockerbie, close to ***Quintinshill***, still brings home the horror of that event.

***

The collision over Charkhi Dadri near New Delhi, India

On Wednesday 11 December 1996, a mid-air collision occurred between a Saudi Arabian Airlines Boeing 747 taking off and an incoming Kazakhstan Airlines Ilyushin Il-76. The accident was determined to have been caused by the Kazakh plane flying too low on approach to the runway, probably as a result of a misunderstanding by the flight crew of the approach altitude instructions from the New Delhi air traffic controller. An important secondary issue was the non-availability of air traffic control radar that produced altitude readings of the flights it monitored, as a result of which air traffic control at New Delhi was unaware of the Kazakh aircraft's intrusion into the lower take-off flight corridor. The main cause of the accident, however, was an Indian Air Force requirement restricting civil air space in the area, which meant that only one flight lane, inbound and outbound separated by altitude, was cleared for use by civilian traffic at this major airport. This remains the worst mid-air collision at the time of writing.

***

The Airbus A310 crash near Novosibirsk, Russia

This crash, on Tuesday 22 March 1994 near Novosibirsk, involved an Airbus A310, Aeroflot flight 593, on its way from Moscow to Hong Kong and was caused by a captain who had his son and daughter with him in the cockpit whilst his 1st Officer was in the first-class cabin. The captain, standing behind the pilot's seats in which his children were seated, was showing them how the autopilot brought the plane back on course if that was altered. As he was doing so, his 16-year-old son, unnoticed, moved the manual steering control handle from left to right for a few seconds, as a result of which the autopilot automatically disconnected (as it was designed to do) from controlling the ailerons but remained in overall control. The plane then unexpectedly banked steeply to port, which made reaching the controls well-nigh impossible for someone standing behind the pilot's seats. With the co-pilot being absent it was impossible to regain control of the aircraft to prevent the crash, which killed all 75 passengers and crew on board. As with the ***Ufa*** explosion, this incident illustrates a somewhat cavalier attitude to basic caution and safety, as evidenced in a number of Russian YouTube videos of outrageously daring acts and absurdly dangerous behaviour. It should be mentioned, however, that despite the foregoing examples, air travel remains the safest mode of transport.

***

The sinking of RMS *Titanic*

The *Titanic* was on its maiden voyage when it hit an iceberg, which had floated uncommonly far south for the period of the year and consequently sank, according to the official inquiries in the US and Britain. For unexplained reasons, however, the vessel was on a more northerly great circle sea lane across the Atlantic than was usual, so ice across this track had to be expected. In fact, the extent of the ice was accurately reported in frequent radio messages that were largely ignored on board the liner. The (assumed) high-speed collision with an iceberg, therefore, followed what cannot be described otherwise than erratic navigation in circumstances of known risk. The iceberg that cut short the career of the *Titanic* almost before it had begun was never identified, incidentally, which led to a number of theories about how and why the ship actually foundered. The fact that both the British and American inquiries were flawed did not help, many relevant questions not being asked, many inconsistencies not being queried and downright impossibilities being accepted.

Whatever the truth of the matter, at 02:20 on Monday 15 April 1912, the *Titanic*, transporting about 2,200 passengers and crew, sank with the loss of 1,523 lives. That was a high casualty rate bearing in mind the fact that the vessel took about 2½ hours to sink. The high death toll was partly due to incompetence in managing the evacuation into the insufficient (albeit officially sanctioned) number of lifeboats, as several of these left the vessel barely occupied.

***

The sinking of RMS *Lusitania*

In the same month that the ***Quintinshill*** rail collision occurred, the *Lusitania* was torpedoed and sunk by a German U-boat in an act of war. That incident happened at 14:10 on Friday 7 May 1915, with the loss of 1,195 people, about 16 miles off the Old Head of Kinsale, Ireland. The sinking of the *Lusitania*, despite being effective as wartime propaganda in the UK and the USA, no doubt negatively influenced the integrity of the Quintinshill inquiry, as the British Government was keen to minimise the impact of yet more disastrous news on public morale. Over the years, misuse of official reports and media by various governments to manipulate popular opinion in times of crisis has been well documented.

***

The capsize of MS *Princess Victoria* off Larne

The 2,694 Gross Tons roll-on/roll-off diesel ferry *Princess Victoria*, launched 27 August 1946 at Wm. Denny & Bros. Ltd. to ply the Irish Sea between Stanraer in Scotland and Larne in Northern Ireland, was the first British example of a breed that in time would take over the short sea connections across the world. Apart from 1,500 passengers and cargo she could load 40 cars from a link-span through a large stern door in her hull, which door, left unprotected through not closing a difficult to manage second guillotine-type sea-door, in the event turned out the cause of her foundering. Departing from Stanraer in Scotland on 31-01-1953, during one of the major storms in early 1953 that later, in co-operation with very high spring tides, would cause flooding havoc with severe casualties in the North Sea coastal regions of Eastern England and The Netherlands, she ran into big seas that eventually broke through her stern door and compromised her watertight integrity. The water ingress on her car deck in turn was not cleared fast enough by her scuppers and pumps, as a result of which she developed a list to port that soon was made worse by the cargo shifting. Despite immediate extensive radio reporting of her problems, for a number of (not in all cases clear) reasons assistance could not timely reach her and as a result 133 people died when she capsized and sunk. Only 40 people were able to get off in lifeboats and survived their ordeal.

Stability problems as a result of open or damaged hull-doors that allow free-flowing sea water on the car decks would cause the following ferries to capsize and sink: MV *Pincess Victoria* 1953 *133* deaths, SS *Heraklion* 1966 *200* deaths, TS *Wahine* 1968 *52* deaths, MV *Straitsman* 1974 *2 human and 1800+ sheep* deaths, MV *Herald of Free Enterprise* 1978 *193* deaths, MV *Jan Heweliusz* 1993 *55* deaths, MV *Estonia* 1994 *852* deaths, MV *Express Samina* 2000 *82* deaths, MV *Al-Salam Boccaccio 98* 2006 *1,020* deaths, MV *Queen of the North* 2006 *2* deaths, MV *Sewol* 2014 *304* deaths.

The two worst peacetime sea accidents of the 20th Century both involved ferries. Worst was the sinking of the heavily overbooked Filippine MV *Dona Paz*. After colliding with the tanker MV *Vector* fire broke out. Rescue was impossible due to unavailability of lifeboats and life jackets, whilst radio to call assistance was not operational. An estimated 4,386 people died, only 24 people survived the sinking. The second worst peacetime sinking was that of MV *Le Joola* off the coast of Gambia during a storm on 26-09-2002. Overloaded with three times the amount of people she was certified to carry she capsized and sank during the night, 1,863 people losing their lives. There were no survivors.

***

The capsize of MS *Herald of Free Enterprise* off Zeebrugge

MS *Herald of Free Enterprise* was one of the many modern, fast and large roll-on/roll-off car ferries that operated in fierce competition between British and Continental ports before the Channel Tunnel opened. On Friday 6 March 1987, 18 minutes after casting off and leaving the Zeebrugge quayside in Belgium en route to Dover, the ship suddenly listed to port just outside the harbour breakwaters before momentarily coming back on an even keel and then capsizing to port whilst turning sharply to starboard and making an almost complete 180° turn. At 19:28 the vessel came to rest on its port-side beam ends, partly sunk against a sandbank just outside the shipping channel off the breakwaters, causing 193 deaths, amongst whom were 38 crew members. It was the worst post-World War 2 British sea disaster.

The cause of the accident was soon determined to be the fact that the bow loading door assembly had been left open and that the ship's forward trim tanks, to lower the bow to connect with the Zeebrugge link-span, had not been discharged. This unbelievable oversight, which went unnoticed on the bridge due to a lack of appropriate monitoring equipment, was caused by incompetent handling of the closure and departure drill. The open bow doors, in combination with both the forward-down trim and the bow connector spade fitted to connect with the link-span, caused the accelerating ship to take in copious amounts of water on its car deck, rendering it unstable and leading to its capsize. In fact, the 'Herald' class of ships had not been conceived for the Dover-to-Zeebrugge run, as a result of which they had to trim their bows down to use the link-span at the Belgian port. After the accident other 'Herald' class ships were examined, complete with departure re-enactments. When loaded and trimmed as the *Herald of Free Enterprise* was that evening, on gaining speed in the outer port of Zeebrugge the run-up of the bow wave against the closed bow loading doors was evident.

A personal connection is that on the night of the accident, my wife and I were on our way from Vlissingen to Sheerness on board an Olau Line ferry. Standing port side astern in the cold evening darkness we wondered what the blaze of light near the Belgian coast was about, and it was not until arrival the following morning that we found out when informed by worried family members. The accident spelled the end for Townsend Thoresen ferry operations, this being one example of how an infringement on prescribed safe working practices caused a successful private transportation business to collapse.

***

The explosion at Halifax, Nova Scotia

On Thursday 6 December 1917, an explosion in the Canadian port of

Halifax initially killed more than 1,600 inhabitants and crew (the final death toll being around 2,000) and grievously injured an estimated 9,000, most being wounded and blinded by flying shards of window glass. The centre of Halifax was comprehensively destroyed. The accident occurred when outbound Norwegian SS *Imo*, sailing to New York on charter for the Belgian war relief organisation, collided in the narrowest point of Chebucto Bay with inbound French SS *Mont Blanc* sailing as an ammunition transport from New York to Bordeaux in war-torn Europe. According to the subsequent inquiry, SS *Imo*, proceeding under its port shore without pilot and without permission from the port authorities towards sea, contravened rules of the road with regard to the right of way through the port area narrows. The collision occurred after a mix-up about the intentions of SS *Imo*, following which a fire erupted in the bow of the *Mont Blanc*, causing the panicking crew to abandon ship and leave it to drift. The inevitable explosion occurred half an hour later at about 09:05 near Pier 6 close to Halifax city centre. It was the largest man-made explosion on earth until nuclear bombs destroyed Hiroshima and Nagasaki in Japan during the next world war.

***

The explosion on Russian submarine *Kursk* in the Barents Sea

This incident was another example of lax Russian regard for basic safety, in this case by accepting a damaged exercise torpedo with a known potentially explosive propulsion system on board submarine K-141 *Kursk*. On Sunday 12 August 2000 at 11:28, an initial explosion of the faulty torpedo sent the vessel to the bottom, setting off a chain reaction of other explosions in the forward torpedo rooms (registered as equalling 3 tonnes of TNT on various seismographs). It caused the loss of the submarine and its entire crew. The slow and shifty initial reaction of the president and government of Russia caused widespread anger throughout the country.

Apart from the Halifax explosion, none of the other maritime accidents described above were anywhere near the worst in history. Even if wartime actions are ignored, there was the Dona Paz disaster in the Filippines during which rather more than 4,000 people lost their lives. The worst death toll following a single ship sinking, however, was the result of belligerent action. It occurred when the German passenger vessel MV *Wilhelm Gustloff* was torpedoed by a Russian submarine on 30 January 1945, while on its way from Gdynia to Kiel through the

Baltic. Approximately 9,400 refugees fleeing the onslaught of the Red Army into Germany lost their lives.

***

The explosion at Chernobyl, Ukraine

On Saturday 26 April 1986, reactor No 4 of the Chernobyl nuclear power station in Ukraine exploded following an uncontrollable power surge. Thirty-one deaths were directly attributed to this event, with a total of 64 confirmed deaths so far, but given the enormous escape of radiation it is impossible to estimate how many people and animals really died (and will still die) from radiation-induced illnesses. The occurrence was triggered by tests to stop the reactor in an emergency. Incompetent handling of the exercise, problems with the construction of the reactor building (e.g. no containment shell) as well as the malfunction of emergency core-cooling features caused the explosion and the uncontrolled spread of radiation over much of the world.

Incidentally, another Soviet nuclear disaster attributable to extreme carelessness, details of which are only now beginning to come to light, was the contamination caused by the Chelyabinsk nuclear facility at Kyshtym in September 1957. In fact, in terms of seriousness that event could even outstrip Chernobyl, the Three Mile Island meltdown in the US in March 1979, and the Fukushima Daiichi disaster in Japan in March 2011.

Bibliography

In most cases the accident research literature as well as the air, maritime, rail and road accident investigation reports, either published as hardcopies or available as PDF files on the internet, were the main sources of information. Other internet information (notably via Wikipedia with its links to further sources) and articles in the news media and a variety of foreign railway magazines as well as in *Modern Railways* and the English language magazine *Today's Railways Europe* were freely used. The magazine *Diesel Railway Traction* proved a mine of information on older diesel traction. I have bound copies of the years between 1956 and 1962, as found in the cellars under Bristol Temple Meads station (for which my sincere thanks go to Ivan Davenport!). A useful find in a charity shop in Lyme Regis was a well-organised scrapbook with cuttings from *The Times* between 1952 and 1958 (from the Harrow & Wealdstone crash via the *Princess Victoria* sinking, see chapter 5, to the Lewisham St Johns accident in terms of this book). From the collection of J. J. Haut I had scrapbook 3189 with excerpts from *Engineering* and *The Engineer* between 1890 and 1908 at my disposal (my sincere thanks to Doug Riden!). Additionally, a motley international collection of signalling books, rulebooks and railway crew instruction manuals were consulted. Reference was also made to the titles listed below for a better understanding of the influence of economics and politics, the history and nature of transport safety, transport operator experience and proficiency issues, the ethics of better safety versus additional cost, public transport technical characteristics and historical information on public transport accidents.

Abdill, G. B., *This Was Railroading: Pacific Northwest* (New York NY, 1963)

Aird, A., *The Automotive Nightmare* (London, 1974)

Aldrich, M., *Death Rode the Rails: American Railroad Accidents and Safety, 1828-1965* (Baltimore MD, 2006)

Allward, M., *Safety in the Air* (London, 1968)
Bailey, C. (ed), *European Railway Signalling* (IRSE publication, London, 1995)
Bibel, G., *Train Wreck: The Forensics of Rail Disasters* (Baltimore MD, 2012)
Bird, M. J., *The Town That Died* (London, 1962)
Bishop, Bill, *Off the Rails* (London, 1984)
Blythe, R., *Danger Ahead: The Dramatic Story of Railway Signalling* (London, 1951)
Bonsall, Th.E., *Great Shipwrecks of the 20th Century* (New York NY, 1988)
Botkin, B. A. & A. F. Harlow (eds), *A Treasury of Railroad Folklore*: (New York NY, 1963)
Bramson, A. E. & N. H. Birch, *Flight Emergency Procedures for Pilots* (London, 1973)
Braun, A., *Signale der deutschen Eisenbahnen* (München, 2000)
Braun, J.W., *Und alle haben überlebt. Flugunfälle: Hintergrunden, Ursachen und Konsequenzen*. (München, 2015)
British Railway Disasters (Shepperton, 1996)
British Railways Diesel Traction Manual for Enginemen (London, 1962)
British Transport Commission Handbook for Railway Steam Locomotive Enginemen (London, 1957)
Broeke, W. van den, et al, *Het Spoor: 150 jaar Spoorwegen in Nederland* (Amsterdam, 1989)
Brown, T & M. Hanlon, *In the Interests of Safety: The Absurd Rules That Blight Our Lives* (London, 2014)
Carpenter, T. G., *The Environmental Impact of Railways* (Chichester, 1994)
Constant, O., *Le TGV: La genèse, les prototypes, les lignes* (Betschdorf, 1998)
Cowsill, M. & J. Hendy, *The Townsend Thoresen Years, 1928-1987* (Kilgetty, 1988)
Craig, P. A., *Pilot in Command* (New York NY, 2000)
Duckworth, C.L.D. & Langmuir, G.E., *Railway and other Steamers* (Prescot, 1968)
Eastlake, K., *Train Disasters* (London, 1998) (plus use of the better German edition)
– *Sea Disasters* (London, 1998)
Edwards, B., *S.O.S., Men against the Sea* (Corfe Mullen, 1994)
Ee, R. & M. van, *Ongevallen op Nederlands Spoor* (Alkmaar, 1997)

Faith, N., *Derail: Why Trains Crash* (London, 2000)
— *Black Box*, revised and updated edition (London, 2012)
— *Mayday: The Perils of the Waves* (London, 1998)
Flemming, D. B., *Explosion in Halifax Harbour* (Halifax NS, 2004)
Flohic, J.-L. (ed), *Le Patrimoine de la SNCF et des Chemins de Fer Français*, second edition, Vols 1 & 2 (Paris, 2000)
Gardiner, R. & van der Vat, D., *The Riddle of the Titanic* (London, 1995)
Glover, J., *Dictionary of Railway Industry Terms* (Hersham, 2005)
Goldsack, P. (ed), *Jane's World Railways & Rapid Transit Systems 1978* (London, 1978)
Guillemin, A., *Les Chemins de Fer* (Paris, 1869, facsimile Le Lavandou, 1979)
Gulík, J. Jr, *Železničné Nehody na Slovensku 3* (Bratislava, 2014)
Hadfield, R.L., *Sea-Toll of our Time* (London, 1935)
Hajt, J., *Das Grosse TEE Buch* (Königswinter, 1997)
Hall, S., *Danger Signals* (Shepperton, 1987)
— *Danger on the Line* (Shepperton, 1989)
— *Railway Detectives* (Shepperton, 1990)
— *ABC of Railway Accidents* (Shepperton, 1997)
— *Hidden Dangers* (Shepperton, 1999)
— *Broad Survey*, Vol 1 (York, 2000)
— *Beyond Hidden Dangers* (Hersham, 2003)
Hall, S. & P. van der Mark, *Level Crossings* (Hersham, 2008)
Hamilton, J. A. B., *Disaster Down The Line* (Poole, 1987)
Harford, T., *The Undercover Economist* (London, 2006)
Harris, J., *Without Trace, The Last Voyages of Eight Ships* (London, 1981)
— *Lost at Sea* (London, 1990)
Hauser, H., *Die letzten Segelschiffe* (Hamburg, 1952)
Hawkins, F. H., *Human Factors in Flight*, second edition (Abcoude, 1993)
Hesselink, H. G., *Hoe het spoor spoor werd*, Parts 1 & 2 (Rotterdam, 1973)
— *Spoorwegen in Nederland 100 jaar geleden* (Brussel, 1998)
Hollingsworth, B., *The Illustrated Directory of Trains of the World* (London, 2004)
Hurst, R. & L. (eds), *Pilot Error: The Human Factors*, second edition (London, 1982)
Independent Commission on Transport, *Changing Directions* (London, 1974)

Institution of Railway Operators (IRO), *Operators' Handbook* 2nd edition. (Stafford 2014)
Jack, I., *The Crash That Stopped Britain* (London, 2001)
Jackson, A. A., *The Railway Dictionary* (Stroud, 1992)
Jennings, Ph. S & Bosek, D., *Shipwrecks* (London, 1992)
Johnston, H. & K. Harris, *Jane's Train Recognition Guide* (London, 2005)
Jongerius, R. T., *Spoorwegongevallen in Nederland, 1839-1993* (Haarlem, 1993)
Kichenside, G., *Great Train Disasters* (Bristol, 1997)
Kichenside G. & A. Williams, *Two Centuries of Railway Signalling* (Oxford, 1998)
Kirsche, H.-J. (ed), *Lexikon der Eisenbahn*, fifth edition (Berlin, 1978)
Kjaer, N., *Bergensbanen 50 år 1909-1959* (Oslo, 1960)
Klaauw, B. van der, *Vliegen: Handboek voor luchtreizigers* (Utrecht, 1979)
Kleindel, F., *Unfälle und Schadensfälle* (Wien, 1980)
Klimbie, B., *Aanvaringen in de Binnenvaart* (Bussum, 1980)
Latten, R., *Spoorwegen Jaarboeken 1979-2010* (Alkmaar, 1979-2010)
Letcher Lyle, K., *Scalded to Death by the Steam* (London, 1985)
Lowell, V. W., *Airline Safety is a Myth*, second edition (New York NY, 1967)
Ludy, L. V., *Air Brakes* (Chicago IL, 1913)
Lundy, D., *The Way of a Ship* (London, 2002)
Mak, G., *In Europa: Reizen door de Twintigste Eeuw* (Amsterdam/Antwerpen, 2004)
Marcus, G., *The Maiden Voyage* (London, 1976)
Marriott, J., *Disaster at Sea* (New York NY, 1987)
Marshall, J., *The Guinness Book of Rail Facts and Feats* (Enfield, 1979)
McCluskie, T; Sharpe, M. & Marriott, L,. *Titanic & her sisters Olympic & Britannic* (London 1998)
Middelraad, P., *Voorgeschiedenis, Ontstaan en Evolutie van het NS-Lichtseinstelsel* (Utrecht, 2000)
Morrison, J., *B.R. North Division: Absolute Block Signalling* (course notes for students)
Nader, R., *Unsafe At Any Speed*, sixth edition (New York, 1965)
Nock, O. S., *Historic Railway Disasters* (London, 1966)
— *The Caledonian Dunalastairs* (Newton Abbot, 1968)
— *The Dawn of World Railways, 1800-1850* (London, 1972)
— *Two Miles a Minute* (London, 1980)

— *World Atlas of Railways* (Paulton, 1983)
— *Fifty Years of Railway Signalling* (Teignmouth, 1999)
Offringa, O., *Raising the Kursk* (Yeovil, 2004)
Pottgiesser, H., *Hauptsignale gestern und heute* (Mainz, 1980)
Preuss, E., *Eisenbahnunfälle in Europa* (Berlin, 1991)
— *Kursbuch des Schreckens* (Stuttgart, 1998)
— *Bahn im Umbruch* (Stuttgart, 2004)
— *Eisenbahnunfälle bei der Deutschen Bahn* (Stuttgart, 2004)
— *Eisenbahner vor Gericht* (Stuttgart, 2011)
Püschel, B., *Historische Eisenbahn-Katastrophen* (Freiburg, 1977)
Ramsden, J. M., *The Safe Airline* (London, 1976)
Reed, R. C., *Train Wrecks* (New York NY, 1968)
Reeuwijk, G. F. van, *De breedspoorlokomotieven van de HIJSM* (Alkmaar, 1985)
Richards, J. & A. Searle, *The Quintinshill Conspiracy* (Barnsley, 2013)
Richter, J.A. & C. Wolf, *Notlandung: Flugunfälle, Hintergrunde, Ursachen und Konsequenzen*. (München, 2010)
Ritzau, H. J., *Schatten der Eisenbahngeschichte: Katastrophen der deutschen Bahnen*, Part 1, 1945-1992 (Pürgen, 1992)
— *Schatten der Eisenbahngeschichte: Ein Vergleich britischer, US- und deutscher Bahnen*, Vol 1 (Pürgen, 1994)
Ritzau, H. J. & J. Hörstel, *Die Katastrophenszene der Gegenwart* (Pürgen, 1983)
Ritzau, H. J., J. Hörstel & T. Wolski, *Schatten der Eisenbahngeschichte 4: Deutsche Eisenbahn-Katastrophen* (Pürgen, 1997)
Rolt, L. T. C., *Red for Danger*, fourth edition (Newton Abbot, 1986)
Sampson, H., *Jane's World Railways* (London, 1951)
Schneider, A. & A. Masé, *Katastrophen auf Schienen* (Zürich, 1968)
Semmens, P., *Railway Disasters of the World* (Sparkford, 1994)
Simpson, P., *The mammoth book of air disasters and near misses* (London, 2014)
Sodenkamp, R., *Scheepsrampen 1945 tot heden* (Bussum, 1974)
— *Tot ondergang gedoemd* (Bussum, 1976)
Solomon, B., *Railroad Signaling* (Minneapolis MN, 2010)
Stenvall, F., *Nordens Järnvägar 1974/75* (Malmö, 1974 & 1975)
Stover, J. F., *American Railroads* (Chicago, 1961)
Treichler, H. P., *The Swiss Railway Saga: 150 Years of Swiss Trains* (Zürich, 1996) (English translation by Dieter Ehrismann)
Trevena, A., *Trains in Trouble: British Rail Accidents*, Vols 1-5

(Redruth, 1980-1984)
Vaughan, A., *Obstruction Danger* (Wellingborough, 1989)
— *Tracks to Disaster* (Hersham, 2003)
Vuchic, V. R., *Urban Public Transportation: Systems and Technology* (Englewood Cliffs NJ, 1981)
Way, F. Jr, *Pilotin' Comes Natural* (New York NY, 1943)
White, J. H. Jr, *A History of the American Locomotive: Its Development 1830-1880* (Baltimore MD, 1968)
— *The American Railroad Passenger Car*, Parts 1 & 2 (Baltimore MD, 1978)
— *The American Railroad Freight Car* (Baltimore MD, 1993)
Wilson, F. E. (ed G. B. Claydon), *British Tramway Accidents* (Brora, 2006)
Wolmar, C., *Broken Rails* (London, 2002)
— *Fire & Steam* (London, 2007)
— *Blood, Iron & Gold* (London, 2009)
— *Engines of War* (London, 2010)
— *The Great Railway Revolution* (London, 2012)
— *To the Edge of the World* (London, 2013)
Wragg, D., *Signal Failure: Politics and Britain's Railways* (Stroud, 2004)
'X', Captain & R. Dodson, *Unfriendly Skies: Revelations of a Deregulated Airline Pilot* (London, 1990)

Index of Accidents

Rail accident locations in alphabetic order

Location names in **bold** are discussed in their own section of Chapter 4. Others are mentioned within the chapter as references.